THE DIFFUSION AND DRIFT

OF ELECTRONS IN GASES

WILEY SERIES IN PLASMA PHYSICS

SANBORN C. BROWN ADVISORY EDITOR
RESEARCH LABORATORY OF ELECTRONICS
MASSACHUSETTS INSTITUTE OF TECHNOLOGY

THE DIFFUSION AND DRIFT OF ELECTRONS IN GASES

L. G. H. HUXLEY

R. W. CROMPTON

The Australian National University

A WILEY-INTERSCIENCE PUBLICATION

JOHN WILEY & SONS, NEW YORK · LONDON · SYDNEY · TORONTO

Library of Congress Cataloging in Publication Data:

Huxley, Leonard George Holden.
  The diffusion and drift of electrons in gases.

  (Wiley series in plasma physics)
  "A Wiley-Interscience publication."
  1. Electrons. 2. Diffusion. 3. Transport theory. 4. Electric discharges through gases.
I. Crompton, Robert Woodhouse, joint author.
II. Title.

QC793.5.E628H89   533'.13   73–12313
ISBN 0–471–42590–7

Printed in the United States of America

10 9 8 7 6 5 4 3 2 1

PREFACE

Early experimental and theoretical studies of the diffusion and drift of ions in gases were among those upon which modern atomic physics was based. However, although the foundations of our subject were laid at the turn of this century, the developments that have occurred in the last twenty years have shifted the emphasis from the measurement of macroscopic phenomena to the use of these studies to investigate collisions between low-energy electrons and gas molecules. This book is chiefly concerned with these modern developments.

The scope of the book is deliberately restricted because we believe that there is now a place for a comprehensive treatment in depth of the theory of electron diffusion and drift and its application to the more fundamental of the so-called swarm experiments. Thus the theoretical sections of the book are concerned with the theory of electron transport and its measurement rather than with the fundamental theory of electronic and atomic collisions. Similarly, in describing experimental work in this field, our aim has been not so much to compare the results of swarm experiments with results obtained by other techniques as to help the reader to gain a better understanding of swarm experiments and their interpretation.

At the outset it was necessary to decide between a compendium containing all recent work and a presentation of what we judge to be the most significant work within our self-imposed terms of reference. The standpoint that we have adopted is of necessity the more personal of the alternatives and, as such, entails some advantages and disadvantages. On

v

the one hand, the book is based upon first-hand experience of a significant proportion of the subject matter. We have therefore been able to make critical assessments of work in the field for the guidance of the reader. On the other hand, a degree of subjectivity in the final choice of material for inclusion is inevitable.

The book is in two parts. In Part 1 a chapter devoted to a historical account of early investigations is followed by two chapters in which the Maxwell-Boltzmann equation is developed and applied to electron diffusion and drift in circumstances in which it is legitimate to neglect the effect of electron density gradients on the velocity distribution function. Up to this point the results derived are well known. In Chapter 4 the complete Maxwell-Boltzmann equation, including spatial derivative terms, is considered in detail. Chapter 5 deals with the application of this theory to experiments designed to measure electron drift velocities and diffusion coefficients; in the absence of readily applicable solutions we have attempted to obtain an approximate form of solution which is useful for the interpretation of experiments of the type discussed in this book. Chapter 6 extends the theory to take account of inelastic encounters, and Chapter 7 contains a short account of the principles of the method of free paths. Part 1 concludes with chapters devoted to the theory of diffusion and drift in combined static electric and magnetic fields, and to electron motion in high-frequency electric fields.

The first two chapters of Part 2 contain surveys of experimental techniques that have been used to measure electron drift velocities and diffusion coefficients, while the third chapter (Chapter 12) deals with techniques for measuring "magnetic" drift velocities and attachment coefficients. Chapter 13 is an account of analytical methods for deriving elastic and inelastic cross sections from the results of swarm experiments. The final chapter contains a compilation of data for electron transport coefficients in some gases of general interest, particular emphasis being given to an assessment of the accuracy of the data. This chapter also contains many of the elastic and inelastic cross sections that have been determined through an analysis of electron transport data.

We are greatly indebted to D. S. Burch, M. T. Elford, D. K. Gibson, J. J. Lowke, H. B. Milloy, and J. A. Rees for their critical comment and advice on parts of the manuscript. We are especially grateful to H. R. Skullerud for fruitful discussions on the subject matter of Chapters 4 and 5, and to T. O. Rhymes for his careful reading of much of the text and for his invaluable assistance in the computation associated with Chapter 14. Above all we wish to acknowledge the work of Mrs. A. E. Duncanson, who typed most of the manuscript with endless patience and meticulous care, and of Mrs. J. Mahon, who gave valuable help with part of it.

Finally we wish to acknowledge the help of the staff of John Wiley & Sons, particularly Miss Beatrice Shube and Mr. William Douglas for their general oversight, Mrs. Ruth Flohn, our copy editor, and Mr. Robert Fletcher who supervised production. To them and others who have given every assistance in the preparation of this book we offer our grateful thanks.

L. G. H. HUXLEY

R. W. CROMPTON

Canberra, Australia
November 1973

CONTENTS

CHAPTER 9.   MOTION IN A HIGH FREQUENCY ELECTRIC FIELD, 250

Part II

PRINCIPAL SYMBOLS

General

e	atomic charge (to be considered positive in all formulae)
m	mass of an electron
M	mass of a molecule
n	number density of electrons
n_0	total number of electrons in a group
N	number density of the molecules of a gas
N_0	Avogadro's number (per mole)
κ	Boltzmann's constant
T	absolute temperature
E	electric field strength
B	magnetic field strength
$\boldsymbol{\omega}$	$= -e\mathbf{B}/m$

Spatial and Kinematic Quantities

$\mathbf{r}$	vector position in configuration space
$d\mathbf{r}$	element of volume in configuration space
$\mathbf{c}$	velocity of an electron
$d\mathbf{c}$	element of volume in velocity space
(c, dc)	shell in velocity space
$\mathbf{C}$	velocity of a molecule

$\mathbf{g}$	velocity of an electron relative to a molecule before an encounter
$\mathbf{g}$	relative velocity after an encounter
$\mathbf{r}$	velocity of an electron relative to the centroid of the electron-molecule pair before an encounter: $\mathbf{r} = [m/(m+M)]\mathbf{g}$
$\mathbf{R}$	velocity of a molecule relative to the centroid
$\epsilon = \frac{1}{2}mc^2$	energy of an electron
$c_{in} = (2\epsilon_{in}/m)^{1/2}$	speed of an electron corresponding to an inelastic change ϵ_{in} of the energy state of a molecule
$\mathbf{W}_{conv}(c)$	convective velocity of electrons whose velocity points lie in the shell (c, dc) (Section 2.6 and equations 3.10)
$\mathbf{W}$	drift velocity due to the electric force $e\mathbf{E}$

Distribution Functions

$f(\mathbf{c}, \mathbf{r}, t)$	velocity distribution function (p. 42)
$f(c, \theta, \mathbf{r}, t)$	velocity distribution function for shell electrons (Section 2.6)
$f_0(c, \mathbf{r}, t)$	isotropic component of $f(c, \theta, \mathbf{r}, t)$ (Section 2.6)
$f_1(c, \mathbf{r}, t)$	polarization component of $f(c, \theta, \mathbf{r}, t)$ (Section 2.6)
$f_0^*(c, t)$	
and $f_1^*(c, t)$	refer to the total group n_0 of electrons (Chapter 4)

Collision Cross Sections, Collision Frequencies, and Related Quantities

$p(\gamma, g)$	differential scattering cross section (Section 2.2)
$q_0(g)$	total scattering cross section (equations 2.1 and 2.3)
$q_m(g)$	cross section for momentum transfer (equation 2.9)
$q_1(g)$	see equation 2.9
subscripts el	
and in	refer, respectively, to elastic and inelastic encounters: thus $q_{0k_{in}}(c)$ denotes the total collision cross section for the kth inelastic process
$\nu_{el}(c) = Ncq_{m_{el}}(c)$	collision frequency for elastic encounters for electrons with speed c
$\nu(c) = Ncq_{m_{el}}(c)$	effective collision frequency (Section 2.7.2d)
$q_m(c) = \nu(c)/Nc$	effective collision cross section for momentum transfer
$\mathbf{V}$	$= e\mathbf{E}/m\nu_{el}(c)$
$\mathbf{V}'$	$= e\mathbf{E}/m\nu(c)$

Point Fluxes

$\sigma_E(c)$ velocity point flux due to $e\mathbf{E}$ (equation 2.19)

$\sigma_{coll}(c)$ velocity point flux from encounters (equations 2.20 and 2.21)

Coefficients of Transport

W drift velocity (Section 3.4.2)

D coefficient of isotropic (or transverse) diffusion (equation 3.27)

D_L coefficient of longitudinal diffusion (equation 4.26)

$\mu = W/E$ electron mobility

D/μ see Section 3.9

λ $= W/2D$

λ_L $= W/2D_L$

k Townsend's energy factor (p. 73)

k_m see equation 3.15

Coefficients of Ionization and Attachment

$\nu_{at}(c)$ collision frequency for attachment of electrons with speed c (Section 5.4)

$\nu_i(c)$ collision frequency for ionization (Section 5.4)

$\bar{\nu}_i$ and $\bar{\nu}_{at}$ mean values taken over all values of c (pp. 146, 147)

$\alpha_i = \bar{\nu}_i/W$ volume coefficient of ionization (equation 5.46)

$\alpha_a = \bar{\nu}_{at}/W$ volume coefficient of attachment (equation 5.46)

$\alpha = \alpha_a - \alpha_i$ net volume coefficient of electron loss (equation 5.47)

α_T Townsend's first ionization coefficient (equation 5.53)

α_{at} attachment coefficient analogous to α_T (equation 5.56)

PHYSICAL DATA

Speed of light *in vacuo**	$c = 2.99793 \times 10^{10}$ cm sec^{-1}
Atomic charge*	$e = 1.60219 \times 10^{-19}$ C
	$= 4.80325 \times 10^{-10}$ esu
Planck's constant*	$h = 6.62620 \times 10^{-34}$ J sec
Avogadro's number*	$N_0 = 6.02217 \times 10^{23}$ mole^{-1}
Rest mass of the electron*	$m = 9.10956 \times 10^{-28}$ gm
Charge/mass of electron	$e/m = 1.7588 \times 10^{8}$ C gm^{-1}
	$= 5.2728 \times 10^{17}$ esu gm^{-1}
One atomic mass unit	$1 \text{ amu} = 1/N_0 = 1.6605 \times 10^{-24}$ gm
Mass of proton (C^{12} scale)	$m_p = 1.00728/N_0$
	$= 1.6726 \times 10^{-24}$ gm
Faraday	$F = N_0 e = 9.64867 \times 10^{4}$ C mole^{-1}
Universal gas constant	$R_0 = 8.3143$ J K^{-1} mole^{-1}
Boltzmann's constant	$\kappa = 1.38062 \times 10^{-23}$ J K^{-1}
	$= 1.38062 \times 10^{-16}$ erg K^{-1}
Charge/mass of proton	$e/m_p = 9.5790 \times 10^{4}$ C gm^{-1}
Mass of proton/mass	
of electron	$m_p/m = 1.8361 \times 10^{3}$
Standard atmosphere (760 torr)	$= 1.01325 \times 10^{6}$ dyne cm^{-2}
0 Celsius	$= 273.155$ K

*See *Scientific American*, October 1970.

Volume of 1 mole
of ideal gas at S.T.P. $= 2.2414 \times 10^4$ cm^3 mole^{-1}

Number of molecules in 1 cc
of an ideal gas at S.T.P. $n_0 = 2.6868 \times 10^{19}$ cm^{-3}

Number of molecules in 1 cc
of an ideal gas at 0 C
at a pressure of 1 torr $= 3.5353 \times 10^{16}$ cm^{-3}

Number of molecules in 1 cc
of an ideal gas at 293 K
at a pressure of 1 torr $= 3.2959 \times 10^{16}$ cm^{-3}

$n^0 e$ at S.T.P. $= 1.2905 \times 10^{10}$ esu cm^{-3}

Energy of 1 electron volt   1 eV $= 1.6022 \times 10^{-12}$ erg
                                                         $= 1.602 \times 10^{-19}$ J

Speed of electron
with energy ϵ eV $= 5.9310 \times 10^7 \, \epsilon^{1/2}$ cm sec^{-1}

Wavelength of ϵ eV electron $= 1.2264 \times 10^{-8} \, \epsilon^{-1/2}$ cm

Energy of electron
with speed v cm sec^{-1} $\epsilon = 2.8428 \times 10^{-16} \, v^2$ eV

Frequency of 1 eV photon $= 2.4180 \times 10^{14}$ Hz
Wavelength of 1 eV photon $= 1.2399 \times 10^{-4}$ cm
Wave number of 1 eV photon $= 8.0655 \times 10^3$ cm^{-1}

κT in eV $\kappa T = 8.6171 \times 10^{-5} \, T$ eV
$\epsilon = \frac{1}{2} m \overline{c^2} = \frac{3}{2} \kappa T$ at 293 K $= 3.7872 \times 10^{-2}$ eV
Temperature equivalent
of the eV (defined as $T = 2\epsilon/3\kappa$) $= 7.7366 \times 10^3$ K

Parameter E/N $= 3.0341 (E/p)$ Td
                                                         (E/p in V cm^{-1} torr^{-1};
                                                         $T = 293$ K)

NUMERICAL DATA

$$\pi = 3.14159$$

$$e = 2.71828$$

$$\Gamma(\tfrac{1}{2}) = \pi^{1/2} = 1.7725$$

$$\Gamma(\tfrac{1}{4}) = 3.6256$$

$$\Gamma(\tfrac{3}{4}) = 1.2254$$

$$\Gamma(\tfrac{5}{4}) = 0.9064$$

$$\Gamma(\tfrac{1}{3}) = 2.679$$

$$\Gamma(\tfrac{2}{3}) = 1.354$$

THE DIFFUSION AND DRIFT

OF ELECTRONS IN GASES

PART 1

1

INTRODUCTORY SURVEY

OF EARLY INVESTIGATIONS

1.1. INTRODUCTION

The years of the Second World War separate the pioneer investigations in which the principles of swarm methods were established from recent work in which the stress is upon precision and the aims are broadened. Although in this book we shall be concerned chiefly with these postwar developments, a summary of the early work serves as an appropriate and useful introduction to recent studies. The summary given in this chapter is restricted to the investigations that have direct relevance to the development of swarm techniques and is not intended to cover the wider developments of gaseous conduction. The period covered extends from the discovery of X-rays in 1895 to the outbreak of the Second World War in 1939. Readers who seek a detailed account with full references are referred to the treatise *Basic Processes of Gaseous Electronics* by L. B. Loeb.*

1.2. GASEOUS CONDUCTION PRODUCED BY X-RAYS

Soon after the publication by Röntgen in December 1895 of his discovery of X-rays several physicists, including Röntgen himself, observed that X-rays render gases electrically conducting. It was later found that a similar conductivity was produced by radiations from radioactive substances. The most significant of the investigations which revealed the nature of this conductivity were those of J. J. Thomson and his colleagues at the Cavendish Laboratory in the years 1896 to 1900. In 1896 J. J. Thomson and E. Rutherford showed that the conductivity produced in air by X-rays was reduced when the air was passed through narrow metal tubes and was removed by passage through glass wool. The conductivity

*See bibliography at the end of the chapter for full citation.

1

was also destroyed when the air moved between electrodes maintained at a sufficiently large potential difference. It was also found that the electric current that flows to the electrodes when the gas between them is traversed by a steady beam of X-rays increases as the electric field is strengthened from small values and reaches a maximum or saturation value but thereafter ceases to increase as the electric field is strengthened, within limits.

The conductivity in a stream of air diminishes progressively downstream from the region traversed by the X-rays. The conductivity therefore decays spontaneously. These properties of gaseous conductivity were explained by Thomson and Rutherford on the hypothesis that the conduction is due to the production of small charged particles in the gas at a constant rate by a constant source of X-rays. These particles or ions were later shown to be molecules or clusters of molecules and to carry the atomic charge. The ions partake in the random thermal motions of the molecules and can be lost by diffusion to a boundary surface. They also drift through the gas in an electric field and are lost to an electrode. The spontaneous disappearance of the conductivity is the result of the recombination of ions carrying charges of opposite sign to produce neutral particles.

1.3. MOBILITIES OF IONS

It was found by Rutherford that gaseous ions drift through the gas in an electric field with a velocity that is proportional to the electric field strength. The velocity of drift in a field of one volt per centimetre in a gas at atmospheric pressure was termed the mobility of the ion. The first measurements of ion mobilities in gases were made by Rutherford in 1897 but were soon superseded by the more accurate measurements of J. Zeleny, who in 1898 discovered that the mobility μ^- of a negative gaseous ion exceeds μ^+, that of a positive ion. In 1900 Zeleny published the results of measurements in various gases at atmospheric pressure. He also observed that in moist gases μ^+ and μ^- tend to equality. The values of μ^+ and μ^- that he found are given in Table 1.1.

The dependence of mobility upon the pressure p of the gas was first investigated by Rutherford in 1898. Since in his method ions were produced through photoelectric emission from a plane cathode, the behaviour of negative ions only was observed. He found that in air in which the pressure p was given values that ranged from 765 to 34 torr the product $\mu^- p$ was effectively constant. In the more extensive experiments of P. Langevin (Paris, 1902) it was found that the mobility of positive ions was inversely proportional to the pressure but that the product $\mu^- p$ increased as p was diminished. J. Frank and R. Pohl (1910) showed that the drift

TABLE 1.1.   Ion mobilities (volt^{-1} cm^2 sec^{-1}) at $T(°C)$

Gas	Dry Gas			Moist Gas		
	μ^+	μ^-	T	μ^+	μ^-	T
Air	1.36	1.87	13.5	1.37	1.51	14.0
Oxygen	1.36	1.80	17.0	1.29	1.52	16.0
Hydrogen	6.70	7.95	20.0	5.30	5.60	20.0
Carbon Dioxide	0.76	0.81	17.5	0.82	0.75	17.0

velocities of negative ions became very large in pure argon and nitrogen but were reduced again to small values when these gases were contaminated with oxygen. R. T. Lattey (1910) and R. T. Lattey and H. T. Tizard (1912) found that in dry gases the drift velocities of negative ions became large in comparison with those of positive ions under the same conditions. The conclusion to be drawn was that the negative ions were transforming to free electrons.

The study of ion mobilities was considerably advanced in precision and reliability by the introduction of time-of-flight methods of measurement independently in 1928 by R. J. Van de Graaff and by A. M. Tyndall, L. H. Starr, and C. F. Powell, but these studies fall outside our field of discussion. We note, however, that time-of-flight methods were successfully adapted by N. E. Bradbury and R. A. Nielsen in 1936 to the measurement of drift velocities of electrons in gases.

1.4.   MEASUREMENT OF COEFFICIENTS OF DIFFUSION OF IONS IN GASES

The difficult problem of how to measure the coefficients of diffusion of positive and negative ions in gases was solved by J. S. Townsend* in 1899. Townsend was able to deduce the coefficients of diffusion from the loss by diffusion to the walls when the ionized gas flowed through narrow metal tubes. In fact, the ratio of the losses in tubes of different length was measured. The somewhat difficult theory of the method was also given by Townsend. The results of his first investigation, at atmospheric pressure, are given in Table 1.2.

Townsend also found that in air the coefficients of diffusion were inversely proportional to the pressure of the gas down to a pressure of 200 torr, as predicted by kinetic theory. In these experiments the gases were ionized by X-rays, but in later (1900) studies he found essentially the same

---

*Townsend referred to himself as J. S. Townsend in most of his publications. His full name, however, was John Sealy Edward Townsend.

TABLE 1.2. Coefficients of Diffusion ($cm^2 sec^{-1}$)

Gas	Dry Gas		Moist Gas	
	Positive Ions	Negative Ions	Positive Ions	Negative Ions
Air	0.028	0.043	0.032	0.035
Oxygen	0.025	0.0396	0.0288	0.0358
Carbon dioxide	0.023	0.026	0.0245	0.0255
Hydrogen	0.123	0.190	0.128	0.142

values when the ions were produced by radioactive radiations, by photo-electric emission from a cathode, or by a point discharge. In 1913 E. Salles used Townsend's method in an extensive investigation and confirmed in essence Townsend's values. Townsend remarked that the coefficients of diffusion of ions in their own gases were much smaller than those of carbon dioxide molecules into these gases and pointed out that this disparity can be explained either by considering ions to be clusters of molecules, or in terms of the enhancement of encounters through electrical attraction between ions and molecules.

1.5. EARLY THEORETICAL INVESTIGATIONS OF DIFFUSION AND DRIFT OF IONS IN GASES

1.5.1. NERNST-TOWNSEND FORMULA. In his 1899 paper (Section 1.4) Townsend also deduced from Maxwell's equations of transport (i.e., from Maxwell-Boltzmann theory and not by the method of free paths) an important formula relating the drift velocity W of a species of gaseous ion and its coefficient of diffusion D. This relation is (with modern symbols)

$$\frac{W/E}{D} = \frac{\mu}{D} = \frac{e}{\kappa T} = \frac{N_0 e}{RT} \qquad \text{(Nernst-Townsend relation),} \qquad (1.1)$$

where e is the charge of the ion and E the electric field strength, both expressed in esu, μ is the mobility for $E = 1$ esu, κ is Boltzmann's constant, T is the absolute temperature of the gas, N_0 is Avogadro's number, and R is the universal gas constant. Townsend made important uses of this formula which are discussed in later sections of this chapter. Townsend's proof of equation 1.1 is of interest in being the first application of Maxwell-Boltzmann theory to the motion of gaseous ions.

The formula had previously been deduced by W. Nernst in 1889 for electrolytic ions, but his theoretical treatment was not strictly applicable to gases. We shall therefore refer to equation 1.1 as the Nernst-Townsend formula. In the literature equation 1.1 is often called, for reasons that are not obvious, the Einstein formula. Einstein certainly employed (1905) the formula in his theoretical investigations of the Brownian movement, but it was already well known, as was in fact acknowledged by Einstein (1908).

An important formula deduced by Einstein (1905) which relates the rate of change of the square of the displacement from an origin of a particle in random motion, to the coefficient of diffusion, is

$$\frac{d}{dt}\overline{r^2} = 6D \tag{1.2}$$

in its three-dimensional form.

This formula is also applicable to the motion of an ion in a gas and serves as the starting point for a derivation of a formula for D by the method of free paths.

1.5.2. P. LANGEVIN'S THEORETICAL INVESTIGATION OF DIFFUSION AND DRIFT. In 1913 Langevin deduced elementary formulae for the mobility and coefficient of diffusion by the method of free paths; these are useful in general discussion but are not accurate. He also considered that Townsend's values for the coefficients of diffusion implied that ions were electrically charged clusters of molecules.

In 1905, however, Langevin published an important theoretical paper with the modest title "Une Formule Fondamentale de Théorie Cinétique," in which he extended the dynamical theory of gases as founded by Maxwell and Boltzmann to derive accurate formulae for the coefficient of interdiffusion of pairs of gases. He then proceeded to derive a formula for the mobility of a gaseous ion on the assumption that a neutral molecule of the gas behaves as a perfectly elastic sphere which becomes polarized in the electrostatic field of the ion. This polarization gives rise to a force of attraction (cgs–esu units) between an ion and a molecule:

$$f = \frac{(\epsilon - 1)e^2}{2\pi N r^5}, \tag{1.3}$$

where ϵ is the dielectric constant of the gas, e the charge of the ion, N the molecular number density, and r the separation of the centres of the ion and the molecule. Langevin also assumed that the random thermal speeds of the ions followed Maxwell's distribution law. Langevin's formula for the mobility μ is (Hassé's notation)

$$\mu = A \left[\frac{1 + (m/M)}{\rho(\epsilon - 1)} \right]^{1/2}, \tag{1.4}$$

where A is a function of a parameter λ, M and m are the masses of the ion and molecule, respectively, and ρ is the density of the gas. Here the mobility μ is the drift velocity in an electric field of 1 esu cm^{-1}. The parameter λ is

$$\lambda = \frac{\sigma^2}{e}\left(\frac{8\pi p}{\epsilon-1}\right)^{1/2}, \tag{1.5}$$

where σ is the sum of the radii of the ion and the molecule, and p is the pressure of the gas (in dyne cm^{-2}). The major task of the theory was to calculate $A(\lambda)$ as a function of λ. This evaluation of $A(\lambda)$ was carried out by Langevin but was recalculated by Hassé. When $\lambda=0$, $A(0)=0.5105$ (Langevin gave 0.505); when $\lambda \to \infty$, $\lambda A \to 0.75$. The value of $A(\lambda)$ reaches a maximum of 0.59 for $\lambda=0.6$. Thus, if λ is calculated from equation 1.5 with an assumed value of σ, the mobility μ can be calculated from equation 1.4 since $A(\lambda)$ is known from λ. The calculated and experimental values of μ can then be compared.

When λ is small (σ small or $\epsilon-1$ large), the force f is more important in determining μ than are direct encounters. In the limit $\lambda \to 0$, $A \to 0.5105$, the mobility

$$\mu \to \mu_0 = 0.5105\left[\frac{1+m/M}{\rho(\epsilon-1)}\right]^{1/2} \tag{1.6}$$

so that equation 1.4 can be written as

$$\mu = \frac{A}{0.5105}\mu_0.$$

When λ is large ($\lambda A \cong 0.75$), the rôle of the attractive force becomes relatively unimportant and ions and molecules behave as colliding spheres traversing rectilinear free paths. If $\epsilon-1$ is eliminated from equation 1.4 by means of equation 1.5, it is found that

$$\mu = \frac{(\lambda A)e}{\sigma^2}\left(\frac{1+m/M}{8\pi p\rho}\right)^{1/2}. \tag{1.7}$$

Since $\rho \propto p$, it follows that the drift velocity $W=\mu E$ is proportional to E/p when λ is large, since λA tends towards the constant value 0.75.

Formula 1.6 is independent of the charge e and the collision cross section σ; consequently μ_0 can be calculated from ρ and $\epsilon-1$ when a value is assumed for m/M. Langevin found that in air the assumption $m=M$ led to a value of $\mu_0 \cong 800$ at atmospheric pressure, whereas the values measured by Zeleny were $\mu^+ = 408$ and $\mu^- = 560$. Langevin concluded that the mass M of the ion exceeded that of the molecule. He wrote, "La seule attraction des ions par les molécules neutres ne suffit pas pour expliquer leur faible mobilité." He estimated, by a procedure not free from objection, that the negative ions comprised two molecules and the positive ions three. He also commented upon the large values, measured by H. A. Wilson (1899) and E. Marx (1900), of the mobilities of negative ions in flames, and by means of equation 1.6 he estimated the mass of the ion. Since this mass proved to be approximately that of an electron, it was evident that in gases at high temperature the negative-ion clusters are replaced by free electrons.

Tyndall and Powell and their colleagues at Bristol made considerable use of Langevin's theory of mobility and diffusion in the interpretation of their accurate measurements of mobilities (1938).

Other important investigations of coefficients of transport by means of the principles propounded by Maxwell and Boltzmann were those of H. R. Hassé (1926), Hassé and W. R. Cook (1931), S. Chapman (1916, 1917), and D. Enskog (1917).

The earliest of the investigations based upon quantum mechanics were those of H. S. W. Massey and C. B. O. Mohr (1934).

1.6. THE DISCOVERY OF THE ELECTRON

As the discovery of the electron is a well known topic, a brief summary will suffice. In 1895 J. Perrin demonstrated that cathode rays carry negative charge. In 1897 E. Wiechert, J. J. Thomson, and W. Kaufmann independently measured the specific charge e/m and velocities v of the cathode particles by different techniques. Of these researches those of Wiechert were the least publicized, those of Kaufmann the most accurate, and those of Thomson the most comprehensive and influential. These experiments showed that e/m was about 2000 times as large as e_i/M_H for the hydrogen ion in an electrolyte. Since the velocities v were large, of the order of one tenth the velocity of light, the view taken was that the charges e and e_i were the same and that the mass m was the order of $1/2000$ of the mass M_H of the hydrogen ion. Kaufmann's value of e/m was 1.67×10^7 emu gm^{-1}, later (1899) refined to 1.77×10^7, which is near the modern value. The same value of e/m was found by P. Lenard and by J. J. Thomson in 1899 for the particles emitted by metals in the photoelectric effect and by heated metals (J. J. Thomson, 1899). J. J. Thomson, after reviewing the evidence (e/m values, velocities, Lenard's mass absorption law for cathode particles, Lorentz's theory of the Zeeman effect) strongly and successfully propounded the view that the cathode corpuscle (electron) was a fundamental constituent of all atoms. In 1906 he derived a formula for the scattering coefficient of matter for X-rays on the assumption that the scattering agents are electrons within the atoms. This formula when applied to the measurements by Barkla of X-ray scattering showed that in light elements the number of electrons in an atom is one half the atomic weight. Consequently the mass of an atom is almost entirely associated with positive charge—a result which is consistent with the measurements of e/m for positive rays.

Thus, although many investigators contributed to the elucidation of the nature of cathode ray corpuscles, it was chiefly through the investigations and advocacy of J. J. Thomson that the electron came to be recognized so rapidly as a fundamental particle of matter.

1.7. THE ATOMIC CHARGE AND THE CHARGE OF A GASEOUS ION

1.7.1. INTRODUCTION. Although it was soon recognized (in 1896) that the ionic hypothesis of gaseous conduction accorded accurately with the measurements obtained, nothing beyond surmise was known about the electric charges carried by gaseous ions. The atomic nature of electricity was evident from Faraday's laws of electrolysis, and the value of the faraday F was accurately known to be 9.65×10^3 emu; according to the ionic theory of electrolytic conduction F was equal to the product $N_0 e$ of N_0, Avogadro's number, and e, the charge of a monovalent electrolytic ion, which was the atomic charge. Thus, if e is expressed in electromagnetic units, $N_0 e = 9.65 \times 10^3$ emu. Evidently, if N_0 were known, the value of e would also be known. However, in 1896 crude estimates of N_0 by use of the kinetic theory were all that were available. J. J. Thomson states in his small book *The Discharge of Electricity in Gases* (1898), based on his Princeton Lectures, that the number of molecules in 1 cubic centimetre of a gas at S.T.P., "lies between the limits 10^{18} and 10^{21}."

Two important experimental problems therefore demanded solution:

1. Determination of the value of e.
2. A demonstration that the charge of a gaseous ion was e or some multiple of e, the atomic charge.

The earliest attempt to measure the atomic charge was made by Townsend and published in February 1897. Later Millikan (1917) wrote of this work, "Townsend's method was one of much novelty and of no little ingenuity. It is also of great interest because it contains all the essential elements of some of the subsequent determinations."

The first attempt to measure directly the charge carried by a gaseous ion produced by X-rays or by means of radiation from radium was made by J. J. Thomson, using a modified form of Townsend's method.

The clear proof that the charge of a gaseous ion is equal to the atomic charge carried by a monovalent ion in an electrolyte was furnished by a totally different method devised by Townsend.

1.7.2.   MEASUREMENT OF THE ATOMIC CHARGE.

Townsend's investigation. Townsend found that when relatively large currents of the order of 10 to 12 A are used the gases evolved in the electrolysis of dilute solutions of sulphuric acid and caustic potash are electrically charged, provided that the temperature of the electrolyte exceeds 20°C, the degree of electrification increasing with the temperature. The oxygen and hydrogen evolved from sulphuric acid both carry a preponderance of positive charge, whereas the gases from caustic potash are negatively charged. The charges are carried by molecular aggregates from the electrolyte, which form efficient centres of condensation for water vapour. Consequently, when the gases were bubbled through water, they formed dense clouds on emergence into the atmosphere. Townsend found that the water in the form of drops in the gas greatly exceeded that in the form of vapour in the saturated gas. Such a cloud when introduced into the lower part of a vessel left the upper part clear. The clearly defined upper boundary fell at a constant rate of about one centimetre in three minutes. In order to determine the size of the droplets Townsend used Stokes's formula for the terminal velocity v of a sphere falling in a viscous medium without turbulence. The radius of the sphere a is then found from the equation $a^2 = \rho\mu v/2g$, and the weight w of a single drop is $w = (4\pi/3)\rho a^3$, where μ is the viscosity of the gas, ρ the density of the drop, and g the acceleration due to gravity. The drops were found to be small with $a \cong 7 \times 10^{-5}$ cm.

The rate of evolution of the gas is known; consequently, on passing a known volume of saturated gas through drying tubes, the weight of droplets W in unit volume of the gas was found from the increase in weight of the drying tubes after correcting for the weight of the water vapour. The number of droplets in unit volume was $n = W/w$. Townsend was also able to measure the total charge Q carried by the droplets in unit volume of the same gas. It follows, therefore, that the charge on each drop was $e = Q/n$.

He made the following observations:

1. The charge Q was proportional to W when different electric currents and temperatures of the electrolytes were used.

2. The same value of e was obtained with negatively charged as with positively charged droplets.

3. The value obtained for e fell within the limits estimated from kinetic theory.

Since the charges came directly from the electrolytes, Townsend made the reasonable assumption that his measured value of e was that of the atomic charge. The first value that he gave (February 1897) was $e = 3 \times 10^{-10}$ esu. However, he recognized that, whereas the gases carried a preponderance of charge of one or the other sign, some droplets carried charges of opposite sign to the majority. Thus, if n_1 and n_2 are the numbers of charged particles with positive and negative charges respectively, the charge in unit volume is $e(n_1 - n_2)$ whereas the total number is $n_1 + n_2$. The ratio n_1/n_2 was determined from measurements of the conductivity of the gas and was found to be equal to 4 in positively charged gases. Thus $(n_1 + n_2)/(n_1 - n_2) = 5/3$, and the value of e was therefore corrected to (February 1898)

$$e = \tfrac{5}{3} \times 3 \times 10^{-10} = 5 \times 10^{-10} \text{ esu.}$$

This result is remarkably close to the currently accepted value for e. Two weaknesses in Townsend's investigation were ignorance of how the charges are injected into the gas from the electrolyte and the assumption that all droplets carry the atomic charge.

J. J. Thomson's investigations. As mentioned above, the charges in Townsend's experiments were carried by small molecular aggregates derived directly from the electrolyte, whereas the gas molecules themselves were uncharged. These experiments therefore provided no information about the magnitudes of the charges carried by ions in gases ionized by X-rays or other ionizing radiations.

In December 1898 J. J. Thomson published the results of his first investigations of the charges carried by gaseous ions. The experimental difficulties were considerable, and his measurements were not capable of yielding reliable or consistent results. The principle of his method was essentially that of Townsend, but the experimental conditions required modification of the procedure. Thomson ionized the gas in a closed space by means of X-rays. The lower boundary of the gas was the surface of a pool of water; the upper, a plane metal electrode. The cloud was formed by adiabatic cooling of the gas by expansion in a connecting chamber, causing condensation upon the ions. The mass w of the individual droplets was found, as in Townsend's experiments, by use of Stokes's law, and the total mass W of the cloud in unit volume of the gas was estimated from the difference in the saturation contents of water vapour in unit volumes of the gas at the temperatures before and after expansion. The number of droplets in unit volume was therefore $W/w = n$. In order to obtain the charge in unit volume of the cloud Thomson measured the current I between the plane electrode and the water surface when the X-ray tube ran continuously. The charge $Q = ne = I/(\mu^+ + \mu^-)E$, where $\mu^\pm$ are the ionic mobilities and E is the electric field. As the quantity $\mu^+ + \mu^-$ was known from Rutherford's measurements, the value of Q was deduced. It follows that $e = Q/n = Qw/W$.

The chief sources of error inherent in this modification of Townsend's procedure lie in the method of estimating W and in the assumption that the value of w did not change through evaporation as the gas temperature rose towards room temperature as the cloud fell. These and other sources of error probably account for the wide variation in Thomson's results. His experiments with X-rays and photoelectric emission gave values for e that ranged from 5.5×10^{-10} esu to 8.4×10^{-10} esu. Later, in 1903, Thomson published another value, $e = 3.4 \times 10^{-10}$ esu, which he considered to be more reliable because of an improved technique for expansion and the use of radium as a source of ionization in place of X-rays. Although these investigations indicated that the charges carried by gaseous ions are of the order of magnitude suggested by kinetic theory for the atomic charge carried by a monovalent ion of an electrolyte, it is difficult not to agree with Millikan's comment: "Although Thomson's experiment was an interesting and important modification of Townsend's, it can scarcely be said to have added greatly to the accuracy of our knowledge of e."

Here we mention briefly later developments of the falling drop method for measurement of the atomic charge.

H. A. Wilson's method. An important improvement in Thomson's technique, introduced by H. A. Wilson in 1903, circumvented the need to estimate W, the weight of droplets in unit volume of the cloud, and also the number of droplets. Wilson formed his cloud by expansion, as in Thomson's experiments, but between a pair of plane metal electrodes between which an electric field could be established. Wilson observed the rate of free fall v_1 under gravity and also the rate v_2 in the presence of a vertical electric field E. It follows that

$$\frac{v_1}{v_2} = \frac{mg}{mg + eE}$$

when the direction of the force eE is parallel to that of the force mg.

Consequently

$$e = \frac{mg(v_2 - v_1)}{Ev_1}.$$

The mass m of a droplet is found from v_1 by the use of Stokes's law as in Townsend's method.

The values found for e varied from 2×10^{-10} to 4×10^{-10} esu with a mean value of $e = 3.1 \times 10^{-10}$ esu.

It was observed that in an electric field the droplets separated into sets which moved at distinctive velocities v_2. The charges on the droplets in the different sets were in the ratio $1:2:3$. This observation provided the first indication of the fact that the charge on a droplet was an atomic charge or a multiple of it. A source of inaccuracy, as in Thomson's experiments, was the change in the mass of a falling drop through evaporation.

R. A. Millikan's measurement of e. Millikan's work on the measurement of the atomic charge is well known; therefore a brief sketch is sufficient. Since, as shown by Wilson, the use of an electric field makes the use of clouds superfluous, Millikan measured the velocities v_1 and v_2 of single droplets. In order to eliminate the errors caused by evaporation he used droplets of mercury or oil. By irradiating the mercury droplet with ultraviolet light he was able to impart several atomic charges to it. Similarly the oil drop acquired various numbers of atomic charges by picking up ions from the gas ionized by X-rays.

Millikan, in a long series of measurements, including an investigation of departures from Stokes's formula for very small droplets, established the atomic nature of electricity and gave an accurate value for the atomic charge. For details of this work the reader is referred to Millikan's book (1947) *Electrons (+ and −), Protons, Photons, Neutrons, Mesotrons and Cosmic Rays.* This is a later edition of his book *The Electron* (1917). In 1917 he gave as the value of the atomic charge $e = 4.774 \times 10^{-10}$ esu, and this value was adopted as standard for many years. However, in the course of time, measurements of Avogadro's number N_0 of comparable or greater reliability than Millikan's direct measurement of e permitted an independent evaluation of e from the accurately known value of the faraday $F = N_0 e$. These values cast doubt on the reliability of Millikan's 1917 result. It was found in 1945 that the value for η, the viscosity of air, that Millikan had adopted was in error. With the accurate value of η the revised value of the atomic charge became $e = 4.807 \times 10^{-10}$ esu, which is close to that derived from the best determinations of N_0 and F.

Some other early determinations of e may be mentioned. In 1901 Planck deduced from radiation measurements and his new quantum theory of radiation the value of $N_0 = 6.175 \times 10^{23}$, which gave $e = 4.69 \times 10^{-10}$ esu, a remarkable result. In 1908 Rutherford and Geiger gave $e = 4.65 \times 10^{-10}$ esu from measurement of the charge on the alpha particle, and in 1909 Regner, using the same principle, found $e = 4.79 \times 10^{-10}$ esu. Rutherford and Boltwood (1909) determined the total mass of helium produced in 1 sec by a known mass of radium, and from the Rutherford-Geiger value for the number of alpha particles emitted in 1 sec from 1 gm of radium they derived a value of N_0 which gave $e = 4.81 \times 10^{-10}$ esu.

The droplet method is now of historic interest only, since the modern values of the fundamental constants are determined by other procedures.

1.8.   FIRST CONCLUSIVE DEMONSTRATION THAT GASEOUS IONS CARRY THE ATOMIC CHARGE

It has been seen (Section 1.7.2) that J. J. Thomson's measurements in 1898 of the value of the charge e were imprecise; therefore it was scarcely possible to assert that the charge of the gaseous ion was the atomic charge, although this assumption was plausible.

This equality was demonstrated beyond doubt by J. S. Townsend in 1899. As described in Section 1.4, the first measurements of coefficients of diffusion of ions in gases were published by Townsend in 1899, and in the same paper he derived the formula for the ratio μ/D of the mobility of an ion to its coefficient of diffusion (Section 1.5.1, equation 1.1). This relation is $\mu/D = N_0 e/RT$. At the end of his 1899 paper Townsend combined his measurements of D with the values of μ found by Rutherford in the corresponding gases to derive the value of $N_0 e$ for gaseous ions. This value proved to be so near the value of the faraday that Townsend was able to assert that the charge of a gaseous ion produced by X-rays is the same as the charge of a monovalent electrolytic ion (the atomic charge). This result was valid for both positive and negative ions.

In the year 1900 Townsend left the Cavendish Laboratory to become Wykeham Professor of Physics at Oxford, where he carried out his well known investigations of electrical discharges in gases and also performed the early studies of electron swarms in gases upon which are based the later experiments that form the chief subject matter of this book.

1.9.   IONIZATION BY COLLISION AND THE PARAMETER E/p

1.9.1.   IONIZATION BY COLLISION. It was remarked in Section 1.2 that when the gas between a pair of plane electrodes is ionized by a steady beam of X-rays the current at first increases with the strength of the electric field established between the plates but later reaches a saturation value when the field strength exceeds the minimum value necessary to remove ions as fast as they are produced. A similar behaviour is observed when the current in the gas is maintained by photoelectric emission from the cathode by use of ultraviolet light. It had been observed, however, by A. Stoletow in 1890 that, when photoelectric currents are maintained between plane electrodes at a fixed separation and the strength E of the electric field is increased progressively, at a certain value E_1 the current begins to increase rapidly beyond the saturation level as E is increased beyond E_1. This phenomenon was also observed by H. Kreusler (1898) and E. von Schweidler (1899) when the currents were maintained by X-rays. Stoletow studied the effect of changing the pressure of the gas on this increased ionization when E and the separation of the electrodes were held constant. He found that with high pressures the currents were small, but as the pressure was reduced the current increased to a maximum and then fell to a fixed value at pressures of a fraction of a torr.

The first attempt to explain this behaviour of the current was in terms of catalytic effects of the ultraviolet light at the surface of the cathode, but the correct explanation was found by Townsend (1900) who, by using an apparatus in which the electrode separation could be varied, showed that when the voltage was adjusted to maintain E at a fixed value the current increased as the separation of the electrodes increased. The increase in current was therefore a volume effect. Townsend obtained an accurate quantitative account of the observations, on the hypothesis that negative ions (which were soon recognized to be free electrons) in drifting through the gas under an electric field of sufficient strength ionize by collision, on the average, α molecules for each centimetre of drift in the direction of the electric field E. The new electrons liberated by collision from molecules behave similarly and ionize molecules at the same linear rate α.

Since ionization could be detected in diatomic gases when the voltage between the electrodes was about 20 V, it followed that the ionization potential of the molecules was less than 20 V. This value was much lower than had hitherto been believed.

The simplest conditions are those in which the currents are maintained by steady photoelectric emission of electrons from the cathode. Townsend found that, if i_1 and i_2 were the currents with electrode separations d_1 and d_2 when E and the pressure p were unaltered, i_1 and i_2 for a wide range of values of d_2/d_1 were given closely by the formula

$$\frac{i_2}{i_1} = \exp{(d_2 - d_1)}\alpha.$$

Thus α could be measured. This formula is what is predicted on the hypothesis that α is the number of molecules ionized by an electron in drifting 1 cm (unit distance) in the direction of the force $e\mathbf{E}$. The coefficient α is known as Townsend's first coefficient of ionization.

1.9.2.   THE PARAMETER E/p. Considerations of mean free path suggested that the dependence of α upon the value of the field E (usually measured in V cm^{-1}) and upon the pressure p of the gas (measured in mm of mercury, that is to say, in torr) should be a functional relationship of the form $\alpha/p = f(E/p)$. Townsend discovered that this relationship is accurately endorsed by experiment. It greatly simplifies the tabulation of the experimental determinations of α since α/p is tabulated in terms of E/p instead of α in terms of E and p separately. In the early years of the century Townsend measured α/p in terms of E/p in many gases. It was found that the same values of α/p in terms of E/p are obtained with X-rays as the ionizing agent as with photoelectric emission. It follows that α for electrons liberated from molecules by X-rays is the same as α for electrons emitted photoelectrically. At the time, this result provided direct evidence that electrons were constituents of gas molecules.

Townsend observed that the simple exponential growth of photoelectric currents in uniform fields, namely, $i = i_0 \exp \alpha d$, is no longer valid when d or E/p is large. In terms of a later symbolism, it was found that when E/p is maintained at a fixed value the growth of current with electrode separation is given accurately by the formula

$$i = \frac{i_0(\exp \alpha d)}{1 - \gamma[(\exp \alpha d) - 1]}. \tag{1.8}$$

This formula reduces to the simple exponential form when $\gamma = 0$. Like α/p, the coefficient γ is also a function of the parameter E/p, that is to say, $\gamma = \phi(E/p)$. The coefficient γ is known as Townsend's second coefficient. It is determined experimentally as a function of E/p along with α/p from measurements of i/i_0 at a number of separations d by use of equation 1.8. The formula itself was derived theoretically by attributing to the positive ions the ability, when E/p is large, to liberate new ions and electrons, either by ionizing molecules or by liberating electrons from the cathode. The coefficient γ is therefore a composite coefficient that includes these actions of the positive ions. Later work has shown that the volume ionizing effect of positive ions is usually unimportant, and that a number of other secondary processes contribute to the coefficient γ.

Equation 1.8 formed the foundation of Townsend's well known theory of electrical breakdown in gases. That subject falls outside the terms of reference of this book, however, and will not be discussed.

The parameter E/p, introduced by Townsend in the context of his coefficients α and γ, plays a wider rôle in the electrical properties of gases, since several important quantities, such as the drift velocities W of electrons and ions, and the ratio μ/D of mobility to diffusion coefficient, are also functions of E/p. The pressure p as used by Townsend in this context was not intended to represent the physical pressure of force per unit area but was instead a measure of the number density N of the gas molecules. It was therefore strictly necessary to postulate the temperature of the gas, but this was not always done. For this reason in much recent work the parameter E/p is replaced by the parameter E/N, which is independent of temperature; when E is expressed in volts per centimetre and N in cubic reciprocal centimetres a unit of 10^{-17} is adopted for this parameter. This unit has been named the townsend. This matter is discussed in Chapter 3.

1.10.   DEVELOPMENT OF SWARM TECHNIQUES TO STUDY THE MOTION OF FREE ELECTRONS WITH SMALL ENERGIES IN GASES

The use by Townsend of the Nernst-Townsend formula (equation 1.1) to show that $N_0 e$ is the same for gaseous ions as for monovalent electrolytic ions was described in Section 1.8. In this use, the value of the ratio $\mu/D = (W/D)/E$ is derived from Rutherford's (and Zeleny's) measurements of μ and Townsend's separate measurements of D.

This procedure was unsuitable, however, for use with gases at greatly reduced pressures, and in 1908 Townsend devised a new method in which W/D is measured as a single physical quantity in an enclosed chamber in a gas at rest, at any appropriate pressure. In this method a stream of ions moving in a steady state of motion in a uniform electric field enters the diffusion chamber through a hole in a plane electrode (cathode with negative ions, anode with positive ions) and progresses in the same uniform field to the receiving electrode, which is also plane (Figure 1.1). The distribution of the number density n of ions in the stream was obtained theoretically in the form of the solution of the steady-state differential equation for n appropriate to the boundary conditions, which Townsend assumed to be as follows: $n =$ constant across the surface of the hole through which the ions enter the chamber, but $n = 0$ everywhere else on that surface.

Let the centre of the hole, whose radius is a, lie at the origin of coordinates and take the axis $+Oz$ parallel to the force $e\mathbf{E}$; then for the

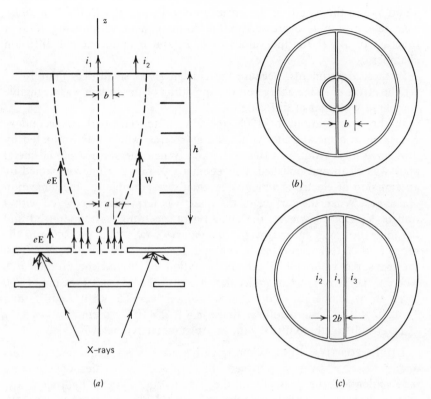

FIG. 1.1. Schematic diagram of Townsend's diffusion apparatus.

steady state the differential equation for n takes the form, since $\mathbf{W}$ is parallel to $+Oz$,

$$\nabla^2 n = \frac{W}{D} \frac{\partial n}{\partial z}. \tag{1.9}$$

It follows that W/D appears as a parameter in the solution $n(x, y, z, W/D)$.

The receiving electrode was divided into a central disk with radius b surrounded by a large annular area such that $n \rightarrow 0$ at its outer boundary. If i_1 is the current to the central disk and i_2 that to the annulus, the total current is $i = i_1 + i_2$. The experimental quantity that is measured is the ratio $R = i_1/i$. This ratio is also found theoretically from the solution $n(x, y, z, W/D)$ of equation 1.9 that satisfies the boundary conditions at the cathode. The ratio R is therefore obtained as a function of W/D, the radii

a and b, and the length of the diffusion chamber h; $R \equiv R(W/D, a, b, h)$. This relationship is expressed in the form of a curve showing R as a function of W/D for the values of a, b, and h of the actual diffusion chamber.

In these experiments, because the current i was small (less than 10^{-12} A), the effect of space charge on the spreading of the stream was negligible compared with that of diffusion.

In the investigations of 1908, the ions were generated by secondary X-rays from a metal plate within the apparatus, which was irradiated by primary X-rays from an external source. When the surface of the metal plate was clean and polished, the secondary X-rays were accompanied by an emission of electrons which were chiefly responsible for the ionization of the gas. When the surface of the plate was tarnished or covered with a film of Vaseline, however, the emission of electrons was suppressed and the ionization was caused by the secondary X-rays. Gas pressures in the range 3 to 25 torr were used. Townsend, instead of determining $N_0 e$ from the measured values of W/D and equation 1.1, found the product $n_0 e$, where n_0 is the number of molecules in 1 cm^3 of a gas at a pressure of 760 torr, at a temperature of 288 K. Since $n_0 = 2.54 \times 10^{19}$ cm^{-3} and $e = 4.802 \times 10^{-10}$ esu, it follows that $n_0 e = 1.22 \times 10^{10}$ esu cm^{-3}.

The results of these investigations are summarized below.

1.10.1. POSITIVE IONS. When positive ions were generated by secondary X-rays from a polished metal plate, the values of $n_0 e$ were independent of the pressure of the gas. In air, oxygen, hydrogen, and carbon dioxide the values (in esu) of $n_0 e \times 10^{-10}$ were, respectively, 1.26, 1.24, 1.26, and 1.32. When the ionization was generated by alpha and beta radiation from radioactive sources, the corresponding values (Haselfoot, 1909) were 1.26 and 1.22. Since the ionization produced by the secondary radiation from the polished plate was due chiefly to the electrons emitted, it was concluded that when positive ions are generated in gases by the action of alpha particles or energetic electrons the positive ions produced carry a single atomic charge.

However, when the gas was ionized directly by the secondary X-rays from a metal plate with a tarnished surface, much larger values of $n_0 e$ were found. In a series of experiments in which the surface of the metal plate was covered with a thin layer of Vaseline, the values obtained in the same gases were 2.03, 1.71, 1.84, and 1.55. It appeared therefore that many of the positive ions carried double charges $2e$.

Investigations by Frank and Westphal (1909) on the mobilities and rates of diffusion of ions at atmospheric pressure and generated by X-rays and by alpha, beta, and gamma rays indicated a difference in the nature of positive ions generated by X-rays and those produced in other ways. In the light of later knowledge it is suggested that the doubly charged ions are the result of the Auger effect.

1.10.2. NEGATIVE IONS. In gases slightly contaminated with water vapour the values of n_0e for negative ions were not influenced by the state, tarnished or bright, of the metal plate from which the secondary radiation was emitted. The values (in esu) of $n_0e \times 10^{-10}$ for air, oxygen, hydrogen, and carbon dioxide were 1.23, 1.23, 1.24, and 1.23, respectively. Similarly, when the ionization was generated by alpha and beta radiation (Haselfoot, 1909), the mean values obtained in a series of measurements were 1.24 and 1.22. In all these experiments where a small amount of water vapour was present the ratio $R = i_1/i$ was independent of the pressure and the nature of the gas. The concordance of the values of n_0e demonstrated that negative ions in slightly moist gases carry each a single atomic charge and move in thermal equilibrium with the gas molecules.

The measurements in dry gases were, on the other hand, strikingly different. At the lower pressures it was found that the ratio $R = i_1/i$ in dry gases not only became pressure dependent but was much less for the same field strength E than at higher pressures or in moist gases. Since the proportion $R = i_1/i$ of the total current that fell on the central disk was greatly diminished, it followed that the stream had spread laterally to an abnormal extent. The values found for W/D and n_0e were consequently also abnormally small. Consider the Nernst-Townsend formula (equation 1.1) in the form

$$\frac{D}{\mu} = \frac{2}{3} \frac{\frac{1}{2}m\overline{c^2}}{e}.$$

From what has been said, D/μ manifested abnormally large values, for the same value of E, compared with those for ions at higher pressures or in moist gases. There were two possible explanations:

1. The implausible one that $\frac{1}{2}m\overline{c^2}$ retains its thermal value $\frac{1}{2}M\overline{C^2}$ $= \frac{3}{2}\kappa T$ whereas e is diminished. This explanation was eliminated by the results of later experiments (Lattey, 1910; Townsend and Tizard, 1913), which showed that in dry gases at pressures of several torr the velocity of drift W and the coefficient of diffusion D become very large in a manner that suggested a reduction in the mass m of the carrier rather than in its charge e.

2. That e remains the same and that $\frac{1}{2}m\overline{c^2}$ assumes values greatly in excess of the value $\frac{3}{2}\kappa T$ appropriate to thermal equilibrium. This change in $\frac{1}{2}m\overline{c^2}$ was associated with a reduction in m.

Thus the entire range of phenomena was consistent with the supposition that the atomic charges e are those of electrons that are able to move freely among the neutral gas molecules. In some gases in which the tendency to form ions by attachment is small, electrons move freely among the molecules even at relatively large pressures provided that the gas is free from impurities.

It follows from the measured values of D/μ and the formula

$$\frac{D}{\mu} = \frac{2}{3} \frac{\frac{1}{2}m\overline{c^2}}{e}$$

that the value of the mean kinetic energy $\frac{1}{2}m\overline{c^2}$ is immediately obtained from

$$\frac{1}{2}m\overline{c^2} = \frac{3e}{2}\frac{D}{\mu}.$$

Townsend wrote

$$k = \frac{\text{mean energy of electron}}{\text{mean energy of translation of a molecule}} = \frac{\frac{1}{2}m\overline{c^2}}{\frac{3}{2}\kappa T}.$$

Townsend's energy factor k is an experimental quantity which was found to be, at a fixed temperature T (15°C), a function of E/p. Townsend and his colleagues determined k experimentally for electrons in many gases.

It is shown in Chapter 3 that the correct expression for the kinetic energy is

$$\frac{1}{2}m\overline{c^2} = \frac{e}{F}\frac{D}{\mu},$$

in which the factor F is determined by the nature of the distribution function $f_0(c)$. When the distribution function is that of Maxwell, $F = \frac{2}{3}$ and Townsend's energy factor k is then simply $(D/\mu)/(\kappa T/e)$.

In general, the distribution function is not that of Maxwell and it is necessary to know the appropriate value of F. The values of k with $F = \frac{2}{3}$ published by Townsend and his colleagues are therefore good approximations to $\left(\frac{1}{2}m\overline{c^2}\right)/(\frac{3}{2}\kappa T)$.

As no methods were available in 1908 for measuring drift velocities W of electrons (which exceed 10^5 cm sec^{-1}), a modified form of diffusion chamber was devised (Townsend and Tizard, 1912, 1913) in which both W/D and W could be measured. In this modification, the electrons for the diffusing stream were supplied by photoelectric emission from a polished metal plate that could be irradiated by ultraviolet light. These electrons were introduced into the diffusion chamber through a narrow slit in the cathode and proceeded through the gas to the anode as a spreading stream. A uniform electric field was maintained between anode and cathode, as in the original apparatus (Figure 1.1a), by use of metal rings at the outer boundary of the chamber. The potential of each ring was maintained at a value Vz/h, where z is the position of the ring, h the length of the chamber, and V the potential of the anode relative to that of the cathode. The receiving electrode (anode) was divided into a central strip, separated by narrow gaps from the remaining and larger portions of the anode (Figure 1.1c). All three portions of the anode were insulated from each other but could be combined as receiving electrodes by external electrical connections.

Suppose the centre of the aperture in the cathode (source of stream) to be at the origin. The anode is the plane $z = h$, and the coordinates of its centre are (O, O, h), which is the centre of symmetry of the central strip of the anode. The edges of the central strip are parallel to those of the aperture in the cathode, which has the form of a narrow rectangle. All these edges are parallel to the axis $+ Oy$. Let $2b$ be the width of the central strip. The coordinates of points in the dividing gaps are therefore ($x = \pm b$, $z = h$, $-\infty \leqslant y \leqslant +\infty$).

The quantity W/D was found, as before, from the ratio $R = i_1/i$, where $i = i_1 + i_2 + i_3$. This ratio, from the appropriate solution of the differential equation for n, was expressed as a function $R(a, b, h, W/D)$, where $2a$ is the width of the slit source. Then W/D was derived from the measured value of R, and thus

$$\tfrac{1}{2}m\,\overline{c^2} = k\left(\tfrac{1}{2}M\,\overline{C^2}\right) = \frac{3e}{2}\frac{D}{\mu},$$

deduced as before.

To obtain W, the central strip was connected electrically to one of the flanking divisions, so that the anode in effect comprised two divisions to which the currents were i_2 and $i_1 + i_3$ (see Figure 1.1c). A uniform magnetic field $\mathbf{B}$ was applied parallel to Oy (i.e., parallel to the edges of the central strip), and its sense and strength were adjusted until the stream of electrons was so deflected that $i_2 = i_1 + i_3$. In this circumstance the central axis of the stream fell on the dividing gap between the currents i_2 and $i_1 + i_3$.

If the direction of the axis of the deflected stream makes an angle θ with $+ Oz$, it follows from the geometry of the apparatus that $\tan \theta = b/h$. It was assumed that $\tan \theta$ was equal to the ratio of the mean lateral force eWB to the onward force eE on an electron. Thus

$$\tan \theta = \frac{BW}{E} = \frac{b}{h}, \qquad W = \left(\frac{b}{h}\right)\frac{E}{B}.$$

TABLE 1.3. Typical values of k and W_M found by Townsend and his colleagues

Gas	E/p	k^a	$W_M \times 10^{-5}$
Hydrogen	1	9.3	11.9
	2	15	16
	20	78	70
Nitrogen	1	21.5	8.7
	2	30.5	13.1
	20	59.5	86
Helium	1	53	8.25
	2	105	12.7
	5	172	30.2
Argon	1.25	320	7.7
	5	310	40
	15	324	82

^{a} At a temperature of $T = 288$ K, the value $k = 27$ corresponds to a value $\frac{1}{2}m\overline{c^2} = 1$ eV on the assumption of a Maxwellian distribution.

The drift velocities W in many gases were determined in this way as functions of E/p, that is, of E/N.

The theory is, however, oversimplified. If $\tan\theta = W_x/W_z$, where W_x and W_z are components of $\mathbf{W}$, and W_M is defined as $(E/B)\tan\theta$, then the true value of W is $W = \psi^{-1}W_M$, where the magnetic deflexion coefficient ψ depends on the nature of the distribution function $f_0(c)$ and on the energy dependence of the momentum transfer cross section. In view of this necessary correction, in recent work the magnetic deflexion of a stream is used to study ψ rather than to find W, which is obtained more accurately and directly from the time of flight of a group of electrons across a known spatial interval.

Townsend and his colleagues derived from W_M and k, which they exhibited as functions of E/p, values of collision cross sections (in the form of mean free paths at a pressure of 1 torr) as functions of the mean speed $\bar{c}$. It is evident that such estimates were not of the precision required for comparison with the predictions of modern theoretical investigations. The work as a whole, however, provided a valuable body of information not only about the macroscopic features of the motions of electrons in many gases, but also about the nature of the encounters between gas molecules and electrons with small energies. In particular the remarkable dependence of the collision cross section upon speed c in argon was strikingly demonstrated (Ramsauer-Townsend effect).

This programme of work carried out by Townsend, Bailey, and their associates in the period between the two World Wars has been fully described (Healey and Reed, 1941; Townsend, 1925, 1947).

To illustrate orders of magnitude, some examples of these measurements are given in Table 1.3. It will be noted that in monatomic gases, where encounters are elastic, the mean energy of agitation is very many times the mean energy of thermal agitation of the molecules.

1.11. THEORETICAL INVESTIGATIONS OF THE MOTION OF FREE ELECTRONS IN GASES

The earliest investigations of the motion of electrons among massive scattering centres were those of H. A. Lorentz of Leiden on electrical conduction in metals. These investigations by Lorentz, reported in his book *The Theory of Electrons*, are important since the theoretical techniques that he introduced were adopted in later investigations of the motions of free electrons in gases.

The essence of his method is the recognition that in the presence of an electric field or of diffusion the velocity distribution function $f(c)$ of the electrons is distorted from the symmetrical Maxwellian form. Lorentz supposed that $f(c)$ was of the form (in the notation of Chapter 2 of this book) $f(c) = f_0(c) + f_1(c)\cos\theta$, where θ is the angle between the velocity c of an electron and the direction of the electric force eE. He assumed that the distortion from the Maxwellian form is small; consequently $f_0(c)$ is Maxwell's distribution function for thermal equilibrium, and $f_1(c) \ll f_0(c)$. The conductivity is then determined by $f_1(c)$, which Lorentz showed to be equal to $-[(eE/m)(l/c)](\partial/\partial c)f_0(c)$, where l is the mean free path, assumed to be independent of c.

The atomic lattice is assumed to be rigidly fixed; consequently no energy is lost by an electron in its encounters with the lattice. The theory is not self-consistent, therefore, since in an electric field the mean energy of the electrons would increase without limit, whereas the postulates that $f_0(c)$ remains Maxwellian and that the mean energy of the electron retains its thermal value imply a transference of energy to the lattice.

We have seen, however, that when electrons move freely in a gas in an electric field the mean energy attains a steady value $k(\frac{3}{2}\kappa T)$, where k is Townsend's energy factor (see Section 1.10). The first theoretical investigations of this property and of other features of the motion of electrons in gases were undertaken by F. B. Pidduck, an Oxford mathematician (1913,

1916). In his first paper Pidduck based the investigation on Maxwell's equations of transport as derived by Maxwell's "second method" in the kinetic theory of gases. Pidduck assumed the "inverse fifth-power" law for the interaction between electrons and molecules and found it necessary to retain the mass M of the molecule as a finite quantity. Although m/M is small, the small losses of energy in encounters ultimately control the value of Townsend's energy factor k. Pidduck dervied the formula $k-1 = W^2/\overline{C}^2$, where W is the drift velocity of the electrons and $\overline{C}^2$ is the mean-square velocity of the molcules. Since $W^2/\overline{C}^2$ can in practice greatly exceed unity, it follows that the energy factor can greatly exceed the value $k=1$ corresponding to thermal equilibrium.

In Pidduck's second paper (1916), which he considered to be the more soundly based, he generalized the method of Lorentz to adapt it to electron motion in gases. He considered two molecular models, inverse fifth-power centres of force and elastic spheres with constant cross section. He found that with molecules regarded as elastic spheres the formula for $k-1$ was virtually unchanged since $(k-1)/(W^2/\overline{C}^2)$ varied from 1.18 to 1.06 as k increased from unity to large values. However, as pointed out in his first paper, a comparison of the measured values of k at known values of E/p and those calculated from the formula using the measured values of $W^2/\overline{C}^2$ for these same values of E/p showed that in diatomic gases the formula greatly overestimated the values of k. Pidduck therefore modified his molecular model by making the spheres inelastic with a coefficient of restitution ϵ. He defined a quantity f as $\frac{1}{2}(1+\epsilon)$ and showed that the formula for $k-1$ is to be replaced by (1916)

$$(k-f) + \frac{kM}{m}(1-f) = \frac{W^2}{\overline{C}^2}f.$$

Since M/m is large, a very small departure of f from the value of unity causes a considerable reduction in the value of k obtained when $f=1$. From the measured values of k and W, since M/m is known, f can be deduced. It is found that even in diatomic gases f is little different from unity. In modern terms nearly all encounters are elastic, but in a small proportion of all encounters electrons lose a large proportion of their energies. In monatomic gases the experimental values of f differ inappreciably from unity.

An important feature of Pidduck's second paper is his derivation of a formula for the isotropic (dominant) term $f_0(c)$ in the velocity distribution function. It was found that $f_0(c)$ was no longer, in general, Maxwellian in form. Pidduck derived the formula

$$f_0(c) = A(p^2 + b)^b \exp -p^2,$$

where A is a constant, $p^2 = hmc^2$, $h^{-1} = 2\kappa T = M\bar{C}^2/3$, $b = 2e^2h^2E^2/3N^2q^2r$, q is the collision cross section (independent of c), $r = m/M$, and N is the molecular number density.

This formula, which incorporates the effects of molecular motion, includes as a special case the well known formula discovered by Druyvesteyn fifteen years later. Twenty years later B. Davydov rediscovered Pidduck's formula, in identical form, except for a change of notation. Pidduck (1916) also investigated electron motion in combined electric and magnetic fields in terms of "the inverse fifth-power" molecular model. It is clear that Pidduck's pioneer investigations were important contributions to the theory of electron motions.

In 1930 M. J. Druyvesteyn deduced a formula for $f_0(c)$ in which the molecules are considered to be at rest:

$$f_0(c) = A \exp\left[-\frac{3m}{M} \frac{c^4}{(eE/mNq)^2} \right].$$

This formula, which, as mentioned, is a special case of Pidduck's formula, is named Druyvesteyn's formula. In 1935 B. Davydov rediscovered Pidduck's formula by a simpler procedure than that employed by Pidduck. It therefore incorporates Druyvesteyn's formula and allows for molecular motion. In 1936 Pidduck drew attention to the relation of the formulae for $f_0(c)$ of Druyvesteyn and Davydov to his earlier formula. All these formulae suppose that the collision cross section q is independent of the electron speed c.

In 1935, P. M. Morse, W. P. Allis, and E. S. Lamar of the Massachusetts Institute of Technology derived the form of $f_0(c)$ when the cross section q is a function $q(c)$ of c but with the limitation that the molecules are at rest. They employed the method of Lorentz as Pidduck had done in his second paper, and they made critical comment on Pidduck's first paper, apparently unaware of the more important second paper in which Pidduck also criticizes his earlier paper. Morse et al. found for $f_0(c)$ the formula (in another notation)

$$f_0(c) = A \exp\left[-\int_0^c \left(\frac{c}{V^2} \right) dc \right],$$

where $V = eE/mNcq(c)$. When $q(c)$ is independent of c, their formula reduces to Druyvesteyn's formula. They also touch upon the more difficult problem of the form of f_0 in the presence of spatial gradients of the electron number density.

In their well known book *The Mathematical Theory of Non-uniform Gases* S. Chapman and T. G. Cowling* deduce for $f_0(c)$ a formula that includes all the formulae discussed above. This formula is, in the notation adopted later in this book,

$$f_0(c) = A \exp\left(-\frac{3m}{M} \int_0^c \frac{c \, dc}{V^2 + \overline{C}^2}\right),$$

where, as before, $V = eE/mNcq(c)$.

Evidently, when $\overline{C}^2 = 0$, the formula of Morse, Allis, and Lamar and that of Druyvesteyn are recovered. In order to recover the formula of Pidduck and Davydov we suppose first that q is independent of c. Next we consider the integrand of the integral. We write

$$\frac{1}{V^2 + \overline{C}^2} \equiv \frac{1 - \left[V^2/(V^2 + \overline{C}^2)\right]}{\overline{C}^2}.$$

Consequently, since q is constant,

$$f_0(c) \propto \exp\left\{-\frac{3m}{M} \int_0^c \left[\frac{c}{\overline{C}^2} - \frac{(eE/mNq)^2 c}{(eE/mNq)^2 + \overline{C}^2 c} \frac{1}{\overline{C}^2}\right] dc\right\}$$

$$= \left[\exp\left(-\frac{\tfrac{1}{2}mc^2}{\kappa T}\right)\right] \exp\left[\left(\frac{eEl}{\kappa T}\right)^2 \frac{M}{6m} \int_0^y \frac{dy}{\tfrac{1}{6}(M/m)(eEl/\kappa T)^2 + y}\right]$$

where $y = mc^2/2\kappa T$ and $l = 1/Nq$.

Thus

$$f_0(c) \propto \left[\exp\left(-\frac{\tfrac{1}{2}mc^2}{\kappa T}\right)\right]\left(\frac{M}{6m}\left(\frac{eEl}{\kappa T}\right)^2\right)$$

*G. H. Wannier (*Am. J. Phys.*, **39**, 281, 1971) draws attention to a defect in the treatment by Chapman and Cowling and presents his own derivation.

$$\times \exp\left\{\log\left[\frac{1}{6}\frac{M}{m}\left(\frac{eEl}{\kappa T}\right)^2 + \frac{mc^2}{2\kappa T}\right] - \log\frac{M}{6m}\left(\frac{eEl}{\kappa T}\right)^2\right\}\right)$$

$$\propto \left[\frac{1}{6}\frac{M}{m}\left(\frac{Eel}{\kappa T}\right)^2 + \frac{mc^2}{2\kappa T}\right]^{(M/6m)(eEl/\kappa T)^2}\left[\exp\left(-\frac{\frac{1}{2}mc^2}{\kappa T}\right)\right],$$

which is the expression given by Davydov and Pidduck.

When the form of the distribution function has been determined, it is a straightforward matter to derive from the theory the appropriate formulae for the mobility and the coefficient of diffusion.

1.12. FURTHER DEVELOPMENTS IN EXPERIMENTAL PROCEDURES

During the remainder of the period before the Second World War the most important advance in the experimental technique of swarm studies, other than the work of Townsend and his associates, was the successful development of time-of-flight methods to measure electron drift velocities W. These methods, which incorporated electrical shutters operated at radio frequencies, were introduced by R. A. Nielsen and N. E. Bradbury in 1936 and provided unambiguous and reliable values of the drift velocities W in terms of E/p (or E/N).

In 1940 L. G. H. Huxley simplified the theory of Townsend's apparatus by making the radius a of the "source" through which electrons enter the diffusion chamber much smaller than the other relevant dimensions. The "hole source" can then be treated as a point source. Since 1950 the designs of apparatuses have in general taken advantage of the flexibility and accuracy implicit in this simpler theory.

In the rest of this book, although reference is made to early work, the chief concern is with the advances in swarm methods that have been made since 1940. There is also a change of outlook, in that much greater stress is laid upon swarm techniques as a tool for measuring collision cross sections than as a means for establishing the macroscopic features of the motion of electrons with small energies in gases.

BIBLIOGRAPHY AND NOTES

ACCOUNTS OF THE EARLY INVESTIGATIONS OF GASEOUS CONDUCTION

1. *Œuvres Scientifiques de Paul Langevin*, Services des Publications du Centre National de la Recherche Scientifique, Paris, 1950.

2. *Œuvres Scientifiques de Jean Perrin*, Services des Publications du Centre National de la Recherche Scientifique, Paris, 1950.
3. *Conduction of Electricity through Gases*, J. J. Thomson and G. P. Thomson, 3rd ed., Cambridge University Press, Vols. 1 and 2, 1928, 1933.
4. *The Life of Sir J. J. Thomson*, Lord Rayleigh, Cambridge University Press, 1943.
5. *J. J. Thomson and the Cavendish Laboratory in his Day*, G. P. Thomson, Nelson, 1964.
6. *Electricity in Gases*, J. S. Townsend, Clarendon Press, Oxford, 1915.
7. *The Electron*, R. A. Millikan, University of Chicago Press, 1917.
8. *Electrons (+ and −), Protons, Neutrons, Mesotrons and Cosmic Rays*, R. A. Millikan, University of Chicago Press, 1937, 1947.

REFERENCES TO SPECIFIC TOPICS

Ionic Theory of Gaseous Conduction

J. Perrin, ref. 2, pp. 27–34.
J. J. Thomson and E. Rutherford, "Passage of Electricity through Gases Exposed to Röntgen Rays," *Phil. Mag.*, **62**, 392, 1896. Also ref. 1, pp. 13–141.

First Measurements of Mobilities of Gaseous Ions

In their 1896 paper (reference above) Thomson and Rutherford indicated a method for measuring the sum of the mobilities of the positive and negative ions.

E. Rutherford (*Phil. Mag.*, November 1897) gave the mean of the values of the mobilities of positive and negative ions for air, oxygen, carbon dioxide, and nitrogen.

J. Zeleny (*Phil. Mag.*, July 1898) found that the mobilities of negative ions exceed those of positive ions in air, oxygen, carbon dioxide, and nitrogen. In a second paper (*Phil. Trans. A*, **195**, 193, 1900) he gave the results of more accurate measurements.

Later and More Reliable Measurements of Ion Mobilities

These are described in the monograph *The Mobility of Positive Ions in Gases*, A. M. Tyndall, Cambridge University Press, 1938.

An account covering the whole period, with full references, is given in Chapter 1 of L. B. Loeb's *Basic Processes of Gaseous Electronics*, University of California Press, 1955.

First Measurements of Coefficients of Diffusion of Gaseous Ions

J. S. Townsend, "The Diffusion of Ions into Gases," *Phil. Trans. A*, **193**, 129, 1899; **195**, 259, 1900. The first of these papers contains the theory of the method of measurement, the deduction of the Nernst-Townsend formula, and its application to demonstrate that gaseous ions carry the atomic charge. Also ref. 6, Chap. V.

The Atomic Charge and the Charge of a Gaseous Ion

General references

R. A. Millikan, "The Electron," ref. 7, Chap. III. Also ref. 8.
"Early History of Determination of Atomic Charge," *Nature*, **131**, 569, 1933.

H. A. Wilson, *Modern Physics*, 3rd ed., Blackie and Son, 1948, Chap. XIII.

G. P. Thomson, ref. 5, Chap. 5.

Specific references

J. S. Townsend, *Proc. Camb. Phil. Soc.*, **9**, 244, February 1897, and *Phil. Mag.*, (5) **45**, 125, February 1898.

J. J. Thomson, *Phil. Mag.*, (5) **46**, 528, December 1898; (5) **48**, 547, December 1899; (6) **5**, 346, 1903.

H. A. Wilson, *Phil. Mag.*, (6) **5**, 429, 1903.

R. A. Millikan, refs. 7 and 8. These books give a full account.

First Demonstration that the Charge of a Gaseous Ion is the Atomic Charge

See references above under "First Measurements of Coefficients of Diffusion of Gaseous Ions."

Ionization and the Parameter E/p

J. S. Townsend, *Nature*, **62**, August 1900.

J. S. Townsend, *The Theory of Ionization in Gases*, Constable and Co., 1910. This small book describes the discovery of ionization by collision and the development of the theory of electrical breakdown in gases.

J. S. Townsend, ref. 6.

Early Swarm Studies by the Method of the Spreading Stream

J. S. Townsend, ref. 6, Chap. 5, pp. 172–187.

J. S. Townsend, *Motion of Electrons in Gases*, address given at Centenary Celebration of the Franklin Institute in Philadelphia, September 1924, Clarendon Press, Oxford, 1925. This monograph gives a good summary of the methods and measurements carried out by Townsend and his colleagues at Oxford.

J. S. Townsend, *Electrons in Gases*, Hutchinson's Scientific and Technical Publications, London, 1947.

R. H. Healey and J. W. Reed, *The Behaviour of Slow Electrons in Gases*, Amalgamated Wireless Ltd., Sydney, 1941. This book covers the work of Townsend and his school at Oxford, as well as the later work of V. A. Bailey and his colleagues at the University of Sydney. It is a valuable survey of the subject up to the year 1939.

Early Measurements of Drift Velocities of Electrons in Gases

Method of magnetic deflexion

See references above under "Early Swarm Studies."

Method of time of flight of an isolated group

N. E. Bradbury and R. A. Nielsen, *Phys. Rev.*, **49**, 388, 1936.

General references to electron drift velocities

L. B. Loeb, *Basic Processes of Gaseous Electronics*, reference above, Chap. III. This chapter provides an extensive coverage of the subject.

Important Theoretical Investigations

The Nernst-Townsend formula

See the first reference under "First Measurements of Coefficients of Diffusion." This proof is of interest, in addition to its importance, in that it is the first application of Maxwell-Boltzmann methods to gaseous ions.

Theory of diffusion and drift of gaseous ions

Paul Langevin's famous investigation was published under the title "Une Formule Fondamentale de Théorie Cinétique," *Ann. Chim. Phys.*, **5**, 245, 1905. The mathematical portion of this paper is reproduced in translation in Earl W. McDaniel's *Collision Phenomena in Ionized Gases*, John Wiley & Sons, 1964.

Theoretical investigations of the motions of free electrons in gases

The following works are referred to in section 1.11:

H. A. Lorentz, *The Theory of Electrons*, G. E. Stechert and Co., New York, 1st ed., 1909, 2nd ed., 1916, p. 267.

R. Becker, *Theorie der Elektrizität*, B. G. Teubner, 1933, Section 36, p. 201.

F. B. Pidduck, *Proc. Roy. Soc. A*, **88**, 296, 1913; *Proc. Lond. Math. Soc.*, **15**, 89, 1916; *Quart. J. Math.*, **7**, 199, 1936.

M. J. Druyvesteyn, *Physica*, **10**, 61, 1930; **1**, 1003, 1934.

P. Davydov, *Phys. Z. Sowjetunion*, **8**, 59, 1935.

B. M. Morse, W. P. Allis, and E. S. Lamar, *Phys. Rev.*, **48**, 412, 1935.

J. S. Townsend, *Phil. Mag.*, (7) **9**, 1145, 1930; (7) **22**, 145, 1936. Townsend discusses the distribution function in terms of the method of free paths.

S. Chapman and T. G. Cowling, *The Mathematical Theory of Non-Uniform Gases*, 2nd ed., Cambridge University Press, 1952, p. 350.

Many authors have written on the theory of electron motion in gases. A general survey of work on distribution functions, with a long list of references, is given in L. B. Loeb's *Basic Processes of Gaseous Electronics*, reference above, Chap. IV.

Theory of diffusion apparatus with a source of negligible dimensions

L. G. H. Huxley, "The Lateral Diffusion of a Stream of Ions in a Gas," *Phil. Mag.*, (7) **30**, 396, 1940. This paper provided the basic theory of the modern form of the Townsend diffusion apparatus.

Experimental Data

The experimental data obtained by the early methods are given as curves or tables in the following publications:

J. S. Townsend, *Motion of Electrons in Gases*, reference above. J. S. Townsend, ref. 6.

R. H. Healey and J. W. Reed, *The Behaviour of Slow Electrons in Gases*, reference above.

S. C. Brown, *Basic Data of Plasma Physics*, John Wiley & Sons. This book also contains many data of a later period than is covered in this chapter.

Einstein's Investigation of the Brownian Movement

A. Einstein, *Investigations on the Theory of the Brownian Movement* (Ed., R. Fürth), Methuen and Co., London, 1926. This book contains five of Einstein's important papers on the subject. The references relating to Section 1.5.1 are pp. 7 and 70.

2

THE MAXWELL-BOLTZMANN EQUATION

AND RELATED TOPICS

2.1. INTRODUCTION

We begin this chapter with a review of the relevant features of the classical dynamics of encounters between moving particles, which in the present context are electrons and molecules with masses m and M, respectively. It is then possible to frame appropriate definitions of the scattering cross section q_0 and of the cross section for momentum transfer q_m for encounters between electrons and molecules.

The fundamental Maxwell-Boltzmann equation is then derived in a general form as a six-dimensional equation of continuity to be satisfied by the quantity $n(\mathbf{r}, t)f(\mathbf{r}, \mathbf{c}, t) \equiv nf$, which is the product of the number density $n(\mathbf{r}, t)$ of the electrons and their velocity distribution function $f(\mathbf{r}, \mathbf{c}, t)$. It is then shown that if f is represented as $f_0(\mathbf{r}, c, t) + f_1(\mathbf{r}, c, t)\cos\theta$ the Maxwell-Boltzmann equation can be replaced by a pair of working equations, a scalar equation representing conservation of matter (or energy) and a vector equation for momentum. The theory of drift and diffusion is then developed from these working equations.

2.2. COLLISION CROSS SECTIONS FOR PARTICLE SCATTERING: CONCEPT OF INTERACTING BEAMS

Consider particles of two kinds with masses m and M between which short-range forces of interaction exist.

Let two uniform but oppositely directed beams, one of particles m, the other of particles M, pass through each other. The particles in each beam

32

are randomly distributed. Let all the particles m move at the same velocity $\mathbf{r}$† and the particles M at velocity $\mathbf{R}$. As a particle m travels through the beam of particles M it will in due course, if the spatial number density of particles M is sufficiently large, pass into the vicinity of a particle M sufficiently close that the mutual interactions change their velocities from $\mathbf{r}$ and $\mathbf{R}$ to $\mathbf{r}'$ and $\mathbf{R}'$ and each is scattered from its beam (Figure 2.1). In this way each beam is attenuated by loss of members through scattering.

We now impose a condition upon the velocities $\mathbf{r}$ and $\mathbf{R}$: let

$$m\mathbf{r} + M\mathbf{R} = 0, \quad \text{that is,} \quad \mathbf{R} = -\frac{m}{M}\mathbf{r}.$$

Because the total momentum is zero before the encounter and the centroid is therefore at rest, it follows from the law of conservation of momentum that the momentum is also zero after the encounter and that the centroid is still at rest. Consequently the velocities $\mathbf{r}'$ and $\mathbf{R}'$ also satisfy the condition

$$m\mathbf{r}' + M\mathbf{R}' = 0, \qquad \mathbf{R}' = -\frac{m}{M}\mathbf{r}'.$$

Conservation of momentum does not imply conservation of kinetic energy, since the encounter may bring about changes in the internal energies of m and M. Therefore we do not postulate that in general $r = r'$, $R = R'$.

Encounters occur when m and M pass in close proximity. As m and M approach their centroid, they are deflected by their interaction as if the centroid were a fixed scattering centre.

Let n be the number of particles m in unit voluume, and N the number of particles M. A particle m is therefore surrounded by particles M distributed with a number density N. Each particle M shares with a particular particle m a centroid such that the distance of the centroid from m is $[M/(M+m)]s$, where s is the separation of m and M. The centroids that surround any particle m are therefore more closely packed than the particles M, and it is easy to see that the centroids associated with a particle m are fixed points distributed with a number density $[(M+m)/M]N$. Each particle m therefore moves at velocity $\mathbf{r}$ through a cloud of fixed centroids associated with it and distributed with number density $[(M+m)/M]N$. Similarly each particle M moves at velocity $\mathbf{R}$ through a

---

†Although the symbol $\mathbf{r}$ is used to mean both the position vector and a relative velocity vector, no confusion arises because the correct meaning is evident from the context.

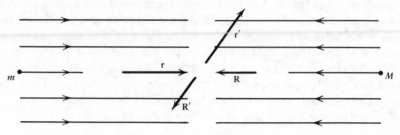

FIG. 2.1.

cloud of centroids associated with the particles m with a number density $[(M+m)/m]n$.

Let us take a cylindrical element of volume whose cross section is the vector area $d\mathbf{S}$ and whose length is $d\mathbf{L}$. Both vectors are parallel to the velocity $\mathbf{r}$. The volume of this elementary cylinder, which we suppose to be immersed in the beam of particles m, is $dS\,dL$. In time dt, $nr\,dS\,dt$ particles m enter the cylinder but fewer than this number, on the average, remain in the beam after emergence since some are scattered within the interval dL. The number scattered, on the average, is proportional to the number entering, to the length dL, and to the number density of centroids; that is, the number of particles m scattered from the elementary cylinder in time dt is proportional to

$$nr\,dS\,dL\,dt\left(\frac{M+m}{M}\right)N.$$

The directions of the velocities $\mathbf{r}'$ of these scattered particles m range over the unit sphere, and the number of them scattered into an elementary solid angle $d\omega=\sin\gamma\,d\gamma\,d\psi$ in the direction (γ,ψ) (Figure 2.2) is proportional to $d\omega$. The factor of proportionality, $p(\gamma,\psi)$, depends on the direction (γ,ψ) and is determined by the nature of the scattering process, which is not our present concern.

We thus arrive at an expression for the total number of particles scattered into $d\omega$ in direction (γ,ψ) in time dt from the volume $dS\,dL$:

$$(dt\,dS\,dL)\left(\frac{M+m}{M}\right)(Nnr)p(\gamma,\psi)\,d\omega.$$

The quantity $p(\gamma,\psi)$ has the physical dimensions of an area and is termed the *differential scattering cross section*. In practice, scattering, as observed

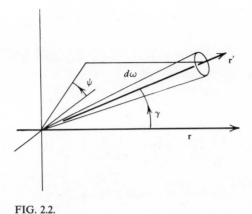

FIG. 2.2.

experimentally, is symmetrical about the direction of **r**; consequently $p(\gamma,\psi)\equiv p(\gamma)$.

Because the velocities **r′** and **R′** satisfy the condition $m\mathbf{r'}+M\mathbf{R'}=0$, particles M are scattered into the solid angle vertically opposite and equal in magnitude to $d\omega$ at the same rate as particles m are scattered into $d\omega$. The velocity of a particle m is **r** relative to a frame of reference in which the centroids are at rest, and that of a particle M is **R**.

Let **g** be the velocity of a particle m relative to a particle M. Then **g** is parallel to **r**, and its magnitude is

$$g=r+R=\left(\frac{M+m}{M}\right)r,$$

that is,

$$\mathbf{r}=\left(\frac{M}{M+m}\right)\mathbf{g}.$$

Similarly the velocity of M relative to m is $-\mathbf{g}$ with $\mathbf{R}=-[m/(m+M)]\mathbf{g}$.

On replacing $[(M+m)/M]r$ and $[(M+m)/m]R$ by g, the number of particles of either kind scattered in time dt into solid angle $d\omega$ from an element of volume $(dS\,dL)$ immersed in both beams is seen to be

$$(dt\,dS\,dL)(nNg)p(\gamma)\,d\omega.$$

The total number of particles scattered over the whole range of directions of **r′** and **R′** is

$$(dt\,dS\,dL)(nNg)\left[\int_\omega p(\gamma)\,d\omega\right].$$

The differential cross section $p(\gamma)$ also depends on the relative speed g and should more strictly be denoted $p(\gamma,g)$.

The integral of $p(\gamma,g)$ over all directions defines the *total scattering cross section* $q_0(g)$: thus

$$q_0(g)=\int_\omega p(\gamma,g)\,d\omega=\int_0^\pi\int_0^{2\pi}p(\gamma,g)\sin\gamma\,d\gamma\,d\psi$$

$$=2\pi\int_0^\pi p(\gamma,g)\sin\gamma\,d\gamma. \tag{2.1}$$

The total number of particles scattered from $dS\,dL$ in time dt is therefore

$$\Delta n_s=(dt\,dS\,dL)(nNg)q_0(g). \tag{2.2}$$

The scattering process is called isotropic when $p(\gamma,g)=p(g)$, that is, the number scattered into $d\omega$ is independent of the direction (γ,ψ) of $d\omega$. In this event,

$$q_0(g)=p(g)\int_\omega d\omega=4\pi p(g).$$

An example of isotropic scattering from classical dynamics is provided by the encounter of two smooth and rigid spheres with radii σ_1 and σ_2. Put $\sigma=\sigma_1+\sigma_2$; then the number of isotropic scattering encounters that occur in $dS\,dL$ in time dt is $nNg(\pi\sigma^2)dS\,dL\,dt$, and it follows that in this example

$$q_0(g)=4\pi p(g)\equiv q_0=\pi\sigma^2,$$

which is indeed the scattering cross section.

It has been assumed that the scattering events are single encounters in which pairs of particles m and M only are involved. If scattering were to involve, for instance, one particle m and two particles M, then the rate of scattering would depend on a term containing N^2 as a factor. Three-body scattering of this kind is assumed to be unimportant when the mean separation of the particles in the beams is much greater than the length $[q_0(g)]^{1/2}$, as occurs in almost all circumstances that we shall consider.

When particles m and M can interact in one of several possible ways during an encounter, each possible kind of encounter is associated with its own differential cross section. Let $p_{el}(\gamma,g)$ be the differential cross section for elastic scattering, and $p_{k_{in}}(\gamma,g)$ that for the kth inelastic process. The expression for the number scattered into $d\omega$ becomes

$$(dt\,dS\,dL)(nNg)\left[\,p_{el}(\gamma,g)+\sum_k p_{k_{in}}(\gamma,g)\right]d\omega,$$

and the total number scattered is

$$(dt\,dS\,dL)(nNg)q_0(g)$$

with

$$q_0(g)=\int_\omega\left[\,p_{el}(\gamma,g)+\sum_k p_{k_{in}}(\gamma,g)\right]d\omega=q_{0_{el}}(g)+\sum_k q_{0k_{in}}(g). \tag{2.3}$$

In all that has preceded it has been assumed that the beams are oppositely directed and such that $m\mathbf{r}+M\mathbf{R}=0$. It is a simple matter to remove these restrictions. Impart to all particles in both beams, and therefore to the centroids of all pairs (m, M), a common velocity $\mathbf{G}$. The velocity of all particles m becomes $\mathbf{c}=\mathbf{r}+\mathbf{G}$ and that of particles M is now $\mathbf{C}=\mathbf{R}+\mathbf{G}$, as is shown vectorially in Figure 2.3a. In Figure 2.3b the velocities $\mathbf{r}'$ and $\mathbf{R}'$ after an encounter and the resultant velocities $\mathbf{c}'$ and $\mathbf{C}'$ are also shown.

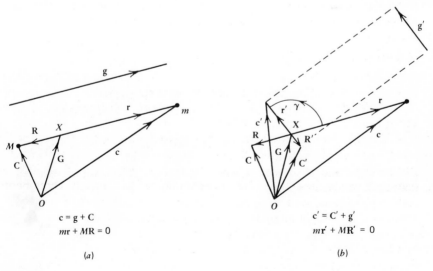

$$c = g + C \qquad\qquad c' = C' + g'$$
$$m\mathbf{r} + M\mathbf{R} = 0 \qquad\qquad m\mathbf{r}' + M\mathbf{R}' = 0$$

(a) $\qquad\qquad\qquad\qquad\qquad$ (b)

FIG. 2.3.

The addition of the common velocity **G** therefore transforms the system into a beam of particles m with velocities **c** and a beam of particles M with velocities **C**, and since in general **c** and **C** are not oppositely directed the beams cross obliquely. The scattering process is not affected by the imposition of the common velocity **G** and may be discussed as before in terms of two oppositely directed beams with $m\mathbf{r} + M\mathbf{R} = 0$ in a system of coordinates moving with velocity **G**.

Conversely, given initially two beams that cross obliquely with respective velocities **c** and **C**, it is necessary to use diagram (a) of Figure 2.3 to deduce the system of velocities **r**, **R**, **g**, and **G** in order to discuss the scattering processes. We note that $\mathbf{g} = \mathbf{c} - \mathbf{C}$ is divided internally at X into **r** and **R** so that $mr = MR$. Then $OX = G$.

The dynamical basis of Figure 2.3b, which is Maxwell's velocity diagram for a simple encounter, is formally discussed in Appendix A.

Some single-collision experiments are used to delineate the differential cross section $p(\gamma, g)$ as a function of γ and g. The associated total cross section $q_0(g)$ is then found by numerical integration of $p(\gamma, g)$ over all directions. In other experiments $q_0(g)$ is measured directly. Difficulties arise when appreciable scattering occurs at small values of the scattering angle γ.

In swarm methods, differential cross sections are not found; instead, total cross sections are determined. No difficulty exists through the presence of appreciable scattering at small angles since the transport properties that are measured in swarm experiments are determined by the scattering at all angles; that is, integration with respect to ω is already physically incorporated in these methods.

2.2.1.   ELASTIC, INELASTIC, AND SUPERELASTIC ENCOUNTERS.   Encounters between electrons and molecules are of three general types: elastic, inelastic, and superelastic, and it is convenient to associate a collision cross section with each type. These matters are next considered.

The total kinetic energy of a pair of particles (m, M) before an encounter is

$$Q = \tfrac{1}{2}mc^2 + \tfrac{1}{2}MC^2.$$

But $\mathbf{c} = \mathbf{G} + \mathbf{r}$ and $\mathbf{C} = \mathbf{G} + \mathbf{R}$, where $\mathbf{G}, \mathbf{R}$, and $\mathbf{r}$ are the velocities already defined. Consequently,

$$Q = \tfrac{1}{2}[(m+M)G^2 + 2\mathbf{G} \cdot (m\mathbf{r} + M\mathbf{R}) + mr^2 + MR^2].$$

Since $m\mathbf{r} + M\mathbf{R} = 0$, this expression reduces to

$$Q = \tfrac{1}{2}(m+M)G^2 + \tfrac{1}{2}mr^2 + \tfrac{1}{2}MR^2. \tag{2.4}$$

Similarly, the kinetic energy of the pair after an encounter is seen to be

$$Q' = \tfrac{1}{2}(m+M)G^2 + \tfrac{1}{2}mr'^2 + \tfrac{1}{2}MR'^2. \tag{2.5}$$

The loss of energy in the encounter is, therefore,

$$\Delta Q = Q - Q' = \tfrac{1}{2}m(r^2 - r'^2) + \tfrac{1}{2}M(R^2 - R'^2)$$

$$= \frac{1}{2}\frac{mM}{m+M}(g^2 - g'^2),$$

where g and g' are speeds of m relative to M before and after the encounter.

We now suppose that m is an electron and M a molecule, and we recognize encounters of three kinds.

(a) Elastic Encounters

In such encounters there is no loss of total kinetic energy of the pair: $\Delta Q = 0$, $g = g'$, although there may be an exchange of energy between the particles. Elastic encounters are the only encounters that take place between electrons and the atoms of monatomic gases when the energies of the electrons do not exceed the threshold values necessary to produce changes in the electronic states or ionization of the atoms.

In encounters with diatomic or polyatomic molecules elastic encounters are usually accompanied by inelastic and superelastic encounters.

(b) Inelastic Encounters

In these encounters the internal energy of the molecule is increased at the expense of the kinetic energy Q and $g' < g$.

All inelastic encounters associated with a particular change in the internal energy state of a molecule involve the same definite loss of kinetic energy, $\Delta Q = \epsilon_k$ for the kth kind of inelastic encounter. Thus

$$\frac{1}{2}\frac{mM}{m+M}(g^2 - g_k'^2) = \epsilon_k,$$

$$g_k'^2 = g^2 - \frac{2(m+M)}{mM}\epsilon_k. \tag{2.6}$$

Consequently, after inelastic encounters in which g is specified and ϵ_k is fixed, all the velocities $\mathbf{g}'$, whatever their directions, possess the same speed

g_k'. This is a convenient property of encounters between electrons and molecules.

(c) Superelastic Encounters

Such encounters may occur when an electron interacts with an atom or molecule already in an excited energy state. The molecule changes to a state of smaller internal energy, and the liberated energy ϵ_k appears as an increment to the kinetic energy of the electron. Thus

$$g_k'^2 = g^2 + \frac{2(m+M)}{mM}\epsilon_k. \tag{2.7}$$

Such encounters are the inverses of inelastic encounters.

When electrons move at random in a state of thermal equilibrium in the absence of an electric field, the effects of inelastic and superelastic encounters are balanced and together give no flow energy, on the average, from electrons to molecules or vice versa.

2.2.2. MOMENTUM TRANSFER CROSS SECTION. Consider interactions between the particles of obliquely crossing beams. One beam consists of particles m with velocity $\mathbf{c}$; the other, of particles M with velocity $\mathbf{C}$; and $\mathbf{g} = \mathbf{c} - \mathbf{C}$ (Figure 2.3a). Let m be associated with electrons, and M with molecules.

The number of electrons scattered in time dt from volume $dS\,dL$ into a solid angle $d\omega$ about the direction (γ, ψ) relative to the direction of $\mathbf{g}$ as polar axis (Figure 2.4) is

$$(dt\,dS\,dL)(nNg)p(\gamma,g)\,d\omega.$$

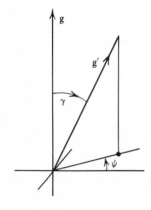

FIG. 2.4.

The momentum of each of these electrons before scattering was $m\mathbf{c}$ $= m(\mathbf{G}+\mathbf{r})$; after scattering it was $m(\mathbf{G}+\mathbf{r}')$. The loss of momentum was therefore $m(\mathbf{r}-\mathbf{r}')$.

The components of this lost momentum along and normal to $\mathbf{g}$ are

$$m(r-r'\cos\gamma), \qquad mr'\sin\gamma\cos\psi, \qquad mr'\sin\gamma\sin\psi.$$

Since, when g (and therefore r) is held fixed in value, r' is constant, and because the scattering is symmetrical about the direction of $\mathbf{g}$, the mean values of $mr'\sin\gamma\cos\psi$ and $mr'\sin\gamma\sin\psi$ are zero, taken over all values of ψ with γ constant. Thus the mean value of the vector sum of the losses of momentum of all these scattered electrons has the direction of $\mathbf{r}$ and the magnitude

$$(dt\,dS\,dL)(nNg)m\int_0^\pi\int_0^{2\pi}p(\gamma,g)(r-r'\cos\gamma)\sin\gamma\,d\gamma\,d\psi.$$

If we adopt the definition:

$$q_1(g)=2\pi\int_0^\pi p(\gamma,g)(\cos\gamma)\sin\gamma\,d\gamma,$$

the expression for lost momentum becomes

$$\Delta\mathbf{P}=(dt\,dS\,dL)(nNg)\left[q_0(g)-\frac{r'}{r}q_1(g)\right]m\mathbf{r}$$

$$=(dt\,dS\,dL)(nNg)q_m(g)\left(\frac{mM}{m+M}\right)\mathbf{g}. \qquad (2.8)$$

The *cross section for momentum transfer* is the quantity $q_m(g)$, whose definition, from what has preceded, is

$$q_m(g)=q_0(g)-\frac{g'}{g}q_1(g)$$

since $r'/r=g'/g$.

When all encounters are elastic, $g'=g$ and

$$q_m(g)\equiv q_{m_{el}}(g)=q_{0_{el}}(g)-q_{1_{el}}(g). \qquad \left.\begin{array}{c}\\[3em]\end{array}\right\} \quad (2.9)$$

2.3.   DISTRIBUTION FUNCTION AND VELOCITY SPACE

The preceding discussion of the scattering from interacting crossed beams of particles m and M provides a useful introduction to the less simple description of the interaction of two sets of particles m and M moving at random. We are specifically concerned to discuss the motion of a set of electrons m, either as an isolated group or in the form of a stream, among the molecules M of a gas in random thermal motion.

We suppose that the molecules of the gas have a number density N which is independent of position and everywhere very much larger than the number density n of the electrons, whose partial pressure is a negligible proportion of the whole. In general, the number density n, unlike N, is a function of position $\mathbf{r}$ and of time t. It is written as $n(\mathbf{r}, t)$ when it is expedient to stress this dependence, but is often denoted simply by n.

An element of volume of configuration space at vector position $\mathbf{r}$ is designated $d\mathbf{r}$; consequently the number of electrons within $d\mathbf{r}$ at time t is $n(\mathbf{r}, t)\, d\mathbf{r}$. The velocities $\mathbf{c}$ of the $n(\mathbf{r}, t)\, d\mathbf{r}$ electrons range widely in magnitude and direction, and in order to depict and discuss the motion of the group of $n(\mathbf{r}, t)\, d\mathbf{r}$ electrons as a whole it is convenient to associate with configuration space a velocity space. In velocity space, velocity $\mathbf{c}$ is represented by a point, its velocity point, which is the end point in this space of the vector $\mathbf{c}$. Thus the motion of the $n(\mathbf{r}, t)\, d\mathbf{r} \equiv n\, d\mathbf{r}$ electrons in configuration space is represented by $n\, d\mathbf{r}$ velocity points in velocity space. An elementary region of velocity space at vector position $\mathbf{c}$ is designated $d\mathbf{c}$.

The number of electrons at time t that lie within the volume element $d\mathbf{r}$ at $\mathbf{r}$, and whose velocity points lie also within the elementary region $d\mathbf{c}$ of velocity space, is designated $[n(\mathbf{r}, t)\, d\mathbf{r}][f(\mathbf{r}, \mathbf{c}, t)\, d\mathbf{c}] \equiv nf\, d\mathbf{r}\, d\mathbf{c}$. The function $f(\mathbf{r}, \mathbf{c}, t)$ is the general velocity distribution function.

Let $d\mathbf{c} = c^2\, dc\, d\omega$, where $d\omega = \sin\theta\, d\theta\, d\phi$; then $nf\, d\mathbf{r}\, d\mathbf{c}$ is the number of the $n\, d\mathbf{r}$ electrons whose velocities at time t are in the small range of directions $d\omega$ about the direction (θ, ϕ) and whose speeds lie in the range dc about c. Thus $nf\, d\mathbf{r}\, d\mathbf{c}$ is the population of the class of electrons with these properties.

2.4.   RESOLUTION OF RANDOM MOTION INTO AN AGGREGATE OF ELEMENTARY BEAMS

The number of electrons that are contained within the element of volume $d\mathbf{r}$ of configuration space at time t, and whose velocities are restricted in magnitude and direction to the range $d\mathbf{c} = c^2\, dc\, d\omega = c^2 \sin\theta\, d\theta\, d\phi\, dc$, has been expressed in the form $n(\mathbf{r}, t)\, d\mathbf{r} f(\mathbf{r}, \mathbf{c}, t)\, d\mathbf{c}$. These electrons are, however, moving through $d\mathbf{r}$ at velocities confined within the narrow range $d\mathbf{c}$ about $\mathbf{c}$ and may therefore be thought to belong to an almost parallel beam of electrons moving with almost the same velocity $\mathbf{c}$. Since the number of particles in $d\mathbf{c}\,d\mathbf{r}$ at time t is $n(\mathbf{r},t)f(\mathbf{c},\mathbf{r},t)\,d\mathbf{c}\,d\mathbf{r}$, the number density of the associated elementary beam (Section 2.2) is $n(\mathbf{r},t)$ $f(\mathbf{c},\mathbf{r},t)\,d\mathbf{c}\equiv nf\,d\mathbf{c}$; moreover $nf\,d\mathbf{c}\,d\mathbf{r}$ is the number of points within $d\mathbf{c}$ in velocity space associated with $d\mathbf{r}$. In the whole group of electrons $\int_{\mathbf{r}} n\,d\mathbf{r}$, the velocities $\mathbf{c}$ range over all directions and speeds c and the total motion may be regarded as an aggregate of elementary beams in which the electrons move with velocities having a small spread about the velocity $\mathbf{c}$.

Since the total number of points in velocity space associated with $d\mathbf{r}$ is equal to the number of electrons $n(\mathbf{r},t)\,d\mathbf{r}$, it follows that

$$n(\mathbf{r},t)\,d\mathbf{r} = n(\mathbf{r},t)\,d\mathbf{r}\int_{\mathbf{c}} f(\mathbf{c},\mathbf{r},t)\,d\mathbf{c},$$

whence

$$\int_{\mathbf{c}} f(\mathbf{c},\mathbf{r},t)\,d\mathbf{c} = 1. \qquad (2.10)$$

The molecules M also move at random with velocities $\mathbf{C}$, and as was done for electrons we define a distribution function $F(\mathbf{C},\mathbf{r},t)$ such that the number of molecules contained within $d\mathbf{r}\,d\mathbf{C}$ at time t when their number density is $N(\mathbf{r},t)$ is

$$N(\mathbf{r},t)\,d\mathbf{r}\,F(\mathbf{C},\mathbf{r},t)\,d\mathbf{C},$$

where $d\mathbf{C}$ is an elementary range of velocities about $\mathbf{C}$.

Since we shall deal only with gases in thermal equilibrium, the directions of the velocities $\mathbf{C}$ are distributed isotropically and the distribution function F and the number density N are independent of both time t and position $\mathbf{r}$. Moreover, F is a function of the speeds C and does not depend on the directions of the velocities $\mathbf{C}$.

We postulate that the number density n of the electrons is everywhere so small compared with N that physical quantities such as momentum and energy imparted to the electrons by an electric force $e\mathbf{E}$, and transferred subsequently to the molecules, have a completely negligible effect on the overall random motion of the molcules. It follows that $N(\mathbf{r},t)\equiv N$; $F(\mathbf{C},\mathbf{r},t)$ $\equiv F(C)$; and the number of molecules within $d\mathbf{r}$ whose velocities are restricted to a range $d\mathbf{C}$ about $\mathbf{C}$ is, at all times,

$$NF(C)\,d\mathbf{C}\,d\mathbf{r} \equiv (N\,d\mathbf{r})F(C)C^2\,dC\,d\Omega,$$

where $d\Omega$ is an elementary solid angle within which the directions of the velocities $\mathbf{C}$ are restricted. It is known from the dynamical theory of gases that $F(C)$ is Maxwell's distribution function.

It follows from what has preceded that we are able to reduce the investigation of the interaction of electrons in random or nearly random motion among the molecules of a gas to that of the pairs of elementary beams into which the motion of the electrons and that of the molecules can be resolved. It follows from equation 2.2 that the number of electrons scattered from $d\mathbf{r}$ in time dt through the interaction of the elementary beam of electrons $n(\mathbf{r},t)f(\mathbf{c},\mathbf{r},t)d\mathbf{c}$ with the elementary molecular beam $NF(C)d\mathbf{C}$ is

$$\Delta n_s = (dt\,d\mathbf{r})n(\mathbf{r},t)f(\mathbf{c},\mathbf{r},t)\,d\mathbf{c}\,NF(C)\,d\mathbf{C}\,gq_0(g)$$

$$\equiv (dt\,d\mathbf{r})(nNg)q_0(g)fF\,d\mathbf{c}\,d\mathbf{C}, \tag{2.11}$$

in which $\mathbf{g} = \mathbf{c} - \mathbf{C}$.

The total number scattered in time dt from $d\mathbf{r}$ is found as the sum of those scattered through the interaction of each elementary beam of electrons with each elementary molecular beam:

$$n_s = (dt\,d\mathbf{r})nN\int_{\mathbf{c}}\int_{\mathbf{C}} gq_0(g)f(\mathbf{c},\mathbf{r},t)F(C)\,d\mathbf{c}\,d\mathbf{C}. \tag{2.12}$$

2.5.   THE MAXWELL-BOLTZMANN EQUATION

2.5.1.   INTRODUCTION. In the theory of electron motion in gases, which in the present context is the theory of diffusion and drift in an electric field, the product $nf \equiv n(\mathbf{r},t)f(\mathbf{r},\mathbf{c},t)$ is fundamental, and we therefore need to determine it in a given physical situation as a function of $\mathbf{r}$, $\mathbf{c}$, and t. Alternatively we may seek to determine the functions $n(\mathbf{r},t)$ and $f(\mathbf{c},\mathbf{r},t)$ individually.

The first step is to formulate a differential equation that is to be satisfied by the function $n(\mathbf{r},t)f(\mathbf{r},\mathbf{c},t)$. This equation is the Maxwell-Boltzmann equation, which is an equation of continuity for the population $nf\,d\mathbf{r}\,d\mathbf{c}$ of the special class of electrons discussed in Section 2.3. We refer to these electrons as the class $nf\,d\mathbf{r}\,d\mathbf{c}$.

2.5.2.   FORMULATION OF THE MAXWELL-BOLTZMANN EQUATION. We consider the various processes that operate to change the population of the class $nf\,d\mathbf{r}\,d\mathbf{c}$. The formal expression for the rate of change of the population is $(\partial/\partial t)(nf)\,d\mathbf{r}\,d\mathbf{c}$, and this rate of change is a consequence of three processes.

(*a*) There is a loss of population from the transport of electrons across the boundary surface of the elementary volume $d\mathbf{r}$ in configuration space. This transport arises from the fact that $d\mathbf{r}$ is immersed in the elementary beam $nf\,d\mathbf{c}$; however, since nf is not constant along or across the beam in general, there is a net loss of electrons in time dt equal to $\mathrm{div}_r\,(\mathbf{c}nf)\,d\mathbf{r}\,d\mathbf{c}\,dt$, where div_r is the divergence operator in configuration space.

(*b*) In the absence of an electric field electrons move at constant velocities between encounters with molecules, and their velocity points remain at rest in velocity space between encounters. In the presence of an electric field $\mathbf{E}$ an acceleration $e\mathbf{E}/m$ is imparted to each of the $n\,d\mathbf{r}$ electrons whose velocity points, as a consequence, all drift in velocity space at the common rate $e\mathbf{E}/m$. Velocity points are therefore lost from the elementary region $d\mathbf{c}$ by drift of points across its surface. Hence the number of points lost in time dt from $d\mathbf{c}$ is $dt\,\mathrm{div}_c\,(nfe\mathbf{E}/m)\,d\mathbf{c}\,d\mathbf{r}$, where div_c is the divergence operator in velocity space. This net loss of points is also a loss of population from the class $nf\,d\mathbf{r}\,d\mathbf{c}$.

(*c*) Points are also lost from $d\mathbf{c}$ because of the quasi discontinuous changes in position $\Delta\mathbf{c}$ in velocity space that occur when an electron encounters a molecule. The discontinuous changes in velocity remove points from $d\mathbf{c}$ to another region of velocity space. The loss of points in time dt from this process is Δn_s of equation 2.11. In addition $d\mathbf{c}$ receives points through displacements $\Delta\mathbf{c}$, due to encounters, from other elements $d\mathbf{c}$ of velocity space. We therefore express the net loss, which is the excess of the loss over the gain in time dt, symbolically by $S\,d\mathbf{c}\,d\mathbf{r}\,dt$.

When the expressions in (*a*) to (*c*) above are assembled to form an equation, we obtain, after removing the common factor $dt\,d\mathbf{r}\,d\mathbf{c}$,

$$\frac{\partial}{\partial t}(nf) + \mathrm{div}_r\,(nf\mathbf{c}) + \mathrm{div}_c\left(nf\frac{e\mathbf{E}}{m}\right) + S = 0. \tag{2.13}$$

Equation 2.13 is the Maxwell-Boltzmann equation to be satisfied by nf. Since $\mathbf{c}$ is independent of $\mathbf{r}$ and $e\mathbf{E}/m$ is independent of $\mathbf{c}$, equation 2.13 may be given the form (e.g., Chapman and Cowling, 1952)

$$\frac{\partial}{\partial t}(nf) + \mathbf{c}\cdot\mathrm{grad}_r\,(nf) + \frac{e\mathbf{E}}{m}\cdot\mathrm{grad}_c\,(nf) + S = 0. \tag{2.14}$$

Although in many theoretical investigations equation 2.14 is adopted as the basic form, we shall, in what follows, find it more convenient to base the theory of diffusion and drift on the fundamental equation 2.13, which is in effect an equation of continuity in time and six-dimensional space.

2.6.   THE MAXWELL-BOLTZMANN EQUATION FOR VELOCITY SHELLS

2.6.1.   VELOCITY SHELLS. We shall find frequent occasion to use a limited region of velocity space which, following Maxwell, we designate a shell. A shell is a thin region of velocity space bounded by spherical surfaces with radii c and $c + dc$ and with the origin as centre. This shell is denoted (c, dc).

The whole of velocity space may therefore be subdivided into shells (c, dc) whose radii c range in magnitude from zero to infinity, although the ultimate shell at the origin is an infinitesimal spherical volume with radius dc. In the absence of an electric force and of a gradient in the number density at the position $\mathbf{r}$, all directions of the velocities $\mathbf{c}$ of the $n\,d\mathbf{r}$ electrons are equally probable, and the $n\,d\mathbf{r}$ velocity points are isotropically distributed in velocity space. In particular the velocity points, on the average, are uniformly distributed within each velocity shell (c, dc). Let $d\mathbf{c} = c^2\,dc\,d\omega$ be an elementary region of the shell (c, dc), where $d\omega = \sin\theta\,d\theta\,d\phi$ and (θ, ϕ) is the direction of a velocity $\mathbf{c}$ whose end point lies in $d\mathbf{c}$. The population of velocity points in $d\mathbf{c}$ is therefore independent of the direction (θ, ϕ) as $d\mathbf{c}$ is displaced over the shell (c, dc), when drift due to $e\mathbf{E}$ and diffusion due to $\mathrm{grad}_r(nf)$ are absent at position $\mathbf{r}$ in configuration space. In other words the vector resultant of the velocities $\mathbf{c}$ associated with the velocity points of a shell (c, dc) is zero, and the $n\,d\mathbf{r}$ electrons move at random but their centroid remains at rest.

When, however, an electric force $e\mathbf{E}$ acts on each electron, or diffusion due to a gradient $\mathrm{grad}_r(nf)$ is present, or both, at position $\mathbf{r}$, the group of $n\,d\mathbf{r}$ electrons moves as a whole through the gas. Their centroid is no longer at rest, and their motion no longer completely random. It follows that in such circumstances the velocities $\mathbf{c}$ associated with a shell (c, dc) have, in general, a resultant $\sum\mathbf{c}$ which is not equal to zero, and the velocity points are no longer uniformly distributed throughout the shell. The total point population of a shell is

$$n_c = (n\,d\mathbf{r})\int_\omega f(\mathbf{c}, \mathbf{r}, t)\,d\mathbf{c} = (n\,d\mathbf{r})(c^2\,dc)\int_\omega f(\mathbf{c}, \mathbf{r}, t)\,d\omega,$$

so that the mean velocity of the electrons associated with the shell (c, dc), that is, the convective velocity, which we denote as $\mathbf{W}_{\mathrm{conv}}(c)$, is

$$\mathbf{W}_{\mathrm{conv}}(c) = \frac{\sum_\mathbf{c}\mathbf{c}}{n_c},$$

and the mean velocity of all the $n\,d\mathbf{r}$ electrons is

$$\mathbf{W}_{\text{conv}} = \frac{\sum_{c} n_c \mathbf{W}_{\text{conv}}(c)}{n\,d\mathbf{r}}.$$

In order to proceed it is now necessary to introduce a postulate about the distribution of velocity points within a shell (c,dc). Hence it is postulated that the distribution of velocity points within (c,dc) is axially symmetrical about the direction of the mean resultant velocity $\mathbf{W}_{\text{conv}}(c)$ of the velocities c of the electrons of the shell. This direction is taken as the polar axis $\theta = 0$ for the elementary solid angles $d\omega = \sin\theta\,d\theta\,d\phi$ of the elements $d\mathbf{c} = c^2\,dc\,d\omega$ of the shell. The population of velocity points in an element $d\mathbf{c}$, that is, $n\,d\mathbf{r}f(\mathbf{c},\mathbf{r},t)\,d\mathbf{c}$, is now $(n\,d\mathbf{r})(c^2\,dc)[f(c,\theta,\mathbf{r},t)\,d\omega]$, since according to the postulate the population within (c,dc) is distributed with axial symmetry about the direction $\theta = 0$ and is therefore independent of ϕ, the angle of azimuth. The distribution function is, consequently, of the form $f \equiv f(c,\theta,\mathbf{r},t)$, where the axis $\theta = 0$ refers to the shell (c,dc) and is in general not a common direction for all shells when drift due to $e\mathbf{E}$ and diffusion due to $\text{grad}_r(nf)$ are both present.

It is next assumed that for a shell (c,dc) the distribution function $f(c,\theta,\mathbf{r},t)$ may be represented by a convergent series of terms as follows:

$$f(c,\theta,\mathbf{r},t) = f_0(c,\mathbf{r},t) + \sum_{k=1}^{\infty} f_k(c,\mathbf{r},t)P_k(\cos\theta), \qquad (2.15)$$

in which $P_k(\cos\theta)$ is the kth-order Legendre polynomial (see Appendix C). It follows from equation 2.15 that:

(a) the population of velocity points of the shell (c,dc) is

$$n_c = (n\,d\mathbf{r})(c^2\,dc)\int_0^\pi \int_0^{2\pi}\left[f_0 + \sum_{k=1}^{\infty} f_k P_k(\cos\theta)\right]\sin\theta\,d\theta\,d\phi$$

$$= (nf_0)(4\pi c^2\,dc)\,d\mathbf{r}, \qquad (2.16)$$

in which $n \equiv n(\mathbf{r},t)$, $f_0 \equiv f_0(c,\mathbf{r},t)$, $f_k \equiv f_k(c,\mathbf{r},t)$; and

(b) the mean velocity $\mathbf{W}_{\text{conv}}(c)$ in the direction $\theta = 0$ has the magnitude

$$W_{\text{conv}}(c) = \frac{1}{n_c}(n\,d\mathbf{r})(c^2\,dc)\int_0^\pi \int_0^{2\pi}\left[f_0 + \sum_{k=1}^{\infty} f_k P_k(\cos\theta)\right](c\cos\theta)\sin\theta\,d\theta\,d\phi$$

$$= \frac{cf_1}{3f_0}.$$

We introduce a vector $\mathbf{f}_1$ with direction that of the axis $\theta = 0$ and magnitude $f_1 \equiv f_1(c, \mathbf{r}, t)$. The velocity $\mathbf{W}_{\text{conv}}(c)$ is therefore

$$\mathbf{W}_{\text{conv}}(c) = \frac{c\mathbf{f}_1}{3f_0}. \qquad (2.17)$$

The notation $\mathbf{W}_{\text{conv}}(c)$ suggests that this velocity is a velocity of convective flow, although it has been introduced as the mean velocity $\sum \mathbf{c}/n_c$ of the n_c electrons whose velocity points lie in the shell (c, dc); that is, it is the instantaneous velocity of the centroid of these n_c electrons. To arrive at the interpretation of $\mathbf{W}_{\text{conv}}(c)$ as a velocity of convective flow, let $d\mathbf{S}$ be an element of surface at position $\mathbf{r}$ in configuration space and let the direction of $d\mathbf{S}$ (i.e., of the normal to dS) be (α, ψ) in a system of coordinates in which the direction of $\mathbf{W}_{\text{conv}}(c)$ is the polar axis. Consider the $(n\,d\mathbf{r})$ $f(\mathbf{c}, \mathbf{r}, t)\,d\mathbf{c}$ points in the element $d\mathbf{c} = c^2\,dc\,d\omega = c^2\,dc\sin\theta\,d\theta\,d\phi$ of the shell (c, dc) that includes the end point of the velocity $\mathbf{c}$ whose direction is (θ, ϕ). The angle θ' between $\mathbf{c}$ and $d\mathbf{S}$ is such that

$$\cos\theta' = \cos\alpha\cos\theta + \sin\alpha\sin\theta\cos(\psi - \phi).$$

Associated with $d\mathbf{c}$ is the elementary beam of electrons with strength at position $\mathbf{r}$ equal to $(nf)\,d\mathbf{c}$. In time dt this beam transports $dt\,(nf\,d\mathbf{c})$ $c\,dS\cos\theta'$ electrons across dS into the region containing the normal. The total net transport of electrons in the sense of the normal by all the elementary beams associated with the shell (c, dc) is

$$dt\,(nc^3\,dc)\,dS\sum f\cos\theta'\,d\omega = dt\,(nc^3\,dc)\,dS\int_0^\pi\int_0^{2\pi}\left[f_0 + \sum_{k=1}^\infty f_k P_k(\cos\theta)\right]$$

$$\times\,[\cos\alpha\cos\theta + \sin\alpha\sin\theta\cos(\psi - \phi)]\sin\theta\,d\theta\,d\phi$$

$$= dt\,(nc^3\,dc)\,dS\frac{4\pi}{3}f_1\cos\alpha = dt\,(4\pi nf_0c^2\,dc)\frac{c\mathbf{f}_1}{3f_0}\cdot d\mathbf{S}$$

$$= dt\,\frac{n_c}{d\mathbf{r}}\mathbf{W}_{\text{conv}}(c)\cdot d\mathbf{S} = dt\,n(c, t)\mathbf{W}_{\text{conv}}(c)\cdot d\mathbf{S}$$

from equations 2.16 and 2.17; $n(c, t) = n_c/d\mathbf{r}$ is the contribution to the number density at $\mathbf{r}$ of the electrons with speeds in the range c to $c + dc$.

It follows that $\mathbf{W}_{\text{conv}}(c)$ is not only the instantaneous velocity of the centroid of the n_c electrons of the shell (c, dc) but also the velocity of convective flow across $d\mathbf{S}$ of electrons with speeds c.

The total convective velocity, which includes electrons of all speeds, is

$$\mathbf{W}_{conv} = \frac{1}{n} \sum_c n(c) \mathbf{W}_{conv}(c) = \frac{1}{n} \sum_c n(c) \frac{c\mathbf{f}_1(c)}{3f_0(c)}. \qquad (2.18)$$

2.7. ADAPTATION OF MAXWELL-BOLTZMANN EQUATION TO A COMPLETE SHELL

2.7.1. THE SCALAR EQUATION. The Maxwell-Boltzmann equation in the form given in equation 2.13 refers to the rate of change of the population of the class $nf\,d\mathbf{r}\,dc$, where dc is the element $c^2\,dc\,d\omega$ of velocity space. Since dc depends on $d\omega = \sin\theta\,d\theta\,d\phi$, its point population, as already remarked, is a function of the direction (θ, ϕ) and, in particular, of the angle θ. The angle θ therefore enters implicitly into equation 2.13. We are interested ultimately, however, not in the directions of the velocities $\mathbf{c}$ but in deriving formulae for the transport coefficients of diffusion and drift and for the coefficients f_0 and f_1 in the expansion for f as given in equation 2.15. It is therefore expedient to transform equation 2.13 at the outset into a form in which it is liberated from dependence on the directions of the velocities $\mathbf{c}$. This transformation is effected by replacing the class $(nf)\,d\mathbf{r}\,dc$ by the class $(nf_0)\,d\mathbf{r}(4\pi c^2\,dc)$, that is, by the electrons that lie in the volume element $d\mathbf{r}$ in configuration space and that move with speeds between c and $c + dc$. The velocity points for these electrons therefore lie somewhere within the complete shell (c, dc).

Members are lost from this extended class by transport across the bounding surface of $d\mathbf{r}$ in configuration space and by loss of velocity points from the shell (c, dc) through two processes: the general drift $e\mathbf{E}/m$ of points in velocity space, which carries points into and out of (c, dc) across its boundaries, and the discontinuous displacements $\Delta\mathbf{c}$ from encounters that remove points from (c, dc) or introduce other points. We therefore proceed as before by separately formulating the contributions of these processes and then assembling them into an equation.

(a) The formal expression for the rate of increase of the population of the class is

$$d\mathbf{r}(4\pi c^2\,dc)\frac{\partial}{\partial t}(nf_0).$$

(b) Let $d\mathbf{S}$ be an element of the bounding surface of $d\mathbf{r}$ with the direction of $d\mathbf{S}$ outwards from the enclosed volume. It follows from the discussion preceding equation 2.18 that the excess of the number of particles that cross $d\mathbf{S}$ outwards in time dt over the number that cross inwards is

$$dt(4\pi c^2 dc)\frac{cn\mathbf{f}_1}{3}\cdot d\mathbf{S}.$$

The net loss of electrons from $d\mathbf{r}$ is therefore

$$dt\,d\mathbf{r}(4\pi c^2 dc)\frac{c}{3}\operatorname{div}_r(n\mathbf{f}_1).$$

(c) We consider next the loss of velocity points from (c, dc) through the common drift eE/m of velocity points.

Take an element $c^2 d\omega$ of the spherical surface with radius c and centre the origin, and let $d\omega = \sin\theta\,d\theta\,d\phi$ as before. Let the direction of $e\mathbf{E}$ be (θ', ϕ') in the system in which the direction of $\mathbf{f}_1$ is the polar axis $\theta = 0$. The angle α between $e\mathbf{E}$ and the velocity $\mathbf{c}$ with direction (θ, ϕ) is such that

$$\cos\alpha = \cos\theta\cos\theta' + \sin\theta\sin\theta'\cos(\phi - \phi').$$

The density of points in velocity space at the surface element $c^2 d\omega$ is

$$(nf)\,d\mathbf{r} \equiv n\,d\mathbf{r}\left[f_0 + \sum_{k=1}^{\infty} f_k P_k(\cos\theta)\right];$$

consequently the rate at which points cross $c^2 d\omega$ outwards is

$$(c^2 d\omega)(d\mathbf{r}\,nf)\frac{eE}{m}\cos\alpha.$$

The total number of points transported outwards across the complete spherical surface in time dt is therefore

$$c^2 d\mathbf{r}\,n\frac{eE}{m}\,dt\int_0^\pi\int_0^{2\pi}\left[f_0 + \sum_{k=1}^{\infty}f_k P_k(\cos\theta)\right]$$

$$\times\left[\cos\theta\cos\theta' + \sin\theta\sin\theta'\cos(\phi - \phi')\right]\sin\theta\,d\theta\,d\phi$$

$$=\left(\frac{4\pi}{3}c^2\frac{eE}{m}f_1\cos\theta'\right)n\,d\mathbf{r}\,dt$$

$$=\left(\frac{4\pi}{3}c^2\frac{eE}{m}\cdot n\mathbf{f}_1\right)d\mathbf{r}\,dt$$

$$=\sigma_E(c)\,d\mathbf{r}\,dt,$$

where

$$\sigma_E(c) = \frac{4\pi}{3} c^2 \frac{e\mathbf{E}}{m} \cdot n\mathbf{f}_1. \tag{2.19}$$

The loss of points from (c, dc) from this cause is therefore

$$dt \, d\mathbf{r} \, dc \frac{\partial}{\partial c} \sigma_E(c).$$

(d) The displacements $\Delta\mathbf{c}$ in velocity space brought about by encounters cause points to cross the spherical surface both inwards and outwards. We represent the excess of points that cross inwards over those that cross outwards in time dt by

$$dt \, d\mathbf{r} \sigma_{\text{coll}}(c).$$

The net gain of points to (c, dc) brought about by encounters in time dt is therefore

$$dt \, d\mathbf{r} \, dc \frac{\partial}{\partial c} \sigma_{\text{coll}}(c).$$

It follows that the net loss of points to (c, dc) from the force $e\mathbf{E}$ and encounters is

$$\text{net loss} = (dt \, d\mathbf{r} \, dc) \frac{\partial}{\partial c} [\sigma_E(c) - \sigma_{\text{coll}}(c)]. \tag{2.20}$$

In general $\sigma_{\text{coll}}(c)$ is not represented by a simple formula, but when all encounters are elastic $\sigma_{\text{coll}}(c)$ is given by

$$\sigma_{\text{coll}}(c) = 4\pi n c^2 \nu_{\text{el}} \left(\frac{m}{M} c f_0 + \frac{\overline{C^2}}{3} \frac{\partial f_0}{\partial c} \right), \tag{2.21}$$

where $\nu_{\text{el}} = N c q_{m_{\text{el}}}(c)$ and $\overline{C^2}$ is the mean-square speed of the molecules. Here N is the molecular density, M the mass of a molecule, and $q_{m_{\text{el}}}(c)$ the momentum transfer cross section (equation 2.9) for elastic encounters. Formula 2.21 is established in Appendix B.

On gathering terms, the following equation, which replaces equation 2.13, is obtained after cancellation of common factors;

$$\frac{\partial}{\partial t}(nf_0) + \frac{c}{3} \operatorname{div}_{\mathbf{r}}(n\mathbf{f}_1) + \frac{1}{4\pi c^2} \frac{\partial}{\partial c} [\sigma_E(c) - \sigma_{\text{coll}}(c)] = 0. \tag{2.22}$$

Because all terms in equation 2.22 are scalar quantities, we shall refer to this equation or its special forms as the *scalar equation* to distinguish it from another equation that we shall establish, the vector equation, in which the terms are vector quantities.

Equation 2.22 is an equation in which two dependent variables, nf_0 and nf_1, appear, and in order to determine either or both a second relation between them is required. This second relationship is the vector equation mentioned above. We proceed to establish it.

2.7.2.   THE VECTOR EQUATION. The scalar equation 2.22 is an expression of the law of conservation of mass or of particle number. It can also be shown to result from an application of the law of conservation of energy to the motion of the electrons.

A second and independent equation is obtained by considering the momentum of the electrons of the shell (c, dc). As was done for the population of the class, we consider in turn the processes that operate to change the momentum.

(a) The total momentum of those of the $n\,d\mathbf{r}$ electrons whose velocity points lie in (c, dc) has the direction of $\mathbf{f}_1$ and the magnitude

$$n\,d\mathbf{r}(c^2\,dc)\,mc \int_\omega (f\cos\theta)\,d\omega$$

$$= n\,d\mathbf{r}(mc^3\,dc) \int_0^\pi \int_0^{2\pi} \left[f_0 + \sum_{k=1}^\infty f_k P_k(\cos\theta) \right] \cos\theta \sin\theta\, d\theta\, d\phi$$

$$= d\mathbf{r}(4\pi c^2\,dc)nm\frac{cf_1}{3}$$

$$= d\mathbf{r}(4\pi nf_0 c^2\,dc)m\frac{cf_1}{3f_0}.$$

The rate of increase of this momentum is

$$dc\,d\mathbf{r}\left(\frac{4\pi}{3}c^3 m \right) \frac{\partial}{\partial t}(n\mathbf{f}_1).$$

(b) We consider next the loss of momentum through loss of electrons from $d\mathbf{r}$ by transport across its boundary.

Take an element of surface $d\mathbf{S}$ which is a portion of a finite closed surface within and on which n is not everywhere equal to zero. In considering the transport of momentum across $d\mathbf{S}$, take for the polar axis of reference the direction of f_1 at dS. Let the direction of a velocity c be (θ, ϕ) and that of dS be (α, ψ). The angle β between dS and c is such that

$$\cos\beta = \cos\alpha \cos\theta + \sin\alpha \sin\theta \cos(\psi - \phi).$$

The elementary beam $nf\,dc$, with velocity c, carries momentum across dS in time dt equal to (with $dc = c^2\,dc\,d\omega = c^2\,dc\sin\theta\,d\theta\,d\phi$)

$$dt\,dc\,nfcmc \cdot dS = dt\,dc\,fnmc^2\,dS\cos\beta.$$

The component of this momentum in the direction of f_1 is

$$dt\,dc\,nfmc^2\,dS\cos\beta\cos\theta$$

$$= dt(c^2\,dc)nmc^2\,dS\left[f_0 + \sum_{k=1}^{\infty} f_k P_k(\cos\theta) \right]$$

$$\times \cos\theta\,[\cos\alpha\cos\theta + \sin\alpha\sin\theta\cos(\psi - \phi)]\,d\omega.$$

On using the formula $\cos^2\theta = [2P_2(\cos\theta) + 1]/3$ and after integration over all directions (θ, ϕ) of the elementary beams, it is found that the total momentum carried across dS in time dt has a component in the direction of f_1 equal to

$$dt(c^2\,dc)nmc^2\,dS\left(\frac{4\pi}{3}f_0 + \frac{4\pi}{5}\frac{2}{3}f_2 \right)\cos\alpha.$$

Since dS can be given any orientation, the angle α can be assigned any arbitrary value. The resolved part in the fixed direction of f_1 of the total momentum carried across dS is thus proportional to $(dS \cdot f_1)/f_1$. The total momentum is therefore in the direction of dS and of magnitude, independent of the orientation of dS, equal to

$$dt(c^2\,dc)nmc^2\,dS\frac{4\pi}{3}(f_0 + \tfrac{2}{5}f_2).$$

The momentum transported in time dt is therefore

$$dt(4\pi c^2\,dc)\left(\frac{mc^2}{3} \right)(nf_0 + \tfrac{2}{5}nf_2)\,dS.$$

It follows that the vector sum of the momenta transported outwards from the closed surface Σ of which dS is an element directed outwards from the enclosed volume is

$$dt(4\pi c^2 \, dc)\left(\frac{mc^2}{3}\right) \iint_\Sigma (nf_0 + \tfrac{2}{5}nf_2)\,d\mathbf{S}.$$

Let λ, μ, and ν be the direction cosines of $d\mathbf{S}$, and $\mathbf{i}_x$, $\mathbf{i}_y$, and $\mathbf{i}_z$ be unit vectors parallel respectively to the axes of coordinates. Consider

$$\iint_\Sigma (nf_0 + \tfrac{2}{5}nf_2)\,d\mathbf{S} = \iint_\Sigma (nf_0 + \tfrac{2}{5}nf_2)(\lambda \mathbf{i}_x + \mu \mathbf{i}_y + \nu \mathbf{i}_z)\,dS.$$

From Green's theorem

$$\iint_\Sigma \lambda(nf_0 + \tfrac{2}{5}nf_2)\,dS = \int\!\!\int\!\!\int_\mathbf{r} \frac{\partial}{\partial x}(nf_0 + \tfrac{2}{5}nf_2)\,d\mathbf{r},$$

and it follows that

$$\iint_\Sigma (nf_0 + \tfrac{2}{5}nf_2)\,d\mathbf{S} = \int\!\!\int\!\!\int_\mathbf{r} \mathrm{grad}_r(nf_0 + \tfrac{2}{5}nf_2)\,d\mathbf{r}.$$

Let Σ shrink to become the surface that encloses $d\mathbf{r}$; then

$$\iint_\Sigma (nf_0 + \tfrac{2}{5}nf_2)\,d\mathbf{S} \to \mathrm{grad}_r(nf_0 + \tfrac{2}{5}nf_2)\,d\mathbf{r}.$$

The loss of momentum from $d\mathbf{r}$ in time dt associated with electrons of the shell (c, dc) is therefore

$$(dt\,dc\,d\mathbf{r})(4\pi c^2)\left(\frac{mc^2}{3}\right)\mathrm{grad}_r(nf_0 + \tfrac{2}{5}nf_2).$$

(c) The force $e\mathbf{E}$ imparts momentum to each electron, and the class of electrons associated with the shell (c, dc) gains momentum from the field.

Let, as in Section 2.7.1c, the direction of the polar axis $\theta = 0$ be that of $\mathbf{f}_1$, that of $e\mathbf{E}$ be (α, ψ), and that of a velocity $\mathbf{c}$ be (θ, ϕ). Consider a surface element $c^2\,d\omega$ of the sphere c at the end of the radius $\mathbf{c}$. The angle between $e\mathbf{E}$ and $\mathbf{c}$ is θ', with $\cos\theta' = \cos\alpha\cos\theta + \sin\alpha\sin\theta\cos(\psi - \phi)$.

The number of points that drift outwards across $c^2\,d\omega$ in time dt is $dt\,d\mathbf{r}nf\cdot(eE/m)\cos\theta'\cdot c^2\,d\omega$, and the momentum associated with these points is $dt\,d\mathbf{r}\,nf\cdot mc(eE/m)\cos\theta'\cdot c^2\,d\omega$, with a component along $\mathbf{f}_1$ equal to

$$dt\,d\mathbf{r}\,eEc^3 n\left[f_0 + \sum_{k=1}^\infty f_k P_k(\cos\theta)\right]\cos\theta\cos\theta'\sin\theta\,d\theta\,d\phi.$$

As before, replace $\cos^2\theta$ by $[2P_2(\cos\theta) + 1]/3$ and integrate over the whole surface of the sphere c, that is, over all directions (θ, ϕ), to obtain the total component along $\mathbf{f}_1$ of the momentum of the electrons whose velocity points cross the sphere c outwards. This component is, in time dt,

$$(dt\, d\mathbf{r})\frac{4\pi c^3}{3}(nf_0 + \tfrac{2}{5}nf_2)eE\cos\alpha.$$

Since eE can be given any direction, it follows that the total momentum itself is parallel to eE and equal to

$$(dt\, d\mathbf{r})\left(\frac{4\pi c^3}{3}\right)(nf_0 + \tfrac{2}{5}nf_2)e\mathbf{E}.$$

Consider the region of velocity space external to the sphere with radius c and centre the origin. The momentum imparted in time dt to the electrons whose velocity points lie in this region is

$$(dt\, d\mathbf{r})n\left(\int_c^\infty 4\pi f_0 c^2\, dc\right)e\mathbf{E}.$$

The total gain of momentum to the region from the force eE directly and from the addition of points by passage across the inner boundary, at the sphere c, is

$$(dt\, d\mathbf{r})(4\pi ne\mathbf{E})\left[\int_c^\infty f_0 c^2\, dc + \frac{c^3}{3}(f_0 + \tfrac{2}{5}f_2)\right].$$

But

$$\int_c^\infty f_0 c^2\, dc = \frac{c^3 f_0}{3}\Big|_c^\infty - \int_c^\infty \frac{c^3}{3}\frac{\partial f_0}{\partial c}\, dc = -\frac{c^3 f_0}{3} - \int_c^\infty \frac{c^3}{3}\frac{\partial f_0}{\partial c}\, dc.$$

The total gain in momentum to this class of electrons is therefore

$$(dt\, d\mathbf{r})\frac{4\pi}{3}\left[-\int_c^\infty c^3\frac{\partial f_0}{\partial c}\, dc + \tfrac{2}{5}c^3 f_2(c)\right]ne\mathbf{E}.$$

The gain in momentum to the electrons associated with the region external to the sphere $(c + dc)$ is

$$(dt\, d\mathbf{r})\frac{4\pi}{3}\left[-\int_{c+dc}^\infty c^3\frac{\partial f_0}{\partial c}\, dc + \tfrac{2}{5}(c+dc)^3 f_2(c+dc)\right]ne\mathbf{E}.$$

It follows that the gain in momentum to the class of electrons whose velocity points lie in the shell (c, dc) is

$$(dt\,d\mathbf{r})\frac{4\pi}{3}\left\{\left(-\int_c^{c+dc} c^3\frac{\partial f_0}{\partial c}\,dc\right)-\tfrac{2}{5}dc\frac{\partial}{\partial c}[c^3f_2(c)]\right\}ne\mathbf{E},$$

and the loss of momentum, which is the negative of this quantity, is

$$(dt\,d\mathbf{r}\,dc)\frac{4\pi}{3}\left\{c^3\frac{\partial f_0}{\partial c}+\frac{2}{5}\frac{\partial}{\partial c}[c^3f_2(c)]\right\}ne\mathbf{E}.$$

(d) There is also a loss of momentum from the transference of momentum to molecules in encounters. It follows from equation 2.8 that the momentum transferred within $d\mathbf{r}$ by the elementary beam with strength $nf\,d\mathbf{c}$ to the molecules of the elementary beam $NF(C)d\mathbf{C}$ in time dt is, with elastic encounters,

$$dt\,d\mathbf{r}\,nf(\mathbf{c})(c^2\,dc\sin\theta\,d\theta\,d\phi)NF(C)\,d\mathbf{C}gq_{m_{el}}(g)\left(\frac{M}{M+m}\right)m\mathbf{g}$$

$$\cong(dt\,d\mathbf{r})nf(\mathbf{c})(c^3\,dc\sin\theta\,d\theta\,d\phi)q_{m_{el}}(c)(m\mathbf{c})NF(C)\,d\mathbf{C}$$

since $m/M\ll1$; $M/(M+m)\cong1$; $g\cong c$. But

$$\int_C F(C)\,d\mathbf{C}=1.$$

Consequently integration over all elementary beams $NF(C)d\mathbf{C}$ gives for the total loss of momentum

$$(dt\,d\mathbf{r})nNq_{m_{el}}(c)(c^3\,dc)f(\mathbf{c})(\sin\theta\,d\theta\,d\phi)m\mathbf{c}.$$

The vector sum of the components of the momentum normal to $\mathbf{f}_1$ transferred to molecules is zero; consequently the momentum transferred is the sum of the components along $\mathbf{f}_1$, which is

$$(dt\,d\mathbf{r}\,dc)[nNcq_{m_{el}}(c)]c^3m\int_0^\pi\int_0^{2\pi}\left[f_0+\sum_{k=1}^\infty f_kP_k(\cos\theta)\right]\cos\theta\sin\theta\,d\theta\,d\phi$$

$$=(dt\,d\mathbf{r}\,dc)[nNcq_{m_{el}}(c)]c^3m\frac{4\pi}{3}\mathbf{f}_1$$

$$=(dt\,d\mathbf{r}\,dc)\frac{4\pi}{3}mc^3v_{el}(n\mathbf{f}_1)=(dt\,d\mathbf{r})(4\pi nf_0c^2\,dc)v_{el}m\mathbf{W}_{conv}(c)$$

with $v_{el}=Ncq_{m_{el}}(c)$.

When inelastic encounters are also significant, the expression for the net loss of momentum to the class is modified. Consider first a single type of inelastic encounter for which the critical energy is $\frac{1}{2}mc_{in}^2$, and the total collision cross section is $q_{0_{in}}(c)I(c-c_{in})$, where I is the unit step function. The additional loss of points to the shell (c,dc) from inelastic encounters is

$$dt\,d\mathbf{r}(4\pi c^2\,dc)(nf_0)\left[Ncq_{0_{in}}(c)I(c-c_{in})\right],$$

and the associated loss of momentum is

$$dt\,d\mathbf{r}\left(\frac{4\pi}{3}c^2\,dc\right)(n\mathbf{f}_1)\left[Ncq_{0_{in}}(c)I(c-c_{in})\right]mc$$

$$=dt\,d\mathbf{r}\,dc\frac{4\pi}{3}m\nu_{0_{in}}I(c-c_{in})c^3n\mathbf{f}_1.$$

Points are also entering the shell (c,dc) as a consequence of inelastic encounters, but the shell from which they originate is not contiguous, in general, to (c,dc) but possesses a radius c_1 such that $c_1^2=c^2+c_{in}^2$. The effective thickness dc_1 of this shell is such that $c_1\,dc_1=c\,dc$. It follows that the momentum contributed to the class is (from equation 2.9)

$$dt\,d\mathbf{r}\frac{4\pi}{3}mNc_1q_{0_{in}}(c_1)c_1{}^2n\mathbf{f}_1(c_1)c_1\,dc_1\left[\frac{(c/c_1)q_{1_{in}}(c_1)}{q_{0_{in}}(c_1)}\right]$$

$$=dt\,d\mathbf{r}\,dc\frac{4\pi}{3}mnNc_1q_{1_{in}}(c_1)\mathbf{f}_1(c_1)c_1c^2.$$

When several forms k of inelastic encounters are active, the net loss of momentum to the shell (c,dc) is

$$dt\,d\mathbf{r}\left(\frac{4\pi}{3}c^2\,dc\right)mc\nu(c)n\mathbf{f}_1(c),$$

where

$$\nu(c)=\nu_{el}(c)+\sum_k\left\{\nu_{0k_{in}}(c)I(c-c_{kin})-\left[Nc_{1k}q_{1k_{in}}(c_{1k})\right]\frac{c_{1k}}{c}\frac{f_1(c_{1k})}{f_1(c)}\right\}$$

and

$$\nu_{0k_{in}}(c)=Ncq_{0k_{in}}(c).$$

The factor $\nu(c)$ appears as an effective or apparent collision frequency, as if all encounters referred to electrons of the shell (c, dc). The effective collision cross section for momentum transfer is

$$q_m(c) = \frac{\nu(c)}{Nc} = q_{m_{el}}(c) + \sum_k \left[q_{0k_{in}}(c) I(c - c_{k_{in}}) - \left(\frac{c_{1k}}{c} \right)^2 q_{1k_{in}}(c_{1k}) \frac{f_1(c_{1k})}{f_1(c)} \right].$$

When inelastic encounters are unimportant, as in many circumstances in monatomic gases, $q_m(c) = q_{m_{el}}(c)$.

Another simple behaviour is that in which inelastic scattering may be regarded as isotropic, in which event $q_{1k_{in}}(c_{1k}) = 0$. The expression for $q_m(c)$ then reduces to

$$q_m(c) = q_{m_{el}}(c) + \sum_k q_{0k_{in}}(c) I(c - c_{k_{in}}).$$

Inelastic encounters are considered in greater detail in Chapter 6.

(e) There is also a small gain of momentum due to the fact that scattering is about the centroid of the system rather than the centre of the molecule. It may be shown that when all encounters are elastic this gain to the shell (c, dc) is less than

$$(dt \, d\mathbf{r} \, dc) \left(\frac{m}{M} \right) \left(\frac{4\pi}{3} mn \right) \frac{d}{dc} \left(\frac{c^4 \nu_{el} q_{1_{el}}}{q_{0_{el}}} \mathbf{f}_1 \right).$$

Since $m/M \cong 10^{-4}$ or 10^{-5}, this term is negligible in comparison with the loss discussed in (d) above. It is therefore ignored.

When all terms (a) to (d) are collected into an equation and common factors are removed, the vector equation is found to be

$$\frac{\partial}{\partial t}(n\mathbf{f}_1) + c \, \mathrm{grad}_r(nf_0 + \tfrac{2}{5}nf_2) + \left[\frac{\partial f_0}{\partial c} + \frac{2}{5} \frac{1}{c^3} \frac{\partial}{\partial c}(c^3 f_2) \right] \frac{ne\mathbf{E}}{m} + \nu n\mathbf{f}_1 = 0.$$

$$(2.23)$$

The presence of the function f_2 is a complication, and f_2 is usually omitted as small for the reason that the drift velocity $\mathbf{W}$ in an electric field is much less in magnitude than the mean speed $\bar{c}$ of the electrons. The motion of the electrons is essentially random, and the coefficients f_0, f_1, and f_2 form a diminishing sequence $f_2 < f_1 \ll f_0$. When f_2 is omitted, the vector equation reduces to

$$\frac{\partial}{\partial t}(n\mathbf{f}_1) + c\operatorname{grad}_r(nf_0) + \frac{e\mathbf{E}}{m}\frac{\partial}{\partial c}(nf_0) + \nu n\mathbf{f}_1 = 0, \qquad (2.24)$$

and this equation, in conjunction with the scalar equation (equation 2.22), is found to give a satisfactory and self-consistent account of electron motion in gases.

2.8. ELECTRON MOTION IN THE PRESENCE OF A MAGNETIC FIELD

The Lorentz force on an electron moving with velocity c in a magnetic field $\mathbf{B}$, assumed to be independent of time t, is

$$\mathbf{F}_L = -e(\mathbf{B}\times\mathbf{c}).$$

The acceleration imparted by $\mathbf{F}_L$ is

$$\frac{d\mathbf{c}}{dt} = -\frac{e}{m}(\mathbf{B}\times\mathbf{c}) = \boldsymbol{\omega}\times\mathbf{c}.$$

Since the charge e of the electron is negative, the vector $\boldsymbol{\omega} = -(e/m)\mathbf{B}$ is parallel to $\mathbf{B}$. When $e\mathbf{E}$ is absent, the acceleration $(d/dt)\mathbf{c} = \boldsymbol{\omega}\times\mathbf{c}$ corresponds in velocity space to uniform motion in a circle in a plane normal to $\mathbf{B}$ and having angular velocity $\boldsymbol{\omega}$.

Adopt a polar axis in velocity space parallel to $\boldsymbol{\omega}$; then the centre of the circle lies on this axis. Thus all velocity points move in circles with their centres on the polar axis $\boldsymbol{\omega}$ and with a common angular velocity $\boldsymbol{\omega}$. The Lorentz acceleration in configuration space is equivalent to a rotation $\boldsymbol{\omega}$ of the frame of reference in velocity space, in which velocity points maintain fixed positions unless displaced by encounters. The presence of a Lorentz acceleration in configuration space thus leaves unaffected the flux of points across a spherical surface in velocity space. Thus, when $e\mathbf{E}$ is zero, the velocity distribution function $f_0(c)$ remains that of Maxwell.

When $e\mathbf{E}$ is present, the acceleration of an electron in configuration space is

$$\frac{d}{dt}\mathbf{c} = (\boldsymbol{\omega}\times\mathbf{c}) + \frac{e\mathbf{E}}{m}.$$

In velocity space the drift of points due to $\boldsymbol{\omega}\times\mathbf{c}$ is tangential to the surface of the sphere c and contributes nothing to the flux of velocity points across the spherical surface, which is due entirely to the drift $e\mathbf{E}/m$ as if $\mathbf{B}$ were absent. The presence of $\mathbf{B}$ contributes no additional term to the scalar equation 2.22, although it influences $\mathbf{f}_1$ in that equation.

The influence of $\mathbf{B}$ appears, however, in the vector equations (equations 2.23 and 2.24) explicitly in the form of an additional term.

Choose, as before, the direction of $\mathbf{f}_1$ as the polar axis. The resultant $\sum[(d/dt)\mathbf{c}]_L$ of the Lorentz accelerations of the $(4\pi f_0 c^2\,dc)n\,d\mathbf{r}$ electrons whose velocity points reside in the shell (c, dc) is

$$d\mathbf{r}\,nc^2\,dc \int_0^\pi \int_0^{2\pi} (\boldsymbol{\omega}\times\mathbf{c})\left[f_0 + \sum_{k=1}^\infty f_k P_k(\cos\theta)\right]\sin\theta\,d\theta\,d\phi.$$

Resolve $\mathbf{c}$ into a component $\mathbf{i}c\cos\theta$ parallel to $\mathbf{f}_1 = \mathbf{i}f_1$ and components $\mathbf{i}_1 c\sin\theta\cos\phi$ and $\mathbf{i}_2 c\sin\theta\sin\phi$ in the plane normal to $\mathbf{f}_1$. The scalar products of the unit vectors are zero, that is, $\mathbf{i}\cdot\mathbf{i}_1 = \mathbf{i}\cdot\mathbf{i}_2 = \mathbf{i}_1\cdot\mathbf{i}_2 = 0$. Since

$$\int_0^{2\pi}\binom{\cos}{\sin}\phi\,d\phi = 0,$$

it follows that the resultant of the Lorentz accelerations is, for the electrons of the shell,

$$\left(\sum\frac{d}{dt}\mathbf{c}\right)_L = d\mathbf{r}\,n(c^3\,dc)(\boldsymbol{\omega}\times\mathbf{i})$$

$$\times\left\{\int_0^\pi\int_0^{2\pi}\left[f_0 + \sum_{k=1}^\infty f_k P_k(\cos\theta)\right]\cos\theta\sin\theta\,d\theta\,d\phi\right\}$$

$$= d\mathbf{r}\left(\frac{4\pi}{3}c^3\,dc\right)(\boldsymbol{\omega}\times n\mathbf{f}_1)$$

$$= d\mathbf{r}\,(4\pi f_0 c^2\,dc)[\boldsymbol{\omega}\times n\mathbf{W}_{\text{conv}}(c)].$$

The gain of momentum by these electrons in time dt through the Lorentz forces is

$$dt\,m\left(\sum\frac{d}{dt}\mathbf{c}\right)_L = (dt\,d\mathbf{r}\,dc)\left(\frac{4\pi}{3}c^3\right)m\,\boldsymbol{\omega}\times n\mathbf{f}_1.$$

The vector equation 2.24 now becomes

$$\frac{\partial}{\partial t}(n\mathbf{f}_1) + c\,\text{grad}_r(nf_0) + \frac{e\mathbf{E}}{m}\frac{\partial}{\partial c}(nf_0) - \boldsymbol{\omega}\times n\mathbf{f}_1 + \nu n\mathbf{f}_1 = 0. \quad (2.25)$$

BIBLIOGRAPHY AND NOTES

FOUNDATIONS OF MAXWELL-BOLTZMANN METHODS

The Scientific Papers of Clerk Maxwell, Cambridge University Press, Vols. I and II, 1890 (reprinted by Dover, New York, 1952). Particular reference: "On the Dynamical Theory of Gases," Vol. II, p. 26.

L. Boltzmann, *Lectures on Gas Theory* (translated by S. G. Brush), University of California Press, 1964.

GENERAL TEXTS

S. Chapman and T. G. Cowling, *The Mathematical Theory of Non-Uniform Gases,* 2nd ed., Cambridge University Press, 1952.

J. H. Jeans, *The Dynamical Theory of Gases,* 4th ed., Cambridge University Press, 1956.

E. H. Kennard, *Kinetic Theory of Gases,* McGraw-Hill Book Co., New York, 1938.

R. D. Present, *Kinetic Theory of Gases,* McGraw-Hill Book Co., New York, 1958.

THEORY OF MOTION OF IONS IN GASES

W. P. Allis, *Motion of Ions and Electrons,"* Handbuch der Physik, Vol. XXI, Springer-Verlag, Berlin, 1956. This useful article is frequently cited. In deriving formulae for drift and diffusion the author considers the electric field to oscillate at high frequency. The formulae appropriate to the motion of electrons in uniform static electric fields are taken to be the special cases of the corresponding formulae with high-frequency fields when the frequency is put equal to zero. It should be noted, however, that this procedure is unsound because an important term implicit in the scalar equation (equation 2.22) averages to zero in a high-frequency field and disappears. This term plays an important role in the theory of motion of electrons in static fields since it is responsible for anisotropic diffusion, which is discussed in Chapter 4 of this book. It is necessary, therefore, to treat motion in static fields separately from motion in high-frequency fields, and this course is adopted in this book.

V. L. Ginsburg and A. V. Gurevič, "Nichtlineare Erscheinungen in einem Plasma, das sich in einem veränderlichen electromagnetischen Feld befindet," *Fortschr. Phys.,* **8**, 97, 1960. This article begins with a helpful discussion of the Maxwell-Boltzmann theory in relation to the motion of electrons in gases. It also includes a discussion of the theory of radio wave interaction.

3

RESTRICTED THEORY

OF DRIFT AND DIFFUSION

3.1. INTRODUCTION

The fundamental equations of the subject were established and discussed in Chapter 2, and it remains to employ them to study the motions of electrons in gases. Since applications of the complete scalar and vector equations are, in general, not simple, there is an advantage in considering first the most straightforward examples so that the understanding of the underlying physical processes is not clouded by mathematical analysis.

The presence in the scalar equation of terms containing $\text{grad}_r(nf_0)$ as a factor and also of the term $(\partial/\partial t)n$ is found to require that nf_0 be of the form $nf_0 \equiv n(\mathbf{r},t)f_0(c,\mathbf{r},t)$ if it is to be a solution of the complete scalar equation. Since in general it is not a simple matter to solve the complete equation in a rigorous manner, in many expositions of the subject simplifying *ad hoc* assumptions are made in order to achieve some progress. It is noted that when the term $(\partial/\partial t)(nf_0)$ and the factor $\text{grad}_r(nf_0)$ vanish everywhere, as in a uniform stream, the residual scalar equation can be readily solved to give $n = \text{constant}$ and $f_0 = f_0(c)$ independently of $\mathbf{r}$ (e.g., equation 3.11). It is then assumed that the *full* scalar equation can be solved in the form $nf_0 \equiv n(\mathbf{r},t)f_0(c)$, in which $n(\mathbf{r},t)$ satisfies a steady-state differential equation (equation 3.29) with constant coefficients whose values are calculated from formulae that depend on the distribution function $f_0(c)$ for the uniform stream. This procedure is adopted in this chapter, but in the final section (Section 3.11) the procedure is subjected to critical comment.

We first consider the uniform and constant stream of electrons and derive the appropriate formulae for the distribution function $f_0(c)$ and the drift velocity **W**. We then make the assumption (invalid in general) that this same distribution function $f_0(c)$ prevails everywhere, even in the presence of density gradients of n, and proceed to establish a formula for the coefficient of diffusion D. The derived quantity $DE/W = D/\mu$ is examined and is shown to be directly related to the mean energy of the electrons. Discussion of the complete scalar equation is undertaken in Chapter 4.

It is convenient to begin with a restatement of the scalar and vector equations that are the actual working equations replacing equation 2.13, the general Maxwell-Boltzmann equation.

3.2.   THE SCALAR EQUATION AND ITS EQUIVALENT FORMS

The scalar equation (equation 2.22) is

$$\frac{\partial}{\partial t}(nf_0) + \frac{c}{3}\,\mathrm{div}_r\,(n\mathbf{f}_1) + \frac{1}{4\pi c^2}\frac{\partial}{\partial c}(\sigma_E - \sigma_{\mathrm{coll}}) = 0, \qquad (3.1)$$

in which, from equation 2.19,

$$\sigma_E(c) = \frac{4\pi}{3}c^2\frac{e\mathbf{E}}{m}\cdot n\mathbf{f}_1. \qquad (3.2)$$

Consequently, equation 3.1 is equivalent to

$$\frac{\partial}{\partial t}(nf_0) + \frac{c}{3}\,\mathrm{div}_r\,(n\mathbf{f}_1) + \frac{1}{3c^2}\frac{\partial}{\partial c}\left(c^2\frac{e\mathbf{E}}{m}\cdot n\mathbf{f}_1\right) - \frac{1}{4\pi c^2}\frac{\partial}{\partial c}\sigma_{\mathrm{coll}}(c) = 0. \qquad (3.3)$$

It is shown in Appendix B that when all encounters are elastic then (equation 2.21)

$$\sigma_{\mathrm{coll}}(c) = 4\pi c^2\nu_{\mathrm{el}}\left[\frac{mc}{M}nf_0 + \frac{\overline{C^2}}{3}\frac{\partial}{\partial c}(nf_0)\right], \qquad (3.4)$$

where

$$\nu_{\mathrm{el}} = Ncq_{m_{\mathrm{el}}}(c). \qquad (3.5)$$

When inelastic encounters are also important, $\sigma_{\mathrm{coll}}(c)$ is represented less simply (Chapter 6).

It can be shown that equation 3.3 is equivalent to the following equation for the kinetic energy associated with the shell (c, dc):

$$\frac{\partial}{\partial t}\left(\tfrac{1}{2}mc^2 nf_0\right) + \frac{c}{3}\,\mathrm{div}_r\left(\tfrac{1}{2}mc^2 n\mathbf{f}_1\right) + \frac{\partial}{\partial c}\left(\frac{c^2 e\mathbf{E}}{6}\cdot n\mathbf{f}_1\right) - \left(\frac{\tfrac{1}{2}mc^2}{4\pi c^2}\right)\frac{\partial}{\partial c}\sigma_{\mathrm{coll}}(c) = 0.$$

$$(3.6)$$

Alternatively,

$$\frac{\partial}{\partial t}\left(\tfrac{1}{2}mc^2 nf_0\right) + \frac{c}{3}\,\mathrm{div}_r\left(\tfrac{1}{2}mc^2 n\mathbf{f}_1\right) + \left(\frac{\tfrac{1}{2}mc^2}{4\pi c^2}\right)\left(\frac{\partial}{\partial c}\left[\sigma_E(c) - \sigma_{\mathrm{coll}}(c)\right]\right) = 0.$$

$$(3.7)$$

3.3.   THE VECTOR EQUATION: RESOLUTION OF $\mathbf{W}_{\mathrm{conv}}(c)$ INTO A DRIFT VELOCITY $\mathbf{W}(c)$ DUE TO $e\mathbf{E}$ AND A CONVECTIVE VELOCITY $\mathbf{W}_G(c)$ DUE TO $\mathrm{grad}_r(nf_0)$

The vector equation (equation 2.25) is

$$\frac{\partial}{\partial t}(n\mathbf{f}_1) + c\,\mathrm{grad}_r(nf_0) + \frac{\partial}{\partial c}(nf_0)\frac{e\mathbf{E}}{m} - \boldsymbol{\omega} \times (n\mathbf{f}_1) + \nu n\mathbf{f}_1 = 0, \quad (3.8)$$

in which ν is the effective collision frequency for momentum transfer (Section 2.7.2d), which, when all encounters are elastic, becomes ν_{el} of equation 3.5. Since $\mathbf{f}_1$ depends on the terms $c\,\mathrm{grad}_r(nf_0)$ and $(\partial/\partial c)(nf_0)$ $(e\mathbf{E}/m)$, which represent independent processes, we may express $\mathbf{f}_1$ as a vector sum:

$$\mathbf{f}_1 = \mathbf{f}_E + \mathbf{f}_G$$

and resolve equation 3.8 into the pair of equations

$$\frac{\partial}{\partial t}(n\mathbf{f}_E) + \frac{\partial}{\partial c}(nf_0)\frac{e\mathbf{E}}{m} - \boldsymbol{\omega} \times (n\mathbf{f}_E) + \nu n\mathbf{f}_E = 0,$$

$$\frac{\partial}{\partial t}(n\mathbf{f}_G) + c\,\mathrm{grad}_r(nf_0) - \boldsymbol{\omega} \times (n\mathbf{f}_G) + \nu n\mathbf{f}_G = 0.$$

$$(3.9)$$

It was shown in Chapter 2 that the velocity $\mathbf{W}_{\mathrm{conv}}(c)$, which is the mean of the vector velocities $\mathbf{c}$ of a shell (c, dc), is $\mathbf{W}_{\mathrm{conv}}(c) = c\mathbf{f}_1/3f_0$, so that on replacing $\mathbf{f}_1$ by $\mathbf{f}_E + \mathbf{f}_G$ it follows that $\mathbf{W}_{\mathrm{conv}}(c) = \mathbf{W}(c) + \mathbf{W}_G(c)$ with

$$\mathbf{W}(c) = \frac{c\mathbf{f}_E}{3f_0} \quad \text{and} \quad \mathbf{W}_G(c) = \frac{c\mathbf{f}_G}{3f_0}.$$

The component $\mathbf{W}(c)$ is determined by $e\mathbf{E}$, and $\mathbf{W}_G(c)$ by $\mathrm{grad}_r(nf_0)$.

The first of equations 3.9, when $\boldsymbol{\omega}=0$ so that $\mathbf{f}_E$ and $e\mathbf{E}$ are parallel, reduces to the following equation relating the magnitudes f_E and eE: †

$$\frac{1}{\nu}\frac{\partial}{\partial t}f_E + \frac{eE}{m\nu}\frac{\partial}{\partial c}f_0 + f_E = 0. \tag{3.9a}$$

It is now shown that, in laboratory experiments on diffusing streams and travelling groups, $(1/\nu)(\partial/\partial t)f_E$ is negligibly small in comparison with the terms $(eE/m\nu)(\partial/\partial c)f_0$ and f_E. For simplicity we assume that ν is constant.

Since the time constant for energy change is much greater than the time constant for momentum change (see Section 9.2), it follows that there are intervals of time $t \gg 1/\nu$ such that within them $(eE/m\nu)(\partial/\partial c)f_0$ changes negligibly and may be considered constant. In what follows we are concerned with such intervals.

Suppose that before time $t=0$ a condition of equilibrium exists between a steady force $e\mathbf{E}_0$ and the vector $\mathbf{f}_{E_0}$, that is, $(1/\nu)(\partial/\partial t)f_{E_0}=0$, so that from equation 3.9a

$$\frac{eE}{m\nu}\frac{\partial}{\partial c}f_0 = -f_{E_0}.$$

Next suppose that at $t=0$ the force becomes $e\mathbf{E}(t)$, with $E(0)=E_0$; f_E then becomes $f_E(c,t)$ and is given by the full equation 3.9a. Divide the first and third terms of equation 3.9a by f_{E_0} and the second term by the equal quantity $-(eE_0/m\nu)(\partial/\partial c)f_0$. It follows that equation 3.9a is equivalent to

$$\frac{1}{\nu f_{E_0}}\left[\frac{\partial}{\partial t}f_E(c,t)\right] - \frac{E(t)}{E_0} + \frac{f_E(c,t)}{f_{E_0}} = 0, \tag{3.9b}$$

whence

$$\frac{f_E(c,t)}{f_{E_0}} = (\exp-\nu t)\left[\nu\int_0^t \frac{E(s)}{E_0}(\exp\nu s)\,ds + 1\right]$$

$$= \frac{E(t)}{E_0} - \frac{\exp-\nu t}{E_0}\int_0^t \frac{d}{ds}E(s)(\exp\nu s)\,ds.$$

Let the interval t, although much greater than $1/\nu$, be so small that within it $(d/ds)E(s)$ may be considered to be constant and equal to $[(d/dt)E(t)]_{t=0}$. It then follows that

†It is assumed that, in effect, n is constant, since changes in n due to diffusion in the time intervals under consideration are negligible.

$$\frac{f_E(c,t)}{f_{E_0}} = \frac{E(t)}{E_0} - \frac{1}{\nu E_0}(1-\exp-\nu t)\left[\frac{d}{dt}E(t)\right]_{t=0}.$$

Suppose, for example, that $\nu t = 6$; then at $t = 6/\nu$, by which time an electron has made about six encounters since $t = 0$, $\exp - \nu t \sim 1/400$. It follows that after an interval t in which an electron has made several encounters, $f_E(c,t)/f_{E_0}$ is given by

$$\frac{f_E(c,t)}{f_{E_0}} \cong \frac{E(t)}{E_0} - \frac{1}{\nu E_0}\left[\frac{d}{dt}E(t)\right]_{t=0}.$$

But

$$\frac{1}{\nu E_0}\left[\frac{d}{dt}E(t)\right]_{t=0}$$

is the proportional change in $E(t)$ in the time equal to the time constant for momentum change $\tau = 1/\nu$. If, therefore, this quantity is negligibly small, then

$$\frac{f_E(c,t)}{f_{E_0}} = \frac{E(t)}{E_0},$$

a result consistent with equation 3.9b only if $(1/\nu)(\partial/\partial t)f_E(c,t)$ is negligibly small.

Thus, when $E(t)$ changes by a negligible proportion of itself in an interval $\Delta t = \tau = 1/\nu$, $f_E(c,t)$ conforms without time lag as if equilibrium were maintained at all times t. The working form of the first of equations 3.9 (with $\boldsymbol{\omega} = 0$) then becomes that in which the term $(1/\nu)(\partial/\partial t)\mathbf{f}_E(c,t)$ is absent. Therefore in laboratory experiments with travelling groups or steady streams of electrons in which the pressure of the gas ranges from a few torr to 100 torr or greater, it follows that an abrupt change in $\mathbf{E}$ is accompanied by an almost instantaneous adjustment of $\mathbf{f}_E$ to the new equilibrium value, since the values of ν are very large (of the order of 10^{10} to 10^{11} sec^{-1}).

When the electric field $\mathbf{E}$ is a continuous function $\mathbf{E}\,(t)$ of time, there are two extreme cases. First suppose that, as above, $\mathbf{E}$ changes by a very small proportion of itself in a time $\Delta t = 1/\nu$. Then in the equation

$$\frac{1}{\nu}\frac{\partial}{\partial t}\mathbf{f}_E + \left(\frac{\partial}{\partial c}f_0\right)\frac{e\mathbf{E}}{m\nu} + \mathbf{f}_E = 0$$

the first term is negligibly small in comparison with the second and third terms; consequently $\mathbf{f}_E$ is a function of t that follows in step with the variation of $\mathbf{E}(t)$ with negligible delay, that is,

$$\mathbf{f}_E(t) \cong - \left(\frac{\partial}{\partial c} f_0 \right) \frac{e\mathbf{E}(t)}{m\nu}.$$

When, however, the relative change in $\mathbf{E}(t)$ is appreciable in time $\Delta t = 1/\nu$, the term $(1/\nu)(\partial/\partial t)\mathbf{f}_E$ is no longer negligible, and it is necessary to employ the complete differential equation from which to derive $\mathbf{f}_E$ as a function of t. Such circumstances are encountered in laboratory experiments with electric fields that oscillate at microwave frequencies and in the ionosphere at radio frequencies because in the ionosphere ν is greatly diminished at the prevailing small gas densities. These matters are treated in detail in Chapter 9.

Consider next the second of equations 3.9. It is observed that a travelling group of electrons in a gas at pressures of several torr or greater has far from completely dispersed in 10^{-4} sec, a time which is many times greater than the time $\Delta t = 1/\nu$. We conclude that in these circumstances $\operatorname{grad}_r(nf_0)$ in the second of equations 3.9 is a slowly varying function of t, judged on the scale with $1/\nu$ as unit. We may therefore omit the term $(\partial/\partial t)(n\mathbf{f}_G)$ and write $n\mathbf{f}_G = -(c/\nu)\operatorname{grad}_r(nf_0)$.

In summary, we have, under laboratory conditions when $\mathbf{E}$ is constant or slowly varying and $\omega = 0$,

$$\mathbf{f}_E = - \frac{e\mathbf{E}}{m\nu} \frac{\partial}{\partial c} f_0;$$

$$\mathbf{f}_G = - \frac{c}{\nu} \frac{1}{n} \operatorname{grad}_r(nf_0);$$

$$\mathbf{W}(c) = - \frac{e\mathbf{E}}{3m\nu} \frac{c}{f_0} \frac{\partial}{\partial c} f_0 = - \frac{\mathbf{V}'c}{3f_0} \frac{\partial}{\partial c} f_0,$$

where

$$\mathbf{V}' = \frac{e\mathbf{E}}{m\nu} \tag{3.10}$$

and ν is the effective collision frequency for momentum transfer, defined in Section 2.7.2d;

$$\mathbf{W}_G(c) = - \frac{c^2}{3\nu f_0} \frac{1}{n} \operatorname{grad}_r(nf_0);$$

$$\mathbf{W}_{\text{conv}}(c) = \mathbf{W}(c) + \mathbf{W}_G(c).$$

In what follows it is supposed in the first instance that $\operatorname{grad}_r(nf_0)$ is equal to zero.

3.4.   UNIFORM STREAM OF ELECTRONS

The stream is considered to proceed through a gas in the direction of the force $e\mathbf{E}$. Suppose that the stream is broad so that in its central regions $\text{grad}_r (nf_0)$ and $\text{div}_r(nf_1)$ are negligible. It is then justifiable to write $\mathbf{f}_G = 0$, $\mathbf{f}_1 = \mathbf{f}_E$. We suppose in addition that the magnetic field is absent so that from equations 3.9 and 3.10 with $\boldsymbol{\omega} = 0$ it follows that

$$\mathbf{f}_1 = \mathbf{f}_E = -\frac{e\mathbf{E}}{vm}\frac{d}{dc}f_0.$$

The scalar equation 3.1 reduces in this case to

$$\frac{d}{dc}(\sigma_E - \sigma_{\text{coll}}) = 0,$$

that is,

$$\sigma_E - \sigma_{\text{coll}} = \text{const.}$$

But both σ_E and σ_{coll} tend to zero as $c \to \infty$ since there are no velocity points as $c \to \infty$. The constant is therefore zero and $\sigma_E = \sigma_{\text{coll}}$, that is (from equation 3.2),

$$\frac{e\mathbf{E}}{m}\cdot\mathbf{f}_1 = \frac{3}{4\pi c^2 n}\sigma_{\text{coll}}.$$

3.4.1.   FORMULA FOR DISTRIBUTION FUNCTION WHEN ALL ENCOUNTERS ARE ELASTIC. With this restriction

$$-\left(\frac{e\mathbf{E}}{m}\right)^2\frac{1}{v_{\text{el}}}\frac{df_0}{dc} = \frac{3}{4\pi c^2 n}\sigma_{\text{coll}} = 3v_{\text{el}}\left(\frac{m}{M}cf_0 + \frac{\overline{C^2}}{3}\frac{df_0}{dc}\right).$$

Write $\mathbf{V} = e\mathbf{E}/mv_{\text{el}}$; it then follows that

$$(V^2 + \overline{C^2})\frac{df_0}{dc} = -3\frac{m}{M}cf_0,$$

the solution of which is

$$f_0(c) = A\exp\left(-\frac{3m}{M}\int_0^c\frac{c\,dc}{V^2 + \overline{C^2}}\right). \tag{3.11}$$

This is the generalized expression for the isotropic term in the distribution function, and it is evident that $A \equiv f_0(0)$. A more general form is

$$f_0(c) = B \exp\left(-\frac{3m}{M} \int_{c_0}^{c} \frac{c \, dc}{V^2 + \overline{C^2}} \right),$$

where $c > c_0$ and $B \equiv f_0(c_0)$. These solutions refer to cases in which $\text{div}_r(n\mathbf{f}_1)$ and $\text{grad}_r(nf_0)$ are negligible and there is no time dependence.

3.4.2. FORMULA FOR DRIFT VELOCITY IN THE GENERAL CASE. From equation 3.10 the instantaneous velocity of the centroid of the $(n \, d\mathbf{r})$ $(4\pi f_0 c^2 \, dc)$ electrons of the shell (c, dc) is, when $\boldsymbol{\omega} = 0$,

$$\mathbf{W}(c) = \frac{c\mathbf{f}_E}{3f_0} = \frac{c\mathbf{f}_1}{3f_0} = -\frac{e\mathbf{E}c}{3mvf_0} \frac{d}{dc} f_0$$

since, as shown above,

$$\mathbf{f}_1 = -\frac{e\mathbf{E}}{vm} \frac{d}{dc} f_0 = -\mathbf{V}' \frac{d}{dc} f_0.$$

The drift velocity $\mathbf{W}$ due to an electric force is the mean value of $\mathbf{W}(c)$ taken over all shells and is therefore ($\boldsymbol{\omega} = 0$):

$$\mathbf{W} = \frac{4\pi}{3} \int_0^\infty \mathbf{f}_1 c^3 \, dc = -\frac{4\pi}{3} \int_0^\infty c^3 \mathbf{V}' \frac{df_0}{dc} dc$$

$$= -\frac{4\pi}{3} \left(\frac{e\mathbf{E}}{m} \right) \int_0^\infty \frac{c^3}{v} \frac{df_0}{dc} dc = -\frac{4\pi}{3} \left(\frac{e}{m} \right) \left(\frac{\mathbf{E}}{N} \right) \int_0^\infty \frac{c^2}{q_m(c)} \frac{df_0}{dc} dc.$$

$$(3.12)$$

We note that $f_0(c)$ in equation 3.12 is not restricted to be $f_0(c)$ as defined in equation 3.11.

Equation 3.12 can be transformed by partial integration as follows:

$$\mathbf{W} = -\frac{4\pi}{3} \left(\frac{e\mathbf{E}}{m} \right) \int_0^\infty \frac{c^3}{v} \frac{df_0}{dc} dc$$

$$= -\frac{4\pi}{3} \left(\frac{e\mathbf{E}}{m} \right) \left[\frac{c^3 f_0}{v} \bigg|_0^\infty - \int_0^\infty \frac{d}{dc} \left(\frac{c^3}{v} \right) f_0 \, dc \right].$$

If $c^3 f_0/\nu \to 0$, both when $c \to 0$ and when $c \to \infty$, then

$$\mathbf{W} = \frac{4\pi}{3}\left(\frac{e\mathbf{E}}{m}\right)\int_0^\infty \frac{d}{dc}\left(\frac{c^3}{\nu}\right)f_0\,dc$$

$$= \left(\frac{e\mathbf{E}}{m}\right)\frac{4\pi}{3}\int_0^\infty\left[c^{-2}\frac{d}{dc}\left(\frac{c^3}{\nu}\right)\right]c^2 f_0\,dc$$

$$= \left(\frac{e\mathbf{E}}{m}\right)\frac{1}{3}\overline{\left[c^{-2}\frac{d}{dc}\left(\frac{c^3}{\nu}\right)\right]}$$

$$= \left(\frac{e}{m}\right)\left(\frac{\mathbf{E}}{N}\right)\frac{1}{3}\overline{\left[c^{-2}\frac{d}{dc}\frac{c^2}{q_m(c)}\right]}. \tag{3.13}$$

The superimposed bar over the factors [ ] indicates that these factors equal the mean value taken over all shells (c, dc) of the quantities in the brackets. Expressed explicitly in terms of $q_m(c)$ and f_0, the formula for $\mathbf{W}$ is

$$\mathbf{W} = \frac{4\pi}{3}\left(\frac{e}{m}\right)\left(\frac{\mathbf{E}}{N}\right)\int_0^\infty \frac{d}{dc}\frac{c^2}{q_m(c)}f_0\,dc. \tag{3.14}$$

The formulae for $\mathbf{W}$ show that W is a function of the parameter E/N.

The physical significance of the integrands in equations 3.12 and 3.14 is discussed in Chapter 7.

The velocity $\mathbf{W}$ discussed above is the convective drift velocity caused by the electric force $e\mathbf{E}$. The drift velocity that is derived in the laboratory from time-of-flight methods is the velocity of the centroid of an isolated group of electrons that moves through the gas in the field $\mathbf{E}$. It is shown in Chapters 4 and 7 that these two velocities are in fact equal.

3.5. MOTION IN MONATOMIC GASES

When electrons move in monatomic gases, there is a wide range of energy within which encounters are elastic. We can therefore apply the theory that has been developed in the preceding sections immediately to electron motion in monatomic gases.

Consider, first, motion in the absence of an electric field. The electrons then move in thermal equilibrium with the gas molecules. In equation 3.11 put $V = eE/mv_{el} = 0$; then

$$f_0(c) = A \exp\left(\frac{-\frac{1}{2}mc^2}{\frac{1}{3}M \overline{C^2}} \right)$$

$$= A \exp\left(\frac{-\frac{1}{2}mc^2}{\kappa T} \right),$$

which is Maxwell's distribution function for electrons in thermal equilibrium with the molecules of the gas. Alternatively, since $\sigma_E = 0$, it follows that $\sigma_{coll} = 0$, and from equation 3.4 that

$$\frac{mc}{M} f_0 + \frac{\overline{C^2}}{3} \frac{df_0}{dc} = 0,$$

the solution of which is $f_0(c)$ as given above.

Next, suppose that $\mathbf{E}$ is not zero. Consider the populations of shells (c, dc) with the same thickness dc but with different radii c. The point population of a shell (c, dc) is $4\pi n \, d\mathbf{r} f_0 c^2 \, dc$, and as c is varied while dc is maintained at the same value the population of a shell attains its maximum value at the value of c for which $(d/dc)[c^2 f_0(c)] = 0$. Let this value be c_m; then

$$2c_m f_0(c_m) + c_m{}^2 \frac{d}{dc} f_0(c)_{c = c_m} = 0,$$

that is,

$$\frac{df_0}{dc} + \frac{2f_0}{c} = 0.$$

We reject the solutions $c = 0$ and $c = \infty$ as these correspond to minima. It follows from equation 3.11 that

$$-\frac{3m}{M} \frac{c_m}{V_m{}^2 + \overline{C^2}} + \frac{2}{c_m} = 0,$$

whence

$$\frac{1}{2}mc_m{}^2 = \frac{M}{3}\left(V_m{}^2 + \overline{C^2} \right),$$

where

$$V_m = \frac{eE}{mNq_{m_{el}}(c_m)c_m}.$$

Let k_m denote the ratio of the energy corresponding to the most probable electron speed to the energy corresponding to the most probable molecular speed, which is $\frac{1}{2}MC_m{}^2 = \frac{1}{3}M\overline{C^2}$ since $F(C)$ is Maxwellian; then

$$k_m = \frac{\frac{1}{2}mc_m{}^2}{\frac{1}{3}M\overline{C^2}} = 1 + \frac{V_m{}^2}{\overline{C^2}}$$

or

$$k_m - 1 = \frac{V_m{}^2}{\overline{C^2}}. \qquad (3.15)$$

In practice, unless E/N is very small, V_m may notably exceed $(\overline{C^2})^{1/2}$, and it follows that the mean energy of the electrons of a group moving in a steady state of motion through a gas in a uniform electric field may greatly exceed the mean energy of translation of the molecules. This fact was discovered experimentally by Townsend (Section 1.10). Formula 3.15 is a generalization of a formula given by Pidduck (Section 1.11) for the case in which $q_{m_{el}}(c) \propto c^{-1}$, in which event, as follows from equation 3.11, $f_0(c)$ becomes Maxwell's distribution function since V is then made independent of c. Equation 3.13 may be given the form

$$\mathbf{W} = \frac{1}{3}\overline{\left[c^{-2}\frac{d}{dc}(c^3\mathbf{V}) \right]};$$

consequently, when V is independent of c $[cq_{m_{el}}(c) = \text{constant}]$,

$$\mathbf{W} = \mathbf{V} = \frac{e\mathbf{E}}{m\nu_{el}}.$$

Also, since in equation 3.15 both distribution functions are Maxwellian, it follows that

$$k_m = \frac{\frac{1}{3}m\overline{c^2}}{\frac{1}{3}M\overline{C^2}} = k = 1 + \frac{W^2}{\overline{C^2}}$$

where $k = \frac{1}{2} m \overline{c^2} / \frac{1}{2} M \overline{C^2}$ is *Townsend's energy factor*. Thus

$$k - 1 = \frac{W^2}{\overline{C^2}}, \tag{3.16}$$

which is Pidduck's formula, valid for the special case $q_{m_{el}}(c) \propto c^{-1}$. Since the measured values of W usually notably exceed $(\overline{C^2})1/2$, Pidduck's formula also indicates that the mean energy of the electrons may be many times the value of the mean translational energy of the molecules.

From the condition for a steady state of motion, namely, $\sigma_E = \sigma_{coll}$ for every spherical surface c in velocity space, it follows from the expressions for σ_E and σ_{coll} in Section 3.2 that

$$\nu_{el} \left[\left(\frac{eE}{m \nu_{el}} \right)^2 \frac{d}{dc} f_0 + \frac{3mc}{M} f_0 + \overline{C^2} \frac{d}{dc} f_0 \right] = 0.$$

With eE fixed, and provided ν_{el} increases without limit as c approaches infinity, the first term tends to zero as c tends to infinity. Consequently the steady-state condition leads to

$$\frac{3mc}{M} f_0 + \overline{C^2} \frac{df_0}{dc} \underset{c \to \infty}{\to} 0,$$

whence

$$f_0(c) \to \text{const} \cdot \exp \left(- \frac{\frac{1}{2} mc^2}{\frac{1}{3} M \overline{C^2}} \right) = \text{const} \cdot \exp \left(- \frac{\frac{1}{2} mc^2}{\kappa T} \right).$$

The distribution function thus tends to the Maxwellian form as c is increased to values notably in excess of c_m, although the normalizing constant is not that of the thermal distribution function. In the special case of $\nu_{el} = \text{constant}$, that is, $q_{m_{el}}(c) \propto c^{-1}$, $f_0(c)$ is of the Maxwellian form for all values of c since $V = eE / m \nu_{el}$ is constant. In this case, neither the normalizing constant nor the parameter in the exponent are those of the thermal distribution function. These properties of the distribution function also follow from equation 3.11.

Maxwell showed that, when m and M repel in proportion to the inverse fifth power of their separation, $q_{m_{el}}(c) \propto c^{-1}$. In practice the dependence of $q_{m_{el}}(c)$ upon c is not normally observed to be of this form in encounters between electrons and molecules.

With the larger values of the experimental parameter E/N the values of

V^2 may greatly exceed $\overline{C^2}$ throughout the effective range of speeds c within which $c^2 f_0(c)/c_m^2 f_0(c_m)$ is not negligible. When such is the case, the formula for $f_0(c)$ (equation 3.11) reduces to

$$f_0(c) = A \exp\left(-\frac{3m}{M}\int_0^c \frac{c\,dc}{V^2}\right)$$

$$= A \exp\left[-\frac{3m}{M}\left(\frac{m}{e}\right)^2\left(\frac{1}{E/N}\right)^2\int_0^c c^3 q_{m_{el}}^2(c)\,dc\right]. \tag{3.17}$$

The coefficient A in both equations 3.17 and 3.11 is determined through the normalizing condition,

$$4\pi\int_0^\infty f_0(c)c^2\,dc = 1.$$

3.6. UNIT OF MEASUREMENT FOR THE PARAMETER E/N

The parameter E/N is an important quantity because the velocity distribution function $f_0(c)$ is determined by E/N through V (equation 3.11), and it follows that other quantities, such as the drift velocity W (equation 3.14) and coefficient of diffusion D (equation 3.27), are functions of E/N or of E/N and N. In practice E is expressed in volts per centimetre, but because N, the molecular concentration, is always an extremely large number (N at a pressure of 1 torr with $T = 293$ K is 3.30×10^{16} cm^{-3}), E/N (in units of V cm^2) is an inconveniently small number of the order of magnitude 10^{-17}. It has therefore proved to be convenient to express E/N in terms of a unit chosen to be 10^{-17} V cm^2. This unit has been named the townsend in honour of the most important pioneer in the field of electrons in gases, who first used a parameter E/p, equivalent to E/N, and stressed its importance (Section 1.9). For convenience of transition from the former parameter (E/p) to E/N expressed in townsends we note the relation $E/N(\text{Td}) = 3.03(E/p)$ when E is in V cm^{-1}, p is in torr, and $T = 293$ K.

3.7. DISTRIBUTION FUNCTION WHEN $q_{m_{el}}(c)$ IS PROPORTIONAL TO A POWER OF c AND $V^2 \gg \overline{C^2}$

We consider first the form assumed by the distribution function of equation 3.17 when $q_{m_{el}}(c)$ is independent of c. The result so established finds many useful applications because $q_{m_{el}}(c)$ may often be regarded as independent of c within the effective range of integration in formulae (such as those for W) for a given value of E/N. With $q_{m_{el}}(c) \equiv q_{m_{el}}$ a constant, equation 3.17 reduces to

$$f_0(c) = A \exp - \left(\frac{c}{\alpha} \right)^4$$

with

$$\alpha^4 = \frac{4M}{3m} \left(\frac{e}{m} \frac{E/N}{q_{m_{el}}} \right)^2$$

(3.18)

which is the Druyvesteyn form of the distribution function.

We now consider the more general case in which we suppose that $q_{m_{el}}(c)$ is proportional to a power of c, that is, $q_{m_{el}}(c) = ac^r$ where a and r are constants. It then follows from equation 3.17 that

$$f_0(c) = A \exp - \left(\frac{c}{\alpha} \right)^s$$

with

$$\alpha = \left[\frac{s}{a^2} \frac{M}{3m} \left(\frac{e}{m} \right)^2 \left(\frac{E}{N} \right)^2 \right]^{1/s}$$

(3.19)

and

$$s = 2r + 4.$$

The value of the coefficient A is chosen to satisfy the requirement

$$4\pi \int_0^\infty f_0(c) c^2 dc = 1$$

whence

$$4\pi A \int_0^\infty \left[\exp - \left(\frac{c}{\alpha} \right)^s \right] c^2 dc = 4\pi \alpha^3 A \int_0^\infty [\exp(-u)^s] u^2 du = 1.$$

To evaluate the integral we employ the standard formula

$$\int_0^\infty [\exp(-v^n)] v^m dv = \frac{1}{n} \Gamma \left(\frac{m+1}{n} \right).$$

(3.20)

It then follows that

$$\frac{4\pi\alpha^3 A}{s}\Gamma\left(\frac{3}{s}\right)=1, \qquad A=\frac{s}{4\pi\alpha^3\Gamma(3/s)}. \tag{3.21}$$

Two special cases of equation 3.19 are particularly useful.

Maxwell's distribution function. Let $s=2$; then

$$f_0(c)=\frac{1}{(\alpha\sqrt{\pi})^3}\exp-\left(\frac{c}{\alpha}\right)^2. \tag{3.22}$$

In this case the most probable speed $c_m=\alpha$.

Druyvesteyn's distribution function. Let $s=4$; then

$$f_0(c)=\frac{1}{\alpha^3\pi\Gamma(\frac{3}{4})}\exp-\left(\frac{c}{\alpha}\right)^4 \tag{3.23}$$

with $\Gamma(\frac{3}{4})=1.2254$. Druyvesteyn's distribution function is important in that it is often, in practice, a close representation of the actual distribution function, as was pointed out in relation to equation 3.18.

3.7.1.   THE MOST PROBABLE SPEED c_m. The most probable speed is that for which $(d/dc)[c^2 f_0(c)]=0$. The solutions $c=0$, $c=\infty$ are not what we require since they correspond to minima. It follows that

$$2c_m f_0(c_m)+c_m^2\left[\frac{d}{dc}f_0(c)\right]_{c=c_m}=0,$$

that is,

$$c_m=-\frac{[2f_0(c_m)]}{[(d/dc)f_0(c)]_{c=c_m}}.$$

Let $f_0(c)=A\exp-(c/\alpha)^s$; then $c_m=2\alpha^s/(sc_m^{s-1})$, that is,

$$c_m=\left(\frac{2}{s}\right)^{1/s}\alpha. \tag{3.24}$$

Thus, when $s=2$ (Maxwell),

$$c_m=\alpha;$$

and when $s=4$ (Druyvesteyn),

$$c_m = 2^{-1/4}\alpha = \frac{\alpha}{1.189} = 0.841\alpha.$$

3.7.2. FORMULA FOR THE MEAN VALUE OF c^x. We adopt, for convenience in discussion, the normalized distribution function (equations 3.19 and 3.21)

$$f_0(c) = \frac{s}{4\pi\alpha^3\Gamma(3/s)} \exp -\left(\frac{c}{\alpha}\right)^s.$$

The value of the mean of the xth power of c is

$$\overline{c^x} = \frac{4\pi s}{4\pi\alpha^3\Gamma(3/s)} \int_0^\infty c^{x+2}\left[\exp -\left(\frac{c}{\alpha}\right)^s\right] dc$$

$$= \frac{s\alpha^x}{\Gamma(3/s)} \int_0^\infty v^{x+2}(\exp - v^s)\, dv.$$

Consequently, from equation 3.20,

$$\overline{c^x} = \left[\frac{\Gamma\left(\dfrac{x+3}{s}\right)}{\Gamma\left(\dfrac{3}{s}\right)}\right]\alpha^x = \left(\frac{s}{2}\right)^{x/s}\left[\frac{\Gamma\left(\dfrac{x+3}{s}\right)}{\Gamma\left(\dfrac{3}{s}\right)}\right]c_m{}^x. \qquad (3.25)$$

We again consider the special cases $s=2$ and $s=4$.
When $s=2$ (Maxwell),

$$c_m = \alpha,$$

$$\bar{c} = \left[\frac{\Gamma(2)}{\Gamma(\frac{3}{2})}\right]\alpha = \frac{2\alpha}{\sqrt{\pi}} = 1.128\alpha,$$

$$\overline{c^2} = \left[\frac{\Gamma(\frac{5}{2})}{\Gamma(\frac{3}{2})}\right]\alpha^2 = \tfrac{3}{2}\alpha^2,$$

$$\overline{c^{-1}} = \left[\frac{\Gamma(1)}{\Gamma(\frac{3}{2})} \right] \alpha^{-1} = \frac{2\alpha^{-1}}{\sqrt{\pi}} = \frac{1.128}{\alpha},$$

since $\Gamma(\frac{1}{2}) = \sqrt{\pi} = 1.7725$. A useful combination of mean speeds is

$$\frac{\bar{c}}{\overline{c^{-1}}} = \alpha^2 = \frac{2}{3}\,\overline{c^2}\,.$$

When $s = 4$ (Druyvesteyn),

$$c_m = \frac{\alpha}{2^{1/4}} = 0.841\alpha,$$

that is,

$$\alpha = 1.189 c_m,$$

$$\alpha^2 = \sqrt{2}\,c_m^{\,2} = 1.414 c_m^{\,2}.$$

Since $\Gamma(\frac{3}{4}) = 1.2254$, it follows that

$$\bar{c} = \left[\frac{\Gamma(1)}{\Gamma(\frac{3}{4})} \right] \alpha = \frac{\alpha}{\Gamma(\frac{3}{4})} = 0.816\alpha = 0.970 c_m,$$

$$\overline{c^2} = \left[\frac{\Gamma(\frac{5}{4})}{\Gamma(\frac{3}{4})} \right] \alpha^2 = \frac{0.9064}{\Gamma(\frac{3}{4})} \alpha^2 = 0.740\alpha^2 = 1.046 c_m^{\,2},$$

$$\overline{c^{-1}} = \left[\frac{\Gamma(\frac{1}{2})}{\Gamma(\frac{3}{4})} \right] \alpha^{-1} = \frac{\pi^{1/2}}{\Gamma(\frac{3}{4})} \alpha^{-1} = \frac{1.772\alpha^{-1}}{\Gamma(\frac{3}{4})} = \frac{1.446}{\alpha} = \frac{1.217}{c_m},$$

$$\frac{\bar{c}}{\overline{c^{-1}}} = \frac{\alpha^2}{\sqrt{\pi}} = 0.564\alpha^2 = 0.763\,\overline{c^2} = 0.798 c_m^{\,2}.$$

3.8.   DIFFERENTIAL EQUATION FOR n AND COEFFICIENT OF DIFFUSION

In swarm techniques we are concerned with the distribution $n(\mathbf{r}, t)$ of the electron number density in a travelling group or with the distribution of electrons within a closed chamber. Before the year 1967 it had generally been assumed that when the electric force $e\mathbf{E}$ is directed parallel to the coordinate axis $+Oz$, the number density $n(\mathbf{r}, t) \equiv n$ was to be found as a solution of an equation of the form

$$\frac{\partial}{\partial t} n - D\nabla^2 n + W \frac{\partial}{\partial z} n = 0,$$ (3.26)

in which D is the isotropic coefficient of diffusion and W is the drift speed. In equation 3.26 the drift velocity is given by equation 3.12, whereas the formula for D is shown below to be

$$D = 4\pi \int_0^\infty \frac{c^2}{3\nu} f_0 c^2 \, dc,$$ (3.27)

where ν is the effective collision frequency for momentum transfer. In these formulae the distribution function f_0 is assumed to be that for a uniform stream, which, when all encounters are elastic, is the generalized function (equation 3.11). Thus f_0 is assumed to be independent of position $\mathbf{r}$. In order to relate equation 3.26 to the scalar equation for nf_0 we multiply each term of equation 3.1 by $4\pi c^2$ and integrate with respect to c from zero to infinity, that is to say, over all shells (c, dc). Equation 3.1 is then transformed to

$$\frac{\partial}{\partial t} n + \mathrm{div}_r \left(n \frac{4\pi}{3} \int_0^\infty c\mathbf{f}_1 c^2 \, dc \right) + (\sigma_E - \sigma_{\mathrm{coll}})|_0^\infty = 0.$$

Since σ_E and σ_{coll} vanish at both limits and because, in the absence of a magnetic field (equation 3.10),

$$n\mathbf{f}_1 = n(\mathbf{f}_G + \mathbf{f}_E) = -\frac{e\mathbf{E}}{\nu m} \frac{\partial}{\partial c} (nf_0) - \frac{c}{\nu} \mathrm{grad}_r (nf_0),$$

it follows that the transformed equation reduces to

$$\frac{\partial}{\partial t} n - \mathrm{div}_r \left[\mathrm{grad}_r \left(\frac{4\pi}{3} n \int_0^\infty \frac{c^2}{\nu} f_0 c^2 \, dc \right) \right]$$

$$- \mathrm{div}_r \left[\frac{4\pi}{3} \int_0^\infty \frac{e\mathbf{E}c}{m\nu} \frac{\partial}{\partial c} (nf_0) c^2 \, dc \right] = 0.$$

We now assume that f_0 is independent of position $\mathbf{r}$. The transformed equation then becomes

$$\frac{\partial}{\partial t} n - D\nabla^2 n + \operatorname{div}_r (n\mathbf{W}) = 0, \qquad (3.28)$$

where D is given by equation 3.27 and $\mathbf{W}$ by equation 3.12. Because $\mathbf{W}$ is independent of $\mathbf{r}$, since $\mathbf{E}$ is uniform, equation 3.28 is equivalent to

$$\frac{\partial}{\partial t} n - D\nabla^2 n + \mathbf{W} \cdot \operatorname{grad}_r (n) = 0, \qquad (3.29)$$

which reduces to equation 3.26 when the direction of $\mathbf{W}$ is that of $+ Oz$. It follows that the validity of equation 3.29 rests on the assumption that f_0 is independent of position. However, in general, the distribution function does in fact depend on position, that is, $f_0 \equiv f_0(c, \mathbf{r}, t)$.

Consider the scalar equation (equation 3.3) and suppose that no magnetic field is present. On replacing $n\mathbf{f}_1$ by its value given above, equation 3.3 becomes

$$\frac{\partial}{\partial t}(nf_0) + \operatorname{div}_r \left\{ -\operatorname{grad}_r \left[\left(\frac{c^2}{3\nu} \right) (nf_0) \right] - \frac{c}{3} \mathbf{V}' \frac{\partial}{\partial c}(nf_0) \right\}$$

$$+ \frac{1}{c^2} \frac{\partial}{\partial c} \left\{ -\frac{c^2}{3} \nu_{el} \left[\frac{c\mathbf{V}'}{\nu_{el}} \cdot \operatorname{grad}_r (nf_0) + VV' \frac{\partial}{\partial c}(nf_0) + \frac{3}{4\pi c^2 \nu_{el}} \sigma_{coll} \right] \right\} = 0,$$

$$(3.30)$$

in which $\mathbf{V}' = e\mathbf{E}/m\nu$. It is shown in Section 3.11 that, when $\operatorname{grad}_r (nf_0)$ is not zero, it is impossible to satisfy equation 3.30 by a form of (nf_0) in which the distribution function f_0 is independent of position $\mathbf{r}$. Consequently equation 3.29 is inexact. Moreover, as shown in Chapter 4, in the presence of an electric force $e\mathbf{E}$ diffusion is no longer isotropic, as implied by the single coefficient of diffusion D, and equation 3.29 is to be replaced by the equation

$$\frac{\partial}{\partial t} n - D\left(\frac{\partial^2}{\partial x^2} + \frac{\partial^2}{\partial y^2} \right) n - D_L \frac{\partial^2}{\partial z^2} n + \mathbf{W} \cdot \operatorname{grad}_r (n) = 0. \qquad (3.31)$$

The coefficients D and W are given, as before, by equations 3.27 and 3.12. The coefficient D_L is the longitudinal coefficient of diffusion and in general is different from D. It is discussed in detail in Chapter 4. The fact that in the presence of an electric force diffusion of electrons is in general anisotropic was discovered experimentally in 1967 by Wagner, Davis, and Hurst. Nevertheless, as shown in Section 5.3, the quantity that is actually measured in the Townsend-Huxley experiment is D/W and not D_L/W. In practice, however, it is more convenient to deal with the ratio D/μ rather than D/W since D/μ is a function of E/N, whereas D/W is of the form [function $(E/N)]/N$. The mobility μ is by definition the ratio W/E.

Because of the practical importance of the experimental quantity D/μ it is discussed in detail in the following section.

3.9. DISCUSSION OF FORMULAE FOR W AND D/μ

For convenience of reference we restate the formulae for D and W as given in equations 3.27 and 3.12:

$$\left.\begin{array}{c} D = 4\pi \displaystyle\int_0^\infty \frac{c^4}{3\nu} f_0 \, dc = \overline{\left(\frac{c^2}{3\nu}\right)}, \\[1em] W = -\dfrac{4\pi}{3} \displaystyle\int_0^\infty c^3 V' \frac{d}{dc} f_0 \, dc. \end{array}\right\} \qquad (3.32)$$

In these formulae $f_0 \equiv f_0(c)$ is taken to be the distribution function appropriate to a steady stream for the given value of E/N, ν is the effective collision frequency for momentum transfer, and $V' = eE/m\nu$. When all encounters are elastic, $\nu = \nu_{el} = Ncq_{m_{el}}(c)$. Although special forms of the distribution function $f_0(c)$ are adopted, it is not assumed that all encounters are elastic. The formulae, with the postulated distribution functions, are therefore applicable both to monatomic and to polyatomic gases.

It follows from equation 3.32 that

$$\frac{W}{D} = -\frac{\displaystyle\int_0^\infty c^3 V' \frac{d}{dc} f_0 \, dc}{\displaystyle\int_0^\infty \frac{c^4}{\nu} f_0 \, dc} = -\frac{\left(\dfrac{e}{m}\dfrac{E}{N}\right)\displaystyle\int_0^\infty \frac{c^2}{q_m(c)} \frac{d}{dc} f_0 \, dc}{\dfrac{1}{N}\displaystyle\int_0^\infty \frac{c^3}{q_m(c)} f_0 \, dc},$$

in which $q_m(c)$ is the effective momentum transfer cross section, which becomes $q_{m_{el}}(c)$ when inelastic encounters are absent (Section 2.7.2d).

It has proved more convenient in studies of the motion of electrons in gases to use what is, in effect, the reciprocal of the ratio defined above. The mobility μ of electrons is defined to be W/E, so that the ratio D/μ is seen to be

$$\frac{D}{\mu} = -\frac{\int_0^\infty \dfrac{c^3}{q_m(c)} f_0\, dc}{\dfrac{e}{m} \int_0^\infty \dfrac{c^2}{q_m(c)} \dfrac{d}{dc} f_0\, dc}. \tag{3.33}$$

When the denominator is transformed by partial integration, as was done to obtain equation 3.13, we find

$$\frac{D}{\mu} = \frac{\overline{[c/q_m(c)]}}{\left(\dfrac{e}{m}\right)\overline{\left[c^{-2}\dfrac{d}{dc}\dfrac{c^2}{q_m(c)}\right]}}, \tag{3.34}$$

in which, as before, the superincumbent bars denote the average values taken over all speeds c of the quantities in the brackets. The physical dimensions of the quantity on the right-hand side that is multiplied by e^{-1} are those of a kinetic energy $\frac{1}{2}mc^2$, and it is therefore convenient to express D/μ in terms of the mean kinetic energy $\frac{1}{2}m\overline{c^2}$ multiplied by e^{-1} and a dimensionless factor, as follows:

$$\frac{D}{\mu} = \left\{ \frac{2\,\overline{[c/q_m(c)]}}{\overline{(c^2)}\,\overline{\left[c^{-2}\dfrac{d}{dc}\dfrac{c^2}{q_m(c)}\right]}} \right\} \frac{\frac{1}{2}m\overline{c^2}}{e} = \frac{F}{e}\tfrac{1}{2}m\overline{c^2}. \tag{3.35}$$

With Maxwell's distribution function $F = \frac{2}{3}$, but when all encounters are elastic and $q_{m_{el}}(c)$ is constant, $F = 0.7628$. Since the dimensionless factor F in practice lies in value between unity and one half, it is evident that De/μ provides a rough estimate of the mean kinetic energy of the electrons. For this reason De/μ has been termed the *characteristic energy* (Frost and Phelps, 1962).

In the cgs system of units W and D are expressed in units of centimetres per second and square centimetres per second, respectively. It follows that the unit in which the ratio D/W is to be expressed is the centimetre. Similarly the force eE is measured in dynes. In practice the electric field is expressed in volts per centimetre, and the mobility μ is defined as the ratio W/E with W measured in centimetres per second and E in volts per centimetre. With the definition of μ it follows that the unit of measurement for D/μ is the volt.

According to equation 3.35 the mean energy $\frac{1}{2}m\overline{c^2}$ associated with a measured value of D/μ is

$$\tfrac{1}{2}m\,\overline{c^2} = \frac{1}{F}e\left(\frac{D}{\mu}\right).$$

Consequently in this system of units both the mean energy and the characteristic energy eD/μ are measured in electron volts.

The formulae (equations 3.32 and 3.33) for W and D/μ are important in two respects. On the one hand, they provide information about the general characteristics of the motion of electrons in gases: the measured values of D/μ are directly related to the mean energies of random motion when electrons move in a steady state determined by the parameter E/N, whereas the velocity W gives the electrical conductivity of a weakly ionized gas. On the other hand, since the theoretical formulae for W and D/μ contain $q_m(c)$[or $q_m(\epsilon)$], the measured values can be used in conjunction with these formulae to derive the momentum transfer cross section as a function of the electron energy. When all encounters are elastic, either one of the measured quantities W or D/μ is sufficient in principle to determine the function $q_{m_{el}}(c)$ (see Chapter 13). When the effects of inelastic encounters are significant, it is necessary to employ the measured values of both W and D/μ to derive $q_{m_{el}}(c)$ together with the associated inelastic cross sections. The accurate determination of $q_{m_{el}}(c)$, even in the case in which all encounters are elastic, requires the use of an electronic computer since the procedure depends on successive approximation by a sequence of trial functions $q_{m_{el}}(c)$ and $q_{0k_{in}}(c)$ within the integral or integrals. The details of these determinations are more appropriately discussed in Chapter 13, which is concerned directly with analytical procedures and results. It is relevant to note here, however, that good preliminary estimates of $q_{m_{el}}(c)$ and first estimates, at least, of $q_{0k_{in}}(c)$ can be made by means of approximate formulae.

3.10.   THE RATIO W/D WHEN E/N IS SMALL

We shall consider weak fields $\mathbf{E}$ such that $V^2 \ll \overline{C^2}$, where V is defined in Section 3.4.1. Then, in the case of a uniform stream of electrons with inelastic encounters, it follows from equation 3.11 that $f_0(c)$ approaches the thermal Maxwellian distribution function, that is,

$$f_0(c) = \frac{1}{(\alpha\sqrt{\pi})^3} \exp\left(-\frac{c^2}{\alpha^2}\right),$$

in which α is the most probable velocity (section 3.7). When elastic encounters also occur, the condition of quasithermal equilibrium is maintained even when the electric field is stronger than that indicated by this criterion. In what follows, it is also assumed that $\text{grad}_r n$ is so small that the distribution function is negligibly perturbed by the presence of the gradient. With these restrictions we are able to relate W to D independently of the collision frequency. Thus:

$$W = -\frac{eE}{m}\frac{4\pi}{3}\int_0^\infty \frac{c^3}{\nu}\frac{d}{dc}f_0\,dc$$

$$= \frac{eE}{m}\frac{4\pi}{3}\frac{2}{\alpha^2}\int_0^\infty \frac{c^4}{\nu}f_0\,dc \quad (\text{since } f_0 \text{ is Maxwellian as above})$$

$$= \frac{eE}{\tfrac{1}{2}m\alpha^2}\frac{4\pi}{3}\int_0^\infty \frac{c^4}{\nu}f_0\,dc = \frac{eE}{\kappa T}\cdot D \quad (\text{from equation 3.27}),$$

or

$$\frac{W}{D} = \frac{eE}{\kappa T} = \frac{N_0 eE}{RT} \quad (\text{Nernst-Townsend relation});$$

otherwise

$$\frac{D}{\mu} = \frac{RT}{N_0 e} = \frac{\kappa T}{e} = \frac{2}{3}\frac{\overline{\tfrac{1}{2}m c^2}}{e} \tag{3.36}$$

in which N_0 is Avogadro's number, R is the universal gas constant, and $\kappa = R/N_0$ is the Boltzmann constant.

3.11.   COMMENT ON THE APPROXIMATE SOLUTION OF THE SCALAR EQUATION

In this chapter we have given an approximate treatment of the scalar equation 3.30, and it is proposed in this section to comment briefly and critically on what has been done. The scalar equation 3.30 when expressed in its primitive form (equation 3.1) is

$$\frac{\partial}{\partial t}(nf_0) + \text{div}_r\left(n\frac{c}{3}\mathbf{f}_1\right) + \frac{1}{4\pi c^2}\frac{\partial}{\partial c}[\sigma_E(c) - \sigma_{\text{coll}}(c)] = 0. \quad (3.37)$$

We recall that $\sigma_E(c)$ and $\sigma_{\text{coll}}(c)$ vanish as $c \to \infty$ and when $c = 0$ (Section 3.8), and it follows that, when the operation $4\pi\int_0^\infty (\quad)c^2\,dc$ is performed on each term of equation 3.37, that equation is replaced by

$$\frac{\partial}{\partial t}n + \text{div}_r\left(n\frac{4\pi}{3}\int_0^\infty c^3\mathbf{f}_1\,dc\right) + [\sigma_E(c) - \sigma_{\text{coll}}(c)]|_0^\infty = 0,$$

that is to say, from equations 3.10,

$$-\frac{\partial}{\partial t}n + \text{div}_r\left[4\pi\int_0^\infty \frac{c^2}{3\nu}\,\text{grad}_r(nf_0)c^2\,dc + \frac{ne\mathbf{E}}{m}4\pi\int_0^\infty \frac{c}{3\nu}\left(\frac{\partial}{\partial c}f_0\right)c^2\,dc\right] = 0.$$

$$(3.38)$$

We note, however, that further progress is impossible because the nature of the distribution function $f_0(c, \mathbf{r}, t)$ is unknown. The course of action often adopted (and, in fact, followed in this chapter) is to cut the Gordian knot by postulating that f_0 is independent of $\mathbf{r}$ and t and is the same as the distribution function $f_0(c)$ for the uniform steady stream. When all encounters are elastic, $f_0(c)$ is then supposed to be given by equation 3.11. Hence equation 3.38 reduces to equation 3.29, but clearly in this form it is not soundly based, as confirmed by the following argument.

Consider the scalar equation 3.30, which for elastic encounters is, when $e\mathbf{E}$ is parallel to $+Oz$ and $\nu = \nu_{\text{el}}$,

$$-\frac{\partial}{\partial t}(nf_0) + \frac{c^2}{3\nu_{\text{el}}}\nabla^2(nf_0) + \frac{cV}{3}\frac{\partial^2}{\partial z\,\partial c}(nf_0)$$

$$+ \frac{1}{3c^2}\frac{\partial}{\partial c}\left\{c^3V\frac{\partial}{\partial z}(nf_0) + c^2\nu_{\text{el}}\left[(V^2 + \overline{C^2})\frac{\partial}{\partial c}(nf_0) + \frac{3mc}{M}(nf_0)\right]\right\} = 0.$$

$$(3.39)$$

Since, when all encounters are elastic, $f_0(c)$ satisfies the differential equation

$$(V^2 + \overline{C^2}) \frac{d}{dc} f_0(c) = -\frac{3mc}{M} f_0(c),$$

it follows that the substitution of $nf_0 \equiv n(\mathbf{r}, t) f_0(c)$ in equation 3.39 leads to the elimination of the final pair of terms on the left-hand side of the equation, but the remaining terms do not sum to zero. Consequently $n(\mathbf{r}, t) f_0(c)$ is not a solution of the complete scalar equation.

In general, the spatially dependent forms of $n(\mathbf{r}, t)$ and $f_0(c, \mathbf{r}, t)$ cannot be found by equating to zero a selected group of terms of the scalar equation and deriving nf_0 as the solution of this latter equation, for the solution so found cannot satisfy the residual scalar equation, which is a different differential equation.

We note, however, that whereas f_0 is in general a function of c, $\mathbf{r}$, and t the number density depends on $\mathbf{r}$ and t only. It would appear, therefore, to be a more promising approach to examine the possibility that $n(\mathbf{r}, t)$ is a solution of a modified form of equation 3.29 such as equation 3.31. The associated form of $f_0(c, \mathbf{r}, t)$ would then be sought. Such is the approach adopted in the next chapter.

BIBLIOGRAPHY AND NOTES

GENERAL TREATISES AND REVIEW ARTICLES

W. P. Allis, "Motions of Ions and Electrons," *Handbuch der Physik*, Vol. XXI, Springer-Verlag, Berlin, 1956.

R. W. Crompton, "The Contribution of Swarm Techniques to the Solution of Some Problems in Low Energy Electron Physics," *Advances in Electronics and Electron Physics*, Vol. 27, Academic Press, New York, 1969.

V. L. Ginsberg and A. V. Gurevič, "Nichtlineare Erscheinungen in einem Plasma, das sich in einem veränderlichen electromagnetischen Feld befindet," *Fortschr. Phys.*, **8**, 97, 1960.

L. G. H. Huxley and R. W. Crompton, in *Atomic and Molecular Processes* (Ed., D. R. Bates) Academic Press, New York, 1962, p. 335.

L. B. Loeb, *Basic Processes of Gaseous Electronics*, University of California Press, Berkeley, 1955, Chaps. III, IV.

E. W. McDaniel, *Collision Phenomena in Ionized Gases*, John Wiley & Sons, New York, 1964, Chap. 11.

J. S. Townsend, *Electrons in Gases*, Hutchinson's Scientific and Technical Publications, London, 1948.

REFERENCES TO SPECIFIC TOPICS

Distribution Function

See bibliography for Chapter 1.

The Generalized Formula (equation 3.11)

This formula is a special case of a formula given by Margenau for motion of electrons in high-frequency electric fields. Equation 3.11 is obtained when the angular frequency in it is put equal to zero.

H. Margenau, *Phys. Rev.,* **69**, 508, 1946; E. A. Desloge, S. W. Matthysse, and H. Margenau, *Phys. Rev.*, **112**, 1437, 1958.

The generalized formula (equation 3.11) is derived for static electric fields by Chapman and Cowling in their treatise:

S. Chapman and T. G. Cowling, *The Mathematical Theory of Non-uniform Gases,* Cambridge University Press, 1952, p.350.

Formulae for Drift Velocity and Diffusion

It was shown that the formula (equation 3.12) for the drift velocity W can, by partial integration, be given an alternative form (equation 3.13), provided that $q_m(c)$ is such that $c^3 f_0/\nu = c^2 f_0/N q_m(c)$ is zero both when $c = 0$ and $c = \infty$. Conversely, equation 3.13 can be transformed back into equation 3.12. The two formulae, with the stated proviso, are therefore equivalent. It is found in practice that Maxwell-Boltzmann procedures lead in the first instance to equation 3.12, as demonstrated in this chapter, whereas the method of free paths gives equation 3.13. Similarly, both the Maxwell-Boltzmann theory and the method of free paths [by use of Einstein's formula (equation 1.2)] lead to equation 3.27 for D. Thus, as remarked by Huxley (1957), "It may be inferred therefore that the supposed limitations of the method of free paths are in this context in many instances attributable to imperfections of application rather than to those of principle."

Expressions for drift velocity and coefficient of diffusion of the forms of equations 3.12 and 3.27 were given by Allis and Allen (reference below) and by Margenau (references above), and the corresponding expression for W, equation 3.13, was given by Davidson (from considerations of balance of momentum) and by Huxley by the method of free paths.

W. P. Allis and H. W. Allen, *Phys. Rev.,* **52**, 703, 1937.

P. M. Davidson, *Proc. Phys. Soc. B,* **67**, 159, 1954.

L. G. H. Huxley, *Aust. J. Phys.,* **10**, 118, 1957; **13**, 718, 1960.

Additional references to free path formulae are more appropriately associated with later chapters.

The Parameter E/N

Although the convenience attending the replacement of E/p by E/N was becoming apparent, the first use of it in the literature appears to be in a paper by Barbara I. H. Hall, *Aust. J. Phys.,* **8**, 551, 1955.

The *townsend* as the unit for E/N was proposed by L. G. H. Huxley, R. W. Crompton, and M. T. Elford, *Brit. J. Appl. Phys.,* **17**, 1237, 1966.

The Characteristic Energy eD/μ

As explained in the context of equation 3.32, the ratio D/μ is in several respects a more convenient quantity than the ratio W/D, which is measured directly. The introduction of D/μ into general usage was due to Frost and Phelps, who termed eD/μ the characteristic energy.

L. S. Frost and A. V. Phelps, *Phys. Rev.*, **127**, 1621, 1962.

The Nernst-Townsend Formula

For some time it was believed that the Nernst-Townsend formula (equations 1.1 and 3.35) was valid in all circumstances and, in particular, was applicable to the motion of electrons. However, in 1937 Townsend showed that the formula is not exact, except in special circumstances, when applied to electrons, since the ratio $(D/\mu)/\frac{1}{2}m\overline{c^2}$ is not constant but is influenced by the form of the velocity distribution function and the nature of the dependence of $q_m(c)$ upon c, as shown in equation 3.34.

J. S. Townsend, *Phil. Mag.*, (7) **23**, 481, 1937.

Dimensionless Factors

The dimensionless factor F (equation 3.34), depending upon a product of mean values of powers of the velocities c in combination with $q_m(c)$, is an example of the use of such factors, which have proved to be convenient for finding numerical coefficients in a number of formulae.

L. G. H. Huxley and A. A. Zaazou, *Proc. Roy. Soc. A,* **196**, 402, 1949.
L. G. H. Huxley, *Aust. J. Phys.*, **10**, 118, 1957; **13**, 718, 1960.

4

GENERAL THEORY

OF DIFFUSION AND DRIFT

IN A UNIFORM AND CONSTANT

ELECTRIC FIELD

4.1. INTRODUCTION

In Chapter 3 formulae for the drift velocity and coefficient of diffusion were derived on the simplifying supposition that the velocity distribution function $f(\mathbf{c})$ is independent of position $\mathbf{r}$ even in the presence of spatial gradients of the number density n. The usual form of the differential equation for n (equation 3.29) was thus established. It was pointed out, however, that in the presence of spatial derivatives of n the distribution function is in fact a function $f(\mathbf{r}, \mathbf{c}, t)$ of position $\mathbf{r}$.

In this chapter, therefore, we treat the more difficult problem of providing a theory of diffusion and drift in the presence of spatial derivatives of n. The exposition follows that given by Huxley (1972) and discussed further by Skullerud (1973), which proceeds as follows. It is first shown that the overall velocity distribution function for an isolated group of n_0 electrons drifting and diffusing in a uniform electric field acquires the form given in Section 3.4, when all encounters are elastic. This overall form is designated $f^*(\mathbf{c})$. It is also the form appropriate everywhere in a homogeneous stream free of gradients in the number density. It is suggested, therefore, that when spatial derivatives are present the velocity distribution function is distorted from the basic form $f^*(\mathbf{c})$ and that this distortion can be represented by a series of terms proportional to the pure and mixed partial spatial derivatives of n. The coefficients of these distorting terms are functions of the speeds c, and in order for the theory to be valid a procedure for determining them must be established. This procedure requires the adoption of a generalized form of the differential equation for n (equation 4.37), which in many circumstances is adequately represented by equation 3.31. The theory is restricted to circumstances in which the velocity distribution function is of the form $f(\mathbf{r}, \mathbf{c}, t) = f_0(\mathbf{r}, c, t) + f_1(\mathbf{r}, c, t) \cos\theta$ with $f_1/f_0 \ll 1$. The generalized equation shows that in the presence of gradients and an electric field diffusion is anisotropic; when the first- and second-order spatial derivatives only are retained in the equation (equation 3.31), diffusion is described by two coefficients D and D_L which refer to directions transverse and parallel, respectively, to the field.

4.2. ISOLATED TRAVELLING GROUP OF ELECTRONS

The group is considered to travel in an extensive region such that its total population $n_0 = \int_{\mathbf{r}} n(\mathbf{r}, t) d\mathbf{r}$ is contained within a closed surface Σ on which $n(\mathbf{r}, t)$ is everywhere zero. The number of the n_0 electrons whose velocity points lie in the element of volume $d\mathbf{c}$ at time t is described by a distribution function $f^*(\mathbf{c}, t)$ such that this number is $n_0 f^*(\mathbf{c}, t) d\mathbf{c}$. But if, as before, the number of electrons within an element of volume $d\mathbf{r}$ in configuration space is $n(\mathbf{r}, t) d\mathbf{r}$, the number of these whose velocity points reside within $d\mathbf{c}$ is $n(\mathbf{r}, t) d\mathbf{r} f(\mathbf{c}, \mathbf{r}, t) d\mathbf{c}$. It follows that

$$n_0 f^*(\mathbf{c}, t) d\mathbf{c} = d\mathbf{c} \int_{\mathbf{r}} n(\mathbf{r}, t) f(\mathbf{c}, \mathbf{r}, t) d\mathbf{r},$$

that is,

$$f^*(\mathbf{c}, t) = \frac{1}{n_0} \int_{\mathbf{r}} n(\mathbf{r}, t) f(\mathbf{c}, \mathbf{r}, t) d\mathbf{r}. \qquad (4.1)$$

Consequently $f^*(\mathbf{c}, t)$ is the mean value of $f(\mathbf{c}, \mathbf{r}, t)$ taken over the whole group. Moreover, if

$$f(\mathbf{c}, \mathbf{r}, t) \equiv \sum_0^\infty f_k(c, \mathbf{r}, t) P_k(\cos\theta),$$

then $f^*(\mathbf{c}, t)$ is of the form

$$f^*(\mathbf{c}, t) = \sum_0^\infty f_k^*(c, t) P_k(\cos\theta),$$

where

$$f_k^*(c, t) = \frac{1}{n_0} \int_{\mathbf{r}} n(\mathbf{r}, t) f_k(c, \mathbf{r}, t) d\mathbf{r}.$$

In order to find the equations to be satisfied simultaneously by f_0^* and f_1^*, we return to the scalar equation 3.1 and the vector equation 3.8. It is assumed that no processes are present that liberate new electrons or remove electrons from the group and that, therefore, $(d/dt)n_0 = 0$.

First we form the integral $\int_r (\ )d\mathbf{r}$ of each term of equations 3.1 and 3.8, and consider the integrals in turn. From the first term of equation 3.1 we have:

$$\frac{\partial}{\partial t} \int_\mathbf{r} n(\mathbf{r},t) f_0(c,\mathbf{r},t) \, d\mathbf{r} = \frac{\partial}{\partial t} [n_0 f_0^*(c,t)] = n_0 \frac{\partial f_0^*}{\partial t},$$

whereas, from the second term,

$$\frac{c}{3} \int_\mathbf{r} \mathrm{div}_r(n\mathbf{f}_1) \, d\mathbf{r} = \frac{c}{3} \int_\Sigma n\mathbf{f}_1 \cdot d\mathbf{S} = 0$$

since $n = 0$ on all elements $d\mathbf{S}$ of Σ.

Finally, we consider the remaining term:

$$\frac{1}{4\pi c^2} \int_\mathbf{r} \frac{\partial}{\partial c} (\sigma_E - \sigma_{\mathrm{coll}}) \, d\mathbf{r}.$$

The state of motion at time t of the whole group of electrons n_0 is represented by the positions in velocity space of n_0 velocity points, of which

$$n_0 f^*(\mathbf{c},t) \, d\mathbf{c} = n_0 \sum_0^\infty f_k^*(c,t) P_k(\cos\theta) \, d\mathbf{c}$$

occupy the element $d\mathbf{c} = c^2 \, dc \sin\theta \, d\theta \, d\phi$. The $n(\mathbf{r},t) \, d\mathbf{r}$ electrons that reside in $d\mathbf{r}$ contribute $n \, d\mathbf{r} \, d\mathbf{c}[\sum_0^\infty f_k(c,t) P_k(\cos\theta)]$ velocity points to the total population of points within $d\mathbf{c}$, which is

$$d\mathbf{c} \, n_0 \sum_0^\infty f_k^* P_k(\cos\theta) = d\mathbf{c} \int_\mathbf{r} n \sum_0^\infty f_k P_k(\cos\theta) \, d\mathbf{r}.$$

The point fluxes across the surface of a sphere c contributed by the $n \, d\mathbf{r}$ electrons transfer in time dt points equal in number to

$$(dt \, d\mathbf{r})\sigma_E \quad \text{outwards} \quad \text{(Section 2.7.1c and equation 3.2)}$$

and

$$(dt \, d\mathbf{r})\sigma_{\mathrm{coll}} \quad \text{inwards} \quad \text{(Section 2.7.1d).}$$

The corresponding numbers of points transferred when the population is n_0 instead of $n\,d\mathbf{r}$ are found by replacing $(n\,d\mathbf{r})f(\mathbf{c},\mathbf{r},t)$ in the equations leading to $\sigma_E\,d\mathbf{r}$ and $\sigma_{\text{coll}}\,d\mathbf{r}$ by $\int_{\mathbf{r}} n(\mathbf{r},t)f(\mathbf{c},\mathbf{r},t)\,d\mathbf{r} = n_0 f^*(\mathbf{c},t)$. The numbers are, then,

$$d t\sigma_E^* = dt \int_{\mathbf{r}} \sigma_E\,d\mathbf{r} \qquad \text{and} \qquad d t\sigma_{\text{coll}}^* = dt \int_{\mathbf{r}} \sigma_{\text{coll}}\,d\mathbf{r}.$$

Consequently the term

$$\frac{1}{4\pi c^2} \int_{\mathbf{r}} \frac{\partial}{\partial c}(\sigma_E - \sigma_{\text{coll}})\,d\mathbf{r}$$

is equal to

$$\frac{1}{4\pi c^2} \frac{\partial}{\partial c}(\sigma_E^* - \sigma_{\text{coll}}^*).$$

Since

$$\sigma_E = \frac{4\pi}{3}c^2 \frac{e\mathbf{E}}{m}\cdot n\mathbf{f}_1,$$

it follows that

$$\sigma_E^* = \frac{4\pi}{3}c^2 \frac{e\mathbf{E}}{m}\cdot n_0\mathbf{f}_1^*.\dagger$$

When all encounters are elastic, σ_{coll} is (equation 3.4)

$$\sigma_{\text{coll}} = 4\pi c^2 \nu_{\text{el}}\left[\frac{m}{M}cnf_0 + \frac{\overline{C^2}}{3}\frac{\partial}{\partial c}(nf_0) \right];$$

consequently the corresponding formula for σ_{coll}^* is,

$$\sigma_{\text{coll}}^* = 4\pi c^2 \nu_{\text{el}} n_0\left(\frac{m}{M}cf_0^* + \frac{\overline{C^2}}{3}\frac{\partial}{\partial c}f_0^* \right).$$

$\dagger \mathbf{f}_1^*$ is parallel to $\mathbf{E}$ since it refers to the group as a whole. Thus σ_E^* could be more simply written as $(4\pi/3)(e/m)c^2 n_0 E f_1^*$.

Therefore the scalar equation for a travelling group is

$$n_0 \frac{\partial}{\partial t} f_0^*(c,t) + \frac{1}{4\pi c^2} \frac{\partial}{\partial c}(\sigma_E^* - \sigma_{\text{coll}}^*) = 0. \tag{4.2}$$

It remains to consider the terms of the vector equation 3.8 when integrated with respect to $\mathbf{r}$. The first term gives

$$\frac{\partial}{\partial t} \int_{\mathbf{r}} n(\mathbf{r},t) f_1(c,\mathbf{r},t)\,d\mathbf{r} = n_0 \frac{\partial}{\partial t} f_1^*(c,t).$$

Consider next

$$c\int_{\mathbf{r}} \text{grad}_r(nf_0)\,d\mathbf{r} = c\int\int\int_{-\infty}^{\infty} \left[\mathbf{i}_x \frac{\partial}{\partial x}(nf_0) + \mathbf{i}_y \frac{\partial}{\partial y}(nf_0) + \mathbf{i}_z \frac{\partial}{\partial z}(nf_0) \right] dx\,dy\,dz.$$

Since

$$\int\int_{-\infty}^{\infty} dy\,dz \left[\int_{-\infty}^{\infty} \frac{\partial}{\partial x}(nf_0)\,dx \right] = \int\int_{-\infty}^{\infty} \left[nf_0 \Big|_{-\infty}^{\infty} \right] dy\,dz = 0,$$

it follows that

$$c\int_{\mathbf{r}} \text{grad}_r(nf_0)\,d\mathbf{r} = 0.$$

The remaining terms are

$$\frac{e\mathbf{E}}{m} \frac{\partial}{\partial c} \int_{\mathbf{r}} nf_0\,d\mathbf{r} = \frac{e\mathbf{E}}{m} n_0 \frac{\partial}{\partial c} f_0^*,$$

$$\boldsymbol{\omega} \times \int_{\mathbf{r}} nf_1\,d\mathbf{r} = \boldsymbol{\omega} \times n_0 \mathbf{f}_1^*,$$

and

$$\nu \int_{\mathbf{r}} nf_1\,d\mathbf{r} = n_0 \nu \mathbf{f}_1^*.$$

The vector equation for the group is, therefore,

$$\frac{\partial}{\partial t} \mathbf{f}_1^* + \frac{e\mathbf{E}}{m} \frac{\partial}{\partial c} f_0^* - (\boldsymbol{\omega} \times \mathbf{f}_1^*) + \nu \mathbf{f}_1^* = 0. \tag{4.3}$$

We note an important feature of equations 4.2 and 4.3, namely, that terms depending on div_r and grad_r do not appear.

Assume that no magnetic field is present ($\boldsymbol{\omega} = 0$) and that the group has travelled for a sufficient time in the field $\mathbf{E}$ that the functions f_0^* and f_1^*

have reached equilibrium and have become independent of the time t. Equation 4.3 then becomes

$$\mathbf{f}_1^* = -\frac{e\mathbf{E}}{mv}\frac{d}{dc}f_0^* = -\mathbf{V}'\frac{d}{dc}f_0^*.$$

The mean velocity of the group $(n\,d\mathbf{r})(4\pi f_0 c^2\,dc)$ of the $n\,d\mathbf{r}$ electrons whose velocity points occupy the shell (c,dc) is $c\mathbf{f}_1/3f_0$; consequently the velocity of the centroid of the $n_0(4\pi f_0^* c^2\,dc)$ electrons that occupy the shell (c,dc) is

$$\frac{\int_{\mathbf{r}}(c\mathbf{f}_1/3f_0)(n\,d\mathbf{r})(4\pi f_0 c^2\,dc)}{n_0 4\pi f_0^* c^2\,dc} = \frac{c\mathbf{f}_1^*}{3f_0^*}.$$

The velocity of the centroid of the whole group is, therefore,

$$\mathbf{W}_{\text{centroid}} = \int_0^\infty \frac{c\mathbf{f}_1^*}{3f_0^*}4\pi f_0^* c^2\,dc = \frac{4\pi}{3}\int_0^\infty c^3\mathbf{f}_1^*\,dc$$

$$= -\frac{e\mathbf{E}}{m}\frac{4\pi}{3}\int_0^\infty \frac{c^3}{v}\frac{d}{dc}f_0^*\,dc = -\frac{e}{m}\frac{\mathbf{E}}{N}\frac{4\pi}{3}\int_0^\infty \frac{c^2}{q_m(c)}\frac{d}{dc}f_0^*\,dc, \quad (4.4)$$

in which q_m is the effective momentum transfer cross section defined by $q_m(c)=v/Nc$.

Alternatively, from equation 3.13,

$$\mathbf{W}_{\text{centroid}} = \frac{e}{m}\frac{\mathbf{E}}{N}\frac{4\pi}{3}\int_0^\infty \left[c^{-2}\frac{d}{dc}\frac{c^2}{q_m(c)}\right]f_0^* c^2\,dc$$

$$= \frac{e}{m}\frac{\mathbf{E}}{N}\frac{1}{3}\overline{\left[c^{-2}\frac{d}{dc}\frac{c^2}{q_m(c)}\right]} = \frac{e}{m}\frac{\mathbf{E}}{3}\overline{\left[c^{-2}\frac{d}{dc}\left(\frac{c^3}{v}\right)\right]}. \quad (4.5)$$

When all encounters are elastic, $q_m(c)=q_{m_{\text{el}}}(c)$, $v=v_{\text{el}}=Ncq_{m_{\text{el}}}(c)$. The scalar equation 4.2 for the group when $\partial/\partial t=0$ becomes

$$\sigma_E^* = \sigma_{\text{coll}}^*$$

or

$$\frac{e\mathbf{E}}{m}\cdot\mathbf{f}_1^* = \frac{3}{4\pi n_0 c^2}\sigma_{\text{coll}}^*,$$

that is,

$$-\left(\frac{eE}{m}\right)^2\frac{1}{\nu}\frac{d}{dc}f_0^* = \frac{3}{4\pi n_0 c^2}\sigma_{\text{coll}}^*.$$ (4.6)

When all encounters are elastic, $\nu = \nu_{\text{el}}$ and

$$\sigma_{\text{coll}}^* = n_0 4\pi c^2 \nu_{\text{el}}\left(\frac{m}{M}cf_0^* + \frac{\overline{C^2}}{3}\frac{d}{dc}f_0^*\right).$$

Equation 4.6 then reduces to

$$(V^2 + \overline{C^2})\frac{d}{dc}f_0^* = -\frac{3m}{M}cf_0^*$$

where

$$V = \frac{eE}{m\nu_{\text{el}}}.$$

Thus

$$f_0^*(c) = A\exp\left(-\frac{3m}{M}\int_0^c\frac{c\,dc}{V^2 + \overline{C^2}}\right).$$ (4.7)

Evidently, the distribution function for the whole group is the same as that for the uniform stream (equation 3.11). The drift velocity $\mathbf{W}_{\text{centroid}}$ is the same as the convective drift velocity $\mathbf{W}$ due to eE in the uniform stream. We therefore write, without confusion, $\mathbf{W}_{\text{centroid}} = \mathbf{W}$.

Since

$$V = \frac{eE}{m\nu_{\text{el}}} = \frac{eE}{mNcq_{m_{\text{el}}}(c)}$$

when all encounters are elastic, we note that, when E/N is specified, $f_0^*(c)$ is a known function of c if the functional dependence of $q_{m_e} \equiv \equiv q_{m_{\text{el}}}(c)$ on c is known. This knowledge of $q_{m_{\text{el}}}(c)$ then serves to determine W through equations 4.4 and 4.5 $[q_m(c) = q_{m_{\text{el}}}(c)]$ since $f_0^*(c)$ is known.

Although the formula (equation 3.11) for the generalized distribution function is restricted to conditions in which all encounters may be considered to be elastic, it should be remarked that the expressions for the drift velocity $\mathbf{W}$ (equations 3.12 and 3.13) are not similarly restricted since the distribution function is the one appropriate to the circumstances. When inelastic encounters are important, the distribution function $f_0(c)$ is not given by the generalized formula; nevertheless the expressions for W are of considerable value in general considerations of the properties of the motion when the specific form of $f_0(c)$ is not of immediate significance. The same comments are applicable also to the expression for the coefficient of diffusion D (equation 3.27).

4.3. SOLUTION OF SCALAR EQUATION IN A GENERAL FORM

From the discussion in the preceding section it is seen that the velocity distribution function $f_0^*(c)$ for the total population n_0 of an isolated group of electrons satisfies the equation

$$n_0 \frac{\partial}{\partial t} f_0^*(c,t) = - \frac{1}{4\pi c^2} \frac{\partial}{\partial c} (\sigma_E^* - \sigma_{\text{coll}}^*),$$

that is to say, if the population of the shell (c,dc) is denoted as $n_{0c} = n_0 4\pi f_0^* c^2 dc$, then

$$\frac{\partial}{\partial t} n_{0c} = - dc \frac{\partial}{\partial c} (\sigma_E^* - \sigma_{\text{coll}}^*).$$

Consider a group n_0 in thermal equilibrium with the gas in the absence of an electric field **E**. In this event $\sigma_E^* = \sigma_{\text{coll}}^* = 0$, $(\partial/\partial t)n_{0c} = 0$. Let the field be applied at $t = 0$; then σ_E^* is established and $(\partial/\partial t)n_{0c} = - dc(\partial/\partial c)\sigma_E^*$ at $t = 0$.

As t increases, the distribution of velocity points over the shells changes in a manner to increase σ_{coll}^* and (in general) to diminish σ_E^* until the distribution function reaches a new equilibrium form for which $(\partial/\partial t)n_{0c} = 0$ for all shells, that is to say, $\sigma_E^* = \sigma_{\text{coll}}^*$ and $(\partial/\partial t)f_0^* = 0$. In the case of elastic collisions this final form is reached in a time of the order of $(M/m)(1/\nu_{\text{el}})$; when inelastic encounters occur this time is greatly reduced (see page 147).

In what follows we are concerned with the theory of the isolated travelling group in which the velocity distribution function $f_0^*(c)$ for the group as a whole has attained its final unchanging form. Although $(\partial/\partial t)[n_0 f_0^*(c)] = 0$, it is evident that $(\partial/\partial t)(nf_0)$ is not zero throughout the group since through the action of diffusion the number density $n(\mathbf{r},t)$ changes. Moreover, as will appear, $f_0(c)$ is also a function $f_0(\mathbf{r},c,t)$ of position within the group. We therefore seek a form of solution of the scalar equation that correctly describes the properties of an isolated group drifting and diffusing in an electric field. Since the theory of the travelling group and of the diffusing stream in a form adapted to the needs of experiment has hitherto been based on equation 3.29 as a starting point, it is a matter of great importance to discover whether this equation or some other for the number density $n(\mathbf{r}, t)$ is consistent with the full scalar equation (equation 3.37). It will be shown that, if $f_0(\mathbf{r}, c, t)$ is given a particular form, a differential equation can still be found for $n(\mathbf{r}, t)$, but one in which diffusion is no longer isotropic.

In the absence of a magnetic field the scalar equation (equation 3.30) when all encounters are elastic is

$$
-\frac{\partial}{\partial t}(nf_0) + \mathrm{div}_r\left[\frac{c^2}{3\nu_{el}}\,\mathrm{grad}_r(nf_0) + \frac{c\mathbf{V}}{3}\frac{\partial}{\partial c}(nf_0)\right]
$$

$$
+\frac{1}{3c^2}\frac{\partial}{\partial c}\left\{c^2\nu_{el}\left[\frac{c\mathbf{V}}{\nu_{el}}\cdot\mathrm{grad}_r(nf_0) + (V^2+\overline{C^2})\frac{\partial}{\partial c}(nf_0) + \frac{3mc}{M}(nf_0)\right]\right\} = 0,
$$

$$(4.8)$$

in which $\mathbf{V} = e\mathbf{E}/m\nu_{el}$ and $\nu_{el} = Ncq_{m_{el}}(c)$.

We adopt a Cartesian system of coordinates with $+Oz$ parallel to $e\mathbf{E}$ and $\mathbf{V}$; consequently $V = V_z$. With these coordinates equation 4.8 is

$$
-\frac{\partial}{\partial t}(nf_0) + \nabla^2\left(\frac{c^2}{3\nu_{el}}nf_0\right) + \frac{cV}{3}\frac{\partial}{\partial z}\left(\frac{\partial}{\partial c}nf_0\right)
$$

$$
+\frac{1}{3c^2}\frac{\partial}{\partial c}\left\{c^2\nu_{el}\left[\frac{cV}{\nu_{el}}\frac{\partial}{\partial z}(nf_0) + (V^2+\overline{C^2})\frac{\partial}{\partial c}(nf_0) + \frac{3mc}{M}(nf_0)\right]\right\} = 0.
$$

$$(4.9)$$

We seek a solution of equation 4.9 that describes an isolated group of n_0 electrons whose centroid travels at constant speed W along the axis $+Oz$, where W is given by equation 4.5. The constraints on the solution that require it to embody the properties of an isolated travelling group are therefore as follows:

(a) $n_0 f_0^* = \int_r nf_0\,d\mathbf{r}$

(b) $1 = 4\pi\int_0^\infty f_0^* c^2\,dc = 4\pi\int_0^\infty f_0(c, \mathbf{r}, t)\,c^2\,dc,$

(c) (nf_0) and its derivatives tend to zero as $r \to \infty$.

We note also that the total population n_0 of a group can be expressed in alternative ways:

$$n_0 = 4\pi \int_0^\infty \left(\int_r n \, d\mathbf{r} \right) f_0 c^2 \, dc = \int_r \left(4\pi \int_0^\infty f_0 c^2 \, dc \right) n \, d\mathbf{r}.$$

We now adopt a moving Cartesian system with the centroid as origin but with the same directions as above for the axes. If (x', y', z') are the coordinates of a point in the moving system and (x, y, z) its coordinates in the system at rest, then $x' = x$, $y' = y$, $z' = z - Wt$. Moreover, if d/dt is the time-differential operator in the moving system,

$$\frac{\partial}{\partial t} = \frac{d}{dt} - W \frac{\partial}{\partial z'}.$$

It follows that equation 4.9 when referred to the moving system of coordinates is

$$-\frac{d}{dt}(nf_0) + \frac{c^2}{3\nu_{el}} \nabla^2(nf_0) + \frac{cV}{3} \frac{\partial}{\partial z'} \left(\frac{\partial}{\partial c} nf_0 \right) + W \frac{\partial}{\partial z'}(nf_0)$$

$$+ \frac{1}{3c^2} \frac{\partial}{\partial c} \left\{ c^2 \nu_{el} \left[\frac{cV}{\nu_{el}} \frac{\partial}{\partial z'}(nf_0) + (V^2 + \overline{C^2}) f_0^* \frac{\partial}{\partial c} \left(\frac{nf_0}{f_0^*} \right) \right] \right\} = 0, \qquad (4.10)$$

in which the pair of terms $(V^2 + \overline{C^2})(\partial/\partial c)(nf_0) + (3mc/M)nf_0$ is replaced by the equivalent single term $(V^2 + \overline{C^2})(nf_0^*)(\partial/\partial c)(f_0/f_0^*)$. In this expression f_0^* satisfies the equation

$$(V^2 + \overline{C^2}) \frac{d}{dc} f_0^* + \frac{3mc}{M} f_0^* = 0$$

and is therefore the distribution function given by equation 4.7.

It is convenient to proceed in stages by first deriving solutions of equation 4.10 in two relatively simple special circumstances before considering unrestricted solutions. We therefore begin with a discussion of the special case in which the electric force $e\mathbf{E}$ is absent.

4.3.1. MOTION WITHOUT AN ELECTRIC FIELD. When $e\mathbf{E}$ is zero, terms containing V and W as factors vanish and equations 4.10 and 4.9 both reduce to

$$-\frac{\partial}{\partial t}(nf_0) + \frac{c^2}{3\nu_{el}} \nabla^2(nf_0) + \frac{n}{3c^2} \frac{\partial}{\partial c} \left[c^2 \nu_{el} \overline{C^2} f_0^* \frac{\partial}{\partial c} \left(\frac{f_0}{f_0^*} \right) \right] = 0. \qquad (4.11)$$

It is clear that f_0 is not f_0^*, which is now Maxwell's distribution function, because $f_0^{*-1} f_0$ would then be equal to unity and $(\partial/\partial c)(f_0^{*-1} f_0)$ would equal zero. Hence equation 4.11 would become

$$\left(-\frac{\partial}{\partial t}n+\frac{c^2}{3\nu_{el}}\nabla^2 n\right)f_0^*=0.$$

This is an impossible relation since it implies that $(\partial/\partial t)n$ is a function of c. We conclude, therefore, that in an isolated diffusing group, whereas the distribution function of the total group of n_0 electrons is f_0^*, the local distribution function f_0 within the group is a function of position.

Since the departure of f_0 from the form f_0^* is attributable to the presence of spatial derivatives of n, we assume for f_0 the form

$$nf_0=nf_0^*+\sum_{k=1}^{\infty}f_0^*a_{2k}(c)(\nabla^2)^k n$$

$$=\sum_{k=0}^{\infty}f_0^*a_{2k}(c)(\nabla^2)^k n \tag{4.12}$$

with $a_0(c)=1$.

When nf_0 is replaced in equation 4.11 by this series, that equation becomes

$$-f_0^*\sum_{k=0}^{\infty}a_{2k}(c)(\nabla^2)^k\frac{\partial}{\partial t}n+\frac{c^2}{3\nu_{el}}f_0^*\sum_{k=0}^{\infty}a_{2k}(c)(\nabla^2)^{k+1}n$$

$$+\frac{1}{3c^2}\frac{d}{dc}\left\{c^2\nu_{el}\overline{C^2}f_0^*\sum_{k=0}^{\infty}\left[\frac{d}{dc}a_{2k}(c)\right]\right\}(\nabla^2)^k n=0. \tag{4.13}$$

In order to eliminate the factor $(\partial/\partial t)n$ we suppose that $n(x,y,z,t)$ is a solution of an equation of the form

$$\frac{\partial}{\partial t}n=\sum_{l=1}^{\infty}D_{2l}(\nabla^2)^l n. \tag{4.14}$$

When $(\partial/\partial t)n$ is replaced in equation 4.13 by its series representation in equation 4.14, it is found that

$$-f_0^*\sum_{k=0}^{\infty}\sum_{l=1}^{\infty}a_{2k}(c)D_{2l}(\nabla^2)^{k+l}n+\frac{c^2}{3\nu_{el}}f_0^*\sum_{k=0}^{\infty}a_{2k}(\nabla^2)^{k+1}n$$

$$+\frac{1}{3c^2}\frac{d}{dc}\left\{c^2\nu_{el}\overline{C^2}f_0^*\sum_{k=0}^{\infty}\left[\frac{d}{dc}a_{2k}(c)\right](\nabla^2)^k n\right\}=0.$$

By rearrangement of terms this equation can be given the form [the term in n vanishes because $a_0 = 1$ and $(d/dc)a_0 = 0$]

$$\left\{ f_0^* \left(-D_2 + \frac{c^2}{3\nu_{el}} \right) + \frac{1}{3c^2} \frac{d}{dc} \left[c^2 \nu_{el} \overline{C^2} f_0^* \frac{d}{dc} a_2(c) \right] \right\} \nabla^2 n$$

$$+ \left\{ f_0^* \left(-D_4 - a_2 D_2 + \frac{c^2}{3\nu_{el}} a_2 \right) + \frac{1}{3c^2} \frac{d}{dc} \left[c^2 \nu_{el} \overline{C^2} f_0^* \frac{d}{dc} a_4(c) \right] \right\} \nabla^4 n$$

$$+ \left\{ f_0^* \left(-D_6 - a_2 D_4 - a_4 D_2 + \frac{c^2}{3\nu_{el}} a_4 \right) + \frac{1}{3c^2} \frac{d}{dc} \left[c^2 \nu_{el} \overline{C^2} f_0^* \frac{d}{dc} a_6(c) \right] \right\} \nabla^6 n$$

$$+ \cdots$$

$$+ \left\{ f_0^* \left(-D_{2k} - a_2 D_{2k-2} - a_4 D_{2k-4} - \cdots - a_{2k-2} D_2 + \frac{c^2}{3\nu_{el}} a_{2k-2} \right) \right.$$

$$\left. + \frac{1}{3c^2} \frac{d}{dc} \left[c^2 \nu_{el} \overline{C^2} f_0^* \frac{d}{dc} a_{2k}(c) \right] \right\} (\nabla^2)^k n + \cdots = 0. \qquad (4.15)$$

It follows, therefore, that nf_0 in the form given in equation 4.12, and subject to the condition imposed on n by equation 4.14, is a solution of equation 4.11 provided that the coefficients of the factors $(\nabla^2)^k n$ in equation 4.15 vanish for all values of k. We thus obtain the following sequence of equations from which to determine the $a_{2k}(c)$ and the D_{2l}:

$$\left. \begin{array}{l} \left(-D_2 + \frac{c^2}{3\nu_{el}} \right) f_0^* + \frac{1}{3c^2} \frac{d}{dc} \left[c^2 \nu_{el} \overline{C^2} f_0^* \frac{d}{dc} a_2(c) \right] = 0, \\[3mm] \left(-D_4 - a_2 D_2 + \frac{c^2}{3\nu_{el}} a_2 \right) f_0^* + \frac{1}{3c^2} \frac{d}{dc} \left[c^2 \nu_{el} \overline{C^2} f_0^* \frac{d}{dc} a_4(c) \right] = 0, \\[3mm] \qquad\qquad\qquad\qquad \vdots \\[3mm] \left(-D_{2k} - a_2 D_{2k-2} - \cdots - a_{2k-2} D_2 + \frac{c^2}{3\nu_{el}} a_{2k-2} \right) f_0^* \\[3mm] \qquad\qquad + \frac{1}{3c^2} \frac{d}{dc} \left[c^2 \nu_{el} \overline{C^2} f_0^* \frac{d}{dc} a_{2k}(c) \right] = 0, \\[3mm] \qquad\qquad\qquad \text{etc.} \end{array} \right\} (4.16)$$

Let us consider the first of equations 4.16. We form the integral $4\pi \int_0^\infty (\ ) c^2 dc$ of each term; then, because

$$4\pi \int_0^\infty f_0^* c^2 dc = 1 \quad \text{and} \quad \left[c^2 \nu_{\text{el}} \overline{C^2} f_0^* \frac{d}{dc} a_2(c) \right]_0^\infty = 0,$$

it follows that

$$D_2 = 4\pi \int_0^\infty \frac{c^2}{3\nu_{\text{el}}} f_0^* c^2 dc = \overline{\left(\frac{c^2}{3\nu_{\text{el}}} \right)}.$$

Thus $D_2 = D$, the coefficient of diffusion defined by equation 3.27.

Since D_2 is known, we can determine the form of $a_2(c)$ by integration of the first of equations 4.16. Thus

$$\frac{d}{dc} a_2(c) = - \frac{3}{c^2 \nu_{\text{el}} \overline{C^2} f_0^*(c)} \int_0^c \left[\frac{x^2}{3\nu_{\text{el}}(x)} - D_2 \right] f_0^*(x) x^2 dx,$$

whence

$$a_2(c) = \text{const} - \phi_2(c),$$

where

$$\phi_2(c) = 3 \int_0^c \frac{1}{y^2 \nu_{\text{el}}(y) f_0^*(y) \overline{C^2}} \left\{ \int_0^y \left[\frac{x^2}{3\nu_{\text{el}}(x)} - D_2 \right] f_0^*(x) x^2 dx \right\} dy.$$

It follows from equation 4.12 that because

$$4\pi \int_0^\infty f_0 c^2 dc = 1 = 4\pi \int_0^\infty f_0^* c^2 dc,$$

then

$$4\pi \int_0^\infty a_{2k}(c) f_0^* c^2 dc = 0. \tag{4.17}$$

The constant in the expression for $a_2(c)$ is therefore

$$\overline{\phi_2(c)} = 4\pi \int_0^\infty \phi_2(c) f_0^* c^2 dc,$$

and consequently

$$a_2(c) = \overline{\phi_2(c)} - \phi_2(c). \tag{4.18}$$

Similarly from the kth of equations 4.16 we find

$$D_{2k} = -4\pi \int_0^\infty [a_2(c)D_{2k-2} + \cdots + a_{2k-2}(c)D_2]f_0^*c^2\,dc$$

$$+ \frac{4\pi}{3} \int_0^\infty \frac{c^2}{\nu_{el}} a_{2k-2}(c)f_0^*c^2\,dc$$

$$= \frac{4\pi}{3} \int_0^\infty \frac{c^2}{\nu_{el}} a_{2k-2}(c)f_0^*c^2\,dc$$

since the first integral vanishes by virtue of equation 4.17. Also

$$a_{2k}(c) = \overline{\phi_{2k}(c)} - \phi_{2k}(c),$$

where

$$\phi_{2k}(c) = 3 \int_0^c \frac{1}{y^2\nu_{el}(y)f_0^*(y)\,\overline{C^2}} \left\{ \int_0^y \left[\frac{x^2}{3\nu_{el}(x)} a_{2k-2}(x) - a_2(x)D_{2k-2} \right. \right.$$

$$\left. \left. - \cdots - a_{2k-2}(x)D_2 \right] f_0^*(x)x^2\,dx \right\} dy.$$

Since D_{2k} and $a_{2k}(c)$ are expressible in terms of the coefficients of lower order and because D_2 and $a_2(c)$ can be found, it follows that, in principle, the coefficients D_{2k} and $a_{2k}(c)$ can be determined for any value of k.

The number density $n(x,y,z,t)$ is therefore to be found as a solution of the equation:

$$-\frac{\partial}{\partial t}n + D\nabla^2 n + \sum_{k=2}^\infty D_{2k}(\nabla^2)^k n = 0, \tag{4.19}$$

in which D is the isotropic coefficient of diffusion and the coefficients D_{2k}

are determined as described above. Since in many cases the gradients will be small enough to ensure that $D_{2k+2}(\nabla^2)^{k+2}n \ll D_{2k}(\nabla^2)^k n$, the normal form of the diffusion equation, namely,

$$-\frac{\partial}{\partial t}n + D\nabla^2 n = 0,$$

is sufficiently accurate.

We next consider a second relatively simple case, that in which a single spatial coordinate is relevant but an electric force $e\mathbf{E}$ is present.

4.3.2.   MOTION IN ONE SPATIAL DIMENSION. We let

$$q \equiv q(z,t) = \int\limits_{-\infty}^{\infty}\!\!\int n \, dx \, dy, \tag{4.20}$$

and form the integral $\int\limits_{-\infty}^{\infty}\!\!\int (\quad) \, dx \, dy$ of each term of equation 4.10, which is thereby transformed to†

$$-\frac{d}{dt}(f_0 q) + \frac{c^2}{3v_{\text{el}}}\frac{\partial^2}{\partial z^2}(f_0 q) + \frac{cV}{3}\frac{\partial^2}{\partial z \, \partial c}(f_0 q) + W\frac{\partial}{\partial z}(f_0 q)$$

$$+ \frac{1}{3c^2}\frac{\partial}{\partial c}\left\{c^2 v_{\text{el}}\left[\frac{cV}{v_{\text{el}}}\frac{\partial}{\partial z}(f_0 q) + (V^2 + \overline{C^2})f_0^*\frac{\partial}{\partial c}\left(\frac{q f_0}{f_0^*}\right)\right]\right\} = 0. \tag{4.21}$$

We assume that $f_0 q$ can be represented by an expansion

$$f_0 q = f_0^* \sum_{k=0}^{\infty} b_k(c)\frac{\partial^k}{\partial z^k}q \tag{4.22}$$

with $b_0(c) \equiv 1$. In addition we let $q(z,t)$ satisfy an equation of the form

$$\frac{d}{dt}q - \sum_{m=2}^{\infty} D_m \frac{\partial^m}{\partial z^m}q = 0, \tag{4.23}$$

which refers to the moving system of axes. We now replace $f_0 q$ and $(d/dt)q$ in equation 4.21 by their expansions as given in equations 4.22 and 4.23.

---

†In the development that follows, which refers to a system of axes moving with velocity $\mathbf{W}$, the variable z' is denoted by z for notational simplicity.

Equation 4.21 is thus transformed to $[f_0^* \equiv f_0^*(c),\, b_k \equiv b_k(c),\, b_0 = 1]$

$$-f_0^* \sum_{k=0}^{\infty} \sum_{m=2}^{\infty} b_k D_m \frac{\partial^{k+m}}{\partial z^{k+m}} q + \frac{c^2}{3\nu_{\mathrm{el}}} \sum_{k=0}^{\infty} f_0^* b_k \frac{\partial^{k+2}}{\partial z^{k+2}} q$$

$$+ \frac{cV}{3} \sum_{k=0}^{\infty} \frac{d}{dc}(f_0^* b_k) \frac{\partial^{k+1}}{\partial z^{k+1}} q + W f_0^* \sum_{k=0}^{\infty} b_k \frac{\partial^{k+1}}{\partial z^{k+1}} q$$

$$+ \frac{1}{3c^2} \frac{\partial}{\partial c} \left\{ c^2 \nu_{\mathrm{el}} \left[\frac{cV}{\nu_{\mathrm{el}}} f_0^* \sum_{k=0}^{\infty} b_k \frac{\partial^{k+1}}{\partial z^{k+1}} q + (V^2 + \overline{C^2}) f_0^* \sum_{k=0}^{\infty} \frac{d}{dc} b_k \frac{\partial^{k}}{\partial z^{k}} q \right] \right\} = 0.$$

Terms can be grouped to give this equation the form

$$\left(\frac{cV}{3} \frac{d}{dc} f_0^* + W f_0^* + \frac{1}{3c^2} \frac{d}{dc} \left\{ c^2 \nu_{\mathrm{el}} \left[\frac{cV}{\nu_{\mathrm{el}}} f_0^* + (V^2 + \overline{C^2}) f_0^* \frac{d}{dc} b_1 \right] \right\} \right) \frac{\partial q}{\partial z}$$

$$+ \left(-f_0^* D_2 + \frac{c^2}{3\nu_{\mathrm{el}}} f_0^* + \frac{cV}{3} \frac{d}{dc}(f_0^* b_1) + W f_0^* b_1 \right.$$

$$\left. + \frac{1}{3c^2} \frac{d}{dc} \left\{ c^2 \nu_{\mathrm{el}} \left[\frac{cV}{\nu_{\mathrm{el}}} f_0^* b_1 + (V^2 + \overline{C^2}) f_0^* \frac{d}{dc} b_2 \right] \right\} \right) \frac{\partial^2}{\partial z^2} q$$

$$+ \cdots$$

$$+ \left(-f_0^* (D_k + b_1 D_{k-1} + \cdots + b_{k-2} D_2) + \frac{c^2}{3\nu_{\mathrm{el}}} f_0^* b_{k-2} \right.$$

$$+ \frac{cV}{3} \frac{d}{dc}(f_0^* b_{k-1}) + W f_0^* b_{k-1}$$

$$\left. + \frac{1}{3c^2} \frac{d}{dc} \left\{ c^2 \nu_{\mathrm{el}} \left[\frac{cV}{\nu_{\mathrm{el}}} f_0^* b_{k-1} + (V^2 + \overline{C^2}) f_0^* \frac{d}{dc} b_k \right] \right\} \right) \frac{\partial^{k}}{\partial z^{k}} q + \cdots = 0.$$

We now equate to zero separately the coefficients of $(\partial/\partial z)q$, $(\partial^2/\partial z^2)$ $q, \ldots, (\partial^k/\partial z^k)q$ in this reordered equation to obtain a sequence of equations from which to derive expressions for the D_m and the $b_k(c)$. This sequence of equations is as follows:

$$\left.\begin{aligned}
&\frac{cV}{3}\frac{d}{dc}f_0^* + Wf_0^* + \frac{1}{3c^2}\frac{d}{dc}\left\{c^2\nu_{el}\left[\frac{cV}{\nu_{el}}f_0^* + (V^2 + \overline{C^2})f_0^*\frac{d}{dc}b_1\right]\right\} = 0, \\[2ex]
&-f_0^*D_2 + \frac{c}{3\nu_{el}}f_0^* + \frac{cV}{3}\frac{d}{dc}(f_0^*b_1) + Wf_0^*b_1 \\[2ex]
&\qquad + \frac{1}{3c^2}\frac{d}{dc}\left\{c^2\nu_{el}\left[\frac{cV}{\nu_{el}}f_0^*b_1 + (V^2 + \overline{C^2})f_0^*\frac{d}{dc}b_2\right]\right\} = 0, \\[1ex]
&\qquad\qquad\qquad\qquad \vdots \\[1ex]
&-f_0^*(D_k + b_1D_{k-1} + \cdots + b_{k-2}D_2) + \frac{c^2}{3\nu_{el}}f_0^*b_{k-2} + \frac{cV}{3}\frac{d}{dc}(f_0^*b_{k-1}) \\[2ex]
&+ Wf_0^*b_{k-1} + \frac{1}{3c^2}\frac{d}{dc}\left\{c^2\nu_{el}\left[\frac{cV}{\nu_{el}}f_0^*b_{k-1} + (V^2 + \overline{C^2})f_0^*\frac{d}{dc}b_k\right]\right\} = 0,
\end{aligned}\right\} \quad (4.24)$$

etc.

The coefficient $b_1(c)$ is obtained directly from the first of equations 4.24 by integration, in which account is taken of the requirement that $4\pi\int_0^\infty b_k(c)f_0^*(c)c^2\,dc = 0$. We find

$$b_1(c) = \overline{\psi_1(c)} - \psi_1(c), \qquad (4.25)$$

where

$$\psi_1(c) = \int_0^c \frac{3}{y^2\nu_{el}(y)(V^2 + \overline{C^2})f_0^*(y)}$$

$$\times \left\{\int_0^y \left[\frac{xV}{3}\frac{d}{dx}f_0^*(x) + Wf_0^*(x)\right]x^2dx\right\}dy + \int_0^c \frac{yV}{(V^2 + \overline{C^2})\nu_{el}(y)}dy,$$

and

$$\overline{\psi_1(c)} = 4\pi\int_0^\infty \psi_1(c)f_0^*(c)c^2\,dc.$$

In order to find D_2, we form the integral $4\pi \int_0^\infty (\quad)c^2\,dc$ of each term of the second of equations 4.24 and note that $4\pi \int_0^\infty b_1(c)f_0^*(c)c^2\,dc = 0$. It is then seen that

$$D_2 = \frac{4\pi}{3}\int_0^\infty \frac{c^2}{\nu_{el}}f_0^*c^2\,dc + \frac{4\pi}{3}\int_0^\infty cV\frac{d}{dc}(b_1 f_0^*)c^2\,dc.$$

Because of the importance of this coefficient we replace the symbol D_2 by D_L and name it the *longitudinal coefficient of diffusion*. Thus

$$D_L = D + \frac{4\pi}{3}\int_0^\infty cV\frac{d}{dc}(b_1 f_0^*)c^2\,dc$$

$$= D - \frac{4\pi}{3}\int_0^\infty b_1 f_0^* \frac{d}{dc}(c^3V)\,dc, \tag{4.26}$$

where D is the isotropic coefficient of diffusion.

In general, from the kth of equations 4.24,

$$D_k = \frac{4\pi}{3}\int_0^\infty \frac{c^2}{\nu_{el}}b_{k-2}(c)f_0^*(c)c^2\,dc$$

$$+ \frac{4\pi}{3}\int_0^\infty cV\frac{d}{dc}[f_0^*(c)b_{k-1}(c)]c^2\,dc$$

and

$$b_k(c) = \overline{\psi_k(c)} - \psi_k(c) \tag{4.27}$$

with

$$\psi_k(c) = \int_0^c \frac{3}{y^2\nu_{el}(y)(V^2 + \overline{C^2})f_0^*(y)}$$

$$\times \left(\int_0^y \left\{\frac{xV}{3}\frac{d}{dx}[f_0^*(x)b_{k-1}(x)]\right.\right.$$

$$+ Wf_0^*(x)b_{k-1}(x) + \frac{x^2}{3\nu_{el}}f_0^*(x)b_{k-2}(x)$$

$$\left.\left. - f_0^*(x)[D_k + b_1(x)D_{k-1} + \cdots + b_{k-2}D_2]\right\}x^2\,dx\right)dy$$

$$+ \int_0^c \frac{yVb_{k-1}(y)}{\nu_{el}(y)(V^2 + \overline{C^2})}\,dy.$$

When the field is absent, $W = V = 0$ and it can be seen that $b_k(c)$ is zero when k is odd, but $b_k(c)$ becomes $a_{2k}(c)$ of equation 4.12 when k is even. Moreover, of the coefficients D_m, only those of even order remain and these are the same as the D_{2l} of equation 4.14.

The coefficients D_m and $b_k(c)$ can thus be found from equations 4.27 in terms of coefficients of lower order, and because $b_1(c)$ and D_2 are calculable at the outset it is possible in principle to find D_m and $b_k(c)$ for any desired orders m and k.

It follows that $q(z,t)$ is a solution of equation 4.23, which we now write as

$$-\frac{d}{dt}q + D_L\frac{\partial^2}{\partial z^2}q + \sum_{m=3}^{\infty} D_m\frac{\partial^m}{\partial z^m}q = 0, \qquad (4.28)$$

where $D_L \equiv D_2$ is given by equation 4.26.

Equation 4.28 refers to a system of coordinates moving at velocity W in the direction $+Oz$. In the coordinate system at rest $q(z,t)$ is therefore a solution of the equation

$$-\frac{\partial}{\partial t}q + D_L\frac{\partial^2}{\partial z^2}q - W\frac{\partial}{\partial z}q + \sum_{m=3}^{\infty} D_m\frac{\partial^m}{\partial z^m}q = 0. \qquad (4.29)$$

It is evident that in general it would be difficult or impossible to obtain a simple closed analytical expression for $q(z,t)$ that satisfies the complete transport equation 4.29. However, if the physical circumstances are such that the sum of terms in spatial derivatives of order greater than the second is small in comparison with the term $D_L(\partial^2/\partial z^2)q$, equation 4.29 may be replaced by the equation:

$$-\frac{\partial}{\partial t}q + D_L\frac{\partial^2}{\partial z^2}q - W\frac{\partial}{\partial z}q = 0, \qquad (4.30)$$

which is a one-dimensional case of equation 3.31. This equation is susceptible to solution in terms of simple closed expressions.

Fortunately in most of the experimental procedures described later in this book large relative gradients are avoided except in the vicinity of boundaries or near a source, and analyses of the experimental data in terms of equation 4.30, or of other forms of the equation in which higher-order derivatives are neglected, lead, in general, to self-consistent results.

The fact that a group of electrons moving in a gas in the presence of an electric force $e\mathbf{E}$ disperses relative to its centroid with an effective coefficient of diffusion D_L parallel and antiparallel to the electric force which is in general not the same as the lateral or isotropic coefficient D long remained unnoticed. The existence of the phenomenon was first revealed by Wagner, Davis, and Hurst (1967), who from time-of-flight experiments obtained coefficients of diffusion different from those measured by the Townsend-Huxley method of the diffusing stream.

Theories of the phenomenon were developed independently and simultaneously by Parker and Lowke (1969), using Fourier transform methods, and by Skullerud (1969), employing a combination of the earlier theories of Wannier and Huxley. The formula for D_L given in equation 4.26 is, in fact, the same as that obtained by Skullerud. The subject has also been discussed by Huxley (1972).

In the special case in which $q_{m_{el}}$ is independent of c, all encounters are elastic, and E/N is sufficiently large that $\overline{C^2}/W^2 \ll 1$, equation 4.25 gives

$$b_1(c) = \frac{\pi^{1/2}}{2\lambda}\left(\frac{\overline{c^2}-c^2}{\alpha^2}\right) + \frac{\pi}{2\lambda}\sum_{n=0}^{\infty}\frac{1}{\Gamma(n+\tfrac{7}{4})}\left[\frac{\overline{c^{4n+3}}-c^{4n+3}}{(4n+3)\alpha^{4n+3}}\right]$$

$$-\frac{\pi}{2\lambda}\sum_{n=0}^{\infty}\frac{1}{\Gamma(n+\tfrac{5}{2})}\left[\frac{\overline{c^{4n+6}}-c^{4n+6}}{(4n+6)\alpha^{4n+6}}\right], \tag{4.31}$$

in which $2\lambda = W/D$ and α is a parameter in the distribution function. The distribution function in this special case is given by

$$f_0^*(c) = \frac{1}{\pi\Gamma(\tfrac{3}{4})\alpha^3}\exp-\left(\frac{c}{\alpha}\right)^4$$

with

$$\alpha^4 = \frac{4M}{3m}\left(\frac{eE}{mNq_{m_{el}}}\right)^2.$$

It follows from equations 4.26 and 4.31 that (Huxley, 1972)

$$\frac{D_L}{D} = 1 - \frac{4\pi}{3D} \int_0^\infty b_1(c) f_0^* \frac{d}{dc} (c^3 V) \, dc$$

$$= 1 - \left[\frac{\Gamma(\tfrac{5}{4}) \pi^{1/2}}{\Gamma(\tfrac{3}{4})} - 1 \right]$$

$$- \pi^{1/2} \left\{ \sum_{n=0}^\infty \frac{1}{\Gamma(n+\tfrac{7}{4})} \left[\frac{\Gamma\left(\dfrac{2n+3}{2}\right) \dfrac{\pi^{1/2}}{\Gamma(\tfrac{3}{4})} - \Gamma\left(\dfrac{4n+5}{4}\right)}{4n+3} \right] \right.$$

$$\left. - \sum_{n=0}^\infty \frac{1}{\Gamma(n+\tfrac{5}{2})} \left[\frac{\Gamma\left(\dfrac{4n+9}{4}\right) \dfrac{\pi^{1/2}}{\Gamma(\tfrac{3}{4})} - \Gamma(n+2)}{4n+6} \right] \right\}. \quad (4.32)$$

This expression gives for D_L/D the value 0.49, in agreement with the value calculated independently by Parker and Lowke (1969) and Skullerud (1969) for this case.

4.3.3. THREE-DIMENSIONAL MOTION IN A UNIFORM ELECTRIC FIELD. The series representation of nf_0 must now include mixed derivatives, and we write accordingly

$$nf_0 = f_0^* \left[n + \sum_{k=1}^\infty a_{2k}(c) (\nabla_{xy}^2)^k n + \sum_{k=1}^\infty b_k(c) \frac{\partial^k}{\partial z^k} n \right.$$

$$\left. + \sum_{r=1}^\infty \sum_{s=1}^\infty c_{rs} \frac{\partial^r}{\partial z^r} (\nabla_{xy}^2)^s n \right], \quad (4.33)$$

in which

$$\nabla_{xy}^2 \equiv \frac{\partial^2}{\partial x^2} + \frac{\partial^2}{\partial y^2}.$$

Similarly we adopt for $(d/dt)n$ in the moving system the matching expansion:

$$\frac{d}{dt}n = \sum_{l=1}^{\infty} D_{2l}(\nabla_{xy}^2)^l n + \sum_{m=2}^{\infty} D_{zm}\frac{\partial^m}{\partial z^m}n + \sum_{i=1}^{\infty}\sum_{j=1}^{\infty} D_{ij}\frac{\partial^i}{\partial z^i}(\nabla_{xy}^2)^j n. \quad (4.34)$$

We now replace nf_0 and $(d/dt)n$ in equation 4.10 by their expansions and reorder terms again to present the equation in the form of a sum of spatial derivatives of n with coefficients independent of the spatial coordinates. However, as the terms become cumbersome, we shall give only the terms of the equation up to the second-order derivatives of n. It is found that equation 4.10 becomes

$$\left(\frac{cV}{3}\frac{d}{dc}f_0^* + Wf_0^* + \frac{1}{3c^2}\frac{d}{dc}\left\{c^2\nu_{el}\left[\frac{cV}{\nu_{el}}f_0^* + (V^2 + \overline{C^2})f_0^*\frac{d}{dc}b_1\right]\right\}\right)\frac{\partial}{\partial z}n$$

$$+\left(-f_0^* D_{z2} + \frac{c^2}{3\nu_{el}}f_0^* + \frac{cV}{3}\frac{d}{dc}(f_0^* b_1) + Wf_0^* b_1\right.$$

$$\left.+\frac{1}{3c^2}\frac{d}{dc}\left\{c^2\nu_{el}\left[\frac{cV}{\nu_{el}}f_0^* b_1 + (V^2 + \overline{C^2})f_0^*\frac{d}{dc}b_2\right]\right\}\right)\frac{\partial^2}{\partial z^2}n$$

$$+\left\{-f_0^* D_2 + \frac{c^2}{3\nu_{el}}f_0^* + \frac{1}{3c^2}\frac{\partial}{\partial c}\left[c^2\nu_{el}(V^2 + \overline{C^2})f_0^*\frac{d}{dc}a_2\right]\right\}\nabla_{xy}^2 n + \cdots = 0.$$

$$(4.35)$$

On equating to zero, as before, the coefficients of the spatial derivatives of n, we find that the first two equations yielded by equation 4.35 are the same as the first pair of equations 4.24 and that third equation is the same as the first of equations 4.16 but with $\overline{C^2}$ replaced by $(V^2 + \overline{C^2})$. We conclude, therefore, that in equations 4.35, $b_1(c)$ is given by equation 4.25, $D_{z2} = D_L$ as in equation 4.26, and that $D_2 = D$, the lateral coefficient of diffusion. Equation 4.34 can therefore be expressed in the form†

$$-\frac{d}{dt}n + D\left(\frac{\partial^2}{\partial x^2}n + \frac{\partial^2}{\partial y^2}n\right) + D_L\frac{\partial^2}{\partial z'^2}n + S = 0, \quad (4.36)$$

†In this final statement of the differential equation for n in the moving system, the superscript is again added to the symbol z to stress that this equation refers to such a system.

in which S is the sum of terms each of which is a spatial derivative of n of order greater than two and multiplied by a constant coefficient.

In the system of coordinates at rest equation 4.36 transforms to

$$-\frac{\partial}{\partial t}n+D\left(\frac{\partial^2}{\partial x^2}n+\frac{\partial^2}{\partial y^2}n\right)+D_L\frac{\partial^2}{\partial z^2}n-W\frac{\partial}{\partial z}n+S=0. \quad (4.37)$$

In most circumstances treated in this book it is justifiable to neglect S; consequently, unless otherwise stated, the basic equation for n is taken to be the conventional equation that results when S is omitted from equation 4.37:

$$-\frac{\partial}{\partial t}n+D\left(\frac{\partial^2}{\partial x^2}n+\frac{\partial^2}{\partial y^2}n\right)+D_L\frac{\partial^2}{\partial z^2}n-W\frac{\partial}{\partial z}n=0. \quad (4.38)$$

4.4.   EXTENSION OF THE THEORY TO A STEADY STREAM

In Chapter 5 we examine in detail the properties of a steady stream. It is therefore convenient at this point to relate the theory of the travelling group to that of the steady stream.

We consider the simplest example, that of a steady stream of electrons from an isolated pole source at the origin of coordinates. The stream moves in the direction of $+Oz$ in a uniform electric field $\mathbf{E}$ and carries current i across any fixed plane normal to that axis.

The current i may be considered as the superposition at a fixed plane of the fluxes of electrons in an infinite sequence of elementary groups continually liberated at the source, each with a population $i\,dt/e$. The time-independent number density $n(x,y,z)$ at a fixed point in the stream is the superposition of the number densities Δn in these travelling groups, whose times of travel t (lifetimes) range from zero to infinity, that is, $n(x,y,z)=\Sigma\,\Delta n(x,y,z,t)$.

Each elementary distribution $\Delta n(x,y,z,t)$ satisfies the special form of equation 4.37 with S assumed to be negligible, that is equation 4.38. When the contributions of these distributions at a fixed point (x,y,z) are summed, the resulting number density $n(x,y,z)$ is independent of t and satisfies equation 4.38 with $\partial n/\partial t=0$, that is

$$D\left(\frac{\partial^2}{\partial x^2}n+\frac{\partial^2}{\partial y^2}n\right)+D_L\frac{\partial^2}{\partial z^2}n-W\frac{\partial}{\partial z}n=0; \quad (4.39)$$

the summation therefore represents the distribution of number density in a steady stream. It is clear that D and D_L are the same for the steady stream as for the isolated travelling group.

To illustrate more specifically the relationship between the travelling group and the steady stream, we first note that, in an elementary group with a time of travel t, the centroid lies at the point $(0,0,Wt)$; consequently the contribution to the number density of this group is given by the appropriate solution of equation 4.38, namely,

$$\Delta n = \frac{i\,dt}{e(4\pi Dt)(4\pi D_L t)^{1/2}}\left[\exp\left(-\frac{\rho^2}{4Dt}\right)\right]\left\{\exp\left[-\frac{(z-Wt)^2}{4D_L t}\right]\right\}.$$

The number density $n(x,y,z)$ in the stream is therefore

$$n(x,y,z) = \frac{i}{e(4\pi D)(4\pi D_L)^{1/2}}\int_0^\infty\left\{\exp\left[-\frac{\rho^2}{4Dt}-\frac{(z-Wt)^2}{4D_L t}\right]\right\}t^{-3/2}\,dt.$$

We put $2\lambda_L = W/D_L$ and $\tau = tW^2/4D_L = \lambda_L Wt/2$; then

$$n(x,y,z) = \frac{i}{e(4\pi D)(4\pi D_L)^{1/2}}$$

$$\times D_L^{1/2}\lambda_L(\exp\lambda_L z)\int_0^\infty\left(\exp\left\{-\tau-\frac{\lambda_L^2[z^2+(D_L/D)\rho^2]}{4\tau}\right\}\right)\tau^{-3/2}\,d\tau.$$

But [Watson, §6.22(15)]

$$\int_0^\infty\left[\exp\left(-\tau-\frac{s^2}{4\tau}\right)\right]\tau^{-(\nu+1)}\,d\tau = 2\left(\frac{2}{s}\right)^\nu K_\nu(s),$$

where $K_\nu(s)$ is a modified Bessel function of the second kind and of order ν.

Consequently

$$n(x,y,z) = \frac{i}{e(4\pi D)(4\pi D_L)^{1/2}} 2^{3/2} (\lambda_L D_L)^{1/2} \left(z^2 + \frac{D_L}{D} \rho^2 \right)^{-1/4}$$

$$\times (\exp \lambda_L z) K_{1/2} \left[\lambda_L \left(z^2 + \frac{D_L}{D} \rho^2 \right)^{1/2} \right]$$

$$= \frac{i}{e(4\pi D)} \frac{\exp -\lambda_L (r' - z)}{r'}, \tag{4.40}$$

where $r' = [z^2 + (D_L/D)\rho^2]^{1/2}$.

4.5. NUMBER DENSITY AND CURRENT

4.5.1. NUMBER DENSITY. It has been shown that the distribution of number density $n(x,y,z,t)$ is to be found in the form of a solution of equation 4.37 appropriate to the boundary conditions. In this equation S represents a sum of terms containing the individual spatial derivatives of n with orders exceeding the second. The boundaries are usually metal surfaces on which it is assumed that $n = 0$ (see Chapter 5). It is not to be expected that analytically simple solutions of the complete equation 4.37 exist, and the usual practice is to conduct measurements in regions where the relative density gradients are so small that it is considered legitimate to omit the term S from equation 4.37, which then becomes equation 4.38. This equation is to be regarded as an asymptotic form of the complete equation 4.37. Solutions $n(x,y,z,t)$ of equation 4.38 are therefore also to be regarded as asymptotic forms of solutions of the complete equation 4.37. Whereas the solutions of equation 4.37 are valid throughout the region, those of equation 4.38 are valid only within a restricted region where the asymptotic equation 4.38 is valid.

4.5.2. CURRENT DENSITY. We refer to equations 3.10 in which the convective velocity $\mathbf{W}_{conv}(c)$ is expressed as the sum

$$\mathbf{W}_{conv}(c) = \mathbf{W}(c) + \mathbf{W}_G(c)$$

with

$$\mathbf{W}(c) = -\frac{\mathbf{V}'c}{3f_0} \frac{\partial}{\partial c} f_0, \qquad \mathbf{V}' = \frac{e\mathbf{E}}{mv}$$

and

$$\mathbf{W}_G(c) = -\frac{c^2}{3vf_0} \frac{1}{n} \operatorname{grad}_r(nf_0).$$

$$\left. \right\} \tag{4.41}$$

The current across a surface element $d\mathbf{S}$ is therefore

$$di = e4\pi \int_0^\infty (nf_0)[\mathbf{W}_{\text{conv}}(c) \cdot d\mathbf{S}]c^2 dc. \tag{4.42}$$

Let $d\mathbf{S}$ be directed parallel to $e\mathbf{E}$, which is itself parallel to the direction $+Oz$. It then follows that

$$di = di_z = e4\pi \, dS \int_0^\infty \left[-\frac{V'c}{3} \frac{\partial}{\partial c}(nf_0) - \frac{c^2}{3\nu} \frac{\partial}{\partial z}(nf_0) \right] c^2 dc.$$

When nf_0 is replaced by its expansion as given in equation 4.33, it is found that

$$di_z = dS \, e4\pi \int_0^\infty \left[-\frac{V'c}{3} n \frac{d}{dc} f_0^* - \frac{V'c}{3} \frac{d}{dc}(b_1 f_0^*) \frac{\partial}{\partial z} n \right.$$

$$\left. -\frac{c^2}{3\nu} f_0^* \frac{\partial}{\partial z} n - (\text{terms in higher-order derivatives}) \right] c^2 dc,$$

whence

$$di_z = dS \, e\left(nW - D_L \frac{\partial}{\partial z} n + S' \right), \tag{4.43}$$

where S' depends on spatial derivatives with order higher than the first. The conventional expression for di_z is

$$di_z = dS \, e\left[nW - D_L \frac{\partial}{\partial z} n \right], \tag{4.44}$$

which implicitly neglects the term S'.

In regions where the asymptotic solution for $n(x,y,z,t)$ is valid the assumption will be made that it is also legitimate to neglect S'; this standard practice is followed throughout this book.

4.5.3. ILLUSTRATIVE EXAMPLE. Consider the distribution $n(x,y,z)$ in a steady stream of electrons originating from a small hole in a metal cathode that occupies the plane $z = 0$. The centre of the hole is at the origin, and the stream travels in the direction $+Oz$ in a uniform electric field. It is shown in Chapter 5 that an expression for $n(x,y,z)$ that satisfies the asymptotic steady state form of equation 4.37 ($S=0$), namely,

$$D\left(\frac{\partial^2}{\partial x^2}n + \frac{\partial^2}{\partial y^2}n\right) + D_L\frac{\partial^2}{\partial z^2}n - W\frac{\partial}{\partial z}n = 0, \tag{4.45}$$

is

$$n = (\exp\lambda_L z) \sum_{k=0}^{\infty} A_k r'^{-1/2}K_{k+1/2}(\lambda_L r')P_k(\mu), \tag{4.46}$$

where $r'^2 = x'^2 + y'^2 + z^2$ with $x' = (D_L/D)^{1/2}x$ and $y' = (D_L/D)^{1/2}y$; $2\lambda_L = W/D_L$; K is a modified Bessel function of the second kind; and $\mu = \cos\theta = z/r'$.

This solution is an asymptotic solution of the full equation 4.37 valid at distances r from the source where spatial gradients have become relatively small. Such a form of solution would not be valid in the vicinity of the source. Since $n=0$ over the plane $z=0$, the asymptotic solution must possess this property; consequently k must be an odd number in order for $P_k(\mu)$ to vanish when $\theta = \pi/2$.

In Chapter 5 a single-term solution ("dipole" solution) for which $k=1$ is adopted. This solution combined with an image solution has been found to give self-consistent results when equation 4.44 is employed to calculate the ratio of the currents to separate portions of the anode. This matter is discussed further in Chapter 11.

4.6.   INELASTIC ENCOUNTERS

In this section we give a brief indication of how the theoretical treatment adopted in this chapter can be modified to include cases in which inelastic encounters are important.

The first effect of inelastic encounters is to modify the momentum transfer cross section from $q_{m_{el}}(c)$, appropriate to elastic encounters, to an effective cross section $q_m(c)$ with an associated collision frequency ν. The vector equation to be employed is equation 3.8 (with $\boldsymbol{\omega}=0$); consequently the relation between $\mathbf{W}_{conv}(c)$ and $\mathbf{f}_1$ is still $\mathbf{W}_{conv}(c) = c\mathbf{f}_1/3f_0$.

The scalar equation retains its basic form of equation 3.30 unchanged, but when inelastic encounters are important the flux term $\sigma_{coll}(c)$ is no longer given by equation 3.4. It follows that in assuming an expansion for nf_0 of the form adopted in equation 4.33 the distribution function $f_0^*(c)$ for the group as a whole is no longer the generalized distribution function (equation 4.7). Although the distribution function may not be known precisely in advance, its form can be deduced (Chapter 13) from the measurements of W and $\lambda = W/2D$.

When all encounters are elastic, then

$$\frac{1}{4\pi c^2} \frac{\partial}{\partial c} [\sigma_E(c) - \sigma_{\text{coll}}(c)]$$

$$= \frac{1}{3c^2} \frac{\partial}{\partial c} \left\{ c^2 \nu_{\text{el}} \left[\frac{cV}{\nu_{\text{el}}} \frac{\partial}{\partial z} (nf_0) + (V^2 + \overline{C^2}) \frac{\partial}{\partial c} (nf_0) + \frac{3mc}{M} (nf_0) \right] \right\}.$$

It is shown in Chapter 6 that the effect of inelastic encounters is to modify slightly the collision frequency ν_{el}, which is then written ν, and to add a further term to $\sigma_{\text{coll}}(c)$, which is also a function of c. Hence we write

$$\frac{1}{4\pi c^2} \frac{\partial}{\partial c} [\sigma_E(c) - \sigma_{\text{coll}}(c)]$$

$$= \frac{1}{3c^2} \frac{\partial}{\partial c} \left\{ c^2 \nu_{\text{el}} \left[\frac{cV'}{\nu_{\text{el}}} \frac{\partial}{\partial z} (nf_0) + (VV' + \overline{C^2}) \frac{\partial}{\partial c} (nf_0) + F(c)(nf_0) \right] \right\},$$

in which $V' = eE/m\nu$ and $F(c) = (3mc/M) + \phi(c)$.

Let f_0^* be defined as

$$f_0^*(c) = A \exp \left[- \int_0^c \frac{F(c)}{VV' + \overline{C^2}} \, dc \right];$$

then we can write

$$\frac{1}{4\pi c^2} \frac{\partial}{\partial c} [\sigma_E(c) - \sigma_{\text{coll}}(c)]$$

$$= \frac{1}{3c^2} \frac{\partial}{\partial c} \left\{ c^2 \nu_{\text{el}} \left[\frac{cV'}{\nu_{\text{el}}} \frac{\partial}{\partial z} (nf_0) + (VV' + \overline{C^2}) f_0^* \frac{\partial}{\partial c} \left(\frac{f_0}{f_0^*} \right) n \right] \right\}. \quad (4.47)$$

The formal treatment of the full scalar equation now proceeds as before with this modified form of f_0^*. The series representation of nf_0 (equation 4.33) is adopted and leads to equation 4.35. We deduce again that $n(\mathbf{r}, t)$ is a solution of equation 4.38 and that currents are to be calculated by means of equations 4.44.

4.7.   EINSTEIN'S FORMULA

Einstein established in 1905 an important formula relevant to the Brownian motion of microscopic but visible particles in a fluid. Consider an isolated group of n_0 particles moving with random thermal motion in a fluid. The group is dispersed by diffusion with a coefficient of diffusion D. Einstein showed that, if (x,y,z) are the coordinates of a typical particle, then

$$\frac{d}{dt}\,\overline{x^2} = 2D, \quad \text{or, alternatively,} \quad \frac{d}{dt}\,\overline{r^2} = 6D,$$

where $r^2 = x^2 + y^2 + z^2$ and

$$\overline{x^2} = \frac{1}{n_0} \int_{\mathbf{r}} nx^2\,d\mathbf{r}$$

is the mean of the squares of the x coordinates of the particles in a stationary coordinate system.

We establish the theorem for the steady-state motion of electrons in a gas in the presence of an electric field. In a system of coordinates moving with the centroid of the group equation 4.36 becomes (with $S=0$)

$$\frac{dn}{dt} = D\left(\frac{\partial^2 n}{\partial x^2} + \frac{\partial^2 n}{\partial y^2}\right) + D_L \frac{\partial^2 n}{\partial z'^2}.$$

The rate of increase of the mean value of the squares of the x coordinates of the n_0 electrons of the group is

$$\frac{1}{n_0}\frac{d}{dt}\int_{\mathbf{r}} x^2 n\,d\mathbf{r} = \frac{1}{n_0}\int_{\mathbf{r}} x^2\frac{\partial n}{\partial t}\,d\mathbf{r}$$

$$= \frac{1}{n_0}\int\!\!\int\!\!\int_{-\infty}^{\infty}\left[D\left(\frac{\partial^2 n}{\partial x^2} + \frac{\partial^2 n}{\partial y^2}\right) + D_L\frac{\partial^2 n}{\partial z'^2}\right]x^2\,dx\,dy\,dz'.$$

Consider the constituent integrals separately.

$$\int\!\!\int\!\!\int_{-\infty}^{\infty}\frac{\partial^2 n}{\partial x^2}x^2\,dx\,dy\,dz' = \int\!\!\int_{-\infty}^{\infty} dy\,dz'\left(\int_{-\infty}^{\infty}\frac{\partial^2 n}{\partial x^2}x^2\,dx\right).$$

The integral

$$\int_{-\infty}^{\infty} \frac{\partial^2 n}{\partial x^2} x^2 \, dx = x^2 \frac{\partial n}{\partial x} \Big|_{-\infty}^{\infty} - 2 \int_{-\infty}^{\infty} \frac{\partial n}{\partial x} x \, dx = 0 - 2 \left(nx \Big|_{-\infty}^{\infty} - \int_{-\infty}^{\infty} n \, dx \right)$$

$$= 2 \int_{-\infty}^{\infty} n \, dx,$$

since n and its derivatives vanish at $r = \infty$. Thus

$$\int_{\mathbf{r}} \frac{\partial^2 n}{\partial x^2} x^2 \, d\mathbf{r} = 2 \int_{\mathbf{r}} n \, d\mathbf{r} = 2n_0.$$

Consider next

$$\int_{\mathbf{r}} \frac{\partial^2 n}{\partial y^2} x^2 \, d\mathbf{r} = \int\int_{-\infty}^{\infty} x^2 \, dx \, dz' \left(\int_{-\infty}^{\infty} \frac{\partial^2 n}{\partial y^2} \, dy \right) = 0$$

since

$$\int_{-\infty}^{\infty} \frac{\partial^2 n}{\partial y^2} \, dy = \frac{\partial n}{\partial y} \Big|_{-\infty}^{\infty} = 0.$$

Similarly,

$$\int_{\mathbf{r}} \frac{\partial^2 n}{\partial z'^2} x^2 \, d\mathbf{r} = 0.$$

It follows that

$$\frac{d}{dt} \overline{x^2} = \frac{1}{n_0} \int_{\mathbf{r}} x^2 \frac{\partial n}{\partial t} \, d\mathbf{r} = 2D.$$

The same procedure gives

$$\frac{d}{dt} \overline{y^2} = 2D \qquad \text{and} \qquad \frac{d}{dt} \overline{z'^2} = 2D_L. \qquad\qquad (4.48)$$

Einstein's formula is therefore applicable to a group of electrons that spreads by diffusion as the centroid moves at speed W in the presence of an electric force.

Since $\overline{\rho^2} = \overline{x^2} + \overline{y^2}$, it follows that $(d/dt)\overline{\rho^2} = 4D$. One application of Einstein's formula is the derivation of the formula $D = \overline{c^2 / 3\nu(c)}$ by the method of free paths.

4.8. THERMODYNAMIC TREATMENT OF ANISOTROPIC DIFFUSION IN AN ELECTRIC FIELD

In a recent paper, Robson (1972) has given approximate relationships between D_L, D, and W which were derived from a general thermodynamical argument. Since the domain of nonequilibrium thermodynamics, on which his derivation is based, is restricted to situations in which the deviations from equilibrium are small, it is difficult to assess the range of validity of the formulae. Nevertheless, when put to practical tests, the formulae have been found to be remarkably successful even in highly nonequilibrium conditions, that is, when $D/\mu \gg \kappa T/e$. For this reason, and because of their extreme simplicity, the formulae are likely to find wide application.

For the case of electrons where only elastic scattering occurs Robson obtained the formula

$$\frac{D_L}{D} = \frac{\partial (\ln W)}{\partial (\ln E)}$$

$$= \frac{\partial (\ln W)}{\partial (\ln E/N)}. \tag{4.49}$$

Thus the ratio can be obtained simply from the slope of the log-log plots of W versus E/N.

When the momentum transfer cross section varies in such a way that it may be represented adequately by $q_{m_{el}} = q_0 c^l$, and inelastic scattering is negligible, it may be shown that (see, e.g., equation 13.36)

$$W = \text{const.} \left(\frac{E}{N} \right)^{1/(l+2)}$$

provided that $D/\mu \gg \kappa T/e$. Consequently equation 4.49 becomes

$$\frac{D_L}{D} = \frac{1}{l+2}.$$

Thus, when $q_{m_{el}} = \text{const}$, $l=0$ and $D_L/D = 0.5$ in agreement with the result quoted at the end of Section 4.3.2. As is to be expected, this formula is considerably more accurate than the semiquantitative result

$$\frac{D_L}{D} = \frac{l+3}{2(l+2)},$$

derived by Parker and Lowke (1969) and used by them to illustrate their work on the formal solution of this problem via the Boltzmann equation.

BIBLIOGRAPHY AND NOTES

SOLUTION OF THE COMPLETE SCALAR EQUATION

There are relatively few theoretical investigations of the diffusion and transport of electrons based on the complete scalar equation, in which spatial derivatives are important and the velocity distribution function is no longer independent of position **r**. As mentioned in Section 1.11, Morse, Allis, and Lamar in 1935 touched upon the problem, but more serious attention was given to it by Allis and Allen, who by a change of independent variables were able to achieve a separation of the scalar equation. In 1963 Parker considered in detail the problem of the distribution of electrons in a stream from a point source that emits monoenergetic electrons. He obtained a solution for the scalar equation for this special case by making use of the change of independent variables discovered by Allis and Allen to separate the equation. His work is of interest in assessing the errors that can arise in the simplified theory of the spreading stream used in measurements of D/μ. He concludes that it appears that in most cases the experiments have not been appreciably affected by use of the simple theory. Parker's paper begins with an interesting assessment of the problem.

W. P. Allis and H. W. Allen, *Phys. Rev.*, **52**, 703, 1937.

J. H. Parker, *Phys. Rev.*, **132**, 2096, 1963.

ANISOTROPIC DIFFUSION

As mentioned in Section 4.3.2, this phenomenon was discovered by Wagner, Davis, and Hurst at Oak Ridge National Laboratory. Theoretical treatments have been presented by Parker and Lowke and by Skullerud. The presentation given in Chapter 4 follows that of a paper by Huxley (1972) as corrected by Skullerud (1973). In this paper the validity of the use of a differential equation for n including higher-order terms is examined.

E. B. Wagner, F. J. Davis, and G. S. Hurst, *J. Chem. Phys.*, **47**, 3138, 1967.

J. H. Parker and J. J. Lowke, *Phys. Rev.*, **181**, 290, 1969.

J. J. Lowke and J. H. Parker, *Phys. Rev.*, **181**, 302, 1969.

H. R. Skullerud, *J. Phys. B*, **2**, 696, 1969.

L. G. H. Huxley, *Aust. J. Phys.*, **25**, 43, 1972.

R. E. Robson, *Aust. J. Phys.*, **25**, 685, 1972.

H. R. Skullerud, personal communication, 1973.

G. N. Watson, *Theory of Bessel Functions*, 2nd ed., Cambridge University Press, 1944.

5

THE THEORY OF SWARM METHODS

FOR MEASUREMENT OF DRIFT VELOCITY

W, CHARACTERISTIC ENERGY eD/μ,

AND RELATED QUANTITIES

5.1. INTRODUCTION

5.1.1. GENERAL. We now consider the application of the general theory of electron motion in gases to some methods for measuring the transport coefficients W, D/μ, D, and related quantities. The principles of other methods for which the theory is relatively simple are discussed in Part 2 in association with the experimental details.

In this chapter we are concerned chiefly with time-of-flight methods for measuring the drift velocity W and with the method of the steady diverging stream for the measurement of the characteristic energy eD/μ (equation 3.34) because of the central importance of these quantities. Before commencing the discussion, it is important to examine an important problem that arises in applying the theory developed in previous chapters to situations of practical importance, that is, the problem of boundary conditions.

5.1.2. BOUNDARY CONDITION AT A METAL SURFACE. In practice the experimental region in which a stream or a travelling group is examined is bounded by metal electrodes. We assume that when any electron of a stream or travelling group comes into contact with a metal surface it is absorbed by the surface and does not return to the gas. Thus the metal surface imposes a boundary condition which must be taken into account when seeking the appropriate solution of the differential equation for n

whether in the unrestricted form of equation 4.37 or the asymptotic form of equation 4.38. Therefore, we now examine this boundary condition.

We consider first the particle flux across an elementary geometrical surface dS in a region where n can be considered to satisfy the differential equation in its asymptotic form (equation 4.38). We take the direction of dS to be that of the coordinate axis $+Oz$ which is also the direction of the drift velocity W. It then follows from equation 4.44 that the net particle flux across dS is

$$dS(W_{conv})_z = dS\left(-D_L\frac{\partial}{\partial z}n + nW\right).$$

This net flux is, however, the difference between two much larger, oppositely directed fluxes:

$$F_+ dS = dS\left[\frac{n\bar{c}}{4} + \frac{1}{2}\left(-D_L\frac{\partial}{\partial z}n + nW\right)\right]$$

$$= dS\left[\frac{n\bar{c}}{4} + \tfrac{1}{2}n(W_{conv})_z\right] \quad \text{in the direction } +Oz$$

and

$$F_- dS = dS\left[\frac{n\bar{c}}{4} - \frac{1}{2}\left(-D_L\frac{\partial}{\partial z}n + nW\right)\right]$$

$$= dS\left[\frac{n\bar{c}}{4} - \tfrac{1}{2}n(W_{conv})_z\right] \quad \text{in the direction } -Oz.$$

The mean random speed $\bar{c}$ of the electrons is much greater than W_{conv}; consequently, the principal contributions to these oppositely directed fluxes are from the terms $(n\bar{c}/4)dS$.

We next consider the flux to an element dS' of the surface of a plane metal anode in the plane $z = h$.

There is a flux of electrons from the gas to the element dS' but no return flux from dS' to the gas. If n' is the number density of electrons in the immediate vicinity of dS' then an elementary free path treatment shows that an approximate expression for the flux of electrons to dS' is

$$F'_+ dS' = \left(\frac{n'\bar{c}}{4} - \frac{l\bar{c}}{3}\frac{\partial}{\partial z}n'\right)dS'$$

where l is the mean free path of an electron. Since the probability that a free path should exceed the value x is $(\exp -x/l)$, it follows that the majority of the $F'_+ \, dS'$ electrons suffer their last encounter with a gas molecule in a layer only several free paths in thickness and contiguous to the metal surface. Since in practice gas pressures of several to 760 torr are commonly used, it follows that this transition layer is very thin. Beyond this layer the electron motion is essentially random, and the net flux is again the difference of two large but oppositely directed fluxes, $F_+ \, dS$ and $F_- \, dS$, namely $n(W_{\text{conv}})_z \, dS$. It follows, with $dS' = dS$, that the flux to dS' on the metal surface is approximately equal to that across dS in the gas a short distance beyond the transition layer but where the number density is n. Consequently,

$$\frac{n'\bar{c}}{4} - \frac{l\bar{c}}{3}\frac{\partial}{\partial z}n' = n(W_{\text{conv}})_z \quad \text{and} \quad \frac{n'\bar{c}}{4} < n(W_{\text{conv}})_z.$$

Since $W_{\text{conv}} \ll \bar{c}$ it follows that $n' \ll n$. It is the practice to suppose that n'/n is so small that n' may be considered to be negligible. The boundary condition at a metal surface is therefore taken to be $n' = 0$. The distribution of n' within the transition layer does not, however, satisfy the asymptotic form of the differential equation; it is in general unknown and thus of no value for the calculation of the flux to an element dS' of an electrode. The procedure usually adopted to calculate the flux to an element dS' of the plane metal anode is to arrange the experimental conditions to ensure that the asymptotic form of the differential equation is valid in the neighbourhood of the anode and to seek an asymptotic solution that makes n zero on the metal boundaries. It is then supposed that this solution is valid right up to the surface of the anode and that the flux across the transition layer to a surface element dS' is

$$dS'\left[-D_L\left(\frac{\partial}{\partial z}n\right)_{\text{anode}}\right],$$

since n is zero at the anode.

We may illustrate the procedure by means of a simple example, that of a uniform stream impinging on a plane metal anode. We suppose that the number density in the stream at a large distance from the anode is n_0 and independent of position. The stream moves in the direction $+Oz$ and is received by a metal anode lying in the plane $z = h$. The differential equation to be satisfied by n is equation 4.39 with n a function of z alone. We therefore require a solution of the equation

$$D_L\frac{\partial^2}{\partial z^2}n - W\frac{\partial}{\partial z}n = 0$$

which gives $n=0$ when $z=h$ and $n \to n_0$ as $z \to -\infty$. The appropriate solution is

$$n = n_0 [1 - (\exp - 2\lambda_L s)]$$

where $2\lambda_L = W/D_L$ and $s = h - z$.

At large distances s from the anode $n \to n_0$, $(\partial/\partial z)n \to 0$, and the flux across a surface element dS is $n_0 W dS$. According to the procedure adopted, the flux to a surface element of the anode, where $n=0$ and $s=0$, is taken to be

$$dS \left[-D_L \left(\frac{\partial}{\partial z} n \right)_{z=h} \right] = dS n_0 D_L 2\lambda_L$$

$$= dS n_0 W.$$

The flux across any element dS normal to the stream is thus the same whether or not dS lies on the electrode, so that the procedure leads to a correct result in this instance. It is assumed to be valid in other examples discussed later in this chapter especially in relation to diatomic gases and helium where small values of the mean free path prevail.

In experiments in which an electron stream originates from a small hole in the cathode, as for example in the Townsend-Huxley experiment, strong gradients of n exist in the vicinity of the source. In this case it may be necessary for the product Nh to be large in order that the asymptotic form of the differential equation for n should accurately describe the distribution of n in the vicinity of the anode. In diatomic gases, in which inelastic collisions result in relatively small values of the time constant τ (see p. 147) the solution is valid for the majority of the values of N and h that are commonly employed. The experiments can therefore be interpreted with sufficient accuracy using the assumptions described in this section. On the other hand, in the heavier monatomic gases in which the ratio M/m is relatively large and, at some energies, the collision cross section becomes relatively small, it is possible that equation 4.39 is invalid at all points in the stream unless unusually large values of Nh are employed. Care is then needed in the choice of the experimental parameters and the interpretation of the results.

5.2. MEASUREMENT OF DRIFT VELOCITY BY TIME-OF-FLIGHT METHODS

In principle these methods are straightforward in so far as they are intended to determine W from the time required for the centroid of a travelling group to traverse a known distance. In practice, however, when high accuracy is important, it is necessary to take into account the effect of diffusive processes when analyzing experimental measurements. Once the nature and causes of such effects are recognized, it is possible to reduce or eliminate them by suitable choice of the parameters of the experiment.

5.2.1. COMBINED EFFECT OF DIFFUSIVE PROCESSES. We examine the combined effect of a number of diffusive processes on the interpretation of a typical time-of-flight experiment.

The mathematical description of the ideal isolated travelling group of n_0 electrons uninfluenced by electrodes is (see page 112)

$$n = \frac{n_0}{(4\pi Dt)(4\pi D_L t)^{1/2}}\left[\exp\left(-\frac{\rho^2}{4Dt}\right)\right]\left\{\exp\left[-\frac{(z-Wt)^2}{4D_L t}\right]\right\}, \quad (5.1a)$$

where $\rho^2 = x^2 + y^2$. Let $q(z,t) = 2\pi\int_0^\infty n\rho\,d\rho$. Then it follows that

$$q(z,t) = \text{const}\cdot t^{-1/2}\exp\left[-\frac{(z-Wt)^2}{4D_L t}\right]. \quad (5.1b)$$

The centroid and the position of the maximum of the number density of the group coincide and travel with velocity **W**. This idealized description of a travelling group provides a useful model of the group whose time of travel across a known spatial interval is measured in time-of-flight methods to find the velocity **W**. In practice, however, the travelling group is not entirely isolated since, in general, it moves away from a cathode at which $n = 0$ everywhere on the plane $z = 0$ at all time $t > 0$ towards an electrode in the plane $z = h$ over which n is always zero. It is necessary, therefore, to modify the description of the travelling group to meet these boundary conditions at the electrodes; that is, it is necessary to incorporate in the mathematical description the contributions of back diffusion to the cathode and forward diffusion to the electrode that terminates the drift space.

As indicated in Section 5.1.2, the presence of the electrodes not only alters the distribution of electron number density in such a way that $n \approx 0$ at the electrodes but also modifies the electron energy distribution function in the vicinity of the electrodes from the form it would have in their absence. In what follows, the description of the isolated travelling group is modified to meet the first condition simply by seeking solutions of the time-dependent differential equation for n (equation 4.38), which satisfy the boundary condition $n=0$ at the electrode surfaces.† While the development in this section represents a considerable oversimplification of the complex effects introduced by the electrodes, it is useful in indicating the order of magnitude of the effects and their dependence on the experimental parameters.

In addition to allowing for the effects of the boundaries it is necessary to examine in detail the way in which the velocity of the travelling group is inferred from the experimental measurements since diffusive effects may modify significantly a straightforward interpretation of the results. In electrical shutter experiments, for example, an electrode placed behind an electrical shutter in the plane $z=h$ collects electrons transmitted by the shutter in the brief interval during which it is open. Hence it is necessary to find the correct relationship between W and the time t_m at which the number of transmitted electrons is a maximum.

We suppose initially that the anode is at such a distance from the cathode that the group can move an appreciable distance uninfluenced by the anode. We are therefore concerned only with the influence of the cathode. In order to reduce n to zero over the plane $z=0$, we add to the simple solution of equation 4.38 a second solution representing a dipole distribution with a strength so chosen that its negative contribution to the number density nullifies the positive contribution of the simple pole group everywhere on the plane $z=0$.

We therefore adopt the following representation as the appropriate solution of the equation for n (Huxley, 1972):

$$n = \frac{n_0}{(4\pi Dt)(4\pi D_L t)^{1/2}} \left[\exp\left(-\frac{\rho^2}{4Dt} \right) \right]$$

$$\times \left\{ \exp\left[-\frac{(z-Wt)^2}{4D_L t} \right] - \frac{2D_L}{W} \frac{\partial}{\partial z} \exp\left[-\frac{(z-Wt)^2}{4D_L t} \right] \right\},$$

---

†The effect of the boundaries on the local electron energy distribution function which results from the perturbation to the electron density gradients is, to first order, accounted for in this development.

which is equivalent to

$$n = \frac{n_0}{(4\pi Dt)(4\pi D_L t)^{1/2}} \left[\exp\left(-\frac{\rho^2}{4Dt} \right) \right] \left[1 + \frac{2D_L}{W} \frac{(z - Wt)}{2D_L t} \right]$$

$$\times \exp\left[-\frac{(z - Wt)^2}{4D_L t} \right]$$

$$= \frac{n_0 \exp(-\rho^2/4Dt)}{(4\pi Dt)(4\pi D_L t)^{1/2}} \frac{z}{Wt} \exp\left[-\frac{(z - Wt)^2}{4D_L t} \right]. \tag{5.2}$$

Thus $n = 0$ when $z = 0$ for all values of $t > 0$. It follows that

$$q = \text{const} \cdot z t^{-3/2} \exp\left[-\frac{(z - Wt)^2}{4D_L t} \right]. \tag{5.3}$$

Consider next the influence of the anode in the plane $z = h$. It is evident that the solution adopted in equation 5.2 is inappropriate since it does not make $n = 0$ when $z = h$. It is necessary, therefore, to add to the solution given above another term that satisfies equation 4.38 and that has the appropriate strength and sign to ensure that the resultant value of n satisfies this boundary condition. The additional term is analytically the same as that for n in equation 5.2 except for its origin of reference, which is the point $(0, 0, 2h)$, so that z is replaced by $z - 2h$. The term is also multiplied by a constant factor whose value is to be determined and which for convenience is given the form $\exp b$. We therefore adopt as the appropriate representation of n the following expression:

$$n = \frac{n_0 \exp(-\rho^2/4Dt)}{(4\pi Dt)(4\pi D_L t)^{1/2}} \frac{1}{Wt} \left\{ z \exp\left[-\frac{(z - Wt)^2}{4D_L t} \right] \right.$$

$$\left. + (z - 2h)(\exp b)\exp\left[-\frac{(z - 2h - Wt)^2}{4D_L t} \right] \right\}.$$

Next the constant b is so determined that when $z = h$ the two exponents are equal. The number density then becomes zero since the coefficient z becomes h and $z - 2h$ becomes $-h$. We require therefore that

$$b - \frac{(-h - Wt)^2}{4D_L t} = \frac{-(h - Wt)^2}{4D_L t},$$

that is,

$$b = \frac{(h + Wt)^2 - (h - Wt)^2}{4D_L t} = \frac{hW}{D_L} = 2\lambda_L h.$$

Consequently

$$n = \frac{n_0 \exp(-\rho^2/4Dt)}{(4\pi Dt)(4\pi D_L t)^{1/2}} \frac{1}{Wt} \left\{ z \exp\left[-\frac{(z - Wt)^2}{4D_L t} \right] \right.$$

$$+ (z - 2h) \exp\left[-\frac{(z - Wt)^2 + 4h(h - z)}{4D_L t} \right] \left. \right\}, \qquad (5.4)$$

and

$$q = \frac{n_0}{(4\pi D_L)^{1/2} Wt^{3/2}}$$

$$\times \left\{ z \exp\left[-\frac{(z - Wt)^2}{4D_L t} \right] + (z - 2h)\left(\exp\frac{hW}{D_L} \right) \exp\left[-\frac{(z - 2h - Wt)^2}{4D_L t} \right] \right\}$$

$$= \text{const} \cdot t^{-3/2} \left\{ z \exp\left[-\frac{(z - Wt)^2}{4D_L t} \right] + (z - 2h)\left(\exp\frac{hW}{D_L} \right) \right.$$

$$\times \exp\left[-\frac{(z - 2h - Wt)^2}{4D_L t} \right] \left. \right\}. \qquad (5.5)$$

It can be seen that $n=q=0$ when $z=h$ for all values of t.†

As a basis for the subsequent discussion of the Bradbury-Nielsen technique (Chapter 10) we now need to find the instantaneous flux of electrons transmitted by an electrical shutter in the small interval of time during which it is open. To do so we assume that the shutter acts like a metal plane when shut but as a geometrical plane when open. The instantaneous transmitted flux $i(h,t)$ is therefore

$$i(h,t) = \left(Wq - D_L \frac{\partial}{\partial z} q \right)_{z=h} = -D_L \left(\frac{\partial}{\partial z} q \right)_{z=h}$$

since $q_{z=h}=0$.

From equation 5.5 it follows that

$$\frac{\partial}{\partial z} q = \text{const} \cdot t^{-3/2} \left\{ \left[1 - \frac{z(z-Wt)}{2D_L t} \right] \exp\left[-\frac{(z-Wt)^2}{4D_L t} \right] \right.$$

$$\left. + \left[1 - \frac{(z-2h)(z-2h-Wt)}{2D_L t} \right] \left(\exp\frac{hW}{D_L} \right) \exp\left[-\frac{(z-2h-Wt)^2}{4D_L t} \right] \right\}$$

and, therefore,

$$i(h,t) = \text{const} \cdot t^{-3/2} \left(\frac{h^2}{D_L t} - 2 \right) \exp\left[-\frac{(h-Wt)^2}{4D_L t} \right].$$

We next seek an expression for the time t at which $i(h,t) \equiv i(t)$ reaches its maximum value. We equate $(d/dt)i$ to zero to find

$$\frac{d}{dt} i = -3t^{-5/2} + \frac{5h^2 t^{-7/2}}{2D_L}$$

$$+ \left(t^{-7/2} - \frac{h^2 t^{-9/2}}{2D_L} \right) \frac{\left\{ 2Wt(h-Wt) + (h-Wt)^2 \right\}}{2D_L} = 0.$$

† Although this representation of n leads to $n=q=0$ when $z=h$, it should be noted that the addition of the image term slightly perturbs the cathode boundary condition $n=0$. In practice the effect is usually negligible. If required, an exact solution can be obtained with an infinite set of images (Huxley, 1972).

After reduction this equation becomes

$$h^2 - W^2 t^2 = \frac{5h^2 - 6D_L t}{(h^2 / 2D_L t) - 1}$$

$$= \frac{5(1 - 6D_L t / 5h^2)2D_L t}{1 - 2D_L t / h^2}.$$

In practice the groups spread little, and the time t_m of maximum current is not greatly different from h/W. Consequently, except in the factor $h - Wt$, we may write $t_m = h/W$. Thus

$$h - Wt_m \cong \frac{5(1 - 6D_L / 5Wh)2D_L h}{2hW(1 - 2D_L / Wh)}$$

$$= \frac{5(1 - (6/5)\beta)\beta h}{1 - 2\beta},$$

that is,

$$W \cong \frac{h}{t_m}(1 - 5\beta), \tag{5.6}$$

where $\beta = D_L / Wh$, and higher-order terms in β have been neglected.

The drift velocity W and ND_L are both functions of E/N; consequently, with a given value of E/N, D_L is diminished as N is increased. The relative error 5β which is introduced if W is taken to be $W = h/t_m$ can therefore be diminished by employing large pressures; it is also reduced by increasing h. With a suitable choice of N and h the relative error resulting from the use of the simple formula can usually be made comparable with, or less than, other experimental errors.

5.2.2. CONTRIBUTIONS OF INDIVIDUAL DIFFUSIVE PROCESSES. In the design of time-of-flight experiments, it is important to identify the factors that contribute to the total correction factor 5β (Lowke, 1962). We therefore examine in turn several special cases in order to discuss these factors separately, since the importance of each factor depends on the kind of experiment being considered, that is, whether the experiment is based on the so-called pulsed Townsend discharge, the use of electrical shutters, or the measurement of the arrival times of individual electrons.

(a) The Decay of the Group

The simplest interpretation of electrical shutter experiments is based on the assumption that the number of electrons transmitted by the shutter terminating the drift space (the sampling shutter), and thus collected by the receiving electrode, is $q(h,t)W\Delta t$, where Δt is the open time of the shutter and $q(h,t)$ is found from equation 5.1. This assumption neglects the influence both of the shutter and of the electrode at which the electron group is formed on the number density within the travelling group. It also neglects electron transport due to diffusion across the plane of the shutter during the time interval Δt.

With such an interpretation it might be expected that t_m would coincide with the arrival of the centre of mass of the travelling group at the shutter and that the drift velocity should therefore be calculated from $W = h/t_m$. However, the simple formula is inexact even in this instance because the time at which $q(h,t)$ is a maximum does not coincide with the time of transit of the centre of mass of the group through the plane $z = h$. The difference between these times is caused by the decay of the number density at the centre of the group due to diffusion. The more exact relation between W and t_m is found by determining from equation 5.1 the value of t for which $q(h,t)$ is a maximum, that is, $(d/dt)q = 0$, whence it may be shown that

$$W \cong \frac{h}{t_m}(1-\beta). \tag{5.7}$$

(b) Flux Across a Geometrical Plane Due to Diffusion

The next approximation follows from the inclusion of the flux due to diffusion across the plane $z = h$. The flux is then given by

$$i(t) = \left(qW - D_L \frac{\partial}{\partial z}q\right)_{z=h},$$

and it may be shown by using equation 5.1 and finding the value of $t = t_m$ for which $(d/dt)i = 0$ that

$$W \cong \frac{h}{t_m}(1-2\beta). \tag{5.8}$$

(c) Back Diffusion to the Cathode

We consider next the effect of the electrode (cathode) at which the electron group is formed, disregarding for the present the influence of the shutter that terminates the drift space. We note in passing that even when the cathode is an electrical shutter it may be regarded as a plane metal boundary since all electrons that arrive at the shutter are absorbed except those that arrive during the brief interval in which the shutter is open. Thus for a shutter, as well as for a metal electrode, the condition $n=0$ at the boundary is assumed to apply when the shutter is closed. Hence the appropriate representations for n and q are those given by equations 5.2 and 5.3, respectively.

With the aid of these equations it can be shown without difficulty that the position of maximum density within the group moves at velocity **W** and lies $2D_L/W$ ahead of the corresponding position for an isolated travelling group. Furthermore, by determining the value of $t=t_m$ for which the flux across a geometrical plane $z=h$ is a maximum, it is found that

$$W \cong \frac{h}{t_m}(1-4\beta).$$ (5.9)

This result is consistent with equation 5.8 and with the fact that the position at which $q(z,t)$ is a maximum is displaced forward by an amount $2D_L/W$ as a consequence of back diffusion to the cathode. It demonstrates that the total correction factor is the algebraic sum of the factors from the several diffusive processes that contribute to the total.

(d) Diffusion to the Second Shutter

We have seen that the correction factor resulting from the use of the formula $W=h/t_m$ is 4β when the effect of the shutter on the distribution of number density within the travelling group is ignored. On the other hand, the analysis of Section 5.2.1, which included this effect, predicted a correction factor of 5β. It follows that the contribution to the total correction factor from diffusion to the second shutter is $\beta=D_L/Wh$.

For convenience the analysis given in this section has been based on an idealized electrical shutter experiment although, as will be seen later, much of it is applicable to other time-of-flight experiments. In practice the total correction factor for such an experiment may differ from the value 5β given by equation 5.6 as a consequence of other factors, some of which have already been alluded to but have not been included here. A further discussion of this experiment and other time-of-flight experiments based on the theory developed in this section will be found in Chapter 10.

We note, in passing, that if a steady stream be supposed to arise from a succession of groups of the kind described by equation 5.2, then on following the analysis presented in Section 4.4 and using the integral

$$\int_0^\infty \left[\exp\left(-\tau-\frac{s^2}{4\tau}\right)\right]\tau^{-(\nu+1)}\,d\tau = 2\left(\frac{2}{s}\right)^\nu K_\nu(s)$$

it will be found that the number density in the steady stream is (Huxley, 1972)

$$n \propto (\exp\lambda_L z)\left(\frac{z}{r'}\right) r'^{-1/2} K_{3/2}(\lambda_L r'),$$

where $r'^2 = x'^2 + y'^2 + z^2$, with $x' = (D_L/D)^{1/2}x$ and $y' = (D_L/D)^{1/2}y$.

This important and useful formula is discussed in detail in what follows in this chapter.

5.3. THE THEORY OF THE STEADY STREAM FROM A POINT SOURCE

The theory of the steady stream from a small source is important through its application to the measurement of the characteristic energy eD/μ by means of a Townsend-Huxley diffusion apparatus.

5.3.1. THEORY OF UNIMPEDED STREAM.
We first derive an expression for the number density $n(\mathbf{r})$ in a steady stream that originates from a point source and proceeds to infinity.

The time-independent equation for n is

$$D\left(\frac{\partial^2 n}{\partial x^2} + \frac{\partial^2 n}{\partial y^2}\right) + D_L\frac{\partial^2 n}{\partial z^2} = W\frac{\partial n}{\partial z}. \tag{5.10}$$

We let $2\lambda_L = W/D_L$ and write equation 5.10 in the form

$$\frac{D}{D_L}\left(\frac{\partial^2 n}{\partial x^2} + \frac{\partial^2 n}{\partial y^2}\right) + \frac{\partial^2 n}{\partial z^2} - 2\lambda_L\frac{\partial n}{\partial z} = 0.$$

Next we introduce new variables: $x' = (D_L/D)^{1/2}x$ and $y' = (D_L/D)^{1/2}y$; then equation 5.10 is equivalent to

$$\frac{\partial^2 n}{\partial x'^2} + \frac{\partial^2 n}{\partial y'^2} + \frac{\partial^2 n}{\partial z^2} = 2\lambda_L\frac{\partial n}{\partial z}. \tag{5.11}$$

Since it is convenient to work with a simpler equation than equation 5.11, we replace n by

$$n(x',y',z) \equiv (\exp\lambda_L z)V(x',y',z). \tag{5.12}$$

It follows that

$$\nabla'^2 V = \lambda_L^2 V \tag{5.13}$$

with

$$\nabla'^2 \equiv \frac{\partial^2}{\partial x'^2} + \frac{\partial^2}{\partial y'^2} + \frac{\partial^2}{\partial z^2}.$$

We let $r'^2 = x'^2 + y'^2 + z^2$ and define angles θ and ϕ through the following relations: $\cos\theta = z/r'$, $x' = r'\sin\theta\cos\phi$, and $y' = r'\sin\theta\sin\phi$. Expressed in terms of the variables r', θ, and ϕ, equation 5.13 becomes

$$\frac{1}{r'^2}\frac{\partial}{\partial r'}\left(r'^2\frac{\partial V}{\partial r'}\right) + \frac{1}{r'^2\sin\theta}\frac{\partial}{\partial\theta}\left(\sin\theta\frac{\partial V}{\partial\theta}\right) + \frac{1}{r'^2\sin^2\theta}\frac{\partial^2 V}{\partial\phi^2} = \lambda_L^2 V. \quad (5.14)$$

Because we are interested in solutions that are independent of ϕ, we set $\partial^2 V/\partial\phi^2 = 0$. It is now assumed that $V(r',\theta)$ has the functional form $V = R_k(r')P_k(\mu)$, in which $R_k(r')$ remains to be determined, $\mu = \cos\theta$, and $P_k(\mu)$ is a Legendre polynomial with order k. Since

$$\frac{1}{\sin\theta}\frac{\partial}{\partial\theta}\left(\sin\theta\frac{\partial V}{\partial\theta}\right) \equiv R_k(r')\frac{d}{d\mu}\left[(1-\mu^2)\frac{dP_k(\mu)}{d\mu}\right]$$

and it is known that

$$\frac{d}{d\mu}\left[(1-\mu^2)\frac{dP_k(\mu)}{d\mu}\right] = -k(k+1)P_k(\mu),$$

it can be seen that equation 5.14 reduces to

$$\frac{1}{r'^2}\frac{d}{dr'}\left(r'^2\frac{dR_k}{dr'}\right) - \left[\frac{k(k+1)}{r'^2} + \lambda_L^2\right]R_k = 0,$$

that is,

$$\frac{d^2R_k}{dr'^2} + \frac{2}{r'}\frac{dR_k}{dr'} - \left[\lambda_L^2 + \frac{k(k+1)}{r'^2}\right]R_k = 0. \quad (5.15)$$

Equation 5.15 is a special case of a more general equation encountered in the study of Bessel functions (F. E. Relton, *Applied Bessel Functions*, Blackie and Sons, p. 103; a still more general form is given in G. N. Watson, *Bessel Functions*, 2nd ed., Cambridge University Press, p. 97). This generalized equation is

$$\frac{d^2y}{dx^2} + \frac{1-2\alpha}{x}\frac{dy}{dx} - \left[(\beta\gamma x^{\gamma-1})^2 + \frac{\nu^2\gamma^2 - \alpha^2}{x^2}\right]y = 0. \quad (5.16)$$

Solutions of this equation, apart from constants of proportionality, are

$$y = x^\alpha I_\nu(\beta x^\gamma) \quad \text{and} \quad y = x^\alpha K_\nu(\beta x^\gamma),$$

where I_ν and K_ν are modified Bessel functions of the first and second kinds, respectively, and of order ν. It can be seen that equation 5.15 is the special case of equation 5.16 in which $\alpha = -\frac{1}{2}$; $\gamma = 1$; $\beta = \lambda_L$. It then follows that $\nu^2 - \frac{1}{4} = k(k+1)$, that is, $\nu = k + \frac{1}{2}$. Solutions of equation 5.15 are therefore

$$R_k(r') = r'^{-1/2} I_{k+1/2}(\lambda_L r') \quad \text{and} \quad R_k(r') = r'^{-1/2} K_{k+1/2}(\lambda_L r').$$

In a solution that represents a steady stream from a point source, $n \to \infty$ as $r' \to 0$ and $n \to 0$ as $r' \to \infty$. These are properties of the $K_{k+1/2}$ functions but not of the $I_{k+1/2}$ functions. We therefore suppose that at distances large compared with the radius of the aperture the number density in the stream is represented in a general manner by

$$n = (\exp \lambda_L z) V = (\exp \lambda_L z) \sum_{k=0}^{\infty} A_k r'^{-1/2} K_{k+1/2}(\lambda_L r') P_k(\mu). \quad (5.17)$$

Equation 5.17 represents a steady stream that proceeds to infinity without interruption. Without a knowledge of the nature of the source it is impossible to predict the relative magnitudes of the coefficients A_k except in the most general manner. For instance, when the source is a circular hole in a metal cathode in the plane $z = 0$, with the centre of the hole at the origin, equation 5.17 can be taken to represent the number density n in the stream at distances $r \gg a$, where a is the radius of the aperture. However, the boundary condition at the cathode requires that $n = 0$ when $z = 0$ and $\rho > a$, since it is assumed that $n = 0$ at a metal surface. This condition requires that every term in equation 5.17 for which the order k is an even number must vanish because $P_k(\mu) \neq 0$ when $\mu = 0$ if k is even. Equation 5.17 now reduces to

$$n = (\exp \lambda_L z) \sum_{k=0}^{\infty} A_{2k+1} r'^{-1/2} K_{2k+3/2}(\lambda_L r') P_{2k+1}(\mu) \quad (5.18)$$

where $\mu = \cos \theta = z/r'$.

On the other hand, with a point source at the origin in the geometrical plane $z = 0$ which is not occupied by a metal surface, the correct representation of the stream is equation 5.17 in which both even- and odd-order terms may be present. The terms in equation 5.17 for which $k = 0$ and $k = 1$ are, respectively, the "pole" and "dipole" terms or sources. Terms for which $k \geqslant 2$ are multipoles.

In what follows we are concerned exclusively with a stream which enters a diffusion chamber through a small hole at the origin in the cathode which lies in the plane $z = 0$. In conformity with the discussion in Section 5.1.2 we assume that $n = 0$ at the metal surface of the cathode. The distribution of number density n at the points in the stream where $r > a$ is therefore represented by equation 5.18. In practice the radius of the aperture a is small (of the order of 0.5 mm); consequently there is a large region in the stream within which $r \gg a$. The coefficients A_k are functions of $\lambda_L a$ to a power that increases progressively with k, so that when $\lambda_L a < 1$ the coefficients diminish progressively as k increases. It is next assumed that in practice equation 5.18 can be effectively represented by the first term on the right, the dipole term, which is

$$n = A_1(\exp\lambda_L z) r'^{-1/2} K_{3/2}(\lambda_L r') \cdot \cos\theta \quad \text{(dipole term)}. \quad (5.19a)$$

We also note that the pole term gives

$$n = A_0(\exp\lambda_L z) r'^{-1/2} K_{1/2}(\lambda_L r') \quad \text{(pole term)}. \quad (5.19b)$$

Although we shall be concerned to develop the theory of the stream from a hole source in terms of equation 5.19a, which makes $n = 0$ at the cathode $z = 0$, except at the origin, it is useful to give equation 5.19b for the pole source since it is a basic solution of equation 5.10 such that the other terms in equations 5.17 and 5.18 can be expressed in terms of it. Moreover, the pole solution has been extensively used in the literature of the subject along with the assumption that diffusion is isotropic. It is shown later (Section 5.3.3) that a successful semiempirical formula based on the assumption of a pole source and isotropic diffusion can be obtained by assuming a dipole source and anisotropic diffusion.

Before proceeding with the discussion of the stream from a small aperture it is convenient to recall some relevant properties of the Bessel K functions.

5.3.2. SOME PROPERTIES OF THE BESSEL K FUNCTIONS. We first examine the behaviour of the Bessel K functions at large values of the arguments.

In equation 5.16 let $\alpha = \frac{1}{2}$, $\gamma = 1$, $\beta = \lambda_L$. The equation then becomes

$$\frac{d^2y}{dx^2} = \left(\lambda_L^2 + \frac{\nu^2 - 1/4}{x^2}\right)y = \lambda_L^2\left[1 + \frac{\nu^2 - 1/4}{(\lambda_L x)^2}\right]y, \quad (5.20)$$

and its solution in terms of K functions is $y = \text{const} \cdot x^{1/2} K_\nu(\lambda_L x)$.

Let x increase until $(v^2-1/4)/(\lambda_L x)^2 \ll 1$; then the differential equation tends to the form $d^2y/dx^2 = \lambda_L^2 y$, and a solution that tends to zero as $x\to\infty$ is $\exp-\lambda_L x$. Thus, as $x\to\infty$, $x^{1/2}K_v(\lambda_L x) \propto \exp-\lambda_L x$. Consequently, as $x\to\infty$, $K_v(\lambda_L x)\to \text{const}\cdot x^{-1/2}(\exp-\lambda_L x)$, which is a form independent of v.

If we confine our attention to the smaller values of k, we can say that $K_{3/2}(\lambda_L x)$ resembles $K_{1/2}(\lambda_L x)$ when $(\lambda_L x)^2 \gg v^2 - \frac{1}{4} = 2$.

The exact representations of the $K_{k+1/2}(x)$ for small values of k are as follows:

$$K_{1/2}(x) = \left(\frac{\pi}{2x}\right)^{1/2}\exp-x; \quad K_{3/2}(x) = \left(\frac{\pi}{2x}\right)^{1/2}\left(1+\frac{1}{x}\right)\exp-x,$$

$$K_{5/2}(x) = \left(\frac{\pi}{2x}\right)^{1/2}\left(1+\frac{3}{x}+\frac{3}{x^2}\right)\exp-x,$$

etc., and in general

$$K_{k+1/2}(x) = \left(\frac{\pi}{2x}\right)^{1/2}\left(\sum_{l=0}^{k}\frac{(k+l)!}{l!(k-l)!}\frac{1}{(2x)^l}\right)\exp-x. \tag{5.21}$$

It is evident that with large values of x all the functions tend to become $(\pi/2x)^{1/2}\exp-x$. For example, $K_{3/2}(x) = [1+(1/x)]K_{1/2}(x)$. When $x = 100$, an error of 1% is committed in equating $K_{3/2}(x)$ to $K_{1/2}(x)$.

Other useful relationships are as follows (Watson, *op. cit.*, p. 79):

$$\left(\frac{1}{x}\frac{d}{dx}\right)^m\left[\frac{K_v(x)}{x^v}\right] = (-1)^m\frac{K_{v+m}(x)}{x^{v+m}},$$

$$\left(\frac{1}{x}\frac{d}{dx}\right)^m[x^v K_v(x)] = (-1)^m x^{v-m}K_{v-m}(x). \tag{5.22}$$

These properties of the Bessel K functions show that equations 5.19 for the pole and dipole distributions are equivalent to

$$\text{pole source:} \quad n \propto \frac{\exp-\lambda_L(r'-z)}{r'} \tag{5.23}$$

and

dipole source: $\quad n \propto (\exp\lambda_L z)\left(\dfrac{z}{r'}\right)\left(1 + \dfrac{1}{\lambda_L r'}\right)\dfrac{\exp -\lambda_L r'}{r'}$

(because $z/r' = \cos\theta$)

$$= \left(\frac{z}{r'}\right)\left(1 + \frac{1}{\lambda_L r'}\right)\frac{\exp -\lambda_L(r' - z)}{r'}. \tag{5.24}$$

Also from equation 5.22

dipole source: $\quad n \propto (\exp\lambda_L z)r'^{-1/2}\left(\dfrac{z}{r'}\right)K_{3/2}(\lambda_L r')$

$$= -(\lambda_L r')^{3/2} r'^{-1/2}\left(\frac{z}{r'}\right)(\exp\lambda_L z)\left(\frac{1}{\lambda_L r'}\right)\frac{d}{d(\lambda_L r')}\left[\frac{K_{1/2}(\lambda_L r')}{(\lambda_L r')^{1/2}}\right]$$

$$\propto -(\exp\lambda_L z)\left(\frac{z}{r'}\right)\frac{d}{dr'}\left(\frac{\exp -\lambda_L r'}{r'}\right)$$

$$= -(\exp\lambda_L z)\frac{\partial}{\partial z}\left(\frac{\exp -\lambda_L r'}{r'}\right), \tag{5.25}$$

since $r'^2 = \rho'^2 + z^2$; $r' \, dr' = z \, dz$ (ρ constant).

5.3.3.   FORMULA FOR THE CURRENT RATIO WITH A CIRCULAR DIVISION OF THE ANODE. We first consider the distribution of the number density n in streams that impinge on an anode at $z = h$ and originate from either a pole or a dipole source. Equations 5.23 and 5.25 refer to streams that proceed from the origin to infinity without interruption. The associated values of V that satisfy equation 5.13 are, respectively,

$$V_{\text{pole}} \propto \frac{\exp -\lambda_L r'}{r'}$$

and

$$V_{\text{dipole}} \propto -\frac{\partial}{\partial z}\left(\frac{\exp -\lambda_L r'}{r'}\right).$$

The expression for V_{dipole} is an illustration of the fact that if V_{pole} is a solution of equation 5.13 then $(\partial^k/\partial z^k)V_{\mathrm{pole}}$ is also a solution. The same sources moved to the point $(0,0,2h)$ give

$$V_{\mathrm{pole}} \propto \frac{\exp-\lambda_L r''}{r''} \qquad \text{and} \qquad V_{\mathrm{dipole}} \propto -\frac{\partial}{\partial z}\left(\frac{\exp-\lambda_L r''}{r''}\right),$$

where $r''^2 = \rho'^2 + (z-2h)^2$. These expressions are also solutions of equation 5.13. It follows also that

$$n_{\mathrm{pole}} \propto (\exp\lambda_L z)\left(\frac{\exp-\lambda_L r''}{r''}\right) \quad \text{and} \quad n_{\mathrm{dipole}} \propto -(\exp\lambda_L z)\frac{\partial}{\partial z}\left(\frac{\exp-\lambda_L r''}{r''}\right)$$

are solutions of equation 5.11.

Since the difference or the sum of two solutions is also a solution, it follows that equation 5.11 is satisfied by

$$n_{\mathrm{pole}} \propto (\exp\lambda_L z)\left(\frac{\exp-\lambda_L r'}{r'} - \frac{\exp-\lambda_L r''}{r''}\right) \tag{5.26a}$$

and

$$n_{\mathrm{dipole}} \propto -(\exp\lambda_L z)\frac{\partial}{\partial z}\left(\frac{\exp-\lambda_L r'}{r'} + \frac{\exp-\lambda_L r''}{r''}\right), \tag{5.26b}$$

where $r'^2 = \rho'^2 + z^2$ and $r''^2 = \rho'^2 + (z-2h)^2$; $\rho'^2 = (D_L/D)\rho^2 = (\lambda/\lambda_L)\rho^2$ and $\rho^2 = x^2 + y^2$; $\lambda_L = W/2D_L$ and $\lambda = W/2D$. These expressions give $n=0$ when $z=h$ for all values of ρ and also positive values of n when $z<h$. Equations 5.26 therefore give the distribution of number density n in a stream from an isolated pole or dipole source in a uniform electric field when the stream falls upon a plane metal electrode.

When the source is isolated in free space at the origin, we do not impose boundary conditions on the behaviour of n at the plane $z=0$. The supplementary source required to give $n=0$ at the plane $z=h$ of the anode also changes the value of n in the plane $z=0$ in such a way as to represent the actual influence of the anode. The complete solution is here given by the original pole source and a single supplementary source.

If, however, the dipole source is employed to simulate a small hole source in a plane cathode, the boundary condition at the cathode, that is, $n=0$ everywhere except at the origin, must be preserved in a precise solution. In order to preserve this condition, and with it the condition at the anode, that is, $n=0$ for all values of ρ when $z=h$, it is necessary to invoke an infinite hierarchy of supplementary dipole sources at the points $(0,0,\pm k\cdot 2h)$, where $k=1,2,3,\ldots$. However, it is found in practice that unless h is small (say 1 cm or less) the system comprising the original dipole source and a single supplementary term, as in equation 5.26b, is adequate.

We proceed to derive a formula for the current ratio R using the assumptions discussed in Section 5.1.2 regarding the influence of the metal boundaries and assuming, therefore, that equation 5.26b is an accurate description of the distribution of the number density n in a stream from a small aperture. The z component of current density J at any plane $z = $ constant with $0 \leqslant z \leqslant h$ is

$$J = e\left(nW - D_L \frac{\partial n}{\partial z}\right) = eD_L\left(2\lambda_L n - \frac{\partial n}{\partial z}\right),$$

where n is given by equation 5.26b. Consequently, if A is the constant of proportionality in equation 5.26b, then

$$J = eD_L\left\{2\lambda_L n - \lambda_L n + A(\exp\lambda_L z)\frac{\partial}{\partial z}\left[\frac{z}{r'}\frac{d}{dr'}\left(\frac{\exp - \lambda_L r'}{r'}\right)\right.\right.$$

$$\left.\left. + \frac{z - 2h}{r''}\frac{d}{dr''}\left(\frac{\exp - \lambda_L r''}{r''}\right)\right]\right\}$$

$$= eD_L\left(\lambda_L n + A(\exp\lambda_L z)\left\{\frac{1}{r'}\frac{d}{dr'}\left(\frac{\exp - \lambda_L r'}{r'}\right) + \frac{z^2}{r'}\frac{d}{dr'}\left[\frac{1}{r'}\frac{d}{dr'}\left(\frac{\exp - \lambda_L r'}{r'}\right)\right]\right.\right.$$

$$\left.\left. + \frac{1}{r''}\frac{d}{dr''}\left(\frac{\exp - \lambda_L r''}{r''}\right) + \frac{(z - 2h)^2}{r''}\frac{d}{dr''}\left[\frac{1}{r''}\frac{d}{dr''}\left(\frac{\exp - \lambda_L r''}{r''}\right)\right]\right\}\right).$$

At the plane $z = h$, $r'' = r'$, and $n = 0$; consequently

$$J_{z=h} = 2eD_L A(\exp\lambda_L h)\left\{\frac{1}{r'}\frac{d}{dr'}\left(\frac{\exp - \lambda_L r'}{r'}\right)\right.$$

$$\left. + \frac{h^2}{r'}\frac{d}{dr'}\left[\frac{1}{r'}\frac{d}{dr'}\left(\frac{\exp - \lambda_L r'}{r'}\right)\right]\right\}_{z=h}.$$

Current ratio for a central disk (*Figure 1.1b*). The current $i_{0,b}$ received by a central disk of radius b of the anode is

$$i_{0,b} = 2\pi \int_0^b J_{z=h}\rho \, d\rho = 2\pi\left(\frac{D}{D_L}\right)\int_0^{b'} J_{z=h}\rho' \, d\rho' = 2\pi\left(\frac{D}{D_L}\right)\int_h^{d'} J_{z=h}r' \, dr'$$

since $\rho'^2 = (D_L/D)\rho^2$, $r'^2 = \rho'^2 + h^2$, $b'^2 = (D_L/D)b^2$, and $d'^2 = b'^2 + h^2$.
It follows that

$$i_{0,b} = A4\pi e D(\exp\lambda_L h) \int_h^{d'} d\left[\frac{\exp-\lambda_L r'}{r'} + \frac{h^2}{r'}\frac{d}{dr'}\left(\frac{\exp-\lambda_L r'}{r'}\right)\right]_{z=h}$$

$$= A4\pi e D(\exp\lambda_L h)\left[-\frac{\exp-\lambda_L h}{h} + \frac{\exp-\lambda_L d'}{d'}\right.$$

$$\left. + (1+\lambda_L h)\frac{\exp-\lambda_L h}{h} - \frac{h^2}{d'^2}(1+\lambda_L d')\frac{\exp-\lambda_L d'}{d'}\right].$$

The total current received by the anode is seen to be

$$i \equiv i_{0,\infty} = A4\pi e D\lambda_L;$$

consequently,

$$i_{0,b} = i\left[1 - \left(\frac{h}{d'} - \frac{1}{\lambda_L h} + \frac{h}{\lambda_L d'^2}\right)\frac{h}{d'}\exp-\lambda_L(d'-h)\right]. \tag{5.27}$$

The current ratio is therefore

$$R = \frac{i_{0,b}}{i} = 1 - \left[\frac{h}{d'} - \frac{1}{\lambda_L h}\left(1 - \frac{h^2}{d'^2}\right)\right]\frac{h}{d'}\exp-\lambda_L(d'-h) \tag{5.28a}$$

in which $\lambda_L = W/2D_L$, $d'^2 = h^2 + b'^2 = h^2 + (D_L/D)b^2 = h^2 + (\lambda/\lambda_L)b^2$, and $\lambda = W/2D$.
The same procedure applied to equation 5.26a gives

$$R_{\text{pole}} = 1 - \frac{h}{d'}\exp-\lambda_L(d'-h). \tag{5.28b}$$

Because the parameter λ_L is in evidence in equations 5.28, it would not be unreasonable to suppose that from the measured values of R the value of λ_L could be immediately derived without ambiguity. It should be recalled, however, that the parameter λ is also present, although concealed within the symbol d'. It is therefore somewhat of a surprise to discover that, in practice, equations 5.28 assume forms that give λ rather than λ_L, thus demonstrating that the spreading of the stream is governed chiefly by D. We proceed to examine this behaviour of equation 5.28a.

Consider first the exponent $-\lambda_L(d'-h)$. Let the radius b of the central disk of the anode be 0.5 cm, and let h be several centimetres so that b/h is notably less than unity. It is readily seen that when b/h is small $\lambda_L(d'-h)$ is accurately represented by the leading terms of an expansion:

$$\lambda_L(d'-h)=\tfrac{1}{2}(\lambda h)\left(\frac{b}{h}\right)^2-\frac{1}{8}\frac{\lambda}{\lambda_L}(\lambda h)\left(\frac{b}{h}\right)^4+\cdots.$$

Similarly $\lambda(d-h)$ is given by

$$\lambda(d-h)=\tfrac{1}{2}(\lambda h)\left(\frac{b}{h}\right)^2-\tfrac{1}{8}(\lambda h)\left(\frac{b}{h}\right)^4+\cdots.$$

Thus

$$\lambda_L(d'-h)-\lambda(d-h)\cong\tfrac{1}{8}\left(1-\frac{\lambda}{\lambda_L}\right)(\lambda h)\left(\frac{b}{h}\right)^4=x.$$

Consider some realistic values of x: let $\lambda=20$ cm^{-1}; $h=10$ cm; $b=0.5$ cm, and assume that $|1-\lambda/\lambda_L|<1$; then $x<1/6400$. If h is reduced to 4 cm, $x<1/410$. It follows that, with these experimental conditions,

$$\exp-\lambda_L(d'-h)\cong[\exp-\lambda(d-h)](1-x),$$

and that negligible error is introduced by replacing $\exp-\lambda_L(d'-h)$ by $\exp-\lambda(d-h)$. Since $\lambda/\lambda_L<2$ for many gases and since the values of λ encountered in practice usually exceed $\lambda=20$ cm^{-1}, this replacement is justifiable in many circumstances.

Next consider the factor $[h/d'-(1-h^2/d'^2)/(\lambda_L h)]h/d'$. It can be shown that when $(\tfrac{1}{8})(\lambda/\lambda_L)^2(b/h)^4\ll1$ this factor can, with negligible error, be replaced by $[1+(\tfrac{1}{2}-\lambda/\lambda_L)(b/d)^2]h/d$. When, therefore, $(\tfrac{1}{8})(\lambda h)(1-\lambda/\lambda_L)(b/h)^4$ and $(\tfrac{1}{8})(\lambda/\lambda_L)^2(b/h)^4$ are both very small quantities, the expression for R in equation 5.28a may be replaced by

$$R=1-\left[1+\left(\frac{1}{2}-\frac{D_L}{D}\right)\left(\frac{b}{d}\right)^2\right]\left(\frac{h}{d}\right)\exp-\lambda(d-h). \qquad (5.29)$$

In a gas in which $q_m(c)$ is constant or varies very slowly with c, as, for instance, the behaviour of $q_m(c)$ in hydrogen, $D_L/D\cong0.5$ unless E/N is small (Lowke and Parker, 1969). With $D_L/D=0.5$, equation 5.29 reduces to

$$R = 1 - \frac{h}{d}\exp-\lambda(d-h). \tag{5.30}$$

In any case when $(b/d)^2$ is small, the term $(\frac{1}{2} - D_L/D)(b/d)^2$ is usually even smaller, and in practice equation 5.30 is an accurate representation of R. As will be demonstrated in Part 2, equation 5.30 accurately describes the structure of an electron stream in hydrogen, even with values of h as small as 2 cm.

It is of interest that equation 5.30, which has been used for many years, was originally derived on the assumptions of a pole source and isotropic diffusion $(D_L = D)$ and is in fact what equation 5.28b becomes with these assumptions (see Chapter 11).

5.3.4. FORMULA FOR THE CURRENT RATIO WITH STRIP DIVISION OF THE ANODE. The anode is here divided as shown in Figure 1.1c into a central strip with width $2b$ and flanking electrodes. Let the centre of the whole anode be the point $(0,0,h)$, and take the direction of the axis Oy to be parallel to the edges of the central strip. The source, as before, is a small aperture with the origin at its centre. The boundary conditions are taken, as before, to be $n=0$ at a metal surface.

Consider the current to a thin geometrical strip of the anode parallel to the axis Oy and with width dx and coordinate x. This strip extends as a chord across the circular anode, which is sufficiently large that the current density J is zero near and at its circumference. Since $n=0$ at the anode, the current received by the thin strip is

$$dx \int_{-\infty}^{\infty} J(x,y)_{z=h}\,dy.$$

But according to the discussion in the preceding section the current density from a dipole source is

$$J(x,y)_{z=h} = 2eD_L A(\exp\lambda_L h)$$

$$\times \left\{ \frac{1}{r'}\frac{d}{dr'}\left(\frac{\exp-\lambda_L r'}{r'}\right) + \frac{h^2}{r'}\frac{d}{dr'}\left[\frac{1}{r'}\frac{d}{dr'}\left(\frac{\exp-\lambda_L r'}{r'}\right)\right]\right\}_{z=h}.$$

The replacement of $(\exp-\lambda_L r')/r'$ by the equivalent $K_{1/2}(\lambda_L r')$ function from equation 5.21, followed by an application of equation 5.22, leads to the following expression:

$$J(x,y)_{z=h} = -2eD_L A\left(\frac{2}{\pi}\right)^{1/2}(\exp\lambda_L h)\lambda_L^3\left[\frac{K_{3/2}(\lambda_L r')}{(\lambda_L r')^{3/2}} - (\lambda_L h)^2\frac{K_{5/2}(\lambda_L r')}{(\lambda_L r')^{5/2}}\right],$$

$$\tag{5.31}$$

in which $r' = (h^2 + x'^2 + y'^2)^{1/2} = [h^2 + (\lambda/\lambda_L)(x^2 + y^2)]^{1/2}$.

The total current received by the thin strip is

$$di_x = dx \int_{-\infty}^{\infty} J(x,y)_{z=h} dy = dx \left(\frac{D}{D_L}\right)^{1/2} \int_{-\infty}^{\infty} J(x',y')_{z=h} dy'. \quad (5.32)$$

This integral can be evaluated by means of the standard formula [Watson, op. cit., §13.47(6)]

$$\int_0^{\infty} \frac{K_\nu\left[a(c^2 + s^2)^{1/2}\right]}{(c^2 + s^2)^{\nu/2}} s^{2\mu+1} ds = \frac{2^\mu \Gamma(\mu+1)}{a^{\mu+1}c^{\nu-\mu-1}} K_{\nu-\mu-1}(ac). \quad (5.33)$$

It is then seen that

$$di_x = dx \left[4eA\lambda_L(DD_L)^{1/2}\right](\exp\lambda_L h)$$

$$\times \left\{ (\lambda_L h)^2 \frac{K_2\left[\lambda_L(h^2 + x'^2)^{1/2}\right]}{\lambda_L(h^2 + x'^2)} - \frac{K_1\left[\lambda_L(h^2 + x'^2)^{1/2}\right]}{(h^2 + x'^2)^{1/2}} \right\}.$$

The current received by a strip with finite width whose edges are $x = b$ and $x = c$ is [since $dx = (\lambda_L/\lambda)^{1/2} dx' = (D/D_L)^{1/2} dx'$]

$$i_{b,c} = 4eAD\lambda_L(\exp\lambda_L h)$$

$$\times \int_{b'}^{c'} \left\{ \frac{(\lambda_L h)^2 K_2\left[\lambda_L(h^2 + x'^2)^{1/2}\right]}{\lambda_L(h^2 + x'^2)} - \frac{K_1\left\{\lambda_L(h^2 + x'^2)^{1/2}\right\}}{(h^2 + x'^2)^{1/2}} \right\} dx',$$

$$(5.34)$$

where $b' = (D_L/D)^{1/2}b$ and $c' = (D_L/D)^{1/2}c$.

The total current i to the whole anode is found by putting $b' = -\infty$ and $c' = +\infty$ in equation 5.34, followed by a further application of equation 5.33. It is then found that $i = 4\pi eDA\lambda_L$ in agreement with equation 5.27. The ratio $R_{b,c}$ of the current to the strip to the total current is $R_{b,c} = i_{b,c}/i$ and is obtained directly from equation 5.34. The most important instance is that of the symmetrically located strip. If in equation 5.34 we replace b

by $-b$ and c by b, the ratio is

$$R_{-b,b} = \frac{i_{-b,b}}{i}$$

$$= \frac{2}{\pi}(\exp\lambda_L h) \int_0^{b'} \left\{ \frac{(\lambda_L h)^2 K_2\left[\lambda_L(h^2+x'^2)^{1/2}\right]}{\lambda_L(h^2+x'^2)} \right.$$

$$\left. - \frac{K_1\left[\lambda_L(h^2+x'^2)^{1/2}\right]}{(h^2+x'^2)^{1/2}} \right\} dx'. \quad (5.35)$$

This formula for $R_{-b,b}$ is the analogue of that for the centrally placed disk as given in equation 5.28a.

In practice, $\lambda_L h$ is large whereas $(b/h)^2 \ll 1$. We may therefore replace the K functions by their asymptotic representations for very large arguments, namely, $K_\nu(s) = (\pi/2s)^{1/2}\exp-s$. Moreover $x'^2 \ll h^2$ within the range of integration, and, except in the exponent, $h^2 + x'^2$ may be replaced by h^2. Equation 5.35 then simplifies to

$$R_{-b,b} \cong \left(\frac{2\lambda_L}{\pi h}\right)^{1/2}\left(1 - \frac{1}{\lambda_L h}\right)\int_0^{b'}\exp\left\{\lambda_L\left[h-(h^2+x'^2)^{1/2}\right]\right\}dx'$$

$$\cong \left(1 - \frac{1}{\lambda_L h}\right)\left(\frac{2\lambda}{\pi h}\right)^{1/2}\int_0^b\left[\exp-\left(\frac{\lambda x^2}{2h}\right)\right]dx$$

$$= \left(1 - \frac{1}{\lambda_L h}\right)\frac{2}{\sqrt{\pi}}\int_0^{b(\lambda/2h)^{1/2}}(\exp-s^2)\,ds.$$

Thus, when $\lambda_L h \gg 1$,

$$R_{-b,b} \cong \text{erf}\left[b\left(\frac{\lambda}{2h}\right)^{1/2}\right] = \text{erf}\left[\left(\frac{b}{h}\right)\left(\frac{\lambda h}{2}\right)^{1/2}\right]. \quad (5.36)$$

As in the limiting form for R in equation 5.30, the measured parameter is λ rather than λ_L. The reason in each case is that, near the central region of the uninterrupted stream, transport from longitudinal diffusion is unimportant in comparison with that due to drift W, when $\lambda_L h$ is large.

5.4.   DISTRIBUTION OF NUMBER DENSITY $n(\mathbf{r}, t)$ IN THE PRESENCE OF A SLIGHT DEGREE OF IONIZATION BY COLLISION AND OF LOSS OF ELECTRONS BY ATTACHMENT TO MOLECULES

5.4.1.   EQUATION FOR $n(\mathbf{r}, t)$. Although an extensive discussion of ionization by collision does not fall within our terms of reference, it is relevant to consider the effects of a slight degree of ionization on the measured values of drift velocity by time-of-flight methods and of D/μ by the method of the diffusing stream. It is also important to be able to assess the consequences of loss of electrons from a group or stream from attachment to molecules to form negative ions. In what follows we modify the theory of the travelling group and of the steady stream to incorporate the influences of these processes.

We stress at the outset that we envisage experimental conditions in which the pressure of the gas is not small but of the order 1 to 10 torr or greater. Since the collision frequency of an electron in a gas at a pressure of 1 torr is of the order 10^9 sec^{-1}, it follows that we are concerned with conditions in which the collision frequency is of the order 10^{10} sec^{-1}.

We consider first the increase in the population of a travelling group of n_0 electrons brought about by the process of ionization by collision. We suppose that the group has attained a steady state of motion, for which the velocity distribution function is $f_0^*(c)$.

An electron cannot ionize a molecule unless its energy, $\epsilon = \frac{1}{2}mc^2$, exceeds the critical energy for ionization, $\epsilon_i = \frac{1}{2}mc_i^2$. The number of electrons of the group n_0 whose velocity points lie within the shell (c, dc) is $n_0 4\pi f_0^*(c)c^2\, dc$. Let the number of molecules ionized in time dt by this subgroup of electrons be $dt\, \nu_i(c)n_0 4\pi f_0^*(c)c^2\, dc$ so that $\nu_i(c)$ is zero when $c < c_i$. The total number of molecules ionized in time dt by the whole group of n_0 electrons is

$$\bar{\nu}_i n_0\, dt, \quad \text{where} \quad \bar{\nu}_i = 4\pi \int_0^\infty \nu_i(c)f_0^*(c)c^2\, dc.$$

The rate of growth of the population $n_0(t)$ is therefore

$$\frac{d}{dt}n_0(t) = \bar{\nu}_i n_0(t),$$

that is to say,

$$n_0(t) = n_0(0)\exp \bar{\nu}_i t,$$

$$\left.\right\} \quad (5.37)$$

where $n_0(0)$ is the population of the group at time $t = 0$.

Similarly, when electrons are lost to the subgroup through the process of attachment to molecules, the number lost in time dt is $dt\,\nu_{at}(c)n_0 4\pi f_0^*(c)\,c^2\,dc$; and the rate of loss to the whole group when attachment alone is operative is

$$\frac{d}{dt}n_0(t) = -\bar{\nu}_{at}n_0(t),$$

where

$$\bar{\nu}_{at} = 4\pi \int_0^\infty \nu_{at}(c)f_0^*(c)c^2\,dc.$$

It follows that

$$n_0(t) = n_0(0)\exp - \bar{\nu}_{at}t.$$

When both processes are active together, then

$$\left.\begin{array}{c} \dfrac{d}{dt}n_0(t) = (\bar{\nu}_i - \bar{\nu}_{at})n_0(t), \\[2mm] n_0(t) = n_0(0)\exp(\bar{\nu}_i - \bar{\nu}_{at})t. \end{array}\right\} \qquad (5.38)$$

Next we enquire whether the addition of electrons through ionization produces a significant effect on the transport coefficients D, D_L, and W of the group.

We first examine approximately the time rate of change of the mean energy of a group Δn of electrons that are not moving in a steady state of motion in a uniform and constant electric field. For simplicity we shall assume that the collision frequency ν_{el} is constant. If $\bar{\epsilon}(t)$ is the mean energy of an electron of the group at time t, and $\bar{\epsilon}(t) \gg \frac{3}{2}\kappa T$, then

$$\frac{d}{dt}\bar{\epsilon}(t) = eEW - \frac{2m}{M}\nu_{el}\bar{\epsilon}(t),$$

in which it is assumed that all encounters are elastic so that the mean energy lost in an encounter is $(2m/M)\bar{\epsilon}(t)$. It follows that

$$\bar{\epsilon}(t) = \frac{eEWM}{2m\nu_{el}}\left[1 - \exp\left(-\frac{t}{\tau}\right)\right] + \bar{\epsilon}(0)\exp\left(-\frac{t}{\tau}\right)$$

where the time constant $\tau = M/2m\nu_{el}$.

Since $W = eE/m\nu_{el}$ when ν_{el} is constant, it follows that †

$$\bar{\epsilon}(t) = \tfrac{1}{2}MW^2\left[1 - \exp\left(\frac{-t}{\tau}\right)\right] + \bar{\epsilon}(0)\exp\left(-\frac{t}{\tau}\right).$$

At a pressure of 1 torr the value of ν_{el} is approximately 10^9 sec^{-1}, and, for example, in helium $2m/M \sim 2 \times 10^{-4}$; consequently at a pressure of 10 torr the time constant $\tau = M/2m\nu_{el} \sim \tfrac{1}{2} \times 10^{-6}$ sec. The time constant τ is of the order of magnitude of the time required for a group Δn of electrons with small energies to attain a steady state of motion in the field E with a velocity distribution function $f_0^*(c)$. In diatomic and polyatomic gases, in which the mean energy lost in an encounter greatly exceeds $(2m/M)\bar{\epsilon}$, the time constant τ is much smaller than in helium.

It follows that at any time t in a travelling group all the electrons produced by ionization are moving in equilibrium with the same distribution function $f_0^*(c)$ as the original $n_0(0)$ electrons of the group, with the exception of the electrons that are associated with ionizing encounters in the short time interval $(t - \tau)$ to t. We now show that the relative number of such electrons is small.

From equation 5.37 it follows that, if we define α_i as $\alpha_i = \bar{\nu}_i/W$, the rate of growth of the group n_0 in terms of the distance $\bar{z}$ travelled by its centroid is

$$n_0(\bar{z}) = n_0(0)\exp\alpha_i\bar{z},$$

where α_i is the volume coefficient of ionization, which represents the average number of molecules ionized by an electron in drifting 1 cm in the direction of the electric force eE.

Consider a situation in which E/N is adjusted so that the population of a group is doubled when its centroid drifts 10 cm through helium at a pressure of 10 torr. It follows that

$$\frac{n_0(\bar{z})}{n_0(0)} = 2 = \exp 10\alpha_i$$

whence

$$\alpha_i = \frac{\ln 2}{10} = 0.07.$$

†We note that the steady-state mean energy, when $k = \tfrac{1}{2}m\overline{c^2}/\tfrac{1}{2}M\overline{C^2} \gg 1$, is $\tfrac{1}{2}MW^2$, in agreement with Pidduck's formula (equation 3.16).

The drift velocities in helium are of the order of magnitude 10^6 cm sec^{-1}; consequently $\bar{v}_i = W\alpha_i \sim 7 \times 10^4 \sim 10^5$ sec^{-1}. Thus, in the time $\tau = 5 \times 10^{-7}$ sec, the approximate equilibration time constant for helium at 10 torr, the number of electrons produced by ionization is

$$\bar{v}_i \tau n_0(t) \sim 10^5 \times 5 \times 10^{-7} n_0(t) \sim 5 \times 10^{-2} n_0(t).$$

At a pressure of 100 torr, τ is reduced by a factor of 10, so that if E/N is reduced to maintain the same multiplication factor, the number of new electrons produced within the group in the equilibration time is only about $\frac{1}{2}\%$ of its population. Thus, unless the pressure is small and α_i relatively large, the distribution function is imperceptibly changed from $f_0^*(c)$ by the presence of a moderate degree of ionization due to collision. Similarly, when electrons are lost by attachment from the group $n_0(t)$ the velocity distribution function acquires some form $f_0^*(c)$ appropriate to a steady state of motion. Thus in the presence of ionization by collision and attachment the scalar equation for the group, whose velocity distribution function has become independent of time, is

$$f_0^* \frac{d}{dt} n_0(t) + (\bar{v}_{at} - \bar{v}_i) n_0(t) f_0^* + \frac{1}{4\pi c^2} \frac{d}{dc} [\sigma_E^*(c) - \sigma_{coll}^*(c)] = 0.$$

When this equation for the time dependence of the population of a shell (c, dc) is integrated over all shells, it reduces to equation 5.38.

In order to determine $f_0^*(c)$, we replace $n_0(t)$ in this time-dependent equation by $n_0(t) = n_0(0) \exp - (\bar{v}_{at} - \bar{v}_i)t$, with the result that the equation reduces to

$$\sigma_E^*(c) - \sigma_{coll}^*(c) = 0.$$

On removal of the common factor $n_0(0)$ a differential equation remains from which to determine $f_0^*(c)$.

When the number n_0 for the group as a whole is replaced by $n \, d\mathbf{r}$, the population of the element of volume $d\mathbf{r}$, then $f_0^*(c)$ must be replaced by $f_0(c, \mathbf{r}, t) \equiv f_0$. The increase in time dt in the population of the subgroup $n \, d\mathbf{r} \, 4\pi f_0 c^2 \, dc$ now becomes $dt \, n \, d\mathbf{r}[v_i(c) - v_{at}(c)] 4\pi f_0 c^2 \, dc$. It can be seen that the scalar equation in the presence of ionization by collision and attachment is modified to take the following form:

$$\frac{\partial}{\partial t}(nf_0) + [v_{at}(c) - v_i(c)](nf_0) + \frac{c}{3} \text{div}_r(n\mathbf{f}_1)$$

$$+ \frac{1}{4\pi c^2} \frac{\partial}{\partial c}[\sigma_E(c) - \sigma_{coll}(c)] = 0. \qquad (5.39)$$

Since, in a state of quasi equilibrium of the momentum, the vector equation is

$$n\mathbf{f}_1 = -\left[\frac{c}{v}\,\text{grad}_r(nf_0) + \frac{e\mathbf{E}}{mv}\frac{\partial}{\partial c}(nf_0)\right], \tag{5.40}$$

the scalar equation becomes

$$\frac{\partial}{\partial t}(nf_0) + [\nu_{\text{at}}(c) - \nu_i(c)](nf_0) - \text{div}_r\left[\frac{c^2}{3v}\,\text{grad}_r(nf_0) + \frac{ce\mathbf{E}}{3mv}\frac{\partial}{\partial c}(nf_0)\right]$$

$$+ \frac{1}{4\pi c^2}\frac{\partial}{\partial c}[\sigma_E(c) - \sigma_{\text{coll}}(c)] = 0. \tag{5.41}$$

This equation, when treated by the method described in Chapter 4, leads to the following equation to be satisfied by $n(\mathbf{r}, t)$:

$$-\frac{\partial}{\partial t}n + [\bar{\nu}_i - \bar{\nu}_{\text{at}}]n + D\left(\frac{\partial^2}{\partial x^2}n + \frac{\partial^2}{\partial y^2}n\right) + D_L\frac{\partial^2}{\partial z^2}n - \mathbf{W}\cdot\text{grad}_r(n) = 0.$$

$$\tag{5.42}$$

This equation, but with $D_L = D$, is that which has been previously used as the basis of theoretical discussions of the steady diffusing stream, the influence of diffusion on the Townsend discharge, and the growth of a travelling group, when either ionization or attachment is present, or both are (Huxley, 1959, 1968; Burch and Huxley, 1967). The consequences of the previously adopted assumption of isotropic diffusion will be discussed in Section 5.4.3.

We now consider some applications of equation 5.42. It is supposed throughout in what follows that $\mathbf{W}$ is parallel to $+0z$ so that $\mathbf{W}\cdot\text{grad}_r(n) \equiv W(\partial/\partial z)n$.

5.4.2.   THE ISOLATED TRAVELLING GROUP. We employ equation 5.42 in the form

$$-\frac{\partial}{\partial t}n + (\bar{\nu}_i - \bar{\nu}_{\text{at}})n + D\left(\frac{\partial^2}{\partial x^2}n + \frac{\partial^2}{\partial y^2}n\right) + D_L\frac{\partial^2}{\partial z^2}n - W\frac{\partial}{\partial z}n = 0. \tag{5.43}$$

Let $n = U(x, y, z, t)\exp[(\bar{\nu}_i - \bar{\nu}_{\text{at}})t]$; then U satisfies the equation

$$-\frac{\partial}{\partial t}U + D\left(\frac{\partial^2}{\partial x^2}U + \frac{\partial^2}{\partial y^2}U\right) + D_L\frac{\partial^2}{\partial z^2}U - W\frac{\partial}{\partial z}U = 0. \tag{5.44}$$

The solution of equation 5.44 appropriate to a travelling group was given in equation 5.1a. It is

$$U = \frac{A}{(4\pi Dt)(4\pi D_L t)^{1/2}} \left[\exp\left(-\frac{\rho^2}{4Dt} \right) \right] \left\{ \exp\left[-\frac{(z-Wt)^2}{4D_L t} \right] \right\}.$$

The appropriate solution of equation 5.43 is therefore

$$n(\mathbf{r},t) = \frac{A \exp[(\bar{\nu}_i - \bar{\nu}_{at})t]}{(4\pi Dt)(4\pi D_L t)^{1/2}} \left[\exp\left(-\frac{\rho^2}{4Dt} \right) \right] \left\{ \exp\left[-\frac{(z-Wt)^2}{4D_L t} \right] \right\},$$

where $\rho^2 = x^2 + y^2$.

The constant A is determined from the condition (equation 5.38)

$$n_0(t) = \int\!\!\int\!\!\int_{-\infty}^{\infty} n(\mathbf{r},t)\, dx\, dy\, dz = n_0(0) \exp[(\bar{\nu}_i - \bar{\nu}_{at})t],$$

whence it is found that $A = n_0(0)$; consequently

$$n(\mathbf{r},t) = \frac{n_0(0) \exp[(\bar{\nu}_i - \bar{\nu}_{at})t]}{(4\pi Dt)(4\pi D_L t)^{1/2}} \left[\exp\left(-\frac{\rho^2}{4Dt} \right) \right] \left\{ \exp\left[-\frac{(z-Wt)^2}{4D_L t} \right] \right\}.$$

$$(5.45)$$

Since the group begins as a highly concentrated group at the origin at time $t = 0$, the distance travelled in time t by the centroid is $\bar{z} = Wt$. We may therefore express the growth factor $\exp[(\bar{\nu}_i - \bar{\nu}_{at})t]$ in terms of the distance travelled by the centroid in the direction of the electric force $e\mathbf{E}$ to obtain the following equation for n:

$$n = \frac{n_0(0) \exp[(\alpha_i - \alpha_a)\bar{z}]}{(4\pi D\bar{z}/W)(4\pi D_L \bar{z}/W)^{1/2}} \left[\exp\left(-\frac{\rho^2 W}{4D\bar{z}} \right) \right] \left\{ \exp\left[-\frac{(z-\bar{z})^2 W}{4D_L \bar{z}} \right] \right\}$$

$$(5.46)$$

where

$$\alpha_i = \frac{\bar{\nu}_i}{W} \quad \text{and} \quad \alpha_a = \frac{\bar{\nu}_{at}}{W}.$$

The coefficient α_i is the volume coefficient of ionization, and α_a is the volume coefficient of attachment. Both coefficients α_i/N and α_a/N are functions of E/N. However, when three-body encounters result in attachment, α_a/N^2 is a function of E/N.

It can be seen from equations 5.45 and 5.46 that neither the position of the centroid nor the relative distribution of number density $n(\mathbf{r}, t)$ has been affected by the presence of ionization and attachment. Consequently time-of-flight methods can be used to measure W in the presence of ionization by collision.

5.4.3. DISTRIBUTION OF NUMBER DENSITY IN A STREAM. Let $(\partial/\partial t)n = 0$ in equation 5.43, which then becomes

$$D\left(\frac{\partial^2}{\partial x^2}n + \frac{\partial^2}{\partial y^2}n\right) + D_L\frac{\partial^2}{\partial z^2}n - W\frac{\partial}{\partial z}n - (\alpha_a - \alpha_i)Wn = 0.$$

This equation is equivalent to

$$\frac{\partial^2}{\partial x'^2}n + \frac{\partial^2}{\partial y'^2}n + \frac{\partial^2}{\partial z^2}n - 2\lambda_L\frac{\partial}{\partial z}n - 2\lambda_L\alpha n = 0, \tag{5.47}$$

where $x' = (D_L/D)^{1/2}x$, $y' = (D_L/D)^{1/2}y$, $\alpha = \alpha_a - \alpha_i$, and $2\lambda_L = W/D_L$.

Let $n = V\exp\lambda_L z$; then from equation 5.47 it follows that

$$\nabla'^2 V = (\lambda_L^2 + 2\lambda_L\alpha)V = \eta^2 V \tag{5.48a}$$

where

$$\eta^2 = \lambda_L^2 + 2\lambda_L\alpha. \tag{5.48b}$$

Equation 5.48a is different from equation 5.13 only in the replacement of λ_L^2 by η^2, and it follows that the solutions appropriate to pole and dipole sources, respectively, are, from equations 5.23 and 5.25,

$$V = A_0\frac{\exp - \eta r'}{r'} \quad \text{(pole term)}$$

and

$$V = -A_1\frac{\partial}{\partial z}\left(\frac{\exp - \eta r'}{r'}\right) \quad \text{(dipole term)}$$

with $A_0 = i_0 / [2\pi e D(\lambda_L + \eta)/\eta]$ and $A_1 = i_0 / 2\pi e D(\lambda_L + \eta)$, where i_0 is the total current from the source.

It follows that the corresponding solutions of equation 5.47 are as follows:

$$n = A_0(\exp\lambda_L z)\frac{\exp - \eta r'}{r'} \qquad \text{(pole term)} \qquad (5.49a)$$

and

$$n = -A_1(\exp\lambda_L z)\frac{\partial}{\partial z}\left(\frac{\exp - \eta r'}{r'}\right) \qquad \text{(dipole term)}. \qquad (5.49b)$$

It is found that the equation analogous to equation 5.27 is

$$\left.\begin{array}{l}\text{(dipole)} \quad i_{0,b} = i\left\{1 - \left[\frac{h}{d'} - \frac{1}{\eta h}\left(1 - \frac{h^2}{d'^2}\right)\right]\frac{h}{d'}\exp - \eta(d' - h)\right\}, \\[4mm] \text{and the corresponding equation for the pole is} \\[4mm] \qquad\qquad \text{(pole)} \qquad i_{0,b} = i\left[1 - \frac{h}{d'}\exp - \eta(d' - h)\right],\end{array}\right\} \ (5.50)$$

where

$$d'^2 = h^2 + b'^2 = h^2 + \frac{D_L}{D}b^2 = h^2 + \frac{\lambda}{\lambda_L}b^2$$

and

$$i = i_0\exp - (\eta - \lambda_L)h, \qquad (5.51)$$

which (since $\eta^2 = \lambda_L{}^2 + 2\lambda_L\alpha$) is equivalent to (Huxley, 1959)

$$i = i_0\exp\left[-\left(\frac{2\lambda_L\alpha}{\eta + \lambda_L}\right)h\right]. \qquad (5.52)$$

Ionization without attachment. Let $\alpha_a = 0$; then equation 5.52 becomes

$$\left.\begin{array}{l}\qquad\qquad i = i_0\exp\left[\left(\frac{2\lambda_L\alpha_i}{\eta + \lambda_L}\right)h\right], \\[4mm] \text{which is the same as Townsend's formula} \\[4mm] \qquad\qquad\qquad i = i_0\exp\alpha_T h\end{array}\right\} \ (5.53)$$

for the current in terms of the electrode separation h when i_0 and E/N are maintained unchanged as h is varied.

The definition of the first Townsend ionizing coefficient α_T, as given by Townsend, neglected the influence of diffusion and is equivalent to the definition of α_i, that is to say, $\alpha_T = \alpha_i$. It can be seen, however, that, when diffusion is small, $\lambda_L = W/2D_L$ becomes very large and the coefficient of the exponent in equation 5.53 approaches α_i in value. This is evident because $\eta^2 = \lambda_L^2 - 2\lambda_L \alpha_i$; consequently, as λ_L increases, $\eta \to \lambda_L$ and $\alpha_T = 2\lambda_L \alpha_i/(\eta + \lambda_L) \to \alpha_i$.

The experimental quantity α_T is important not only in the interpretation of electrical breakdown in gases but also for calculating the mean rate of ionization per electron $\alpha_i W$. To a first approximation $\alpha_i = \alpha_T$, but a more accurate value is obtained through the relation $\alpha_T = 2\lambda_L \alpha_i/(\eta + \lambda_L)$, that is, $\alpha_i = (\eta + \lambda_L)\alpha_T/2\lambda_L$. This matter has been discussed by Huxley (1959), Crompton (1967), Burch and Huxley (1967), and Huxley (1968).

From equation 5.50 the proportion of the total current that is received by the central disk with radius b can be shown to be

$$R = \frac{i_{0,b}}{i} = 1 - \left[\frac{h}{d'} - \frac{1}{\eta h}\left(1 - \frac{h^2}{d'^2}\right) \right] \frac{h}{d'} \exp - \eta(d' - h). \quad (5.54)$$

If this expression is simplified to assume a form analogous to equation 5.29, we have

$$\eta(d' - h) = (\lambda_L - \alpha_T)(d' - h)$$

$$\cong (\lambda_L - \alpha_T)\frac{\lambda}{\lambda_L}(d - h) \qquad [\text{since } \lambda_L(d' - h) \cong \lambda(d - h)]$$

$$= \lambda\left(1 - \frac{\alpha_T}{\lambda_L}\right)(d - h).$$

Thus

$$R \cong 1 - \left[1 + \left(\frac{1}{2} - \frac{D_L}{D}\right)\left(\frac{b}{d}\right)^2\right] \frac{h}{d} \exp\left[-\lambda\left(1 - \frac{\alpha_T}{\lambda_L}\right)(d - h)\right]. \quad (5.55)$$

It is possible to derive $\lambda = W/2D$ from the measured values of R and α_T. The term $\lambda\alpha_T/\lambda_L$ is small and has the nature of a correction to λ, derived on the basis of equation 5.29 when ionization by collision is present to a small degree.

Attachment without ionization. Put $\alpha_i = 0$; then $\eta^2 = \lambda_L^2 + 2\lambda_L \alpha_a$, $\eta - \lambda_L = 2\alpha_a \lambda_L / (\eta + \lambda_L)$. Equation 5.52 then assumes the form

$$i = i_0 \exp - \left[\left(\frac{2\lambda_L \alpha_a}{\eta + \lambda_L} \right) h \right]$$

$$= i_0 \exp - \alpha_{at} h, \tag{5.56}$$

and the current carried to the central disk with radius b by electrons is

$$i_{0,b} = R_{0,b} i = i \left\{ 1 - \left[\frac{h}{d_b'} - \frac{1}{\eta h} \left(1 - \frac{h^2}{d_b'^2} \right) \right] \frac{h}{d_b'} \exp - \eta (d_b' - h) \right\}, \tag{5.57}$$

where $d_b'^2 = h^2 + b'^2 = h^2 + (D_L / D) b^2$. This expression for $i_{0,b}$ could be expressed in an approximate form in terms of λ, α_{at} and the dimensions of the apparatus, using the procedure employed to transform equation 5.54 to 5.55. However, because it is sometimes necessary to use experimental conditions which may invalidate the approximations employed to derive equation 5.55, the analysis which follows will be developed without this simplification.

When electron attachment occurs, negative ions are formed in the region above the source hole of the lateral diffusion apparatus, resulting in a mixture of unknown composition entering the diffusion chamber through the source. This difficulty is overcome experimentally by collecting all these ions on the central disk of a multiply divided anode, the disk remaining earthed throughout the measurement (see Section 11.4). It is therefore necessary to calculate the ratio of the current received by the annulus immediately surrounding the central disk to the total current arriving at the anode outside the disk, that is, the ratio $R = (i_{c,b} + I_{c,b}) / (i_{c,\infty} + I_{c,\infty})$, where i and I refer to electrons and ions, respectively, and c and b are the inner and outer radii of the annulus. We may rewrite R as

$$R = 1 - \frac{i_{b,\infty} + I_{b,\infty}}{i_{c,\infty} + I_{c,\infty}},$$

and therefore we require expressions for $i_{b,\infty}$, $i_{c,\infty}$, $I_{b,\infty}$, and $I_{c,\infty}$.

The current $i_{0,c}$ can be calculated from equation 5.57 after b has been replaced with c. The expressions for $i_{0,b}$ and $i_{0,c}$ can then be used together with equation 5.51 to obtain the following expressions for $i_{b,\infty}$ and $i_{c,\infty}$:

$$
\begin{aligned}
i_{b,\infty} = i - i_{0,b} &= \left\{ \left[\frac{h}{d'_b} - \frac{1}{\eta h}\left(1 - \frac{h^2}{d'^2_b}\right) \right] \frac{h}{d'_b} \exp - \eta(d'_b - h) \right\} i \\
&= \left\{ \left[\frac{h}{d'_b} - \frac{1}{\eta h}\left(1 - \frac{h^2}{d'^2_b}\right) \right] \frac{h}{d'_b} \exp - (\eta d'_b - \lambda_L h) \right\} i_0, \\
i_{c,\infty} &= \left\{ \left[\frac{h}{d'_c} - \frac{1}{\eta h}\left(1 - \frac{h^2}{d'^2_c}\right) \right] \frac{h}{d'_c} \exp - (\eta d'_c - \lambda_L h) \right\} i_0.
\end{aligned}
\tag{5.58}
$$

It remains to calculate the current carried by the negative ions formed by attachment between the cathode and the anode. Let n_I be the number density of the ions. The differential equation to be satisfied by $n_I(\mathbf{r})$ in the steady stream is†

$$
D_I \nabla^2 n_I = W_I \frac{\partial}{\partial z} n_I - \alpha_a W n, \tag{5.59}
$$

where D_I and W_I refer to the negative ions, and $n(\mathbf{r})$ is the electron number density. A solution of equation 5.59 that has been given by Hurst and Huxley (1960) is the following:

$$
n_I = \frac{\lambda \alpha_a}{\lambda_I - \lambda} \frac{D}{D_I}\left(\exp\frac{\lambda \alpha_a z}{\lambda_I - \lambda}\right) \int_{-\infty}^{z}\left[\exp-\left(\frac{\lambda \alpha_a}{\lambda_I - \lambda}\right)z'\right] n\, dz', \tag{5.60}
$$

where $\lambda_I = W_I/2D_I$. In practice, however, this expression for n_I is inconvenient, and we therefore employ an approximation. We take advantage of the fact that the ions move with thermal velocities, whereas the mean energy of the electrons is usually notably in excess of the thermal value. Thus it is commonly the case that $\lambda_I = W_I/2D_I$ greatly exceeds $\lambda = W/D$. The consequence is that the ions diffuse laterally to a very small extent in comparison with the electrons in travelling from their points of origin to the anode. A good approximation to $n_I(\mathbf{r})$ is therefore obtained by neglect-

†Since the ions are virtually in thermal equilibrium with the gas, diffusion is assumed to be isotropic.

ing the term $D_I \nabla^2 n_I$ in equation 5.59 and thus obtaining for n_I the expression

$$n_I(x,y,z) = \left(\alpha_a \frac{W}{W_I} \right) \int_0^z n \, dz'.$$

The current carried by negative ions to an element dS of the electrode is

$$dI = e n_I(x,y,h) W_I \, dS = e \alpha_a W \left(\int_0^h n \, dz' \right) dS.$$

But the expression for n which satisfies equation 5.42 (with $\partial n / \partial t$ and $\bar{v}_I$ equal to zero) and the boundary conditions is, from equations 5.26b and 5.49b, †

$$n = -\left(\frac{i_0}{4\pi e D \eta} \right) (\exp \lambda_L z) \frac{\partial}{\partial z} \left(\frac{\exp - \eta r'}{r'} + \frac{\exp - \eta r''}{r''} \right);$$

consequently

$$dI = -\left(\frac{i_0}{2\pi} \right) \left(\frac{\lambda}{\eta} \alpha_a \right) \left[\int_0^h (\exp \lambda_L z) \frac{\partial}{\partial z} \left(\frac{\exp - \eta r'}{r'} + \frac{\exp - \eta r''}{r''} \right) dz \right] dS.$$

It follows that the required ionic currents are

$$I_{b,\infty} = -i_0 \left(\frac{\lambda}{\eta} \alpha_a \right) \int_b^\infty \left[\int_0^h (\exp \lambda_L z) \frac{\partial}{\partial z} \left(\frac{\exp - \eta r'}{r'} + \frac{\exp - \eta r''}{r''} \right) dz \right] \rho \, d\rho$$

$$= -i_0 \left(\frac{\lambda_L}{\eta} \alpha_a \right) \int_{b'}^\infty \left[\int_0^h (\exp \lambda_L z) \frac{\partial}{\partial z} \left(\frac{\exp - \eta r'}{r'} + \frac{\exp - \eta r''}{r''} \right) dz \right] \rho' \, d\rho' \qquad \left. \begin{matrix} \\ \\ \\ \\ \\ \\ \\ \\ \\ \end{matrix} \right\} (5.61)$$

since $\rho'^2 = (\lambda / \lambda_L) \rho^2$, and

$$I_{c,\infty} = -i_0 \left(\frac{\lambda_L}{\eta} \alpha_a \right) \int_{c'}^\infty \left[\int_0^h (\exp \lambda_L z) \frac{\partial}{\partial z} \left(\frac{\exp - \eta r'}{r'} + \frac{\exp - \eta r''}{r''} \right) dz \right] \rho' \, d\rho'.$$

† When the anode boundary condition is met, as here, by the addition of the image term, the constant of proportionality in the equation analogous to equation 5.26b is found to be $i_0 / 4\pi e D \eta$.

Using equations 5.58 and 5.61, we obtain the following formula for the current ratio:

$$R = 1 - \frac{[(h/d'_b) - (1/\eta h)(1 - h^2/d'^2_b)](h/d'_b)[\exp - (\eta d'_b - \lambda_L h)] - (\lambda_L/\eta)\alpha_a \int_{b'}^{\infty}\left[\int_0^h (\exp \lambda_L z) V \, dz\right]\rho' \, d\rho'}{[(h/d'_c) - (1/\eta h)(1 - h^2/d'^2_c)](h/d'_c)[\exp - (\eta d'_c - \lambda_L h)] - (\lambda_L/\eta)\alpha_a \int_{c'}^{\infty}\left[\int_0^h (\exp \lambda_L z) V \, dz\right]\rho' \, d\rho'}$$

(5.62)

where

$$V = \frac{\partial}{\partial z}\left(\frac{\exp - \eta r'}{r'} + \frac{\exp - \eta r''}{r''}\right).$$

It can be seen that R is a function of the geometry of the apparatus and of the parameters λ, λ_L, and α_a. By making measurements of R with the same value of N and E/N, but with three different values of h, three equations relating R with λ, λ/λ_L, and α_a are obtained which, in principle, may be solved to give the three parameters. The details of the procedure used in actual practice will be described in Chapter 11.

A discussion of the principles of other methods for measuring attachment coefficients is postponed to Chapter 12 since these are comparatively straightforward.

We consider next the theoretical basis of a different type of experiment in which D is measured directly.

5.5. DIRECT MEASUREMENT OF THE COEFFICIENT OF DIFFUSION

Cavalleri (1969) has devised a direct method for measuring D at a known gas number density. In Cavalleri's method the gas in a cylindrical chamber with conducting walls and with insulated and circular electrodes that form the ends of the chamber is ionized by a flash of X-rays, and the electrons rapidly diffuse to fill the whole chamber. The intensity of the flash is such that the space charge from the presence of positive ions is negligible. The electron temperature is maintained above the thermal value to any desired level by a high-frequency field, and the essence of the method, details of which are given in Chapter 11, is the observation of the time constant of decay of electron number density brought about by diffusion to the walls. We therefore require a theory of the decay of the population of electrons in a gas contained in a cylindrical chamber when the mean energy of the electrons is maintained at a fixed value by a high-frequency electric field. For the sake of generality we shall suppose that small degrees of ionization by collision and attachment are present, and that a constant electric field is maintained between the electrodes, which adds to the loss by diffusion a loss due to drift to one of the electrodes.

We let the radius of the cylinder be a and its length h, and take the axis $+Oz$ as that of the cylinder. The origin is taken as the centre of one of the circular electrodes. We may therefore adopt equation 5.43 without modification. In equation 5.43 we let $n = U(x', y', z) \exp\{[(\bar{\nu}_i - \bar{\nu}_{at}) - \beta^2 D_L] t + \lambda_L z\}$, where $\bar{\nu}_i = W\alpha_i$, $\bar{\nu}_{at} = W\alpha_a$, $2\lambda_L = W/D_L$, and β^2 remains to be determined. Then

$$\frac{\partial^2}{\partial x'^2} U + \frac{\partial^2}{\partial y'^2} U + \frac{\partial^2}{\partial z^2} U = -k^2 U, \qquad (5.63)$$

where $x' = (D_L/D)^{1/2} x$, $y' = (D_L/D)^{1/2} y$, and $k^2 = \beta^2 - \lambda_L^2$.

When expressed in cylindrical coordinates (ρ', θ, z), where $\rho' = (x'^2 + y'^2)^{1/2}$, equation 5.63 takes the form

$$\frac{\partial^2 U}{\partial \rho'^2} + \frac{1}{\rho'} \frac{\partial U}{\partial \rho'} + \frac{1}{\rho'^2} \frac{\partial^2 U}{\partial \theta^2} + \frac{\partial^2 U}{\partial z^2} = -k^2 U.$$

We assume axial symmetry; consequently $\partial^2 U / \partial \theta^2 = 0$, and

$$\frac{\partial^2 U}{\partial \rho'^2} + \frac{1}{\rho'} \frac{\partial U}{\partial \rho'} + \frac{\partial^2 U}{\partial z^2} = -k^2 U. \qquad (5.64)$$

Let $U = R(\rho')Z(z)$; then

$$\frac{d^2 Z}{dz^2} = -\gamma^2 Z, \qquad \frac{d^2 R}{d\rho'^2} + \frac{1}{\rho'} \frac{dR}{d\rho'} + (k^2 - \gamma^2) R = 0, \qquad (5.65)$$

where γ^2 is the constant of separation.

It follows from equation 5.65 that

$$Z \propto \begin{bmatrix} \cos \\ \sin \end{bmatrix} \gamma z \qquad \text{and} \qquad R \propto J_0 \left[(k^2 - \gamma^2)^{1/2} \rho' \right].$$

The boundary conditions are as follows: $n=0$ when $z=0$ and $z=h$ for all values of ρ', and $n=0$ when $\rho'=a'$ for all values of z, where $a' =(D_L/D)^{1/2}a$. Consequently

$$Z \propto \sin\frac{m\pi}{h}z, \quad m=1,2,3\ldots; \qquad \gamma=\frac{m\pi}{h},$$

$$R \propto J_0\left(\frac{c_l\rho'}{a'}\right), \quad \text{where } c_l \text{ is the } l\text{th root of } J_0(x)=0.$$

It follows that

$$k^2-\gamma^2=k^2-\left(\frac{m\pi}{h}\right)^2=\left(\frac{c_l}{a'}\right)^2,$$

and, since $k^2=\beta^2-\lambda_L^2$,

$$\beta^2=\left(\frac{m\pi}{h}\right)^2+\lambda_L^2+\left(\frac{c_l}{a'}\right)^2.$$

A general form of solution appropriate to the present application is

$$n(\rho',z,t)=(\exp\lambda_L z)[\exp(\bar{v}_i-\bar{v}_{\mathrm{at}})t]$$

$$\times \sum_{m=1}^{\infty}\sum_{k=1}^{\infty} A_{ml}\left\{\exp-D_L\left[\left(\frac{m\pi}{h}\right)^2+\lambda_L^2+\left(\frac{c_l}{a'}\right)^2\right]t\right\}J_0\left(\frac{c_l\rho'}{a'}\right)\sin\frac{m\pi}{h}z.$$

$$(5.66)$$

The coefficients A_{ml} are to be determined in terms of $n(\rho',z,t)$ at $t=0$, but because this is unknown we must be content with a relative rather than a specific solution. We note that the effective coefficients are

$$A_{ml}\exp-D_L\left[\left(\frac{m\pi}{h}\right)^2+\lambda_L^2+\left(\frac{c_l}{a'}\right)^2\right]t$$

and that the decay factor with $m=1$, $l=1$ diminishes much more slowly than the decay factors associated with higher modes; consequently the number density rapidly attains a distribution that is closely represented by the lowest mode for which $m = l = 1$. When this condition is attained, then

$$n(\rho',z,t) \cong A_{11}[\exp(\bar{\nu}_i - \bar{\nu}_{at})t]\left\{\exp - D_L\left[\left(\frac{\pi}{h}\right)^2 + \lambda_L^2 + \left(\frac{c_1}{a'}\right)^2\right]t\right\}$$

$$\times (\exp \lambda_L z) J_0\left(\frac{c_1 \rho'}{a'}\right)\sin\frac{\pi z}{h}.$$

We consider the special case in which $\bar{\nu}_i$ and $\bar{\nu}_{at}$ are zero and also $W = 0$, so that $D_L = D$, $\lambda_L = 0$, $\rho' = \rho$, and $a' = a$. Then

$$n(\rho,z,t) \cong A_{11}\left[\exp\left(-\frac{t}{\tau_{11}}\right)\right]J_0\left(\frac{c_1 \rho}{a}\right)\sin\frac{\pi z}{h}, \qquad (5.67)$$

where $c_1 = 2.405$ and the time constant of decay is

$$\tau_{11} = \frac{1}{D\left[(\pi/h)^2 + (c_1/a)^2\right]}.$$

It follows that D can be deduced immediately from the measured value of τ_{11} in the absence of ionization, attachment, and a steady drift W.

BIBLIOGRAPHY AND NOTES

TRAVELLING GROUPS AND STEADY STREAMS

L. G. H. Huxley, *Aust. J. Phys.*, **25**, 523, 1972.

THEORY OF CORRECTIONS TO MEASUREMENTS OF W BY TIME-OF-FLIGHT METHODS

R. A. Duncan, *Aust. J. Phys.*, **10**, 54, 1957.
J. J. Lowke, *Aust. J. Phys.*, **15**, 39, 1962.
The discussion in Section 5.2 carries the subject forward from the findings of these authors.

STRUCTURE OF A STREAM FROM A SMALL SOURCE

The theory of the stream from a point source and its practical importance were first given in 1940 by Huxley, who proposed the formula for the current ratio R given in equation 5.28b with $d' = d$ since the distinction between λ and λ_L was unknown at that time. This formula was erroneously associated with a dipole source, and the correct interpretation was given by Huxley and Crompton in 1955. An extension of the theory to include the effects of a slight degree of ionization by collision and loss of electrons by attachment to neutral molecules was given by Huxley (1959) and a more precise theory by Hurst and Huxley. More recently the topic has been considered by Lowke and Parker and by Lowke.

L. G. H. Huxley, *Phil. Mag.*, **30**, 396, 1940.

L. G. H. Huxley and R. W. Crompton, *Proc. Phys. Soc. B*, **68**, 381, 1955.

J. J. Lowke and J. H. Parker, *Phys. Rev.*, **181**, 302, 1969.

J. J. Lowke, *Proceedings of the Tenth International Conference on Phenomena in Ionized Gases*, Oxford, 1971.

L. G. H. Huxley, *Aust. J. Phys.*, **25**, 43, 1972.

EXPERIMENTAL TESTS OF THE THEORY OF THE SPREADING STREAM OF ELECTRONS

R. W. Crompton and R. L. Jory, *Aust. J. Phys.*, **15**, 451, 1962.

R. W. Crompton, M. T. Elford, and J. Gascoigne, *Aust. J. Phys.*, **18**, 409, 1965.

THEORY OF STREAM IN PRESENCE OF IONIZATION BY COLLISION AND ATTACHMENT

L. G. H. Huxley, *Aust. J. Phys.*, **12**, 171, 1959.

C. A. Hurst and L. G. H. Huxley, *Aust. J. Phys.*, **13**, 21, 1960.

L. G. H. Huxley, *Aust. J. Phys.*, **21**, 761, 1968.

D. S. Burch and L. G. H. Huxley, *Aust. J. Phys.*, **20**, 625, 1967.

R. W. Crompton, *J. Appl. Phys.*, **38**, 4093, 1967.

DIRECT MEASUREMENT OF THE COEFFICIENT OF DIFFUSION

G. Cavalleri, *Phys. Rev.*, **179**, 186, 1969.

6

THE INFLUENCE OF INELASTIC

ENCOUNTERS

6.1. INTRODUCTION

Although the contributions of inelastic encounters are not ignored in the preceding chapters, references to them are made in general rather than in specific terms in contrast to the rôle of elastic encounters, which has received detailed discussion.

The presence of inelastic encounters requires not that the form of the scalar equation (equation 3.3) or of the vector equation (equation 3.8) should be changed, but only that modifications should be made to their constituent terms. There are two such modifications.

1. The expression 3.4 that represents the contribution of elastic encounters to $\sigma_{coll}(c)$ is now supplemented by terms that are the contributions from inelastic or superelastic encounters.

2. The momentum transfer cross section $q_m(c)$ is no longer that for elastic encounters but is a composite cross section representing loss of momentum both in elastic encounters and in encounters that are not elastic. This generalization of $q_m(c)$ is indicated in equations 3.8 and 3.9 by the use of an effective collision frequency ν. In practice it is usually assumed that the effective cross section is little different from the momentum transfer cross section $q_{m_{el}}(c)$ for elastic encounters since in many useful circumstances this proves to be much larger than the associated inelastic cross sections.

An inelastic encounter is one in which the internal energy state of the molecule is changed from one of smaller to one of greater energy. The energy of the impacting electron must therefore exceed a threshold value equal to the difference in the internal energies of the molecule associated with its energy states before and after the encounter. In molecular gases

163

electrons may bring about by impact changes in the states of rotational energy, of vibrational energy, and of the energy of electron configuration in the molecule. In monatomic gases the only inelastic encounters are those which effect a change in the electron configuration.

At room temperatures not all of the molecules of molecular gases are in the ground state of rotational energy, and superelastic as well as elastic and inelastic encounters occur. In general, inelastic encounters of many kinds contribute to σ_{coll}, but in order to avoid the complexities of a complete treatment we shall in the first instance consider a hypothetical model gas whose molecules possess a single energy state other than the ground state and therefore a single threshold energy for an inelastic encounter. We shall also assume at first that all molecules are in the ground state so that superelastic encounters are absent. When the threshold energy is relatively large, this model gas is a good representation of a monatomic gas when there are sufficient electrons with energies above the threshold to bring about transitions from the ground state to the next state in the electronic configuration, but no other transitions. When the transition energy is small, there are exceptional cases only in which the model represents an actual gas. One example is parahydrogen at reduced temperatures, where single transitions in the rotational energy from the ground state $(J=0, v=0)$ to $(J=2, v=0)$ can occur over a limited range of the electron mean energy. Here J is the rotational and v the vibrational quantum number. The threshold energy for this transition is 0.044 eV.

We proceed to derive an expression for σ_{coll} in the model gas.

6.2. EXPRESSIONS FOR $\sigma_{coll}(c)$ AND $f_0(c)$ IN THE MODEL GAS

In order to avoid the need to consider the effects of diffusion we consider that $\sigma_{coll}(c)$ and $f_0(c)$ relate to a whole group of n_0 electrons as in Section 4.2, but here for convenience the asterisk is omitted. Consequently, unless otherwise stated, in this chapter $\sigma_{coll}(c)$ is the same as $\sigma^*_{coll}(c)$ and $f_0(c)$ and $\mathbf{f}_1(c)$ have the meanings of $f_0^*(c)$ and $\mathbf{f}_1^*(c)$ of Section 4.2.

The velocity of the centroid of the shell (c, dc) of the group is now

$$\mathbf{W}(c) = \frac{c}{3} \frac{\mathbf{f}_1}{f_0}$$

whereas that of the whole group is

$$\mathbf{W} = \frac{4\pi}{3} \int_0^\infty c^3 \mathbf{f}_1 \, dc. \tag{6.1}$$

When $f_0(c)$ for the group as a whole has reached its equilibrium form, it follows from the first paragraph of Section 4.3 that whether or not inelastic encounters are present

$$\sigma_E(c) = \sigma_{\text{coll}}(c). \tag{6.2}$$

We therefore first seek an expression for $\sigma_{\text{coll}}(c)$ when inelastic encounters are present, since $\sigma_E(c)$ is still given by

$$\sigma_E(c) = \frac{4\pi}{3} n_0 c^2 \frac{e\mathbf{E}}{m} \cdot \mathbf{f}_1. \tag{6.3}$$

The contribution to $\sigma_{\text{coll}}(c)$ from elastic encounters is, as before,

$$\sigma_{\text{coll}_{\text{el}}}(c) = 4\pi n_0 c^2 \nu_{\text{el}} \left(\frac{m}{M} c f_0 + \frac{\overline{C^2}}{3} \frac{d}{dc} f_0 \right), \tag{6.4}$$

where $\nu_{\text{el}} = N c q_{m_{\text{el}}}(c)$ with $q_{m_{\text{el}}}(c)$ the elastic cross section for momentum transfer.

We require an additional term representing the contribution of inelastic encounters in the model gas. Let ϵ_{in} be the threshold energy for inelastic encounters in the model gas, and $q_{0_{\text{in}}}(c)$ be the total inelastic collision cross section. The speed of an electron with energy ϵ_{in} is

$$c_{\text{in}} = \left(\frac{2\epsilon_{\text{in}}}{m} \right)^{1/2}. \tag{6.5}$$

Since $q_{0_{\text{in}}}(c)$ is zero when the speed c of an electron is less than c_{in}, we express $q_{0_{\text{in}}}(c)$ in the form $q_{0_{\text{in}}}(c) I(c - c_{\text{in}})$, where I is the unit step function, and we shall use either form according to convenience. The number of inelastic encounters made in time dt by electrons whose velocity points lie in the shell (c, dc) is $(4\pi n_0 f_0 c^2 dc)[N c q_{0_{\text{in}}}(c)] dt$. The associated inelastic cross section for momentum transfer is, from equation 2.9,

$$q_{m_{\text{in}}}(g) = q_{0_{\text{in}}}(g) - \frac{g'}{g} q_{1_{\text{in}}}(g), \tag{6.6}$$

and the associated collision frequency is $\nu_{\text{in}} = N g q_{m_{\text{in}}}(g)$. In what follows we make the usual assumption that no significant error is made by replacing g with c.

Consider Figure 6.1, which represents spherical surfaces in velocity space. Let the speeds of an electron before and after an inelastic encounter

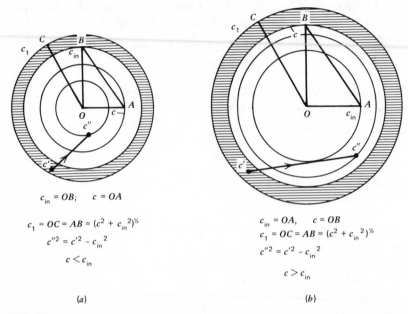

$c_{in} = OB;$ $c = OA$

$c_1 = OC = AB = (c^2 + c_{in}^2)^{1/2}$

$c''^2 = c'^2 - c_{in}^2$

$c < c_{in}$

(a)

$c_{in} = OA,$ $c = OB$

$c_1 = OC = AB = (c^2 + c_{in}^2)^{1/2}$

$c''^2 = c'^2 - c_{in}^2$

$c > c_{in}$

(b)

FIG. 6.1.

be c' and c''; then, if the change in the translational energy of the molecule is neglected, it follows that $\frac{1}{2}m(c'^2 - c''^2) = \epsilon_{in}$, that is,

$$c''^2 = c'^2 - \frac{2\epsilon_{in}}{m} = c'^2 - c_{in}^2, \qquad c' > c_{in}.$$

The necessary condition for a velocity point to move from a position outside to a position inside a sphere c is that $c'' \leqslant c \leqslant c'$. Two cases arise.

(a) $c < c_{in}$ (Figure 6.1a)

For a velocity point to be displaced from outside to inside the sphere c, its initial position must lie within the region bounded by the spheres with radii c_{in} and $c_1 = (c^2 + c_{in}^2)^{1/2}$, and it follows that the number of points that cross the surface c in time dt is

$$dt \cdot 4\pi n_0 N \int_{c_{in}}^{c_1} f_0(x) x^3 q_{0_{in}}(x)\, dx,$$

Consequently the contribution to $\sigma_{coll}(c)$ from inelastic encounters is, with case (a),

$$\sigma_{coll_{in}}(c) = 4\pi n_0 N \int_{c_{in}}^{c_1} f_0(x) x^3 q_{0_{in}}(x)\, dx. \qquad (6.7)$$

(b) $c > c_{in}$ (Figure 6.1b)
It can be seen that in this case the expression for $\sigma_{coll_{in}}(c)$ is

$$\sigma_{coll_{in}}(c) = 4\pi n_0 N \int_{c}^{c_1} f_0(x) x^3 q_{0_{in}}(x)\, dx. \qquad (6.8)$$

It follows from the form of $q_{0_{in}}(c)$ that equations 6.7 and 6.8 are included in the single expression

$$\sigma_{coll_{in}}(c) = 4\pi n_0 N \int_{c}^{c_1} f_0(x) x^3 q_{0_{in}}(x) I(x - c_{in})\, dx. \qquad (6.9)$$

The total flux is therefore

$$\sigma_{coll}(c) = \sigma_{coll_{el}}(c) + \sigma_{coll_{in}}(c)$$

$$= 4\pi n_0 N \left[c^3 q_{m_{el}}(c) \left(\frac{m}{M} c f_0 + \frac{\overline{C^2}}{3} \frac{d}{dc} f_0 \right) + \int_{c}^{c_1} f_0(x) x^3 q_{0_{in}}(x) I(x - c_{in})\, dx \right].$$
$$(6.10)$$

The differential equation for $f_0(c)$ (from equations 6.3 and 6.4) is, for the steady state,

$$-\frac{eE}{m\nu_{el}} \cdot \mathbf{f}_1 + \overline{C^2} \frac{df_0}{dc} + \frac{3m}{M} c f_0 + \frac{3}{c^3 q_{m_{el}}(c)} \int_{c}^{c_1} f_0(x) x^3 q_{0_{in}}(x) I(x - c_{in})\, dx = 0$$
$$(6.11)$$

with $\nu_{el} = N c q_{m_{el}}(c)$.

It was mentioned above that in the relation $\frac{1}{2}m(c'^2 - c''^2) = \epsilon_{in}$ the change in the translational energy of the molecule was neglected. Consider equation B.4 (Appendix B), which gives

$$\epsilon_{in} = \tfrac{1}{2}m(c'^2 - c''^2)$$

$$= m\left[\left(\frac{m}{M}g'^2 + Cg'\cos\beta \right)\left(1 - \frac{g''}{g'}\cos\gamma \right) - Cg''\sin\beta\sin\gamma\cos\psi \right] + \frac{m}{2}(r'^2 - r''^2).$$

(6.12)

Since

$$\frac{m}{2}(r'^2 - r''^2) = \frac{m}{2}\left(\frac{M}{M+m} \right)^2 (g'^2 - g''^2) \cong \frac{m}{2}(g'^2 - g''^2),$$

this term is itself a close approximation to $\epsilon_{in} = (m/2)(c'^2 - c''^2)$, from which c_1 is defined. The remaining terms on the right-hand side of equation 6.12 are small in comparison with $(m/2)(r'^2 - r''^2)$, and they account for a small contribution to $\sigma_{coll_{in}}(c)$ equal to

$$4\pi n_0 N c_1^3 q_{m_{in}}(c_1)\left[\frac{m}{M}c_1 f_0(c_1) + \frac{\overline{C^2}}{3}\frac{d}{dc_1}f_0(c_1) \right].$$

The velocity points that are concerned with this flux start within a shell (c_1, dc_1), where $dc_1 \sim (m/M)c_1$ and the flux is therefore omitted as negligible.

6.3.   THE FUNCTION $\mathbf{f}_1(c)$

In order to elucidate the equilibrium of the momentum content of the electrons of the shells (c, dc) we first set out the processes that increase or diminish this momentum.

Loss of momentum through inelastic encounters (see also Section 2.7.2d). We distinguish between the cases $c > c_{in}$ and $c < c_{in}$. When $c < c_{in}$ electrons of the shell make no inelastic encounters since their energies are less than $\tfrac{1}{2}mc_{in}^2$, but when $c > c_{in}$ such encounters are possible. Through an inelastic encounter electrons are removed from the shell (c, dc) to another region of velocity space. The momentum lost in time dt by the group, whether $c > c_{in}$ or $c < c_{in}$, is expressible as

$$dt\left(\frac{4\pi}{3}n_0 c^3 dc \right)m\mathbf{f}_1(c)Ncq_{0_{in}}(c)I(c - c_{in}).$$

Gain of momentum through inelastic encounters (see also Section 2.7.2d). Velocity points are transferred by inelastic encounters from the shell (c_1, dc_1) to the shell (c, dc). The number of points lost to the shell (c_1, dc_1) in time dt from this cause is $dt[4\pi n_0 c_1^2 f_0(c_1) dc_1] N c_1 q_{0_{in}}(c_1)$ with an associated momentum $dt[(4\pi/3)n_0 c_1^2 dc_1]\mathbf{f}_1(c_1)Nc_1 q_{0_{in}}(c_1)mc_1$ (see Section 2.7.2d). But, from the discussion in Section 2.2.2, the fraction of the momentum retained by a group of electrons with momentum mc after scattering is, when $m \ll M$, approximately $(c'/c)[q_1(c)/q_0(c)]$ where c' is the final velocity. It follows that the momentum retained by the electrons scattered inelastically from the shell (c_1, dc_1) in time dt and therefore transferred to the shell (c, dc) is

$$dt\left(\frac{4\pi}{3}n_0 c_1^2 dc_1\right)\mathbf{f}_1(c_1)Nc_1 q_{0_{in}}(c_1)mc_1\left(\frac{c}{c_1}\right)\frac{q_{1_{in}}(c_1)}{q_{0_{in}}(c_1)}.$$

Since $c_1^2 = c^2 + c_{in}^2$, it follows that $dc_1 = (c/c_1)dc$; consequently the latter expression reduces to

$$dt\left(\frac{4\pi}{3}n_0 m\right)[Nc_1 q_{1_{in}}(c_1)]\mathbf{f}_1(c_1)c_1 c^2 dc.$$

Gain of momentum from the force $e\mathbf{E}$. It was shown in Chapter 2 that this gain is

$$-dt\frac{4\pi}{3}n_0 e\mathbf{E}c^3\frac{df_0}{dc}dc.$$

Equation for $\mathbf{f}_1(c)$. On assembling the gains and losses of momentum, the following equation for the state of equilibrium is obtained:

$$-\frac{e\mathbf{E}}{m}\frac{d}{dc}f_0 = [\nu_{el} + \nu_{0_{in}}I(c - c_{in})]\mathbf{f}_1(c) - \left[Nc_1 q_{1_{in}}(c_1)\frac{c_1}{c}\mathbf{f}_1(c_1)\right] \quad (6.13)$$

with $\nu_{el} = Ncq_{m_{el}}(c)$ and $\nu_{0_{in}} = Ncq_{0_{in}}(c)$.

6.4.   THE SCALAR AND VECTOR EQUATIONS FOR A TRAVELLING GROUP

6.4.1.   THE SCALAR EQUATION. Equation 6.11 is the scalar equation for the model gas, which it is convenient to restate for ease in reference:

$$-\frac{e\mathbf{E}}{mv_{el}}\cdot\mathbf{f}_1(c)+\overline{C^2}\,\frac{d}{dc}f_0(c)+\frac{3m}{M}cf_0(c)$$

$$+\frac{3}{c^3q_{m_{el}}(c)}\int_c^{c_1}f_0(x)x^3q_{0_{in}}(x)I(x-c_{in})\,dx=0. \tag{6.14}$$

6.4.2. THE VECTOR EQUATION. Equation 6.13 is equivalent to

$$\mathbf{f}_1(c)=-\frac{e\mathbf{E}}{m\big[\nu_{el}+\nu_{0_{in}}I(c-c_{in})\big]}\,\frac{d}{dc}f_0(c)$$

$$+\frac{q_{1_{in}}(c_1)}{q_{m_{el}}(c)+q_{0_{in}}(c)I(c-c_{in})}\Big(\frac{c_1}{c}\Big)^2\mathbf{f}_1(c_1). \tag{6.15}$$

The presence of the term depending on c_1 is a complication, and the simplifying assumption usually made is that the term may be omitted as small on the grounds that $q_{m_{el}}(c)$ is much greater than $q_{0_{in}}(c)$ and $q_{1_{in}}(c)$ over a useful range of speeds c. This procedure would require justification, however, according to the particular circumstances. A case in which the omission is fully justified occurs when the inelastic scattering is isotropic, in which event $q_{1_{in}}(c)=0$ since $q_{1_{in}}(c)=q_0(c)\,\overline{\cos\gamma}$. In any event the first term on the right-hand side of equation 6.15 is the predominant term.

When we treat equation 6.13 in accordance with the procedure followed in Section 2.7 we may write, without approximation,

$$\mathbf{f}_1(c)=-\mathbf{V}'\frac{d}{dc}f_0(c), \tag{6.16}$$

where

$$\mathbf{V}'=\frac{e\mathbf{E}}{mv}, \tag{6.17}$$

and v is defined in Section 2.7.2d.

We note that when inelastic scattering is isotropic

$$v=\nu_{el}+\nu_{0_{in}}I(c-c_{in}). \tag{6.18}$$

It follows from equation 6.16 that, in the general case, the scalar equation becomes

$$(V'V + \overline{C^2})\frac{d}{dc}f_0 + \frac{3mc}{M}f_0 + \frac{3}{c^3 q_{m_{el}}(c)}\int_c^{c_1} f_0(x)x^3 q_{0_{in}}(x)I(x - c_{in})\,dx = 0$$

(6.19)

with $\mathbf{V} = e\mathbf{E}/mv_{el}$.

A formal solution of equation 6.19 is

$$f_0(c) = \left(\exp - \frac{3m}{M}\int_0^c \frac{y\,dy}{V'V + \overline{C^2}}\right)\left(A + \int_0^c \left\{\left[\exp\frac{3m}{M}\int_0^z \frac{y\,dy}{VV' + \overline{C^2}}\right]\right.\right.$$

$$\left.\left. \times \frac{3}{z^3 q_{m_{el}}(z)(V'V + \overline{C^2})}\int_z^{z_1} f_0(x)x^3 q_{0_{in}}(x)I(x - c_{in})\,dx\right\}dz\right) \quad (6.20)$$

where A is a constant of integration and $z_1^2 = z^2 + c_{in}^2$. In monatomic gases c_{in} is large and, over a wide range of mean energies $\frac{1}{2}m\,\overline{c^2}$, an inconsiderable proportion of the electrons move with speeds c that exceed c_{in}. Thus, in effect, $I(x - c_{in})$ vanishes, and, for $c < c_{in}$, equation 6.20 reduces to

$$f_0(c) = A\exp\left(-\frac{3m}{M}\int_0^c \frac{y\,dy}{V^2 + \overline{C^2}}\right) \quad (6.21)$$

which is the generalized distribution function (equation 4.7), but with a modified value of A. In the model gas which resembles parahydrogen, the distribution function is appreciably modified by the contribution of inelastic encounters since c_{in} is not taken to be large.

We return to equations 6.13 and 6.15. We have already remarked that when scattering is isotropic $q_{1_{in}}(c_1)$ vanishes and v reduces to the expression given in equation 6.18, that is to say, $q_m(c) = q_{m_{el}}(c) + q_{0_{in}}(c)I(c - c_{in})$. In another circumstance in which a simplification is found, the excitation energy $\frac{1}{2}mc_{in}^2$ is a small proportion of the energies $\frac{1}{2}mc^2$ of the majority of the electrons. In this event $c_1^2 = c^2(1 + c_{in}^2/c^2) \cong c^2$, so that, in equations 6.13 and 6.15, $f_1(c_1) \cong f_1(c)$, $c_1/c \cong 1$, and $q_{1_{in}}(c_1) \cong q_{1_{in}}(c)$. Equation 6.15 then becomes

$$\mathbf{f}_1(c) = -\frac{e\mathbf{E}}{mv_1}\frac{d}{dc}f_0(c), \quad (6.22)$$

where

$$\nu_1 = Nc\left[q_{m_{el}}(c) + q_{0_{in}}(c) - q_{1_{in}}(c)\right].$$

Thus $q_m(c)$ has here become $q_m(c) = q_{m_{el}}(c) + q_{0_{in}}(c) - q_{1_{in}}(c)$.

6.5. CONTRIBUTION OF SUPERELASTIC ENCOUNTERS

It has been assumed in what has preceded that all molecules are in the ground state of internal energy. This assumption is justified when the smallest threshold energy for inelastic encounters greatly exceeds κT as in monatomic gases. In molecular gases, however, the threshold energies for changes in the rotational energies of molecules are comparable in magnitude with κT at room temperatures, and the number of molecules in rotational states other than the ground state is an appreciable proportion of the whole. We again consider the model gas with a single state of internal energy other than the ground state, with the threshold energy designated, as before, by $\epsilon_{in} = \tfrac{1}{2}mc_{in}^2$.

Let N_0 be the number density of molecules in the ground state; then the number in the state with the threshold energy ϵ_{in} is $N_0\exp(-\epsilon_{in}/\kappa T)$. In some encounters between electrons and molecules in the excited state, the molecules revert to the ground state and the energy ϵ_{in} is communicated to the electrons. Such encounters are superelastic (collisions of the second kind). Let c' and c'' be, respectively, the speeds of an electron before and after a superelastic encounter; then $\tfrac{1}{2}m(c''^2 - c'^2) = \epsilon_{in} = \tfrac{1}{2}mc_{in}^2$. Since, in such an encounter, the velocity point of the electron is displaced away from the origin of velocity space, these encounters give rise to an outward flux of points across the sphere c in contrast to inelastic encounters, which maintain an inward flux.

Consider the sphere c and let $c_2^2 = c^2 - (2\epsilon_{in}/m) = c^2 - c_{in}^2$. Then two cases occur.

(a) $c < c_{in}$.

We express the number of superelastic encounters made by the $4\pi n_0 f_0(x)x^2\,dx$ electrons of the shell (x, dx) in time dt as $dt\,4\pi n_0[Nxq_{0_s}(x)]$ $f_0(x)x^2\,dx$, where $q_{0_s}(x)$ is the total superelastic collision cross section. It follows that the outward flux of points across the sphere c is

$$4\pi n_0 N \int_0^c f_0(x)x^3 q_{0_s}(x)\,dx.$$

(b) $c > c_{in}$.

The outward flux from superelastic encounters across the surface of the sphere c is now

$$4\pi n_0 N \int_{c_2}^{c} f_0(x) x^3 q_{0_s}(x)\, dx.$$

There is also a small term, as with inelastic encounters, that contributes an additional flux:

$$4\pi n_0 N c_2^{3} q_{m_s}(c_2) \left[\frac{m}{M} c_2 f_0(c_2) + \frac{\overline{C^2}}{3} \frac{d}{dc_2} f_0(c_2) \right],$$

which is omitted as negligibly small.

Equation 6.10 for the net inward flux now becomes

$$\sigma_{coll} = 4\pi n_0 \left\{ N c^3 q_{m_{el}}(c) \left(\frac{m}{M} c f_0 + \frac{\overline{C^2}}{3} \frac{d}{dc} f_0 \right) \right.$$

$$\left. + N \left[\int_{c}^{c_1} f_0(x) x^3 q_{0_{in}}(x) I(x - c_{in})\, dx - \int_{0 \text{ or } c_2}^{c} f_0(x) x^3 q_{0_s}(x)\, dx \right] \right\} \qquad (6.23)$$

in which the lower limit in the second integral is 0 when $c < c_{in}$ but c_2 when $c > c_{in}$.

The steady state of motion of the group is characterized by the condition $\sigma_E(c) = \sigma_{coll}(c)$ for every spherical surface in velocity space with the origin as centre. In the absence of an electric field, $\sigma_E(c)$ vanishes and the condition reduces to $\sigma_{coll}(c) = 0$. However, according to the dynamical theory of gases and to statistical mechanics, the distribution function $f_0(c)$ is Maxwell's function

$$f_0(c) = \frac{1}{(\alpha \sqrt{\pi})^3} \exp - \frac{c^2}{\alpha^2},$$

where $\frac{1}{2} m \alpha^2 = \frac{1}{3} M \overline{C^2} = \kappa T$ and α is the most probable speed. It follows that the factor

$$\left[\frac{mc}{M} f_0(c) + \frac{\overline{C^2}}{3} \frac{d}{dc} f_0(c) \right] = 0$$

in equation 6.23. Therefore the condition $\sigma_{\text{coll}}(c)=0$ reduces to

$$\int_c^{c_1} f_0(x)x^3 q_{0_{\text{in}}}(x) I(x-c_{\text{in}})\,dx = \int_{0 \text{ or } c_2}^c f_0(x)x^3 q_{0_s}(x)\,dx. \quad (6.24)$$

Since $f_0(x)$ in equation 6.24 is Maxwell's distribution function, it follows that a relationship exists between $q_{0_{\text{in}}}(c)$ and $q_{0_s}(c)$. This relationship is usually found by application of the principle of detailed balancing, but we shall here proceed to determine it directly from equation 6.24.

We first transform the integrals in equation 6.24 and consider in turn the circumstances $c>c_{\text{in}}$ and $c<c_{\text{in}}$. When $c>c_{\text{in}}$, equation 6.24 becomes

$$\int_c^{c_1} f_0(x)x^3 q_{0_{\text{in}}}(x)\,dx = \int_{c_2}^c f_0(y)y^3 q_{0_s}(y)\,dy,$$

where x and y are dummy variables, $c_1^2 = c^2 + c_{\text{in}}^2$, $c_2^2 = c^2 - c_{\text{in}}^2$.

We write $y^2 = x^2 - c_{\text{in}}^2$; then $y\,dy = x\,dx$. When $y=c_2$, then $x=c$; when $y=c$, then $x=c_1$. The integral on the right is therefore equivalent to

$$\int_c^{c_1} f_0\left(\sqrt{x^2-c_{\text{in}}^2}\,\right)\cdot(x^2-c_{\text{in}}^2)\cdot q_{0_s}\left(\sqrt{x^2-c_{\text{in}}^2}\,\right)x\,dx,$$

and it follows that

$$\int_c^{c_1}\left[f_0(x)x^2 q_{0_{\text{in}}}(x) - f_0\left(\sqrt{x^2-c_{\text{in}}^2}\,\right)\cdot(x^2-c_{\text{in}}^2)\cdot q_{0_s}\left(\sqrt{x^2-c_{\text{in}}^2}\,\right)\right]x\,dx=0.$$

$$(6.25)$$

When $c<c_{\text{in}}$, equation 6.24 becomes

$$\int_{c_{\text{in}}}^{c_1} f_0(x)x^3 q_{0_{\text{in}}}(x)\,dx = \int_0^c f_0(y)y^3 q_{0_s}(y)\,dy.$$

Again let $y^2 = x^2 - c_{\text{in}}^2$. Then when $y=0$, $x=c_{\text{in}}$; when $y=c$, $x=c_1$; also $x\,dx = y\,dy$. Consequently

$$\int_{c_{\text{in}}}^{c_1}\left[f_0(x)x^2 q_{0_{\text{in}}}(x) - f_0\left(\sqrt{x^2-c_{\text{in}}^2}\,\right)\cdot(x^2-c_{\text{in}}^2)\cdot q_{0_s}\left(\sqrt{x^2-c_{\text{in}}^2}\,\right)\right]x\,dx=0.$$

$$(6.26)$$

Thus, in order for equations 6.25 and 6.26 to hold in general, the integrand, which is the same in each, must be equal to zero for all values of $x < c_{in}$. We deduce therefore that

$$f_0(c)c^2 q_{0_{in}}(c) = f_0\left(\sqrt{c^2 - c_{in}^2}\right) \cdot (c^2 - c_{in}^2) \cdot q_{0_s}\left(\sqrt{c^2 - c_{in}^2}\right).$$

Since

$$f_0(c) = \frac{1}{(\alpha\sqrt{\pi})^3} \exp\left(-\frac{\frac{1}{2}mc^2}{\kappa T}\right),$$

it follows that

$$f_0\left(\sqrt{c^2 - c_{in}^2}\right) = f_0(c) \exp\left(\frac{\epsilon_{in}}{\kappa T}\right)$$

and that

$$(c^2 - c_{in}^2) q_{0_s}\left(\sqrt{c^2 - c_{in}^2}\right) = c^2 q_{0_{in}}(c) \exp\left(-\frac{\epsilon_{in}}{\kappa T}\right). \qquad (6.27)$$

Thus, when $c = c_{in}$, $q_{0_{in}}(c_{in}) = 0$, as it should.

We return to equation 6.23, which may now be expressed as

$$\sigma_{coll}(c) = 4\pi n_0 N \left(c^3 q_{m_{el}}(c) \left(\frac{m}{M} c f_0 + \frac{\overline{C^2}}{3} \frac{d}{dc} f_0 \right) \right.$$

$$\left. + \left\{ \int_c^{c_1} \left[f_0(x) - f_0\left(\sqrt{x^2 - c_{in}^2}\right) \exp\left(-\frac{\epsilon_{in}}{\kappa T}\right) \right] x^3 q_{0_{in}}(x) I(x - c_{in}) \, dx \right\} \right).$$

$$(6.28)$$

The condition for equilibrium in the presence of the field $\mathbf{E}$ is $\sigma_E(c) - \sigma_{coll}(c) = 0$, that is to say (compare equation 6.11),

$$-c^2 \frac{e\mathbf{E}}{3m} \cdot \mathbf{f}_1(c) + N c^3 q_{m_{el}}(c) \left(\frac{m}{M} c f_0 + \frac{\overline{C^2}}{3} \frac{d}{dc} f_0 \right)$$

$$+ N \int_c^{c_1} \left[f_0(x) - f_0\left(\sqrt{x^2 - c_{in}^2}\right) \exp\left(-\frac{\epsilon_{in}}{\kappa T}\right) \right] x^3 q_{0_{in}}(x) I(x - c_{in}) \, dx = 0.$$

$$(6.29)$$

6.5.1. FORMULA FOR $\mathbf{f}_1(c)$ WHEN SUPERELASTIC ENCOUNTERS ARE SIGNIFICANT. The loss of momentum in time dt by electrons of the shell

(c, dc) in encounters with excited molecules is

$$dt\, n_0 mc \left(\frac{4\pi}{3} \mathbf{f}_1 c^2\, dc \right)(Ncq_{0_s}).$$

The gain of momentum to the class of electrons $n_0 4\pi f_0(c) c^2\, dc$ in time dt from superelastic encounters is zero when $c < c_{in}$ since the velocity points pass beyond the sphere c. When $c > c_{in}$, the momentum gained by the class from electrons of the shell (c_2, dc_2) is (cf. Section 6.3)

$$dt\, n_0 \left[\frac{4\pi}{3} \mathbf{f}_1(c_2) c_2^2\, dc_2 \right] mc_2 \left[Nc_2 q_{0_s}(c_2) \right] \left[\left(\frac{c}{c_2} \right) \frac{q_{1_s}(c_2)}{q_{0_s}(c_2)} \right],$$

where $c_2^2 = c^2 - c_{in}^2$ and $c_2\, dc_2 = c\, dc$.
The gain of momentum is therefore

$$dt \left(\frac{4\pi}{3} n_0 m \right) \left[Nc_2 q_{1_s}(c_2) \right] \mathbf{f}_1(c_2) c_2 c^2\, dc\, I(c - c_{in}). \tag{6.30}$$

It can be seen that when the contributions of superelastic encounters are included equation 6.13 takes the form

$$-\frac{eE}{m} \frac{d}{dc} f_0 = \left(\nu_{el} + \nu_{0_{in}} I(c - c_{in}) + \nu_{0_s} \right) \mathbf{f}_1(c)$$

$$-N \left[\left(\frac{c_1^2}{c} \right) q_{1_{in}}(c_1) \mathbf{f}_1(c_1) + \left(\frac{c_2^2}{c} \right) q_{1_s}(c_2) \mathbf{f}_1(c_2) I(c - c_{in}) \right], \tag{6.31}$$

in which $\nu_{el} = Ncq_{m_{el}}(c)$, $\nu_{0_{in}} = Ncq_{0_{in}}$, and $\nu_{0_s} = Ncq_{0_s}$.
This equation can also be represented in the form of equation 6.16. When the inelastic and superelastic scattering is isotropic, the cross sections $q_{1_{in}}(c_1)$ and $q_{1_s}(c_2)$ vanish, as already remarked. Alternatively they may be small compared with $q_{m_{el}}(c)$. In either event, equation 6.31 is reduced to the form 6.16 but with ν in equation 6.18 replaced by $\nu = \nu_{el} + \nu_{0_{in}} I(c - c_{in}) + \nu_{0_s}$.

6.6.   FORMULATION FOR MOLECULES WITH SEVERAL OR MANY ENERGY STATES

In actual gases we are concerned in general not with a single kind of inelastic encounter, as in the model gas discussed above, but with a variety of inelastic encounters associated with transitions between many pairs of energy states. In the formulation of the scalar and vector equations it is necessary, therefore, to take account of this fact.

The scalar equation 6.29 is modified to read

$$-c^2\frac{e\mathbf{E}}{3m}\cdot\mathbf{f}_1+Nc^3q_{m_{el}}(c)\left(\frac{mc}{M}f_0+\frac{\overline{C^2}}{3}\frac{d}{dc}f_0\right)+N\sum_k I_k=0, \quad (6.32)$$

where

$$I_k=\int_c^{c_{1k}}\left[f_0(x)-f_0\left(\sqrt{x^2-c_{k_{in}}^2}\right)\exp\left(-\frac{\epsilon_{k_{in}}}{\kappa T}\right)\right]x^3q_{0k_{in}}(x)I(x-c_{k_{in}})\,dx,$$

$\epsilon_{k_{in}}$ is the threshold energy required to excite the kth inelastic transition, $c_{k_{in}}^2=2\epsilon_{k_{in}}/m$, $c_{1k}^2=c_{k_{in}}^2+c^2$, and $q_{0k_{in}}(c)$ is the collision cross section for the kth inelastic transition. The summation over k implies that the relevant inelastic encounters associated with changes in the rotational, vibrational, and electronic changes of state are included.

The modified form of the vector equation 6.31 is

$$-\frac{e\mathbf{E}}{m}\frac{d}{dc}f_0=\left\{\nu_{el}+\sum_k\left[\nu_{0k_{in}}I(c-c_{k_{in}})+\nu_{0k_s}(c)\right]\right\}\mathbf{f}_1(c)$$

$$-N\left\{\sum_k\left[\left(\frac{c_{1k}^2}{c}\right)q_{1k_{in}}(c_{1k})\mathbf{f}_1(c_{1k})+\left(\frac{c_{2k}^2}{c}\right)q_{1k_s}(c_{2k})\mathbf{f}_1(c_{2k})I(c-c_{k_{in}})\right]\right\}.$$

$$(6.33)$$

Equation 6.33 can be expressed in a simplified form in terms of an effective collision frequency for momentum transfer ν (cf. equations 6.13 and 6.16). Thus

$$\mathbf{f}_1(c)=-\frac{e\mathbf{E}}{m\nu}\frac{d}{dc}f_0=-\mathbf{V}'\frac{d}{dc}f_0.$$

When the cross sections $q_{1k_{in}}$ and q_{1k_s} are zero or negligible, the effective collision frequency ν becomes

$$\nu=\nu_{el}+\sum_k\left[\nu_{0k_{in}}I(c-c_{k_{in}})+\nu_{0k_s}\right],$$

$$(6.34)$$

and the effective cross section for momentum transfer is

$$q_m(c)=q_{m_{el}}(c)+\sum_k\left[q_{0k_{in}}(c)I(c-c_{k_{in}})+q_{0k_s}(c)\right].$$

When this expression for $\mathbf{f}_1(c)$ is used in equation 6.32, the scalar equation becomes

$$(V'V + \overline{C^2})\frac{d}{dc}f_0 + \frac{3mc}{M}f_0 + \frac{3}{c^3 q_{m_{el}}(c)}\sum_k I_k = 0, \qquad (6.35)$$

which is the differential equation satisfied by the velocity distribution function $f_0(c)$ for the group of n_0 electrons as a whole.

6.7.   THE SCALAR AND VECTOR EQUATIONS FOR nf_0 AND nf_1 AT AN ELEMENT OF VOLUME $d\mathbf{r}$

We readily obtain the scalar and vector equations for nf_0 and nf_1 relative to the $n\,d\mathbf{r}$ electrons within the element of volume $d\mathbf{r}$ by changing the interpretation of $f_0(c)$ and $\mathbf{f}_1(c)$ from that of distribution functions of the group as a whole to that of local distribution functions $f_0(c,\mathbf{r},t)$ and $\mathbf{f}_1(c,\mathbf{r},t)$ relevant to the electrons $n\,d\mathbf{r}$. Similarly the flux terms $\sigma_E(c)$ and $\sigma_{\text{coll}}(c)$ now become $\sigma_E(c,\mathbf{r})$ and $\sigma_{\text{coll}}(c,\mathbf{r})$. In addition it is necessary to restore the term $(c/3)\,\text{div}_r(n\mathbf{f}_1)$, which was removed by integration over the whole group. When all these actions are taken, the scalar and vector equations for nf_0 and nf_1 are seen to be as follows (compare equations 3.3 and 3.8).

6.7.1.   THE SCALAR EQUATION.

$$\frac{\partial}{\partial t}(nf_0) + \frac{c}{3}\,\text{div}_r(n\mathbf{f}_1) + \frac{1}{c^2}\frac{\partial}{\partial c}\left(\frac{c^2}{3}\frac{e\mathbf{E}}{m}\cdot n\mathbf{f}_1 - \frac{1}{4\pi}\sigma_{\text{coll}}\right) = 0, \quad (6.36)$$

where

$$\frac{1}{4\pi}\sigma_{\text{coll}} = c^2 \nu_{\text{el}}\left[\frac{m}{M}cnf_0 + \frac{\overline{C^2}}{3}\frac{\partial}{\partial c}(nf_0)\right] + nN\sum_k I_k$$

with $\nu_{\text{el}} = Ncq_{m_{el}}(c)$ and

$$I_k = \int_c^{c_{1k}}\left[f_0(x) - f_0\left(\sqrt{x^2 - c_{k_{in}}^2}\right)\exp\left(-\frac{\epsilon_{k_{in}}}{\kappa T}\right)\right]x^3 q_{0k_{in}}(x)I(x - c_{k_{in}})\,dx.$$

6.7.2.   THE VECTOR EQUATION AND THE WORKING SCALAR EQUATION.
The vector equation is seen to be

$$\frac{\partial}{\partial t}(n\mathbf{f}_1) + c\operatorname{grad}_r(nf_0) + \frac{e\mathbf{E}}{m}\frac{\partial}{\partial c}(nf_0)$$

$$= -Nc\left\{q_{m_{\mathrm{el}}}(c) + \sum_k \left[q_{0k_{\mathrm{in}}}(c)I(c-c_{k_{\mathrm{in}}}) + q_{0k_s}(c)\right]\right\}n\mathbf{f}_1(c)$$

$$+ N\left\{\sum_k\left[\frac{c_{1k}^{\,2}}{c}q_{1k_{\mathrm{in}}}(c_{1k})n\mathbf{f}_1(c_{1k})\right.\right.$$

$$\left.\left.+\left(\frac{c_{2k}^{\,2}}{c}\right)q_{1k_s}(c_{2k})n\mathbf{f}_1(c_{2k})I(c-c_{k_{\mathrm{in}}})\right]\right\}, \quad (6.37)$$

which, for the condition of quasi equilibrium, may be written as

$$-n\mathbf{f}_1 = \frac{c}{\nu}\operatorname{grad}_r(nf_0) + \frac{e\mathbf{E}}{m\nu}\frac{\partial}{\partial c}(nf_0), \quad (6.38)$$

where ν is the effective collision frequency for momentum transfer.

We next replace $n\mathbf{f}_1$ in equation 6.36 by its value from equation 6.38 and
obtain (with $V' = eE/m\nu$) the working scalar equation

$$\frac{\partial}{\partial t}(nf_0) - \operatorname{div}_r\left[\frac{c^2}{3\nu}\operatorname{grad}_r(nf_0) + \frac{c}{3}\mathbf{V}'\frac{\partial}{\partial c}(nf_0)\right]$$

$$-\frac{1}{c^2}\frac{\partial}{\partial c}\left\{\frac{c^3}{3}\mathbf{V}'\cdot\operatorname{grad}_r(nf_0) + \frac{c^2\nu_{\mathrm{el}}}{3}VV'\frac{\partial}{\partial c}(nf_0)\right.$$

$$\left.+ c^2\nu_{\mathrm{el}}\left[\frac{m}{M}c(nf_0) + \frac{\overline{C^2}}{3}\frac{\partial}{\partial c}(nf_0)\right]\right\} - \frac{nN}{c^2}\sum_k\left(\frac{\partial}{\partial c}I_k\right) = 0,$$

which is equivalent to

$$\frac{\partial}{\partial t}(nf_0) - \text{div}_r\left[\frac{c^2}{3\nu}\,\text{grad}_r(nf_0) + \frac{c}{3}\,\mathbf{V}'\frac{\partial}{\partial c}(nf_0)\right]$$

$$-\frac{1}{3c^2}\frac{\partial}{\partial c}\left\{c^2\nu_{\text{el}}\left[\frac{c}{\nu_{\text{el}}}\mathbf{V}'\cdot\text{grad}_r(nf_0) + (VV' + \overline{C^2})\frac{\partial}{\partial c}(nf_0) + \frac{3m}{M}c(nf_0)\right]\right\}$$

$$-\frac{nN}{c^2}\sum_k\left(\frac{\partial}{\partial c}I_k\right) = 0. \qquad (6.39)$$

Equation 6.39 is the form assumed by the scalar equation when inelastic encounters are important. It is different from the scalar equation 4.8 when all encounters are elastic through the addition of the last term and the introduction of V'.

The term $(\partial/\partial c)I_k$ is

$$\frac{\partial}{\partial c}\int_c^{c_{1k}}\left[f_0(x) - f_0\left(\sqrt{x^2 - c_{k_{\text{in}}}^2}\right)\exp\left(-\frac{\epsilon_{k_{\text{in}}}}{\kappa T}\right)\right]x^3 q_{0k_{\text{in}}}(x)I(x - c_{k_{\text{in}}})\,dx$$

$$= \frac{c}{c_{1k}}\left[f_0(c_{1k}) - f_0(c)\exp\left(-\frac{\epsilon_{k_{\text{in}}}}{\kappa T}\right)\right]c_{1k}^3 q_{0k_{\text{in}}}(c_{1k})$$

$$-\left[f_0(c) - f_0\left(\sqrt{c^2 - c_{k_{\text{in}}}^2}\right)\exp\left(-\frac{\epsilon_{k_{\text{in}}}}{\kappa T}\right)\right]c^3 q_{0k_{\text{in}}}(c)I(c - c_{k_{\text{in}}})$$

$$= c\left\{\left[c_{1k}^2 q_{0k_{\text{in}}}(c_{1k})f_0(c_{1k}) - c^2 q_{0k_{\text{in}}}(c)f_0(c)I(c - c_{k_{\text{in}}})\right]\right.$$

$$+\left[c^2 q_{0k_{\text{in}}}(c)f_0\left(\sqrt{c^2 - c_{k_{\text{in}}}^2}\right)I(c - c_{k_{\text{in}}})\right.$$

$$\left.-c_{1k}^2 q_{0k_{\text{in}}}(c_{1k})f_0(c)\right]\exp\left(-\frac{\epsilon_{k_{\text{in}}}}{\kappa T}\right)\right\}. \qquad (6.40)$$

Equation 6.39 is usually presented in a form (but in another notation with $\epsilon = \frac{1}{2}mc^2$ as independent variable; see Section 6.8) in which $(\partial/\partial c)I_k$ is expanded as in equation 6.40. The condensed form presented in equation 6.39 is, however, more convenient.

6.8. FORMULAE AND EQUATIONS WITH THE ENERGY $\epsilon = \frac{1}{2}mc^2$ AS INDEPENDENT VARIABLE

In many investigations of the motion of electrons in gases the energy $\epsilon = \frac{1}{2}mc^2$ of an electron is employed as an independent variable in place of c. We therefore summarize the more important formulae and equations in this alternative notation.

6.8.1. DISTRIBUTION FUNCTION $f(\epsilon)$.

The point population of a shell (c, dc) of velocity space is $n\,d\mathbf{r}(4\pi f_0 c^2\,dc)$, which is also the number of electrons with energies in the range ϵ to $\epsilon + d\epsilon$, where

$$\epsilon = \tfrac{1}{2}mc^2 \quad \text{and} \quad d\epsilon = mc\,dc = (2m\epsilon)^{1/2}\,dc.$$

Since $c^2\,dc = (2\epsilon/m^3)^{1/2}\,d\epsilon$, it follows that the population of a shell (c, dc) is proportional to $\epsilon^{1/2}\,d\epsilon$. A distribution function $f(\epsilon)$ is so defined that the point population of the shell is also represented as $(n\,d\mathbf{r})\epsilon^{1/2}f_0(\epsilon)\,d\epsilon$, that is to say, $\epsilon^{1/2}f_0(\epsilon)\,d\epsilon = 4\pi f_0(c)c^2\,dc$. Thus $f_0(\epsilon) = Af_0(c)$, where $A = 4\pi\sqrt{2/m^3}$.
Integration over all shells gives

$$\int_0^\infty \epsilon^{1/2}f_0(\epsilon)\,d\epsilon = 4\pi \int_0^\infty f_0(c)c^2\,dc = 1.$$

Also

$$\frac{d}{dc} = \left(\frac{d}{dc}\epsilon\right)\frac{d}{d\epsilon} = (2m\epsilon)^{1/2}\frac{d}{d\epsilon}, \qquad c\frac{d}{dc} = 2\epsilon\frac{d}{d\epsilon}.$$

As written above, $f_0(\epsilon)$ and $\mathbf{f}_1(\epsilon)$ in fact imply $f_0(\epsilon, \mathbf{r}, t)$ and $\mathbf{f}_1(\epsilon, \mathbf{r}, t)$.

6.8.2. THE MEAN VALUE OF A FUNCTION $\phi(\epsilon)$.

Let this mean value be denoted as $\overline{\phi(\epsilon)}$; then

$$\overline{\phi(\epsilon)} = \int_0^\infty \epsilon^{1/2}\phi(\epsilon)f(\epsilon)\,d\epsilon.$$

6.8.3. FORMULAE FOR **W** AND D. The mean of the velocities of the electrons of the shells (c, dc) is

$$\mathbf{W} = -\frac{e\mathbf{E}}{mN}\frac{4\pi}{3}\int_0^\infty \frac{c^2}{q_m(c)}\frac{df_0}{dc}\,dc.$$

When we replace $4\pi c^2\,dc\,f_0(c)$ by $\epsilon^{1/2}\,d\epsilon\,f_0(\epsilon)$ and d/dc by $(2m\epsilon)^{1/2}(d/d\epsilon)$, we obtain

$$\mathbf{W} = -\frac{e\mathbf{E}}{3N}\left(\frac{2}{m}\right)^{1/2}\int_0^\infty \frac{\epsilon}{q_m(\epsilon)}\frac{d}{d\epsilon}f_0(\epsilon)\,d\epsilon. \tag{6.41}$$

Similarly the formula for the lateral coefficient of diffusion,

$$D = \frac{4\pi}{3N}\int_0^\infty \frac{c}{q_m(c)}f_0(c)c^2\,dc,$$

becomes

$$D = \frac{1}{3N}\left(\frac{2}{m}\right)^{1/2}\int_0^\infty \frac{\epsilon}{q_m(\epsilon)}f_0(\epsilon)\,d\epsilon. \tag{6.42}$$

It follows that

$$\frac{eD}{\mu} = -\left\{\frac{\int_0^\infty [\epsilon/q_m(\epsilon)]f_0(\epsilon)\,d\epsilon}{\int_0^\infty [\epsilon/q_m(\epsilon)](d/d\epsilon)f_0(\epsilon)\,d\epsilon}\right\}. \tag{6.43}$$

In equations 6.41 to 6.43, $q_m(\epsilon)$ is the effective momentum transfer cross section.

6.8.4. TRANSFORMATION OF EQUATIONS.

Vector equation. Equation 6.38 becomes

$$-n\mathbf{f}_1(\epsilon) = \frac{1}{Nq_m(\epsilon)}\operatorname{grad}_r[nf_0(\epsilon)] + \frac{e\mathbf{E}}{Nq_m(\epsilon)}\frac{\partial}{\partial\epsilon}[nf_0(\epsilon)]. \tag{6.44}$$

Scalar equation. Equation 6.39 becomes

$$\frac{\partial}{\partial t}[nf_0(\epsilon)] - \tfrac{1}{3}\operatorname{div}_r\left\{\left(\frac{2\epsilon}{m}\right)^{1/2}\frac{1}{Nq_m(\epsilon)}\operatorname{grad}_r[nf_0(\epsilon)] + 2\epsilon\,\mathbf{V}'(\epsilon)\frac{\partial}{\partial \epsilon}[nf_0(\epsilon)]\right\}$$

$$-\frac{1}{3}\left(\frac{m^3}{2\epsilon}\right)^{1/2}\frac{\partial}{\partial \epsilon}\left(\left(\frac{2\epsilon}{m}\right)^{3/2}Nq_{m_{el}}(\epsilon)\left\{\frac{1}{Nq_{m_{el}}(\epsilon)}\mathbf{V}'(\epsilon)\cdot\operatorname{grad}_r[nf_0(\epsilon)]\right.\right.$$

$$+\left.\left.\left[V(\epsilon)V'(\epsilon)+\overline{C^2}\right](2m\epsilon)^{1/2}\frac{\partial}{\partial \epsilon}[nf_0(\epsilon)]+\frac{3m}{M}\left(\frac{2\epsilon}{m}\right)^{1/2}nf_0(\epsilon)\right\}\right)$$

$$-\frac{nNm}{2\epsilon}A\sum_k(2m\epsilon)^{1/2}\frac{\partial}{\partial \epsilon}I_k(c)=0, \qquad (6.45)$$

where $A=4\pi\sqrt{2/m^3}$. In equation 6.45

$$V=\frac{eE}{mNq_{m_{el}}(c)c}=\frac{eE}{N(2m\epsilon)^{1/2}q_{m_{el}}(\epsilon)}, \qquad V'=\frac{eE}{N(2m\epsilon)^{1/2}q_m(\epsilon)}.$$

Consider

$$I_k(c)=\int_c^{c_{1k}}\left[f_0(x)-f_0\left(\sqrt{x^2-c_{k_{in}}^2}\right)\exp\left(-\frac{\epsilon_{k_{in}}}{\kappa T}\right)\right]x^3 q_{0k_{in}}(x)I(x-c_{k_{in}})\,dx.$$

It follows that, since $c_{1k}^2=c^2+c_{k_{in}}^2$, or $\epsilon_{1k}=\epsilon+\epsilon_{k_{in}}$,

$$I_k(c)=\frac{1}{A}\frac{2}{m^2}\int_\epsilon^{\epsilon_{1k}}\left[f_0(y)-f_0(y-\epsilon_{k_{in}})\exp\left(-\frac{\epsilon_{k_{in}}}{\kappa T}\right)\right]I(y-\epsilon_{k_{in}})q_{0k_{in}}(y)y\,dy$$

$$=\frac{1}{A}I_k(\epsilon), \qquad (6.46)$$

where $y = \frac{1}{2}mx^2$ and $I(y - \epsilon_{k_{in}})$ is the unit step function. It follows that

$$\frac{m^2}{2}\frac{\partial}{\partial \epsilon}I_k(\epsilon) = \left[f_0(\epsilon + \epsilon_{k_{in}}) - f_0(\epsilon)\exp\left(-\frac{\epsilon_{k_{in}}}{\kappa T}\right)\right](\epsilon + \epsilon_{k_{in}})q_{0k_{in}}(\epsilon + \epsilon_{k_{in}})$$

$$- \left[f_0(\epsilon) - f_0(\epsilon - \epsilon_{k_{in}})\exp\left(-\frac{\epsilon_{k_{in}}}{\kappa T}\right)\right]\epsilon q_{0k_{in}}(\epsilon)I(\epsilon - \epsilon_{k_{in}}). \quad (6.47)$$

We note that equation 6.27 expressed in terms of ϵ is, for the kth transition,

$$(\epsilon - \epsilon_{k_{in}})q_{0k_s}(\epsilon - \epsilon_{k_{in}}) = \epsilon q_{0k_{in}}(\epsilon)\exp\left(-\frac{\epsilon_{k_{in}}}{\kappa T}\right).$$

Also,
$$\left.\begin{array}{c}\\ \\ \\ \end{array}\right\} \quad (6.48)$$

$$\epsilon q_{0k_s}(\epsilon) = (\epsilon + \epsilon_{k_{in}})q_{0k_{in}}(\epsilon + \epsilon_{k_{in}})\exp\left(-\frac{\epsilon_{k_{in}}}{\kappa T}\right).$$

It can be seen, therefore, that an equivalent formula for $(m^2/2)(\partial/\partial\epsilon)I_k(\epsilon)$ is

$$\frac{m^2}{2}\frac{\partial}{\partial \epsilon}I_k(\epsilon) = \left[(\epsilon + \epsilon_{k_{in}})q_{0k_{in}}(\epsilon + \epsilon_{k_{in}})f_0(\epsilon + \epsilon_{k_{in}}) - \epsilon q_{0k_{in}}(\epsilon)f_0(\epsilon)I(\epsilon - \epsilon_{k_{in}})\right]$$

$$- \left[\epsilon q_{0k_s}(\epsilon)f_0(\epsilon) - (\epsilon - \epsilon_{k_{in}})q_{0k_s}(\epsilon - \epsilon_{k_{in}})I(\epsilon - \epsilon_{k_{in}})\right]. \quad (6.49)$$

When each term of equation 6.45 is multiplied by $(m\epsilon/2)^{1/2}$ and V and V' are replaced by their equivalent expressions in terms of ϵ, the equation takes the form

$$\left(\frac{m\epsilon}{2}\right)^{1/2}\frac{\partial}{\partial t}[nf_0(\epsilon)] - \frac{\epsilon}{3Nq_m(\epsilon)}\text{div}_r\left\{\text{grad}_r[nf_0(\epsilon)] + \frac{\partial}{\partial \epsilon}[nf_0(\epsilon)eE]\right\}$$

$$- \frac{1}{3}\frac{\partial}{\partial \epsilon}\left\{\frac{\epsilon}{Nq_m(\epsilon)}eE\cdot\text{grad}_r[nf_0(\epsilon)]\right.$$

$$+ \left[(eE)^2\frac{\epsilon}{Nq_m(\epsilon)} + \frac{6m}{M}\kappa T\epsilon^2 Nq_{m_{el}}(\epsilon)\right]\frac{\partial}{\partial \epsilon}[nf_0(\epsilon)]$$

$$\left. + \frac{6m}{M}\epsilon^2 Nq_{m_{el}}(\epsilon)nf_0(\epsilon)\right\} - Nn\frac{\partial}{\partial \epsilon}\sum_k\frac{m^2}{2}I_k(\epsilon) = 0, \quad (6.50)$$

in which the $(m^2/2)(\partial/\partial\epsilon)I_k(\epsilon)$ have either of the forms presented in equations 6.47 and 6.49.

The form of the scalar equation when inelastic encounters are important has been investigated in depth by Holstein (1946) and by Frost and Phelps (1962), and the form taken by the group of terms that represent the influence of inelastic encounters in equation 6.50 is (in terms of equation 6.49) entirely equivalent to the corresponding terms found by the latter investigators. In the equivalent of equation 6.50 was the basis of Parker's and Lowke's investigations on longitudinal diffusion referred to in Chapter 4 (bibliography). Equation 6.50 is consistent with the results of Holstein (1946) if terms representing the influence of inverse encounters are removed [i.e., $q_{0k_s}(\epsilon)$ are all zero].

The generalized distribution function. The formula

$$f_0(c) = \text{const} \cdot \exp\left(-\frac{3m}{M} \int_0^c \frac{c\,dc}{V^2 + \overline{C^2}} \right)$$

becomes

$$f_0(\epsilon) = \text{const} \cdot \exp\left\{ -\int_0^\epsilon \frac{d\epsilon}{\left[e^2 E^2 M / 6N^2 m\epsilon q_{m_{\text{el}}}^2(\epsilon) \right] + \kappa T} \right\}. \quad (6.51)$$

6.9. MOTION OF ELECTRONS WHEN THE MOLECULAR ENERGY STATES ARE CLOSELY SPACED AND THE $\epsilon_{k_{\text{in}}}$ ARE SMALL

Equation 6.28 for $\sigma_{\text{coll}}(c)$ is

$$\sigma_{\text{coll}}(c) = 4\pi n_0 N \left(c^3 q_{m_{\text{el}}}(c) \left(\frac{m}{M} cf_0 + \frac{\overline{C^2}}{3} \frac{d}{dc} f_0 \right) \right.$$

$$+ \left\{ \int_c^{c_1} \left[f_0(x) - f_0\left(\sqrt{x^2 - c_{\text{in}}^2} \right) \exp\left(-\frac{\epsilon_{\text{in}}}{\kappa T} \right) \right] x^3 q_{0_{\text{in}}}(x) I(x - c_{\text{in}})\, dx \right\} \right)$$

$$(6.52)$$

when only a single kind of inelastic loss, represented by ϵ_{in}, is significant.

We here consider the integral term in equation 6.52 in the particular circumstances in which the mean energy $\frac{1}{2}m\overline{c^2}$ of the electrons moving in an electric field is much greater than $\frac{3}{2}\kappa T$ and, as in diatomic gases, the rotational transition energies are comparable with κT. In these circumstances, if c_{in} refers to a rotational transition, then in the integrand $x^2 \gg c_{in}^2$ and $f_0(\sqrt{x^2 - c_{in}^2}) \rightarrow f_0(x)$. The integrand then simplifies to the form

$$f_0(x) q'_{0_{in}}(x) x^3,$$

where $q'_0(x) = q_{0_{in}}(x)[1 - \exp(-\epsilon_{in}/\kappa T)]$ and $I(x - c_{k_{in}}) = 1$ for almost all the electrons. Moreover, $c_1 = (c^2 + c_{in}^2)1/2 \cong c + c_{in}^1/2c$ with $c_{in}^2/2c \ll c$. The integral term may therefore be replaced by

$$\frac{4\pi n_0 N f_0(c) q'_{0_{in}}(c) c^3 c_{in}^2}{2c};$$

consequently

$$\sigma_{coll}(c) \cong 4\pi n_0 N c^3 q_{m_{el}}(c) \left\{ \left[\frac{mc}{M} + \frac{q'_{0_{in}}(c)}{q_{m_{el}}(c)} \frac{c_{in}^2}{2c} \right] f_0 + \frac{\overline{C^2}}{3} \frac{d}{dc} f_0 \right\}. \quad (6.53)$$

Although equation 6.53 refers to a single rotational level, in practice in a diatomic gas other than hydrogen and hydrides, a range of rotational states exists at room temperatures with J values of the most highly populated states approximately equal to 8. There is thus a spectrum of rotational energy loss by electrons with velocity c in encounters with molecules. We therefore replace $q'_{0_{in}}(c) c_{in}^2/2q_{m_{el}}(c)$ by a mean value taken over all rotational energy levels. We designate this mean value $h(c, T)$. Equation 6.53 now becomes

$$\sigma_{coll}(c) = 4\pi n_0 N c^3 q_{m_{el}}(c) \left\{ \left[\frac{mc}{M} + \frac{h(c,T)}{c} \right] f_0 + \frac{\overline{C^2}}{3} \frac{d}{dc} f_0 \right\}.$$

The steady-state condition $\sigma_E(c) = \sigma_{coll}(c)$ now gives

$$4\pi n_0 N c^3 q_{m_{el}}(c) \left\{ \left[\frac{mc}{M} + \frac{h(c,T)}{c} \right] f_0 + \frac{VV' + \overline{C^2}}{3} \frac{d}{dc} f_0 \right\} = 0,$$

whence

$$f_0(c) = \text{const} \cdot \exp\left[-3 \int_0^c \frac{(mc/M) + h(c,T)/c}{VV' + \overline{C^2}} dc \right]. \quad (6.54)$$

When $h(c,T)/c$ greatly exceeds mc/M and also $VV' \gg \overline{C^2}$, then

$$f_0(c) \to \text{const} \cdot \exp\left[-3\left(\frac{mN}{eE}\right)^2 \int_0^c h(c,T) q_m(c) q_{m_{el}}(c) c\, dc \right]. \quad (6.55)$$

Equation 6.55 resembles Maxwell's distribution if, within the effective range of values of c, the factor $h(c,T)q_m(c)q_{m_{el}}(c)$ varies slowly with c. In terms of the energy $\epsilon = \frac{1}{2}mc^2$ equation 6.54 assumes the form (cf. equation 6.51)

$$f_0(\epsilon) = \text{const} \cdot \exp\left\{ - \int_0^\epsilon \frac{1 + H(\epsilon,T)/\epsilon}{(eE/N)^2 [M/6m\epsilon q_m(\epsilon) q_{m_{el}}(\epsilon)] + \kappa T} d\epsilon \right\},$$

$$(6.56)$$

where $H(\epsilon,T) = Mh(c,T)/2$. When κT can be neglected, this result is essentially that found by Frost and Phelps (1962).

BIBLIOGRAPHY AND NOTES

The earliest attempt to allow for the contributions of inelastic encounters was made by F. B. Pidduck, who attributed to the molecules a coefficient of restitution less than unity (see bibliography for Chapter 1.)

An early theoretical investigation of the effects of inelastic encounters, based on the association of inelastic encounters with changes in the internal energy state of an atom or molecule, was undertaken by Harriet W. Allen in 1937. She first investigated the influence of inelastic encounters which produce excitation and ionization, on the form of the generalized distribution function (molecules considered to be at rest) in helium, argon, and neon.

Similar, but more extensive, investigations were made by J. A. Smit (1936), whose calculated values of drift velocities were in reasonable agreement with the measured values in helium.

The modification of the Boltzmann equation (scalar equation) to accomodate inelastic losses was first undertaken by T. R. Holstein (1946). Terms of the form of the inelastic terms in equation 6.50 were obtained but no account was taken of superelastic encounters. D. Barbiere (1951) used Holstein's theory to determine the electron velocity distribution functions in helium, neon, and argon.

The problem of solving the scalar equation when inelastic encounters are included has been transformed by the advent of the electronic computer. The first thorough and extensive analyses based on the use of high-speed electronic computers were those of L. S. Frost and A. V. Phelps in 1962. This work is discussed in detail in Chapter 13.

H. W. Allen, *Phys. Rev.*, **52**, 707, 1937.

J. A. Smit, *Physica*, **3**, 543, 1936.

T. R. Holstein, *Phys. Rev.*, **70**, 367, 1946.

D. Barbiere, *Phys. Rev.*, **84**, 653, 1951.

L. S. Frost and A. V. Phelps, *Phys. Rev.*, **127**, 1621, 1962.

R. W. Crompton, *Advances in Electronics and Electron Physics*, Vol. 7, Academic Press, Inc., New York, 1969, pp. 1–18.

7

THE METHOD OF FREE PATHS

7.1. INTRODUCTION

In all that has preceded, the discussion of the motion of electrons in gases has been based on the scalar and vector equations 3.1 and 3.8, which are themselves derived from the Maxwell-Boltzmann equation 2.13. In this approach attention is directed to the rate of change of the population of the class $n(\mathbf{r}, t) d\mathbf{r} f(\mathbf{r}, \mathbf{c}, t) d\mathbf{c}$, with particular consideration of the cases in which $d\mathbf{c}$ is the shell (c, dc) of velocity space. Formulae for the velocity of drift $\mathbf{W}$, the coefficient of diffusion Ch attention is directed to the rate of change of the population of the class d, and other quantities may be derived as the sums of the contributions of the populations $d\mathbf{r}(nf_0) 4\pi c^2 dc$ of the shells (c, dc) at a particular instant. For instance, the drift velocity $\mathbf{W}$ is

$$\mathbf{W} = 4\pi \int_0^\infty \mathbf{W}(c) f_0(c) c^2 dc,$$

where $\mathbf{W}(c)$ is the mean velocity of the electrons associated with the shell (c, dc).

In the alternative method of free paths, attention is directed not to the instantaneous contributions of a class $nf d\mathbf{r} d\mathbf{c}$ whose members are continually changing even when its mean population does not vary, but rather to the behaviour in time of a group of the same electrons selected at $t = 0$. This group is in practice the group $d\mathbf{r}(nf_0) 4\pi c^2 dc$ at $t = 0$. For instance, the drift velocity is now found as the distance through which the centroid of the group of the same electrons is displaced in unit time. If this drift velocity is $\mathbf{W}(c)_{FP}$, the velocity of the centroid of the $n d\mathbf{r}$ electrons is

$$\mathbf{W}_{FP} = 4\pi \int_0^\infty \mathbf{W}(c)_{FP} f_0(c) c^2 dc.$$

Although **W** and **W**$_{FP}$ are the same, it should not be inferred from this fact that **W**(c) and **W**$_{FP}(c)$ are identical; in fact these velocities, although related, are not equal. Similarly, the procedures used to derive the coefficient of diffusion D are dissimilar in the two methods; in the Maxwell-Boltzmann approach the basic concept is particle flux in configuration space, but in the method of free paths the squares of the displacements of electrons between successive encounters are the important quantities.

The formulae derived later in this chapter are, for the sake of simplicity, restricted to the case of elastic scattering. Moreover, it is not proposed to derive anew all the formulae already established in the preceding chapters since it is sufficient, in order to illustrate the principles of the method of free paths, to treat in detail the single example of drift in an electric field in the absence of a magnetic field. It is not implied, however, that the method of free paths is restricted by its nature to simple examples. We shall also comment on the velocity distribution function, which is here derived by an alternative procedure.

Before it is possible to undertake this investigation it is necessary to establish some basic formulae relating to free paths.

7.2.   FREE TIMES AND FREE PATHS

Although formulae for scattering cross sections are given in Section 2.2, it is convenient to recall the following definitions:

$$\left.\begin{aligned}
q_0(g) &\cong q_0(c) = 2\pi \int_0^\pi p(\gamma,c)\sin\gamma\,d\gamma, \\[2ex]
q_1(c) &= 2\pi \int_0^\pi p(\gamma,c)\cos\gamma\sin\gamma\,d\gamma, \\[2ex]
q_m(c) &= q_0(c) - \left(\frac{g'}{g}\right)q_1(c) = 2\pi \int_0^\pi p(\gamma,c)(1-\cos\gamma)\sin\gamma\,d\gamma, \\[2ex]
\overline{\cos\gamma} &= \frac{q_1(c)}{q_0(c)},
\end{aligned}\right\} \quad (7.1)$$

in which γ is the angle of scattering of the relative velocity **g**$'$. When scattering is elastic, $g' = g$ and

$$1 - \overline{\cos\gamma} = \frac{q_m(c)}{q_0(c)}.$$

The number of particles scattered in time dt from an element of volume $d\mathbf{r}$ through the interaction of the elementary beams $nf\,d\mathbf{c}$ and $NF\,d\mathbf{C}$ is $(dt\,d\mathbf{r})[nNcq_0(c)]f\,d\mathbf{c}F\,d\mathbf{C}$, which, when summed for all molecular velocities $\mathbf{C}$ and all directions of the velocities $\mathbf{c}$, becomes, since $\int_{\mathbf{C}} F\,d\mathbf{C} = 1$,

$$(dt\,d\mathbf{r})(4\pi f_0 c^2\,dc)[nNcq_0(c)].$$

Consider a finite group of electrons, $n_0 = \int_{\mathbf{r}} n(\mathbf{r},t)\,d\mathbf{r}$. The number of these electrons whose velocity points lie in $d\mathbf{c}$ is $\int_{\mathbf{r}} n(\mathbf{r},t)[f(\mathbf{c},\mathbf{r},t)\,d\mathbf{c}]\,d\mathbf{r}$. But this number may be expressed in terms of a distribution function for the whole group, as follows (Section 4.2):

$$n_0 f^*(\mathbf{c},t)\,d\mathbf{c} = d\mathbf{c}\int_{\mathbf{r}} n(\mathbf{r},t)f(\mathbf{c},\mathbf{r},t)\,d\mathbf{r},$$

that is to say,

$$f^*(\mathbf{c},t) = \frac{1}{n_0}\int_{\mathbf{r}} n(\mathbf{r},t)f(\mathbf{c},\mathbf{r},t)\,d\mathbf{r}$$

and, in particular, if

$$f(\mathbf{c},\mathbf{r},t) = \sum_{k=0}^{\infty} f_k(c,\mathbf{r},t)P_k(\cos\theta)$$

and

$$f^*(\mathbf{c},t) = \sum_{k=0}^{\infty} f_k^*(c,t)P_k(\cos\theta),$$

then

$$f_k^*(c,t) = \frac{1}{n_0}\int_{\mathbf{r}} n(\mathbf{r},t)f_k(c,\mathbf{r},t)\,d\mathbf{r}.$$

The number of electrons of the group n_0 with velocity points in the shell (c,dc) is

$$n_{0c} = n_0 4\pi f_0^*(c,t)c^2\,dc.$$

The number of encounters made in time dt by members of the group n_{0c} is

$$n_{0c}[Ncq_0(c)]\,dt.$$

We select the group $n_{0c} = 4\pi n_0 f_0^*(c,t)c^2\,dc$ at $t=0$ and assume that f^* has effectively become independent of t, that is, $f_0^* \equiv f_0^*(c)$. As t increases, progressively more members of the group will have made at least one encounter with a molecule. We consider the class which comprises those electrons of the group n_{0c} which have not made an encounter since $t=0$. The population of this class is continually diminished by encounters, and in a sufficiently long time the class will, in effect, have disappeared. At time t we let the membership be $n_{0c}(t)$. The rate of change of the class is, at time t,

$$\frac{d}{dt} n_{0c}(t) = -Nc'q_0(c')n_{0c}(t), \qquad (7.2)$$

in which $c' \equiv c'(t)$ is the speed of an electron that makes an encounter in the interval dt at time t. When no electric force eE acts on the electrons, c' is equal at all times t to its value c at $t=0$. It follows from equation 7.2 that

$$n_{0c}(t) = n_{0c} \exp\left[-\int_0^t Nc'q_0(c') \, dt \right]. \qquad (7.3)$$

7.2.1. THE MEAN VALUE OF A FUNCTION $\phi(t)$. Let $\phi(t)$ be a quantity associated with the motion of an electron such that $\phi(t)$ is a function of the time of free flight following a selected instant $t=0$. We seek the mean value of $\phi(t)$, taken over all the n_{0c} first encounters made by the members of the group n_{0c}.

The number of encounters made in the interval dt at time t is given by $-(d/dt)[n_{0c}(t)] \, dt$, and the value of ϕ at each of these encounters is $\phi(t)$. Since the total number of encounters, spread over a long interval, is n_{0c}, it follows that the mean value of ϕ at an encounter is

$$\overline{\phi(t)} = -\frac{1}{n_{0c}} \int_0^\infty \phi(t) \frac{d}{dt}[n_{0c}(t)] \, dt. \qquad (7.4)$$

Suppose the function $\phi(t)$ to be such that

$$n_{0c}(t)\phi(t) \to 0 \quad \text{as } t \to \infty, \quad \text{and} \quad \phi(0) = 0.$$

Equation 7.4 is equivalent to

$$\overline{\phi(t)} = \frac{1}{n_{0c}} \left[-\phi(t)n_{0c}(t) \Big|_0^\infty + \int_0^\infty \frac{d\phi}{dt} n_{0c}(t) \, dt \right].$$

Consequently, when $\phi(t)$ possesses these properties, the first term on the right vanishes at both limits, and

$$\overline{\phi(t)} = \frac{1}{n_{0c}} \int_0^\infty \frac{d\phi}{dt} n_{0c}(t)\, dt,$$

that is, from equation 7.3,

$$\overline{\phi(t)} = \int_0^\infty \frac{d\phi}{dt} \left(\exp\left\{ -N \int_0^t c'(t) q_0[c'(t)]\, dt \right\} \right) dt. \tag{7.5}$$

7.2.2.   SPECIAL CASES.

(a) $c' = c = Constant$ along a Free Path—No Electric Field

Equation 7.5 simplifies to the form

$$\overline{\phi(t)} = \int_0^\infty \frac{d\phi}{dt} \exp\left(-\frac{t}{\tau_0} \right) dt \tag{7.6}$$

with $1/\tau_0 = Ncq_0 = \nu_0$.
 Let $\phi(t) = t^\alpha$; then

$$\overline{t^\alpha} = \alpha \int_0^\infty t^{\alpha-1} \exp\left(-\frac{t}{\tau_0} \right) dt = \alpha \Gamma(\alpha) \tau_0^\alpha. \tag{7.7}$$

It follows that the *mean free time* $\overline{t} = \tau_0$, $\overline{t^2} = 2\tau_0^2$, $\overline{t^3} = 6\tau_0^3$, and so forth.
 The distance travelled at constant speed c during a free time t is the free path $s = ct$. The *mean free path* is $\bar{s} = l_0 = c\tau_0$. Similarly, $\overline{s^2} = \overline{(ct)^2} = 2(c\tau_0)^2 = 2l_0^2$, $\overline{s^3} = 6l_0^3$, and so forth.

(b) $c' = c + \Delta c(t)$ with $\Delta c/c \ll 1$

In equation 7.5 replace $c'(t)q_0[c'(t)]$ in the integrand of the exponent by $cq_0(c) + (d/dc)[cq_0(c)]\Delta c$. Equation 7.5 is then replaced by the relation

$$\overline{\phi(t)} \cong \int_0^\infty \frac{d\phi}{dt} \left[1 - \frac{d}{dc}\left(\frac{1}{\tau_0} \right) \int_0^t \Delta c(t')\, dt' \right] \exp\left(-\frac{t}{\tau_0} \right) dt \tag{7.8}$$

with $1/\tau_0 = Ncq_0(c)$.

Consider the action of the force $e\mathbf{E}$, where $\mathbf{E}$ is uniform and independent of time, in changing c to c'. Let the Cartesian components of c be u, v, w with the component w in the direction of $e\mathbf{E}$. It follows that

$$c'^2 = u^2 + v^2 + \left(w + \frac{eE}{m}t\right)^2 = (u^2 + v^2 + w^2) + 2w\frac{eE}{m}t + \left(\frac{eE}{m}t\right)^2,$$

that is,

$$c' = c\left[1 + 2\frac{eEt}{mc}\cos\theta + \left(\frac{eE}{mc}t\right)^2\right]^{1/2} \qquad (7.9)$$

with $\cos\theta = w/c$.

Let $(eEt/mc) \ll 1$, that is, the speed acquired in the course of a free path be small compared with c. This is not true for free times that are many times greater than τ_0, but these form a negligible proportion of the total of n_{0c} paths. We may write therefore

$$c' \cong c + \left(\frac{eE}{m}t\right)\cos\theta = c + \Delta c(t)$$

with

$$\Delta c(t) = \left(\frac{eE}{m}t\right)\cos\theta,$$

and we assume that this expression is accurate for all free times t. It follows that equation 7.8 is equivalent to

$$\overline{\phi(t)} = \int_0^\infty \frac{d\phi}{dt}\left[1 - \frac{d}{dc}\left(\frac{1}{\tau_0}\right)\left(\frac{eE}{m}\cos\theta\right)\frac{t^2}{2}\right]\exp\left(-\frac{t}{\tau_0}\right)dt. \quad (7.10)$$

Equation 7.10 relates to an electron with velocity c and a component $w = c\cos\theta$ at $t = 0$. If $\phi \equiv \phi(\theta, t)$, the mean value of ϕ taken with respect to θ and t is

$$\overline{\phi(\theta,t)} = \int_0^\pi \int_0^\infty \frac{d\phi}{dt}\left[1 - \frac{d}{dc}\left(\frac{1}{\tau_0}\right)\left(\frac{eE}{m}\cos\theta\right)\frac{t^2}{2}\right]\left[\exp\left(-\frac{t}{\tau_0}\right)\right]\frac{\sin\theta\,d\theta}{2}\,dt.$$

$$(7.11)$$

(c) Unrestricted Case

It is necessary to employ the exact equation 7.9, which expresses c' as a function of the time. It follows from equation 7.5 that

$$\overline{\phi(\theta,t)} = \int_0^\pi \int_0^\infty \frac{d\phi}{dt}\left(\exp\left\{-N\int_0^t c'(t)q_0[c'(t)]\,dt\right\}\right)\frac{\sin\theta\,d\theta}{2}\,dt$$

with   (7.12)

$$c' = \left[c^2 + 2\frac{eEt}{m}c\cos\theta + \left(\frac{eEt}{m}\right)^2\right]^{1/2}.$$

The speed c' may be expressed in terms of powers of eEt/mc or of mc/eEt in the following way. When $eEt/mc = h < 1$,

$$c' = c(1 + 2h\cos\theta + h^2)^{1/2}$$

$$= c\left[1 + \frac{h^2}{3} + \sum_{k=0}^\infty (-1)^k\left(\frac{1}{2k+1} - \frac{h^2}{2k+5}\right)h^{k+1}P_{k+1}(\cos\theta)\right].\quad (7.13)$$

Equation 7.12 yields equation 7.10 when terms involving powers of h greater than unity are rejected.

When $mc/eEt = l < 1$,

$$c' = \frac{eEt}{m}(1 + 2l\cos\theta + l^2)^{1/2}$$

$$= \frac{eEt}{m}\left[1 + \frac{l^2}{3} + \sum_{k=0}^\infty (-1)^k\left(\frac{1}{2k+1} - \frac{l^2}{2k+5}\right)l^{k+1}P_{k+1}(\cos\theta)\right].\,(7.14)$$

7.3. DERIVATION OF FORMULA FOR DRIFT VELOCITY **W** BY THE METHOD OF FREE PATHS

7.3.1. THE INTERPRETATION OF **W** AS THE VELOCITY OF THE CENTROID.
It was shown in Chapter 4 that with the assumptions stated, the number density n in an isolated travelling group of n_0 electrons with the final distribution function $f_0^*(c)$ satisfies equation 4.38, which is

$$\frac{\partial}{\partial t}n = D\left(\frac{\partial^2 n}{\partial x^2} + \frac{\partial^2 n}{\partial y^2}\right) + D_L\frac{\partial^2 n}{\partial z^2} - W\frac{\partial n}{\partial z}.\quad (7.15)$$

In this equation $\mathbf{W}$ ($=\mathbf{i}_z W$) is the mean velocity $\bar{\mathbf{c}}$ of the electrons of the group. The expression for $\mathbf{W}$ given in equation 4.5 was derived by regarding $\mathbf{W}$ as the mean of the resultants $c\mathbf{f}_1^*/3f_0^*$ of the velocities of the electrons of all the shells (c, dc) taken at any instant of time. The interpretation of $\mathbf{W}$ as the velocity of the centroid is also consistent with equation 7.15, as will now be shown.

The vector position of the centroid of the group is

$$\mathbf{r}_0 = \frac{1}{n_0} \int_\mathbf{r} n(\mathbf{r}, t) \mathbf{r} \, d\mathbf{r}. \tag{7.16}$$

It is supposed as before that the group is isolated in space, that is, there is some large closed surface Σ enclosing the whole group and such that n and its derivatives are zero everywhere on Σ.

It follows from equation 7.16 that the velocity of the centroid of the group is

$$\frac{d}{dt} \mathbf{r}_0 = \frac{1}{n_0} \frac{d}{dt} \int_\mathbf{r} n(\mathbf{r}, t) \mathbf{r} \, d\mathbf{r} = \frac{1}{n_0} \int_\mathbf{r} \frac{\partial}{\partial t}(n) \mathbf{r} \, d\mathbf{r}.$$

We replace $(\partial/\partial t)n$ in the integrand by its equivalent from equation 7.15 and obtain

$$n_0 \frac{d}{dt} \mathbf{r}_0 = \int_\mathbf{r} \left[D\left(\frac{\partial^2 n}{\partial x^2} + \frac{\partial^2 n}{\partial y^2} \right) + D_L \frac{\partial^2 n}{\partial z^2} - W \frac{\partial n}{\partial z} \right] \mathbf{r} \, d\mathbf{r}$$

$$= \int\int\int_{-\infty}^{\infty} \left[D\left(\frac{\partial^2}{\partial x^2} n + \frac{\partial^2}{\partial y^2} n \right) + D_L \frac{\partial^2 n}{\partial z^2} - W \frac{\partial}{\partial z} n \right]$$

$$\times (\mathbf{i}_x x + \mathbf{i}_y y + \mathbf{i}_z z) \, dx \, dy \, dz.$$

Consider

$$\int\int\int_{-\infty}^{\infty} \left(\frac{\partial^2}{\partial x^2} n \right) x \, dx \, dy \, dz = \int\int_{-\infty}^{\infty} \left(x \frac{\partial}{\partial x} n - n \right)\Big|_{-\infty}^{\infty} dy \, dz = 0$$

since $\partial n/\partial x$ and n are zero at $x = \pm\infty$. Similarly,

$$\int\int\int_{-\infty}^{\infty} \left(\frac{\partial^2}{\partial x^2} n \right) y \, dx \, dy \, dz = \int\int_{-\infty}^{\infty} \left(\frac{\partial}{\partial x} n \Big|_{-\infty}^{\infty} \right) y \, dy \, dz = 0.$$

It is evident that all integrals of terms with D or D_L as coefficients are zero.

Consider

$$\int\int\int_{-\infty}^{\infty}\left(\frac{\partial}{\partial z}n\right)y\,dx\,dy\,dz = \int\int_{-\infty}^{\infty}\left(n\Big|_{-\infty}^{\infty}\right)y\,dy\,dx = 0$$

and

$$\int\int\int_{-\infty}^{\infty}\left(\frac{\partial}{\partial z}n\right)z\,dx\,dy\,dz = \int\int_{-\infty}^{\infty}\left[\int_{-\infty}^{\infty}\left(\frac{\partial}{\partial z}n\right)z\,dz\right]dx\,dy$$

$$= \int\int_{-\infty}^{\infty}\left(zn\Big|_{-\infty}^{\infty} - \int_{-\infty}^{\infty}n\,dz\right)dx\,dy = -n_0.$$

Thus

$$-W\int\int\int_{-\infty}^{\infty}\left(\frac{\partial}{\partial z}n\right)(\mathbf{i}_x x + \mathbf{i}_y y + \mathbf{i}_z z)\,dx\,dy\,dz = n_0(\mathbf{i}_z W) = n_0\mathbf{W},$$

and therefore

$$\frac{d}{dt}\mathbf{r}_0 = \mathbf{W}. \tag{7.17}$$

Thus **W** is the temporal rate of change of the vector position of the centroid, that is, **W** is the velocity of the centroid.

In the method of free paths this interpretation of the drift velocity is the basis of the derivation of a formula for **W**. It is evident that the formula derived in this manner must be equivalent to that derived from the scalar and vector equations since equation 7.15 is common to both methods. It will in fact be shown that the formula which is obtained is the alternative form of the expression for **W** in equation 4.5, which is

$$\mathbf{W} = \frac{e}{m}\frac{\mathbf{E}}{N}\frac{1}{3}\overline{\left[c^{-2}\frac{d}{dc}\frac{c^2}{q_m(c)}\right]}.$$

7.3.2. DERIVATION OF FORMULA FOR **W**.

Outline of procedure. As indicated in the preceding section, the derivation of a formula for the drift velocity **W** by means of the method of free paths resolves itself into finding the displacement of the centroid of a group of electrons during a specified interval of time of sufficient length that each electron of the group makes very many encounters with molecules. We begin by considering a group of n_{0c} electrons, all with velocity points located in a shell (c, dc) of velocity space at time $t = 0$. For convenience it is assumed that the resultant of their velocities c is zero at $t = 0$ so that the centroid of the group n_{0c} is initially at rest. We shall proceed in stages to calculate the mean displacement in the direction of the electric force eE of an electron of the group in a time T in which each electron traverses many free paths. It is assumed that all encounters are elastic.

A formula for W is first derived on the assumption that scattering is isotropic but that the collision cross section $q_0(c)$ is a function of the speed c. We next consider the scattering at encounters to be anisotropic and then allow for the fact that the mean value of the cosine of the scattering angle, namely, $\overline{\cos\gamma}$, may itself be a function of the speed c of an electron.

Drift velocity when scattering is isotropic. We postulate that the free paths are short and the electric force eE is of such strength that, if the difference in the speeds at the beginning and the end of a free path is $\Delta c(t) = c' - c$, then $\Delta c(t)/c \ll 1$ [case (b), Section 7.2.2]. Therefore, according to equation 7.9, we write

$$\Delta c(t) \cong \frac{eE}{m} t \cos\theta \ll c.$$

Consider an electron of the group whose velocity c at $t = 0$ has a direction lying within the elementary solid angle $d\omega = \sin\theta\, d\theta\, d\phi$. The spatial displacement that this electron receives in the direction of eE $(+Oz: \theta = 0)$ during the course of a free path in time t following $t = 0$ is

$$\Delta z_1 = wt + \frac{1}{2} \frac{eE}{m} t^2,$$

with $w = c \cos\theta$. The mean displacement of such electrons, whose points are within dc at $t = 0$, is found by replacing $\phi(t)$ by Δz_1 in equation 7.10; thus

$$\overline{\Delta z_1} = \int_0^\infty \left(c\cos\theta + \frac{eE}{m} t \right) \left[1 - \frac{d}{dc}\left(\frac{1}{\tau_0} \right)\left(\frac{eE}{m} \frac{t^2}{2} \cos\theta \right) \right] \exp\left(-\frac{t}{\tau_0} \right) dt$$

because $(d/dt)\Delta z_1 = c\cos\theta + (eE/m)t$.

The mean value of $\overline{\Delta z_1}$ taken with respect to θ is

$$\overline{\overline{\Delta z_1}} = \int_0^\pi \int_0^\infty \left(c\cos\theta + \frac{eE}{m}t \right)\left[1 - \frac{d}{dc}\left(\frac{1}{\tau_0}\right)\left(\frac{eE}{m}\frac{t^2}{2}\cos\theta\right) \right]$$

$$\times \left[\exp\left(-\frac{t}{\tau_0}\right) \right]\frac{\sin\theta\, d\theta}{2}\, dt.$$

It is convenient to integrate first with respect to θ, thus eliminating terms in $\cos\theta$, and then with respect to t. It is then found that

$$\overline{\overline{\Delta z_1}} = -\frac{c}{3}\frac{eE}{m}\frac{d}{dc}\left(\frac{1}{\tau_0}\right)\tau_0^3 + \frac{eE}{m}\tau_0^2$$

$$= \left(\frac{eE}{m}\tau_0\right)\left(\frac{c}{3}\frac{d}{dc}\tau_0 + \tau_0\right) = \left(\frac{eE}{m}\tau_0\right)\frac{1}{3c^2}\frac{d}{dc}(c^3\tau_0).$$

The mean displacement in the course of any free path begun at speed c has this value. The mean free time is found by replacing $\phi(t)$ by t in equation 7.11, whence

$$\overline{\overline{t}} = \int_0^\pi \int_0^\infty \left[1 - \frac{d}{dc}\left(\frac{1}{\tau_0}\right)\frac{eE}{m}\frac{t^2}{2}\cos\theta \right]\left[\exp\left(-\frac{t}{\tau_0}\right) \right]\frac{\sin\theta\, d\theta}{2}\, dt = \tau_0.$$

In a time $T \gg \tau_0$, but such that electrons in elastic encounters essentially retain their speeds c throughout, the number of encounters made by an electron on the average is T/τ_0, and the mean aggregate displacement of an electron in time T is $(T/\tau_0)\overline{\overline{\Delta z_1}}$. Thus, if the displacement Δz_1 were the only one to be considered, the velocity of the centroid of the group n_{0c} would be

$$W_{\text{centroid}}(c) = \frac{(T/\tau_0)\overline{\overline{\Delta z_1}}}{T} = \frac{\overline{\overline{\Delta z_1}}}{\tau_0},$$

that is,

$$W_{\text{centroid}}(c) = \frac{eE}{m}\frac{1}{3c^2}\frac{d}{dc}(c^3\tau_0)$$

$$= \frac{eE}{mN}\frac{1}{3c^2}\frac{d}{dc}\frac{c^2}{q_0(c)}. \tag{7.18}$$

This is the expression for $W_{centroid}(c)$ when electrons are scattered isotropically ($\overline{\cos\gamma} = 0$) in all encounters.

Drift velocity when scattering is not isotropic; $\overline{\cos\gamma} \neq 0$. In calculating the mean displacement, it was assumed that all directions (θ,ϕ) of the velocities of the electrons on beginning their free paths were equally probable, that is, that the mean velocity of electrons immediately following the encounters that terminate their first free path was zero. In order to calculate correctly the mean displacement when scattering is anisotropic and this assumption is, therefore, no longer valid, we must add to this displacement the sum of the mean displacements during succeeding free paths because, on the average, the mean residual velocity of an electron after the encounters that terminate its free paths is not zero. Since velocities possessed by the electrons immediately prior to their first encounters are not completely randomized until each electron of the group has made several encounters, it is necessary to sum the displacements due to the residual velocities over a succession of free paths.

Consider electrons that at time $t=0$ possess velocities c with directions (γ,ϕ) relative to Oz. The mean displacement of these electrons in the times of free flight before their first encounter is, as before, $\overline{\Delta z_1}$. At the encounters that terminate the free paths, the electrons are scattered in all directions γ with respect to their initial direction (θ,ϕ). When scattering is isotropic, the mean residual velocity after scattering is zero, but in the present instance ($\overline{\cos\gamma} \neq 0$), the mean residual velocity in the direction of Oz is $[c\,\overline{\cos\gamma}\cos\theta+(eE/m)\tau_0\,\overline{\cos\gamma}]$ since the mean velocity of the electrons at the end of their first free path is $[c+(eE/m)\tau_0]$. After the encounters that terminate the second free paths, the mean residual velocity in the direction of Oz is $[c(\overline{\cos\gamma})^2\cos\theta+(eE/m)\tau_0(\overline{\cos\gamma})^2]$, and after the encounters that terminate the kth free paths it is $[c\cos\theta+(eE/m)\tau_0]$ $(\overline{\cos\gamma})^k$. It follows from equation 7.11, in which $d\phi/dt$ is given the value $[c\cos\theta+(eE/m)\tau_0](\overline{\cos\gamma})^k$, that the mean displacement in the $(k+1)$th free paths is

$$\left(\overline{\cos\gamma}\right)^k\left(\frac{eE}{m}\tau_0\right)\left[-\frac{c}{3}\frac{d}{dc}\left(\frac{1}{\tau_0}\right)\tau_0^2+\tau_0\right]$$

$$=\left(\overline{\cos\gamma}\right)^k\left(\frac{eE}{m}\tau_0\right)\frac{1}{3c^2}\frac{d}{dc}\left(c^3\tau_0\right),$$

when a term involving $(eE/m)^2$ is omitted as small. The total displacement on the average parallel to Oz is therefore

$$\overline{\Delta z} = \overline{\Delta z_1} + \sum_{k=1}^{\infty} (\overline{\cos\gamma})^k \left(\frac{eE}{m}\tau_0\right)\frac{1}{3c^2}\frac{d}{dc}(c^3\tau_0)$$

$$= \left(\frac{eE}{m}\tau_0\right)\frac{1}{3c^2}\frac{d}{dc}(c^3\tau_0)\sum_{k=0}^{\infty}(\overline{\cos\gamma})^k$$

$$= \left(\frac{eE}{m}\tau_0\right)\frac{1}{3c^2}\frac{d}{dc}(c^3\tau_0)\left(\frac{1}{1-\overline{\cos\gamma}}\right).$$

Effect of dependence of $\overline{\cos\gamma}$ *upon the speed c.* Yet another displacement remains to be considered. The factor $(\overline{\cos\gamma})^k$ has been given the value appropriate to a speed at impact equal to c, whereas the actual speed is $c+(eE/m)t\cos\theta$. Since $\Delta c(t)/c = eEt/mc \ll 1$, we replace $(\overline{\cos\gamma})^k$ by

$$(\overline{\cos\gamma})^k + \frac{d}{dc}(\overline{\cos\gamma})^k\left(\frac{eE}{m}\cos\theta\right)t$$

with the average value

$$(\overline{\cos\gamma})^k + \left(\frac{eE}{m}\tau_0\cos\theta\right)\frac{d}{dc}(\overline{\cos\gamma})^k.$$

The displacement during the $(k+1)$th free path in time t after the kth encounter now becomes

$$t\left(c\cos\theta + \frac{eE}{m}\tau_0\right)\left[(\overline{\cos\gamma})^k + \left(\frac{eE}{m}\tau_0\cos\theta\right)\frac{d}{dc}(\overline{\cos\gamma})^k\right].$$

The mean value of the additional contribution to the displacement is

$$\frac{d}{dc}(\overline{\cos\gamma})^k \int_0^\pi \int_0^\infty \left(c\cos\theta + \frac{eE}{m}\tau_0\right)\left(\frac{eE}{m}\tau_0\cos\theta\right)\left[1 - \frac{d}{dc}\left(\frac{1}{\tau_0}\right)\frac{eE}{m}\frac{t^2}{2}\cos\theta\right]$$

$$\times \left[\exp\left(-\frac{t}{\tau_0}\right)\right]\frac{\sin\theta\,d\theta}{2}\,dt,$$

which reduces, when powers of $(eE/m)\tau_0$ greater than unity are omitted, to

$$\frac{d}{dc}(\overline{\cos\gamma})^k\left(\frac{eE}{m}\tau_0\right)\left(\frac{c\tau_0}{3}\right).$$

The sum of such displacements for all values of k, starting with unity, is

$$\left(\frac{eE}{m}\tau_0\right)\left(\frac{c\tau_0}{3}\right)\frac{d}{dc}\left(\frac{\overline{\cos\gamma}}{1-\overline{\cos\gamma}}\right).$$

Thus, when $\overline{\cos\gamma}\neq0$, an additional term $\overline{\Delta z_2}$ must be added to the mean displacement $\overline{\Delta z_1}$, where

$$\overline{\Delta z_2}=\left(\frac{eE}{m}\tau_0\right)\frac{1}{3c^2}\left[\frac{d}{dc}(c^3\tau_0)\left(\frac{\overline{\cos\gamma}}{1-\overline{\cos\gamma}}\right)+c^3\tau_0\frac{d}{dc}\left(\frac{\overline{\cos\gamma}}{1-\overline{\cos\gamma}}\right)\right]$$

$$=\left(\frac{eE}{m}\tau_0\right)\frac{1}{3c^2}\frac{d}{dc}\left(\frac{c^3\tau_0\overline{\cos\gamma}}{1-\overline{\cos\gamma}}\right).$$

The total mean displacement of an electron in the direction of $e\mathbf{E}$ that results from the action of $e\mathbf{E}$ on it in the course of a single free path together with the persistence of velocity is

$$\overline{\Delta z}=\overline{\Delta z_1}+\overline{\Delta z_2}$$

$$=\left(\frac{eE}{m}\tau_0\right)\frac{1}{3c^2}\frac{d}{dc}\left(c^3\tau_0+\frac{c^3\tau_0\overline{\cos\gamma}}{1-\overline{\cos\gamma}}\right)$$

$$=\left(\frac{eE}{m}\tau_0\right)\frac{1}{3c^2}\frac{d}{dc}(c^3\tau),$$

where $\tau=\tau_0/(1-\overline{\cos\gamma})=1/[Ncq_{0_{el}}(1-\overline{\cos\gamma})]=1/Ncq_{m_{el}}(c)$. The total displacement of the centroid of the group of n_{0c} electrons in time $T\gg\tau_0$ is $(T/\tau_0)[(eE/m)\tau_0](1/3c^2)(d/dc)(c^3\tau)$; consequently the mean velocity of the centroid is

$$\mathbf{W}_{centroid}(c)=\frac{e\mathbf{E}}{m}\frac{1}{3c^2}\frac{d}{dc}(c^3\tau)$$

$$=\frac{e\mathbf{E}}{mN}\frac{1}{3c^2}\frac{d}{dc}\frac{c^2}{q_{m_{el}}(c)}.\tag{7.19}$$

The velocity of the centroid of the whole group of n_0 electrons is

$$\mathbf{W}_{centroid}=\mathbf{W}=4\pi\int_0^\infty\mathbf{W}_{centroid}(c)f_0^*(c)c^2\,dc$$

$$= \frac{e\mathbf{E}}{mN} \frac{4\pi}{3} \int_0^\infty \frac{d}{dc}\left(\frac{c^2}{q_{m_{\text{el}}}(c)}\right) f_0^*(c)\, dc. \tag{7.20}$$

This expression for $\mathbf{W}_{\text{centroid}}$ is in agreement with equation 4.5. It is of interest that the expression for **W** derived from the vector equation is

$$\mathbf{W} = -\frac{e\mathbf{E}}{mN} \frac{4\pi}{3} \int_0^\infty \frac{c^2}{q_{m_{\text{el}}}(c)} \frac{d}{dc} f_0^*(c)\, dc$$

and that this expression can be transformed into equation 7.20 by partial integration and vice versa.

Equation 7.20 can also be expressed in the form

$$\mathbf{W}_{\text{centroid}} = \frac{e\mathbf{E}}{mN} \frac{1}{3} \overline{\left[c^{-2} \frac{d}{dc} \frac{c^2}{q_{m_{\text{el}}}(c)} \right]}. \tag{7.21}$$

As already remarked in Section 7.1, although $\mathbf{W}_{\text{centroid}}$ and **W** are equal, the velocities $\mathbf{W}_{\text{centroid}}(c)$ and $\mathbf{W}(c)$ are, in general, not equal. The velocity $\mathbf{W}_{\text{centroid}}(c)$ is the velocity of the centroid of the travelling group of n_{0c} electrons, selected according to the criterion that their velocity points were within the shell (c, dc) at $t = 0$. Then $\mathbf{W}_{\text{centroid}}(c)$ was derived from the displacements in space of these electrons subsequent to $t = 0$. On the other hand, $\mathbf{W}(c)$ is found as the instantaneous mean velocity $c\mathbf{f}_1^*(c)/3f_0^*(c)$ of the electrons whose velocity points lie within the shell (c, dc), that is to say, $\mathbf{W}(c)$ is the mean velocity of a class of electrons at a given instant. There is thus no a priori reason to suppose that $\mathbf{W}_{\text{centroid}}(c)$ and $\mathbf{W}(c)$ should be the same for all values of c, although the discussion in Section 7.3.1 shows that their mean values taken over all shells (c, dc) should be equal.

On the other hand, the distribution function $f_0^*(c)$ is the same in each case since it determines the populations of the shells (c, dc) and of the selected groups n_{0c}.

The group n_{0c} selected at $t = 0$ has been discussed as if its mean velocity were zero from the outset; in other words, the shell (c, dc) was regarded as being uniformly populated. In fact, the distribution of point density in the shell is of the form $n_0 \sum_0^\infty f_k^*(c) P_k(\cos\theta)$ with the first term $f_0^*(c)$ predominant. The mean velocity of the group n_{0c} at $t = 0$ is therefore $c\mathbf{f}_1^*(c)/3f_0^*(c)$, and after each electron has covered, on the average, k free paths this velocity has been reduced to $[c\mathbf{f}_1^*(c)/3f_0^*(c)](\overline{\cos\gamma})^k$ with a contribution to the displacement during the $(k+1)$th free path equal to $[c\mathbf{f}_1^*(c)/3f_0^*(c)]\tau_0(\overline{\cos\gamma})^k$. The total contribution to the displacement that is associated with the dissipation of the initial mean velocity of the group is

$$\frac{cf_1^*(c)}{3f_0^*(c)} \frac{\tau_0}{1 - \overline{\cos\gamma}} = \left[\frac{cf_1^*(c)}{3f_0^*(c)} \right] \tau.$$

This is a fixed magnitude which does not change with time; consequently over a large number of free paths, $T/\tau_0 \gg 1$, this quantity when divided by T in the final expression for $W(c)$ contributes a term that can be made negligible when T is sufficiently large, that is to say, $T/\tau_0 \gg 1$.

7.4. THE MOMENTUM OF THE ELECTRONS OF A SHELL (c, dc) IN THE STEADY STATE OF MOTION

In what has preceded it is shown that in the method followed the initial mean velocity $cf_1^*/3f_0^*$ of the group $n_{0c} = n_0 4\pi f_0^* c^2 dc$ electrons whose velocity points occupy the shell (c, dc) at $t = 0$ need not be taken into account. This initial mean velocity of the shell, which is dependent on the term $f_1^*(c)$, is of importance and interest, however, and we shall therefore seek to determine its value by the method of free paths.

Consider the subgroup of the n_{0c} electrons constituted by the electrons whose velocity points at $t = 0$ lie within the element $dc = c^2 dc d\omega = c^2 dc \sin\theta\, d\theta\, d\phi$. The momentum of an electron of the subgroup is mc. If the electric force eE is removed at $t = 0$, these electrons continue to move with relatively small reduction in speed, even when each has made a number of encounters, since it is assumed that all encounters are elastic. On the other hand, the momentum $n_{0c} mc$ of the group is dissipated after each electron has made relatively few encounters since the directions of travel of the electrons have then become dispersed. The average value of the residual momentum per electron over the second free path is $mc\, \overline{\cos\gamma}$, and over the $(k+1)$th free path it is $mc(\overline{\cos\gamma})^k$. Since $\overline{\cos\gamma} < 1$, the momentum of the subgroup is progressively reduced. For instance, if $\overline{\cos\gamma} = \frac{1}{5}$, after four encounters the momentum of an electron is, on the average, $mc/625$, whereas its energy (say in helium) is reduced merely by an amount that is of the order of magnitude

$$\frac{2m}{M}\left(\tfrac{1}{2}mc^2\right) \times 4 \cong 10^{-3} \times \tfrac{1}{2}mc^2.$$

The speeds c of the electrons therefore remain effectively constant over many encounters. Thus there are intervals $T \gg \tau_0$ such that the momentum of the group vanishes but the energy undergoes negligible change. The distribution function becomes isotropic with the form $f_0^*(c)$ of the isotropic term of $f^*(c)$ in the presence of $e\mathbf{E}$, since the velocity points of the group n_{0c} remain within the same shell (c,dc) but are distributed within it at uniform density.

Next, suppose that the force $e\mathbf{E}$ is restored. The group n_{0c} begins immediately to acquire mean momentum. In the first free path the average momentum acquired is $e\mathbf{E}\tau_0$, and of this $e\mathbf{E}\tau_0 \overline{\cos\gamma}$ carries over into the second free path on the average. The momentum at the second encounter is $e\mathbf{E}\tau_0(1 + \overline{\cos\gamma})$; after it, $e\mathbf{E}\tau_0[\overline{\cos\gamma} + (\overline{\cos\gamma})^2]$. Thus, after the kth encounter, the momentum is

$$e\mathbf{E}\tau_0(\overline{\cos\gamma} + \overline{\cos\gamma}^2 + \cdots + \overline{\cos\gamma}^k) = e\mathbf{E}\tau_0 \frac{\overline{\cos\gamma}}{1 - \overline{\cos\gamma}}(1 - \overline{\cos\gamma}^k),$$

which tends, as k increases, to the limiting value

$$\left(\frac{e\mathbf{E}\tau_0}{1 - \overline{\cos\gamma}}\right)\overline{\cos\gamma}.$$

This is the mean momentum of the electrons in the group n_{0c} immediately after an encounter when the group has acquired a steady momentum. When the group n_{0c} is selected from velocity space at time $t = 0$, its members will have travelled for an average time τ_0 since their last encounters, during which additional momentum equal, on the average, to $e\mathbf{E}\tau_0$ is acquired. The total momentum of the group at $t = 0$ is therefore

$$n_{0c}e\mathbf{E}\tau_0\left(1 + \frac{\overline{\cos\gamma}}{1 - \overline{\cos\gamma}}\right) = \frac{n_{0c}e\mathbf{E}\tau_0}{1 - \overline{\cos\gamma}} = n_{0c}e\mathbf{E}\tau$$

and the mean velocity of the group is

$$\frac{e\mathbf{E}}{m}\tau = \frac{e\mathbf{E}}{m}\frac{1}{Ncq_{0_{el}}(1 - \overline{\cos\gamma})} = \frac{e\mathbf{E}}{mNcq_{m_{el}}(c)}.$$

Thus the final result of the presence of the field is to give the electrons of the shell (c,dc) a mean velocity $(e\mathbf{E}/m)\tau$. The shell, in a steady state of motion, is thus displaced as a whole with its centre now at the vector position $\mathbf{V} = (e\mathbf{E}/m)\tau$. The displacement of the shell is illustrated in Figure 7.1. In a steady state of motion the mean momentum of an electron after an encounter is $e\mathbf{E}\tau \overline{\cos\gamma}$, and at the end of the free path it is increased to $e\mathbf{E}\tau$ by the addition of momentum $e\mathbf{E}\tau_0$ by the field. Thus

$$e\mathbf{E}(\tau \overline{\cos\gamma} + \tau_0) = e\mathbf{E}\tau_0\left(\frac{\overline{\cos\gamma}}{1 - \overline{\cos\gamma}} + 1\right) = e\mathbf{E}\tau$$

as required.

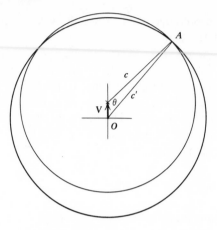

$(V/c) \ll 1, \quad f_0^*(c) = f_0^*(c') + f_1^*(c') \cos \theta$

FIG. 7.1

The value of the mean equilibrium momentum of an electron of a shell may be found directly as follows. Let it be $m\mathbf{V}$; then immediately after its last encounter the average momentum of an electron is $m\mathbf{V} \overline{\cos \gamma}$. At any instant the electrons of the shell (c, dc) will have travelled for an average time τ_0 since the last encounter, during which the momentum is increased by $e\mathbf{E}\tau_0$ to the value $m\mathbf{V}$ that the shell possesses. Consequently,

$$m\mathbf{V} \overline{\cos \gamma} + e\mathbf{E}\tau_0 = m\mathbf{V},$$

whence

$$m\mathbf{V} = \frac{e\mathbf{E}\tau_0}{1 - \overline{\cos \gamma}} = e\mathbf{E}\tau,$$

that is,

$$\mathbf{V} \equiv \mathbf{V}(c) = \frac{e\mathbf{E}}{m}\tau. \tag{7.22}$$

7.4.1. ALTERNATIVE DERIVATION OF FORMULA FOR $W(c)$. We have seen that, if we begin with a group of $n_{0c} = n_0 4\pi f_0^*(c)c^2 dc$ electrons whose velocity points are distributed with constant number density within the shell (c, dc), that is, their centroid is at rest, then after each electron has made relatively few encounters with molecules the mean momentum of the group will have acquired a fixed value $n_{0c}m\mathbf{V} = n_{0c}e\mathbf{E}/\nu_{el}$, which is independent of time. The velocity points of the electrons now occupy a shell (c, dc) whose centre is no longer the origin but rather the end point of the vector $\mathbf{V}$ (Figure 7.1). The velocity points are still distributed uniformly within this displaced shell. On the other hand, the number density of points in the undisplaced shell (c, dc) centred on the origin is not constant in the steady state but is $n_0[f_0^*(c) + f_1^*(c) \cos\theta]$, where n_0 is the population of the complete group. The distribution of points in velocity space when equilibrium is achieved in a field $\mathbf{E}$ therefore has this property: the number density of points within a shell (c, dc) centred on the origin is given by $n_0[f_0^*(c) + f_1^*(c) \cos\theta]$, whereas in a displaced shell, whose centre is at the end of the vector $\mathbf{V} = e\mathbf{E}/m\nu_{\text{el}}$ and whose radius is also c, the number density is constant and equal to $n_0 f_0^*(c)$.

In Figure 7.1 let A be a point on the surface of the displaced sphere with radius c and centre the end of the vector $\mathbf{V}$, and let c' be the distance of A from the origin O. Then

$$c'^2 = c^2 + 2cV\cos\theta + V^2;$$

consequently, if $(V/c) \ll 1$, then $c' \cong c + V\cos\theta$.

Since A is a point on the displaced sphere, the number density at A is $n_0 f_0^*(c)$. But A is also the point $c'(\theta)$ relative to the origin O, and in these terms the number density is given by $n_0[f_0^*(c') + f_1^*(c') \cos\theta]$. It follows that

$$f_0^*(c) = f_0^*(c') + f_1^*(c') \cos\theta$$

$$\cong f_0^*(c) + (V\cos\theta)\frac{d}{dc}f_0^*(c) + f_1^*(c)\cos\theta + \text{small terms},$$

and that

$$f_1^*(c) = -V\frac{d}{dc}f_0^*(c) = -\frac{eE}{m\nu_{\text{el}}}\frac{d}{dc}f_0^*(c), \qquad (7.23)$$

where $\nu_{\text{el}} = Ncq_{m_{\text{el}}}(c)$. This expression for $f_1^*(c)$ is the one already found from the vector equation when integrated to refer to a spatially distributed group n_0. It is valid for the subgroup $n(\mathbf{r})d\mathbf{r}$, for which the distribution function is

$$f(\mathbf{r}, c) = f_0(\mathbf{r}, c) + f_1(\mathbf{r}, c)\cos\theta + \cdots,$$

with

$$f_1 = -\frac{eE}{m}\tau\frac{d}{dc}f_0.$$

The magnitude of the mean velocity of the electrons of the shell (c, dc) at any instant is

$$W(c) = \frac{2\pi\left[\int_0^\pi (c^3 \cos\theta)(f_0^* + f_1^* \cos\theta)\sin\theta\, d\theta\, dc\right]}{4\pi f_0^* c^2\, dc}$$

$$= \frac{cf_1^*}{3f_0^*} = -\frac{ceE\tau}{3mf_0} \frac{d}{dc} f_0^*.$$

Thus

$$\mathbf{W}(c) = -\frac{eE}{3mNq_{m_{el}}(c)} \frac{1}{f_0^*} \frac{d}{dc} f_0^*,$$

and the mean velocity of the electrons in the whole group n_0 is

$$\mathbf{W} = \frac{n_0 \int_0^\infty 4\pi f_0^*(c) c^2 \mathbf{W}(c)\, dc}{n_0}$$

$$= -\left(\frac{4\pi}{3}\right)\left(\frac{eE}{mN}\right)\int_0^\infty \frac{c^2}{q_{m_{el}}(c)} \frac{d}{dc} f_0^*(c)\, dc. \qquad (7.24)$$

This expression is the standard formula for the drift velocity, which we have frequently discussed.

7.5. THE VELOCITY DISTRIBUTION FUNCTION $f_0^*(c)$

In Chapter 4 the isotropic component $f_0^*(c)$ of the velocity distribution function for an isolated group of n_0 electrons in a steady state of motion was found through the requirement that the total flux of velocity points be zero across the surface of a sphere c with the origin as centre. This criterion led to the equation

$$\sigma_E^*(c) = \sigma_{coll}^*(c) \qquad (7.25)$$

since σ_E^* is an outwardly directed and σ_{coll}^* an inwardly directed flux. Expressions for these fluxes were derived on the supposition that the number density of velocity points on the sphere c whose centre is the origin is $n_0[f_0^*(c) + f_1^*(c)\cos\theta]$ with $f_1 \ll f_0$. It was found that

$$\sigma_E^*(c) = n_0 \frac{4\pi}{3} c^2 \frac{eE}{m} f_1^*(c)$$

and that, when all encounters are elastic,

$$\sigma_{coll}^*(c) = n_0 4\pi c^2 \nu_{el}\left(\frac{mc}{M} f_0^* + \frac{\overline{C^2}}{3} \frac{d}{dc} f_0^*\right).$$

$\left.\begin{array}{c} \\ \\ \end{array}\right\} \qquad (7.26)$

According to the assumption made above, the $n_0 4\pi f_0^*(c) c^2 dc$ velocity points in a shell (c, dc) are not distributed with constant number density, and they represent the motion of the associated subgroup of electrons relative to a coordinate system in which the gas is at rest, that is to say, has no mass motion.

We note, however, that the criterion for statistical equilibrium—that the total flux of points across a closed surface be zero—is not restricted to the spherical surface with the origin of velocity space as centre but is true of any closed surface in velocity space. This spherical surface was chosen for convenience since the components σ_E^* and σ_{coll}^* could be readily calculated and the criterion of zero total flux then expressed in the form of equation 7.25, from which $f_0^*(c)$ can be found.

It is natural to enquire whether some other closed surface can also conveniently serve this purpose, and a suitably displaced sphere is found to be equally convenient. Consider the spherical surface with radius c in velocity space but with its centre at the end of the vector $\mathbf{V} = (e\mathbf{E}/m)\tau$. From what was discussed in the preceding section it is seen that the number density of points is constant over the surface of this sphere, the displaced sphere, because the component $f_1^*(c)\cos\theta$ of the distribution function is zero on it. The number density of velocity points is constant within the displaced shell (c, dc), but the velocity of the centroid of the associated $n_0 4\pi f_0^*(c) c^2 dc$ electrons is $\mathbf{V}$ relative to the gas. Although the total flux of velocity points across the surface of the displaced sphere is still zero, it no longer comprises a component dependent on $e\mathbf{E}$ and analogous to $\sigma_E^*(c)$ in equation 7.25. This follows because the drift of velocity points $e\mathbf{E}/m$ is the same everywhere throughout velocity space and because the number density of points is constant over the surface. Thus points enter and leave the sphere at the same rate. We note alternatively that in the expression for σ_E^* in equation 7.26 the factor f_1^* is now equal to zero. The flux is therefore entirely a consequence of encounters between electrons and molecules. Hence, we denote the total flux over the surface of the displaced sphere by $\sigma_{\text{coll}_D}^*(c)$ so that the criterion for equilibrium becomes

$$\sigma_{\text{coll}_D}^*(c) = 0. \tag{7.27}$$

We have noted that the centroid of the electrons associated with the displaced shell possesses a velocity $\mathbf{V}$ relative to the gas; consequently, if the gas is given a mass motion $-\mathbf{V}$, the electric force $e\mathbf{E}$ is able to hold the centroid of these electrons at rest. The molecules now possess velocities $\mathbf{C} - \mathbf{V}$, and since f_1^* is zero the electrons of the shell move at random among them. The square of the speed of a molecule is $(\mathbf{C} - \mathbf{V})^2 = C^2 - 2\mathbf{C} \cdot \mathbf{V} + V^2$, and the mean square speed is $\overline{C^2} + V^2$. The discussion in Appendix B shows that $\sigma^*_{\text{coll}_D}(c)$ is to be derived from $\sigma^*_{\text{coll}}(c)$ of equation 7.26 by replacing $\overline{C^2}$ by $\overline{C^2} + V^2$. It follows therefore that equation 7.27 is equivalent to

$$\frac{3mc}{M} f_0^*(c) + (\overline{C^2} + V^2) \frac{d}{dc} f_0^*(c) = 0,$$

which as before gives the generalized distribution function

$$f_0^*(c) = A \exp\left(-\frac{3m}{M} \int_0^c \frac{c\,dc}{V^2 + \overline{C^2}} \right).$$

We remark that $V(c)$ is a function of c, so that the same value of V is not in general shared by shells with different radii c. Thus, although eE can hold the centroid of the electrons of the particular shell (c, dc) at rest in a gas with a mass motion $-V$, the centroids of electrons of other shells will in general move under the combined action of eE and the molecular stream. An exception to this behaviour occurs when $q_{m_{\text{el}}}(c) \propto c^{-1}$, so that $\tau = 1/\nu_{\text{el}}$ is constant, in which event the velocity V is the same for all shells and the force eE can hold all shell electrons at rest in a gas with mass motion $-V$, which now becomes the drift velocity W. In this quasi thermal state of equilibrium the distribution function becomes Maxwellian, but the mean energy of an electron is $\frac{1}{2}m\overline{c^2} = \frac{1}{2}M(V^2 + \overline{C^2})$, in accordance with Pidduck's formula (equation 3.16). It has been supposed throughout this section that $(V/c) \ll 1$.

7.6. GENERAL EXPRESSION FOR DRIFT VELOCITY IN THE UNRESTRICTED CASE

We remove the restriction that $(eE/m)\tau \ll c$; that is to say, the speed added by the force eE in the course of a free path need not be small in comparison with c.

Equation 7.12 is

$$\overline{\phi(\theta, t)} = \int_0^\pi \int_0^\infty \frac{d\phi}{dt} \left(\exp\left\{ -N \int_0^t c'(t) q_0[c'(t)]\,dt \right\} \right) \frac{\sin\theta\,d\theta}{2}\,dt$$

with

$$c' = \left[c^2 + 2\frac{eE}{m} tc\cos\theta + \left(\frac{eE}{m}t\right)^2 \right]^{1/2},$$

for which expansions are given in equations 7.13 and 7.14.

The displacement in the direction of $e\mathbf{E}$ in the course of a free path of duration t is

$$tc\cos\theta + \frac{eE}{m}\frac{t^2}{2} = \Delta z.$$

If we put $\phi(\theta,t) = \Delta z$, it follows that

$$\overline{\Delta z} = \int_0^\pi \int_0^\infty \left(c\cos\theta + \frac{eE}{m}t\right)\left(\exp\left\{-N\int_0^t c'(t)q_0[c'(t)]\,dt\right\}\right)\frac{\sin\theta\,d\theta}{2}\,dt.$$

The mean free time $\overline{t}$ is

$$\overline{t} = \int_0^\pi \int_0^\infty \left(\exp\left\{-N\int_0^t c'(t)q_0[c'(t)]\,dt\right\}\right)\frac{\sin\theta\,d\theta}{2}\,dt.$$

The expression for $W(c)$ when the displacements Δz are the only type to be considered, that is, when scattering is isotropic, is

$$W_{\text{isotropic}}(c) = \frac{\displaystyle\int_0^\pi \int_0^\infty \left(c\cos\theta + \frac{eE}{m}t\right)\left(\exp\left\{-N\int_0^t c'(t)q_0[c'(t)]\,dt\right\}\right)\frac{\sin\theta\,d\theta}{2}\,dt}{\displaystyle\int_0^\pi \int_0^\infty \left(\exp\left\{-N\int_0^t c'(t)q_0[c'(t)]\,dt\right\}\right)\frac{\sin\theta\,d\theta}{2}\,dt}$$

(7.28)

with

$$c'(t) = \left[c^2 + 2\frac{eE}{m}tc\cos\theta + \left(\frac{eE}{m}t\right)^2\right]^{1/2}.$$

Expression 7.28 is mathematically equivalent to a formula given by G. Cavalleri and G. Sesta (1968) and (apart from notation) becomes the same as theirs when the numerator and denominator are transformed by partial integration with respect to t. This equivalence derives, in fact, from the equivalence of equations 7.4 and 7.5. These authors also consider a distribution function which relates to the distribution of velocities c immediately after encounters. This form of function is chosen because of its isotropic nature when the collision cross section is isotropic: $q_1(c) = 0$, $q_0 = q_m$. Revived interest in the method of free paths is shown by the appearance of a number of recent papers, listed in the bibliography.

BIBLIOGRAPHY AND NOTES

The earliest free path formulae were those for D and W proposed by P. Langevin (1903):

$$D = \tfrac{1}{3}lc \quad \text{and} \quad W = \frac{eE}{m}\frac{l}{c},$$

in which $l = 1/Nq_{0_{el}}$ is the mean free path, and c was taken to be the root mean square of the speeds c of the ions (electrons). The ions (electrons) were assumed to interact as hard smooth spheres, that is, $q_{0_{el}}$ was independent of c and the scattering was isotropic. As no account was taken of the change brought about by the force eE in the intervals between successive encounters (free times), the formulae in fact refer more nearly to the circumstances in which $q_{0_{el}} \propto c^{-1}$. The formulae are therefore not self-consistent and are also inexact. Nevertheless they are consistent with the Nernst-Townsend equation since $W/D = 3eE/mc^2$ and c^2 is assumed to be the mean square speed.

J. S. Townsend (1915) extended Langevin's formulae to include the effect of "persistence of velocity after a collision," that is to say, nonisotropic scattering. He showed that, in the formulae for D and W, l is to be replaced by $l + \lambda$, where λ is the same in each formula. The Nernst-Townsend equation is still recovered.

In none of these investigations is any account taken of the distribution of the velocities of the ions (electrons) or of possible dependence of $q_{0_{el}}$ (or $q_{m_{el}}$) on the speeds c.

In 1936, Townsend deduced an accurate formula for the drift velocity W of electrons applicable to the special case in which $l = 1/Nq_{0_{el}}$ is independent of c and scattering is isotropic. He found

$$\mathbf{W} = \frac{2}{3}\frac{e\mathbf{E}}{m}l\,\overline{c^{-1}} \quad \text{and} \quad D = \tfrac{1}{3}l\bar{c},$$

where $\overline{c^{-1}}$ is the mean of the reciprocal of the speeds, and $\bar{c}$ the mean speed. The form of the distribution function was not specified. Townsend (1937) then recognized that these formulae showed that the Nernst-Townsend relationship is not in general exact when applied to the motion of electrons but needs to be corrected by incorporation of a factor dependent on the form of the distribution function $f_0(c)$ (e.g., equation 3.35), even when $q_{0_{el}}$ is independent of c.

In 1923, a formula for electron mobilities in terms of E/p was derived by K. T. Compton on the assumption of elastic encounters, isotropic scattering, and constant collision cross section. This formula is now chiefly of historic interest.

The generalization of the free path formulae to include the dependence of $q_{0_{el}}$ on c and nonisotropic scattering was made by Huxley (1957, 1960), using the approach adopted in Section 7.3. Huxley also pointed out that the formula for drift velocity derived by the method of free paths and that found by Maxwell-Boltzmann theory are transformed, the one to the other, by integration by parts, as remarked in Section 7.3. He also made use of the displaced sphere in velocity space (Section 7.4.1) in this context. These matters have also been considered by G. L. Braglia (1970).

The fact that $f_0(c)$, rather than $(d/dc)f_0(c)$, appears in the integrand of formulae for drift derived by the method of free paths makes it possible to give formulae in terms of means of the powers of the speeds or in terms of dimensionless factors involving mean values of the powers of the speeds c. In this way the effect of the change in form of the distribution function and of the dependence of $q_{m_{el}}$ on c can be readily assessed. Examples are the formulae of equations 3.13 and 3.35 (Huxley, 1957, 1960).

More recently there has been a revival of interest in free path methods by G. Cavalleri, G. L. Braglia, and others in Italy.

Œuvres Scientifiques de Paul Langevin, Services des Publications du Centre National de la Recherche Scientifique, Paris 1950, pp. 37, 43.

P. Langevin, *Ann. Chim. Phys.,* **28**, 289, 1903.

J. S. Townsend, *Electricity in Gases*, Clarendon Press, Oxford, 1915, pp. 82–91.

J. S. Townsend, *Electrons in Gases*, Hutchinson's Scientific and Technical Publications, London, 1947, Sections 7 and 8.

J. S. Townsend, *Phil. Mag.,* **22**, 145, 1936.

J. S. Townsend, *Phil. Mag.,* **23**, 481, 1937.

K. T. Compton, *Phys. Rev.,* **22**, 333, 432, 1923.

L. G. H. Huxley, *Aust. J. Phys.,* **10**, 118, 1957.

L. G. H. Huxley, *Aust. J. Phys.,* **13**, 578, 718, 1960.

G. Cavalleri and G. Sesta, *Phys. Rev.,* **170**, 286, 1968.

R. Ballerio, R. Bonalumi, and G. Cavalleri, *Energ. Nucl.,* **16**, 455, 1969.

G. L. Braglia, *Nuovo Cimento,* **70B**, 169, 1970.

GENERAL REFERENCES

R. H. Healey and J. W. Reed, *The Behaviour of Slow Electrons in Gases,* Amalgamated Wireless Ltd., Sydney, 1941.

L. B. Loeb, *Basic Processes of Gaseous Electronics*, University of California Press, Berkeley, 1955.

L. G. H. Huxley and R. W. Crompton, "The Motions of Slow Electrons in Gases," *Atomic and Molecular Processes* (Ed., D. R. Bates) Academic Press, New York, 1962.

8

DIFFUSION AND DRIFT IN THE

PRESENCE OF A STEADY

AND UNIFORM MAGNETIC FIELD

8.1. INTRODUCTION

In the preceding chapters we have considered in detail the features of electrons in gases in circumstances where a magnetic field is absent or uninfluential. There are, however, many other circumstances where the presence of a magnetic field directly affects important features of the motion and in particular the drift velocity and the coefficient of diffusion.

The chief concern of this chapter is to investigate the consequences of the presence of a magnetic field for the characteristic properties of electron motion represented by formulae for drift velocity and the coefficient of diffusion, as well as for the differential equation to be satisfied by the number density $n(\mathbf{r}, t)$.

We begin by considering the derivation of formulae for drift and diffusion on the basis of the vector equation, postponing for the moment consideration of the distribution function. In order to treat drift and diffusion individually, it is convenient to separate the vector equation by expressing the vector $\mathbf{f}_1$ as the sum of two vectors $\mathbf{f}_E + \mathbf{f}_G = \mathbf{f}_1$, where $\mathbf{f}_E$ depends upon $e\mathbf{E}$ and $\mathbf{f}_G$ upon $\mathrm{grad}_r(nf_0)$. Moreover the term $-\boldsymbol{\omega} \times (n\mathbf{f}_1)$ is now included in the vector equation to represent the contribution of the magnetic field. Drift and diffusion are considered in turn, both in a general sense and in special cases of particular importance.

The scalar equation for nf_0 is then treated by the method of series employed in Chapter 4, and an equation which is to be satisfied by the number density $n(\mathbf{r}, t)$ is deduced. The rest of the chapter is devoted to the application of the theory to experiments on the properties of the diffusing stream from a small source and the measurement of the magnetic deflexion coefficient ψ and the component W_x of the drift normal to the magnetic field.

8.2. DRIFT IN UNIFORM AND CONSTANT ELECTRIC AND MAGNETIC FIELDS

It was shown in Chapter 2 (Section 2.8) that when a magnetic field $\mathbf{B}$ is present, whereas the scalar equation retains its original form, the vector equation includes a term explicitly dependent on $\mathbf{B}$ (equation 2.25). These two equations are as follows.

Scalar equation (equation 2.22):

$$\frac{\partial}{\partial t}(nf_0) + \frac{c}{3}\operatorname{div}_r(nf_1) + \frac{1}{4\pi c^2}\frac{\partial}{\partial c}[\sigma_E(c) - \sigma_{\text{coll}}(c)] = 0, \qquad (8.1)$$

Vector equation (equation 2.25):

$$\frac{\partial}{\partial t}(nf_1) + c\operatorname{grad}_r(nf_0) + \frac{e\mathbf{E}}{m}\frac{\partial}{\partial c}(nf_0) - \boldsymbol{\omega}\times(nf_1) + \nu nf_1 = 0, \quad (8.2)$$

where $\boldsymbol{\omega} = -(e/m)\mathbf{B}$.

It will be assumed, as before, that when $e\mathbf{E}$ is constant the term $(\partial/\partial t)(nf_1)$ in equation 8.2 can be omitted as negligibly small in comparison with the other terms (Section 3.3); that is to say, a condition of quasi equilibrium is assumed to prevail. It is also supposed that ν is the effective collision frequency ν of Section 2.7.2d.

We first consider the vector equation 8.2, and, as in equation 3.9, we write $\mathbf{f}_1 = \mathbf{f}_E + \mathbf{f}_G$ so that with $(\partial/\partial t)(nf_1)$ omitted equation 8.2 resolves into the two equations (see Section 3.3)

$$\nu n\mathbf{f}_E - \boldsymbol{\omega}\times(n\mathbf{f}_E) = -\frac{\partial}{\partial c}(nf_0)\frac{e\mathbf{E}}{m},$$

$$\nu n\mathbf{f}_G - \boldsymbol{\omega}\times(n\mathbf{f}_G) = -c\operatorname{grad}_r(nf_0). \tag{8.3}$$

Since the convective velocity is $\mathbf{W}_{\text{conv}}(c) = c\mathbf{f}_1/3f_0$, it can be expressed as the sum of two vectors: $\mathbf{W}_{\text{conv}}(c) = \mathbf{W}_E(c) + \mathbf{W}_G(c)$, in which $\mathbf{W}_E(c)$ and $\mathbf{W}_G(c)$ imply $\mathbf{W}_E(c,\mathbf{r},t)$ and $\mathbf{W}_G(c,\mathbf{r},t)$, and

$$\mathbf{W}_E(c) = \frac{c\mathbf{f}_E}{3f_0} \qquad \text{and} \qquad \mathbf{W}_G(c) = \frac{c\mathbf{f}_G}{3f_0}. \tag{8.4}$$

Equations 8.3 are therefore equivalent to

$$vnf_0\mathbf{W}_E(c) - \boldsymbol{\omega} \times [nf_0\mathbf{W}_E(c)] = -\frac{c}{3}\frac{\partial}{\partial c}(nf_0)\frac{e\mathbf{E}}{m},$$

$$vnf_0\mathbf{W}_G(c) - \boldsymbol{\omega} \times [nf_0\mathbf{W}_G(c)] = -\frac{c^2}{3}\operatorname{grad}_r(nf_0).$$

(8.5)

Consider the first of equations 8.5, which, in terms of its vector components, is equivalent to

$$vnf_0W_{E_x}(c) + \omega_z nf_0W_{E_y}(c) - \omega_y nf_0W_{E_z}(c) = -\frac{c}{3}\frac{\partial}{\partial c}(nf_0)\frac{e}{m}E_x,$$

$$-\omega_z nf_0W_{E_x}(c) + vnf_0W_{E_y}(c) + \omega_x nf_0W_{E_z}(c) = -\frac{c}{3}\frac{\partial}{\partial c}(nf_0)\frac{e}{m}E_y, \quad (8.6)$$

$$\omega_y nf_0W_{E_x}(c) - \omega_x nf_0W_{E_y}(c) + vnf_0W_{E_z}(c) = -\frac{c}{3}\frac{\partial}{\partial c}(nf_0)\frac{e}{m}E_z.$$

Let [ ] denote a column matrix and | | a square matrix; then in the notation of matrices equations 8.6 are equivalent to

$$nf_0|M|[W_E(c)] = -\frac{c}{3}\frac{\partial}{\partial c}(nf_0)\left[\frac{e}{m}E\right],$$

where

$$|M| \equiv \begin{vmatrix} v & \omega_z & -\omega_y \\ -\omega_z & v & \omega_x \\ \omega_y & -\omega_x & v \end{vmatrix}, \qquad [W_E(c)] \equiv \begin{bmatrix} W_{E_x}(c) \\ W_{E_y}(c) \\ W_{E_z}(c) \end{bmatrix}, \qquad (8.7)$$

and

$$\left[\frac{eE}{m}\right] \equiv \frac{e}{m}\begin{bmatrix} E_x \\ E_y \\ E_z \end{bmatrix}.$$

The reciprocal of the matrix $|M|$ is

$$|M|^{-1} = \frac{1}{v(v^2+\omega^2)}\begin{vmatrix} v^2+\omega_x^2 & \omega_x\omega_y - v\omega_z & \omega_x\omega_z + v\omega_y \\ \omega_y\omega_x + v\omega_z & v^2+\omega_y^2 & \omega_y\omega_z - v\omega_x \\ \omega_z\omega_x - v\omega_y & \omega_z\omega_y + v\omega_x & v^2+\omega_z^2 \end{vmatrix}, \quad (8.8)$$

where $\omega^2 = \omega_x^2 + \omega_y^2 + \omega_z^2$.

Consequently, from equation 8.7,

$$(nf_0)[W_E(c)] = -\frac{c}{3}\frac{\partial}{\partial c}(nf_0)|M|^{-1}\left[\frac{e}{m}E\right], \tag{8.9}$$

from which n may be removed.

The drift velocity of electrons at position $\mathbf{r}$ is

$$[W_E] = 4\pi\int_0^\infty [W_E(c)]f_0 c^2\,dc$$

$$= |\mu|[E], \tag{8.10}$$

where the mobility matrix $|\mu|$ is

$$|\mu| = \begin{vmatrix} \mu_{xx} & \mu_{xy} & \mu_{xz} \\ \mu_{yx} & \mu_{yy} & \mu_{yz} \\ \mu_{zx} & \mu_{zy} & \mu_{zz} \end{vmatrix},$$

whose elements are as follows:

$$\mu_{xx,yy,zz} = -\frac{4\pi e}{3m}\int_0^\infty \frac{c^3\left(\nu^2 + \omega_{x,y,z}^2\right)}{\nu(\nu^2 + \omega^2)}\frac{\partial}{\partial c}f_0\,dc,$$

$$\mu_{xy,yx} = -\frac{4\pi e}{3m}\int_0^\infty \frac{c^3\left(\omega_y\omega_x \mp \nu\omega_z\right)}{\nu(\nu^2 + \omega^2)}\frac{\partial}{\partial c}f_0\,dc,$$

$$\mu_{xz,zx} = -\frac{4\pi e}{3m}\int_0^\infty \frac{c^3\left(\omega_x\omega_z \pm \nu\omega_y\right)}{\nu(\nu^2 + \omega^2)}\frac{\partial}{\partial c}f_0\,dc,$$ \quad (8.11)

$$\mu_{yz,zy} = -\frac{4\pi e}{3m}\int_0^\infty \frac{c^3\left(\omega_y\omega_z \mp \nu\omega_x\right)}{\nu(\nu^2 + \omega^2)}\frac{\partial}{\partial c}f_0\,dc.$$

The velocity $\mathbf{W}$ of the centroid of a group of n_0 electrons is

$$[W] = \frac{1}{n_0}4\pi\int_0^\infty \int_{\mathbf{r}}[W_E(c)]nf_0 c^2\,dc\,d\mathbf{r} = 4\pi\int_0^\infty [W_E(c)]f_0^* c^2\,dc,$$

where, as in Chapter 4,

$$f_0^* = \frac{1}{n_0}\int_{\mathbf{r}} nf_0\,d\mathbf{r}.$$

But from equation 8.9

$$\frac{1}{n_0} 4\pi \int_0^\infty \int_\mathbf{r} [W_E(c)] n f_0 c^2 \, dc \, d\mathbf{r}$$

$$= -\frac{4\pi}{n_0} \int_0^\infty \int_\mathbf{r} \frac{\partial}{\partial c} (n f_0) \, d\mathbf{r} \, \frac{c}{3} |M|^{-1} \left[\frac{eE}{m} \right] c^2 \, dc$$

$$= -4\pi \int_0^\infty \frac{c}{3} \left(\frac{d}{dc} f_0^* \right) |M|^{-1} \left[\frac{eE}{m} \right] c^2 \, dc,$$

that is,

$$[W] = -\frac{4\pi}{3} \int_0^\infty |M|^{-1} \left[\frac{eE}{m} \right] c^3 \frac{d}{dc} f_0^* \, dc. \tag{8.12}$$

Thus W is found by replacing f_0 by f_0^* in equation 8.9 and then forming the integral $4\pi \int_0^\infty (\quad) c^2 \, dc$ of each side of the equation. When $\mathbf{B} = 0$, $\boldsymbol{\omega} = 0$, then

$$|M|^{-1} = |M| = \frac{1}{\nu} \begin{vmatrix} 1 & 0 & 0 \\ 0 & 1 & 0 \\ 0 & 0 & 1 \end{vmatrix} = \frac{\mathbf{I}}{\nu},$$

and from equation 8.12

$$[W] = -\left(\frac{4\pi}{3} \int_0^\infty \frac{c^3}{\nu} \frac{d}{dc} f_0^* \, dc \right) \left[\frac{eE}{m} \right]$$

in agreement with equation 4.4.

8.2.1. SPECIAL CASE—CROSSED ELECTRIC AND MAGNETIC FIELDS. We consider the important special case in which $\mathbf{B}$ is directed normal to $e\mathbf{E}$.

Let the direction of $e\mathbf{E}$ be that of $+Oz$, whereas $\boldsymbol{\omega}$ is directed along $+Oy$. Then $E = E_z$ and $\omega_x = \omega_z = 0$, $\omega = \omega_y$. The matrix $|M|^{-1}$ reduces to

$$|M|^{-1} = \frac{1}{\nu(\nu^2 + \omega^2)} \begin{vmatrix} \nu^2 & 0 & \nu\omega \\ 0 & \nu^2 + \omega^2 & 0 \\ -\nu\omega & 0 & \nu^2 \end{vmatrix}, \tag{8.13}$$

and $[eE/m]$ becomes

$$\begin{bmatrix} 0 \\ 0 \\ \dfrac{eE}{m} \end{bmatrix}.$$

Equation 8.9 now gives

$$(nf_0)\,W_E(c)_x = -\left(\frac{\omega v}{v^2+\omega^2}\right)\frac{cV'}{3}\frac{\partial}{\partial c}(nf_0),$$

$$W_E(c)_y = 0, \tag{8.14}$$

$$(nf_0)\,W_E(c)_z = -\left(\frac{v^2}{v^2+\omega^2}\right)\frac{cV'}{3}\frac{\partial}{\partial c}(nf_0),$$

where $V' = eE/mv$ and $W_E(c)_x \equiv W_{E_x}(c,\mathbf{r},t)$, $W_E(c)_z \equiv W_{E_z}(c,\mathbf{r},t)$.
It follows from equation 8.14 that

$$W_{E_x} = 4\pi\int_0^\infty W_E(c)_x f_0 c^2\,dc = -\frac{4\pi}{3}\int_0^\infty\left(\frac{\omega v}{v^2+\omega^2}\right)c^3 V'\frac{\partial}{\partial c}f_0\,dc,$$

$$\tag{8.15}$$

$$W_{E_z} = 4\pi\int_0^\infty W_E(c)_z f_0 c^2\,dc = -\frac{4\pi}{3}\int_0^\infty\left(\frac{v^2}{v^2+\omega^2}\right)c^3 V'\frac{\partial}{\partial c}f_0\,dc,$$

in which $W_{E_x} \equiv W_{E_x}(\mathbf{r},t)$ and so forth.

When f_0 is replaced in equations 8.15 by f_0^*, then W_{E_x} and W_{E_z} become the components W_x and W_z of the velocity of the centroid of a freely travelling group of electrons.

8.2.2. MAGNETIC DEFLEXION COEFFICIENT ψ. If at any time the centroid of a travelling group lies in the plane xOz, it continues as it travels to remain in that plane since $W_y = 0$.

The velocity $\mathbf{W}$ makes an angle $\theta = \tan^{-1}W_x/W_z$ with the direction of $+Oz$, that is to say, with $e\mathbf{E}$. It follows from equations 8.15 that

$$W_z = \frac{4\pi}{3}\frac{e}{m}EI_1,$$

$$\tag{8.16}$$

$$W_x = \frac{4\pi}{3}\frac{e}{m}E\omega I_2,$$

where

$$I_1 = -\int_0^\infty \frac{c^3 \nu}{\nu^2 + \omega^2} \frac{d}{dc} f_0^* \, dc,$$

$$I_2 = -\int_0^\infty \frac{c^3}{\nu^2 + \omega^2} \frac{d}{dc} f_0^* \, dc.$$

$$(8.17)$$

The ratio W_x / W_z is a quantity that can be measured directly.

A velocity W_M (magnetic drift velocity) is defined as follows:

$$W_M \equiv \frac{E}{B} \frac{W_x}{W_z} = \frac{E}{B} \frac{\omega I_2}{I_1}. \qquad (8.18)$$

The ratio W_x / W_z was originally used by Townsend (Section 1.10) to find the drift velocity of electrons when time-of-flight techniques were impracticable in the then-extant state of technology. Townsend considered W_M to be the same as W and determined it from measurement of W_x / W_z. In general W_M is approximately equal to W, and the relation between them is written as

$$W = \psi^{-1} W_M, \qquad (8.19)$$

in which ψ is named the magnetic deflexion coefficient. It follows from equation 8.18 that ψ depends on the integrals I_1 and I_2. In laboratory experiments on diffusing streams of electrons the pressure of the gas and the value of B are such that $\omega/\nu \ll 1$, and it is therefore legitimate to employ simplified forms of the expressions for I_1 and I_2. When $\omega/\nu \ll 1$, then

$$I_1 \cong -\int_0^\infty \frac{c^3}{\nu} \frac{d}{dc} f_0^* \, dc,$$

$$I_2 \cong -\int_0^\infty \frac{c^3}{\nu^2} \frac{d}{dc} f_0^* \, dc;$$

and from equation 8.16

$$W_z \cong -\frac{4\pi}{3} \frac{eE}{m} \int_0^\infty \frac{c^3}{\nu} \frac{d}{dc} f_0^* \, dc = W,$$

$$W_x \cong -\frac{4\pi}{3} \frac{eE}{m} \int_0^\infty \frac{\omega}{\nu} \frac{c^3}{\nu} \frac{d}{dc} f_0^* \, dc \ll W.$$

$$(8.20)$$

Thus $W_x/W_z \ll 1$. It follows that

$$W\left(\frac{W_z}{W_x}\right) \cong \frac{W_z{}^2}{W_x}$$

$$= -\left(\frac{4\pi}{3}\frac{eE}{m\omega}\right)\frac{I_1{}^2}{I_2} = \frac{4\pi}{3}\frac{E}{B}\frac{I_1{}^2}{I_2},$$

that is,

$$W \cong \frac{4\pi}{3}\frac{E}{B}\frac{I_1{}^2}{I_2}\frac{W_x}{W_z}.$$

Consequently, from equation 8.18,

$$W \cong \frac{4\pi}{3}\frac{I_1{}^2}{I_2}W_M$$

and

$$\psi \cong \frac{3}{4\pi}\frac{I_2}{I_1{}^2}. \tag{8.21}$$

Since, when $\omega/\nu \ll 1$, the integrals I_1 and I_2 can be written as mean values,

$$I_1 \cong \frac{1}{4\pi}\left[\overline{c^{-2}\frac{d}{dc}\left(\frac{c^3}{\nu}\right)}\right] \qquad I_2 \cong \frac{1}{4\pi}\left[\overline{c^{-2}\frac{d}{dc}\left(\frac{c^3}{\nu^2}\right)}\right],$$

it follows that

$$\psi \cong \frac{3\left[\overline{c^{-2}(d/dc)(c^3/\nu^2)}\right]}{\left\{\left[\overline{c^{-2}(d/dc)(c^3/\nu)}\right]\right\}^2}. \tag{8.22}$$

Also

$$\tan\theta = \frac{W_x}{W_z} = \frac{\omega\left\{\overline{c^{-2}(d/dc)[c^3/(\nu^2+\omega^2)]}\right\}}{\left\{\overline{c^{-2}(d/dc)[c^3\nu/(\nu^2+\omega^2)]}\right\}}. \tag{8.23}$$

Formulae equivalent to equations 8.22 and 8.23 were first given by Huxley (1960), Frost and Phelps (1962), and Engelhardt, Phelps, and Risk (1964) and have been used by Jory (1965) in experiments on the properties of electron streams.

It can be seen from equation 8.22 that, when v is independent of c (i.e., $q_m \propto c^{-1}$), ψ has the value unity and $W_M = W$. When q_m is constant, the formula gives, since $v = Ncq_m$,

$$\psi = \frac{3}{4} \frac{\overline{c^{-2}}}{\left(\overline{c^{-1}}\right)^2}.$$

When q_m is constant, $\overline{C^2}/W^2 \ll 1$, and all encounters are elastic, the distribution function f_0^* tends to the form $f_0^* = \text{const} \cdot [\exp - (c/\alpha)^4]$ and the mean value of the xth power of the speeds c is (equation 3.25)

$$\overline{c^x} = \frac{\alpha^x \Gamma[(x+3)/4]}{\Gamma(\frac{3}{4})}.$$

Consequently

$$\frac{\overline{c^{-2}}}{\left(\overline{c^{-1}}\right)^2} = \frac{\Gamma(\frac{1}{4})\Gamma(\frac{3}{4})}{\Gamma(\frac{1}{2})^2} = \frac{\pi}{\pi \sin \pi/4} = 2^{1/2},$$

whence

$$\psi = \frac{3\sqrt{2}}{4} = 1.06, \qquad W_M = 1.06 W.$$

Evidently ψ is determined by the nature of the function $q_m(c)$.

8.2.3. SPECIAL CASE: ω PARALLEL TO eE. Let $\omega_x = \omega_y = 0$, $\omega = \omega_z$. It follows that $|M|^{-1}$ (equation 8.8) reduces to

$$|M|^{-1} = \frac{1}{v(v^2+\omega^2)} \begin{vmatrix} v^2 & -v\omega & 0 \\ v\omega & v^2 & 0 \\ 0 & 0 & v^2+\omega^2 \end{vmatrix}.$$

When $E = E_z$, it follows from equation 8.9 that

$$(nf_0) \begin{bmatrix} W_E(c)_x \\ W_E(c)_y \\ W_E(c)_z \end{bmatrix} = -\frac{1}{v(v^2+\omega^2)} \begin{bmatrix} v^2 & -v\omega & 0 \\ v\omega & v^2 & 0 \\ 0 & 0 & v^2+\omega^2 \end{bmatrix} \begin{bmatrix} 0 \\ 0 \\ \frac{e}{m}E \end{bmatrix} \frac{c}{3} \frac{\partial}{\partial c}(nf_0),$$

whence

$$W_E(c)_x = W_E(c)_y = 0, \qquad W_E(c)_z = -\frac{cV'}{3f_0}\frac{\partial}{\partial c}f_0$$

with $V' = eE/mv$.

When f_0 is replaced by f_0^*, it follows that

$$W_z = -\frac{4\pi}{3}\int_0^\infty c^3 V' \frac{d}{dc}f_0^* \, dc = W.$$

Thus the centroid moves in the direction of $e\mathbf{E}$, unaffected by $\mathbf{B}$.

8.3. DIFFUSION IN THE PRESENCE OF A MAGNETIC FIELD

The second of equations 8.5 is equivalent to the matrix relationship

$$(nf_0)\,|M|\,[\,W_G(c)\,] = -\frac{c^2}{3}\begin{bmatrix} \frac{\partial}{\partial x}(nf_0) \\[2mm] \frac{\partial}{\partial y}(nf_0) \\[2mm] \frac{\partial}{\partial z}(nf_0) \end{bmatrix}, \qquad (8.24)$$

where $|M|$ is the matrix defined in equation 8.7, whose reciprocal $|M|^{-1}$ is given by equation 8.8. It follows that

$$(nf_0)\,[\,W_G(c)\,] = -|M|^{-1}\frac{c^2}{3}\begin{bmatrix} \frac{\partial}{\partial x}(nf_0) \\[2mm] \frac{\partial}{\partial y}(nf_0) \\[2mm] \frac{\partial}{\partial z}(nf_0) \end{bmatrix}$$

$$= -\frac{c^2}{3\nu(\nu^2+\omega^2)}\begin{vmatrix} \nu^2+\omega_x^2 & \omega_x\omega_y-\nu\omega_z & \omega_x\omega_z+\nu\omega_y \\ \omega_y\omega_x+\nu\omega_z & \nu^2+\omega_y^2 & \omega_y\omega_z-\nu\omega_x \\ \omega_z\omega_x-\nu\omega_y & \omega_z\omega_y+\nu\omega_x & \nu^2+\omega_z^2 \end{vmatrix}\begin{bmatrix} \frac{\partial}{\partial x}(nf_0) \\[2mm] \frac{\partial}{\partial y}(nf_0) \\[2mm] \frac{\partial}{\partial z}(nf_0) \end{bmatrix}.$$

$$(8.25)$$

From equation 8.25 we derive

$$[nW_G] = -\frac{4\pi}{3} \int_0^\infty c^2 |M|^{-1} \begin{bmatrix} \dfrac{\partial}{\partial x}(nf_0) \\ \dfrac{\partial}{\partial y}(nf_0) \\ \dfrac{\partial}{\partial z}(nf_0) \end{bmatrix} c^2\, dc. \qquad (8.26)$$

We define coefficients of diffusion as mathematical quantities by replacing $f_0 \equiv f_0(c,\mathbf{r},t)$ in equation 8.26 by $f_0^* \equiv f_0^*(c)$, where f_0^* is the distribution function for a group as a whole, although diffusion is a local phenomenon. Equation 8.26 can now be written as

$$[nW_G] = -|D| \begin{bmatrix} \dfrac{\partial}{\partial x}n \\ \dfrac{\partial}{\partial y}n \\ \dfrac{\partial}{\partial z}n \end{bmatrix}, \qquad (8.27)$$

where $|D|$ is a square matrix whose elements are

$$D_{xx,yy,zz} = \frac{4\pi}{3} \int_0^\infty \frac{c^2}{\nu(\nu^2+\omega^2)}(\nu^2+\omega_{x,y,z}^2)f_0^* c^2\, dc,$$

$$D_{xy,yx} = \frac{4\pi}{3} \int_0^\infty \frac{c^2}{\nu(\nu^2+\omega^2)}(\omega_x\omega_y \mp \nu\omega_z)f_0^* c^2\, dc,$$

$$D_{xz,zx} = \frac{4\pi}{3} \int_0^\infty \frac{c^2}{\nu(\nu^2+\omega^2)}(\omega_x\omega_z \pm \nu\omega_y)f_0^* c^2\, dc,$$

$$D_{yz,zy} = \frac{4\pi}{3} \int_0^\infty \frac{c^2}{\nu(\nu^2+\omega^2)}(\omega_y\omega_z \mp \nu\omega_x)f_0^* c^2\, dc.$$

$$\left. \right\} \qquad (8.28)$$

When $\omega = 0$, then

$$D_{jk} = 0, \qquad D_{jj} = \frac{4\pi}{3} \int_0^\infty \frac{c^2}{\nu}f_0^* c^2\, dc = D.$$

8.3.1. SPECIAL CASE: $\boldsymbol{\omega}$ DIRECTED ALONG $+Oy$. The matrix $|M|^{-1}$ takes the form shown in equation 8.13. It now follows from equation 8.25 that

$$(nf_0)\, W_G(c)_x = -\frac{c^2}{3(\nu^2+\omega^2)}\left[\nu\frac{\partial}{\partial x}(nf_0)+\omega\frac{\partial}{\partial z}(nf_0)\right],$$

$$(nf_0)\, W_G(c)_y = -\frac{c^2}{3\nu}\frac{\partial}{\partial y}(nf_0), \qquad\qquad (8.29)$$

$$(nf_0)\, W_G(c)_z = -\frac{c^2}{3(\nu^2+\omega^2)}\left[-\omega\frac{\partial}{\partial x}(nf_0)+\nu\frac{\partial}{\partial z}(nf_0)\right].$$

We deduce from these equations that

$$\mathrm{div}_r[nf_0\mathbf{W}_G(c)] = -\frac{c^2\nu}{3(\nu^2+\omega^2)}\left[\frac{\partial^2}{\partial x^2}(nf_0)+\frac{\partial^2}{\partial z^2}(nf_0)\right] - \frac{c^2}{3\nu}\frac{\partial^2}{\partial y^2}(nf_0).$$

$$(8.30)$$

When we replace $f_0(c,\mathbf{r},t)$ in equation 8.30 by $f_0^*(c)$, form the integral $4\pi\int_0^\infty (\quad)c^2\,dc$ of each side, and write

$$\mathbf{W}_G = 4\pi\int_0^\infty \mathbf{W}_G(c)f_0^*c^2\,dc,$$

we find that

$$\mathrm{div}_r(n\mathbf{W}_G) = -\left[D_\perp\left(\frac{\partial^2}{\partial x^2}n+\frac{\partial^2}{\partial z^2}n\right)+D\frac{\partial^2}{\partial y^2}n\right], \qquad (8.31)$$

where $D_\perp = D_{xx} = D_{zz}$ and $D = D_{yy}$ are the values that these quantities in equation 8.28 assume when $\omega_x = \omega_z = 0$ and $\omega_y = \omega$. We remark that $D_{xy,yx}, D_{yz,zy}$ all vanish and

$$D_{xz,zx} = \pm\frac{4\pi}{3}\int_0^\infty \frac{c^2\omega}{\nu^2+\omega^2}f_0^*c^2\,dc.$$

However, both D_{xz} and D_{zx} are eliminated from equation 8.31.

8.4. DISTRIBUTION FUNCTION AND DIFFERENTIAL EQUATION FOR n

Further progress is contingent upon finding how nf_0 depends on $c, \mathbf{r}$, and t and in particular upon discovering the equation satisfied separately by n. It is necessary therefore to consider equation 8.1, the scalar equation. Since $\mathbf{W}_{conv}(c) = c\mathbf{f}_1/3f_0$, equation 8.1 is equivalent to

$$\frac{\partial}{\partial t}(nf_0) + \text{div}_r[nf_0\mathbf{W}_{conv}(c)] + \frac{1}{4\pi c^2}\frac{\partial}{\partial c}[\sigma_E(c) - \sigma_{coll}(c)] = 0, \quad (8.32)$$

which in view of equations 8.4 can be written as

$$\frac{\partial}{\partial t}(nf_0) + \text{div}_r[nf_0\mathbf{W}_G(c)] + \text{div}_r[nf_0\mathbf{W}_E(c)] + \frac{1}{4\pi c^2}\frac{\partial}{\partial c}[\sigma_E(c) - \sigma_{coll}(c)] = 0.$$

$$(8.33)$$

We shall suppose that $e\mathbf{E}$ is directed along $+Oz$ and $\boldsymbol{\omega}$ along $+Oy$. We are then able to replace $\text{div}_r[nf_0\mathbf{W}_G(c)]$ by its representation in terms of nf_0 given in equation 8.30. We also require an expression for $\text{div}_r[nf_0\mathbf{W}_E(c)]$. It follows from equation 8.14 that

$$\text{div}_r[nf_0\mathbf{W}_E(c)] = -\left(\frac{\omega\nu}{\nu^2 + \omega^2}\right)\frac{cV'}{3}\frac{\partial}{\partial c}\left[\frac{\partial}{\partial x}(nf_0)\right]$$

$$-\left(\frac{\nu^2}{\nu^2 + \omega^2}\right)\frac{cV'}{3}\frac{\partial}{\partial c}\left[\frac{\partial}{\partial z}(nf_0)\right]. \quad (8.34)$$

The term $\sigma_E(c)$ is, according to equation 2.19 and since $E = E_z$,

$$\sigma_E(c) = \frac{4\pi}{3}c^2\frac{e\mathbf{E}}{m}\cdot n\mathbf{f}_1 = \frac{4\pi}{3}c^2\frac{eE}{m}nf_{1z} = 4\pi c\nu_{el}V[nf_0W_{conv_z}(c)].$$

Consequently, from equations 8.4, 8.14, and 8.29

$$\sigma_E(c) = 4\pi c\nu_{el}V\left\{-\frac{c^2}{3(\nu^2 + \omega^2)}\left[-\omega\frac{\partial}{\partial x}(nf_0) + \nu\frac{\partial}{\partial z}(nf_0)\right]\right.$$

$$\left. -\left(\frac{\nu^2}{\nu^2 + \omega^2}\right)\frac{cV'}{3}\frac{\partial}{\partial c}(nf_0)\right\}.$$

Equation 8.33 therefore becomes, for the case of crossed electric and magnetic fields,

$$-\frac{\partial}{\partial t}(nf_0)+\frac{c^2\nu}{3(\nu^2+\omega^2)}\left[\frac{\partial^2}{\partial x^2}(nf_0)+\frac{\partial^2}{\partial z^2}(nf_0)\right]+\frac{c^2}{3\nu}\frac{\partial^2}{\partial y^2}(nf_0)$$

$$+\left(\frac{\omega\nu}{\nu^2+\omega^2}\right)\frac{cV'}{3}\frac{\partial}{\partial c}\frac{\partial}{\partial x}(nf_0)+\left(\frac{\nu^2}{\nu^2+\omega^2}\right)\frac{cV'}{3}\frac{\partial}{\partial c}\frac{\partial}{\partial z}(nf_0)$$

$$+\frac{1}{3c^2}\frac{\partial}{\partial c}\left\{\frac{c^3\nu_{el}V}{\nu^2+\omega^2}\left[-\omega\frac{\partial}{\partial x}(nf_0)+\nu\frac{\partial}{\partial z}(nf_0)\right]\right.$$

$$\left.+\left(\frac{\nu^2}{\nu^2+\omega^2}\right)c^2\nu_{el}VV'\frac{\partial}{\partial c}(nf_0)+\frac{3}{4\pi}\sigma_{coll}(c)\right\}=0. \quad (8.35)$$

When all encounters are elastic,

$$\frac{3}{4\pi}\sigma_{coll}(c)=c^2\nu_{el}\left[\frac{3mc}{M}nf_0+\overline{C^2}\frac{\partial}{\partial c}(nf_0)\right],$$

but when inelastic encounters are significant and the inelastic thresholds are closely spaced we replace $3mc/M$ by some function $h'(c)$ where $h'(c)=3[mc/M+h(c,T)/c]$ as appears in equation 6.54.

The last group of terms on the right-hand side of equation 8.35, $(1/3c^2)(\partial/\partial c)\{ \ \}$, can now be written as

$$\frac{1}{3c^2}\frac{\partial}{\partial c}\left(c^2\nu_{el}\left\{\left(\frac{\nu^2}{\nu^2+\omega^2}\right)\frac{cV}{\nu}\left[-\frac{\omega}{\nu}\frac{\partial}{\partial x}(nf_0)+\frac{\partial}{\partial z}(nf_0)\right]\right.\right.$$

$$\left.\left.+\left(\frac{\nu^2VV'}{\nu^2+\omega^2}+\overline{C^2}\right)\frac{\partial}{\partial c}(nf_0)+h'(c)(nf_0)\right\}\right)$$

$$=\frac{1}{3c^2}\frac{\partial}{\partial c}\left(c^2\nu_{el}\left\{\left(\frac{\nu^2}{\nu^2+\omega^2}\right)\frac{cV}{\nu}\left[-\frac{\omega}{\nu}\frac{\partial}{\partial x}(nf_0)+\frac{\partial}{\partial z}(nf_0)\right]\right.\right.$$

$$\left.\left.+\left(\frac{\nu^2VV'}{\nu^2+\omega^2}+\overline{C^2}\right)(nf_0^*)\frac{\partial}{\partial c}\left(\frac{f_0}{f_0^*}\right)\right\}\right),$$

where

$$f_0^* = \text{const} \cdot \exp\left\{ - \int_0^c \frac{h'(c)\,dc}{[\nu^2 VV'/(\nu^2+\omega^2)] + \overline{C^2}} \right\}, \qquad (8.36)$$

in which $h'(c) = 3mc/M$ when all encounters are elastic. The distribution function f_0^* is that for a travelling group as a whole. We therefore express equation 8.35 in the form

$$-\frac{\partial}{\partial t}(nf_0) + \left(\frac{\nu^2}{\nu^2+\omega^2}\right)\frac{c^2}{3\nu}\left(\frac{\partial^2}{\partial x^2} + \frac{\partial^2}{\partial z^2}\right)(nf_0) + \frac{c^2}{3\nu}\frac{\partial^2}{\partial y^2}(nf_0)$$

$$+ \left(\frac{\nu^2}{\nu^2+\omega^2}\right)\left(\frac{cV'}{3}\right)\left(\frac{\omega}{\nu}\frac{\partial}{\partial x} + \frac{\partial}{\partial z}\right)\frac{\partial}{\partial c}(nf_0)$$

$$+ \frac{1}{3c^2}\frac{\partial}{\partial c}\left\{ c^2 \nu_{el}\left[\left(\frac{\nu^2}{\nu^2+\omega^2}\right)\frac{cV}{\nu}\left(-\frac{\omega}{\nu}\frac{\partial}{\partial x} + \frac{\partial}{\partial z}\right)(nf_0) \right.\right.$$

$$+ \left. \left. \left(\frac{\nu^2 VV'}{\nu^2+\omega^2} + \overline{C^2}\right)f_0^* \frac{\partial}{\partial c}\left(\frac{nf_0}{f_0^*}\right) \right] \right\} = 0. \qquad (8.37)$$

Equation 8.37 reduces to equation 4.9 when $\omega = 0$ and $h'(c) = 3mc/M$. In many laboratory experiments the factor $\nu^2/(\nu^2+\omega^2)$ that appears in equation 8.37 is negligibly different from unity.

We now express equation 8.37 in a system of coordinates moving with the centroid of the isolated travelling group. This transformation is achieved by adding to the left-hand side the sum of terms (compare with equation 4.10)

$$W_x \frac{\partial}{\partial x}(nf_0) + W_z \frac{\partial}{\partial z}(nf_0),$$

where W_x and W_z are given by equation 8.15. With f_0 replaced by f_0^* in equations 8.15, we obtain

$$W_x = -4\pi \int_0^\infty \left(\frac{\nu^2}{\nu^2+\omega^2}\right)\left(\frac{\omega}{\nu}\right)\frac{c^3 V'}{3}\frac{d}{dc}f_0^*\,dc,$$

$$W_z = -4\pi \int_0^\infty \left(\frac{\nu^2}{\nu^2+\omega^2}\right)\frac{c^3 V'}{3}\frac{d}{dc}f_0^*\,dc. \qquad (8.38)$$

Equation 8.37 is therefore replaced in the moving system by

$$-\frac{d}{dt}(nf_0) + \left(\frac{v^2}{v^2+\omega^2}\right)\left(\frac{c^2}{3v}\right)\left(\frac{\partial^2}{\partial x^2} + \frac{\partial^2}{\partial z^2}\right)(nf_0) + \frac{c^2}{3v}\frac{\partial^2}{\partial y^2}(nf_0)$$

$$+\left(\frac{v^2}{v^2+\omega^2}\right)\left(\frac{cV'}{3}\right)\left(\frac{\omega}{v}\frac{\partial}{\partial x} + \frac{\partial}{\partial z}\right)\frac{\partial}{\partial c}(nf_0) + W_x\frac{\partial}{\partial x}(nf_0) + W_z\frac{\partial}{\partial z}(nf_0)$$

$$+\frac{1}{3c^2}\frac{\partial}{\partial c}\left\{c^2 v_{el}\left[\left(\frac{v^2}{v^2+\omega^2}\right)\left(\frac{cV}{v}\right)\left(-\frac{\omega}{v}\frac{\partial}{\partial x} + \frac{\partial}{\partial z}\right)(nf_0)\right.\right.$$

$$\left.\left.+\left(\frac{v^2}{v^2+\omega^2}VV' + \overline{C^2}\right)f_0^*\frac{\partial}{\partial c}\left(\frac{nf_0}{f_0^*}\right)\right]\right\} = 0. \qquad (8.39)$$

We next assume that nf_0 is represented as a series analogous to that in equation 4.33 but including x derivatives of all orders to accord with the presence of terms in equation 8.39 that depend on first order x derivatives. We suppose that

$$nf_0 = f_0^*\left[n + \sum_{k=1}^{\infty} a_k\frac{\partial^k}{\partial x^k}n + \sum_{k=1}^{\infty} b_k\frac{\partial^k}{\partial z^k}n + \sum_{k=1}^{\infty} \beta_{2k}\left(\frac{\partial^2}{\partial y^2}\right)^k n\right.$$

$$+\sum_{l=1}^{\infty}\sum_{m=1}^{\infty} a_{lm}\frac{\partial^l}{\partial x^l}\frac{\partial^m}{\partial z^m}n + \sum_{l=1}^{\infty}\sum_{m=1}^{\infty} b_{lm}\frac{\partial^l}{\partial x^l}\left(\frac{\partial^2}{\partial y^2}\right)^m n$$

$$\left.+\sum_{l=1}^{\infty}\sum_{m=1}^{\infty} c_{lm}\frac{\partial^l}{\partial z^l}\left(\frac{\partial^2}{\partial y^2}\right)^m n\right], \qquad (8.40)$$

in which the coefficients $a_k, b_k, \ldots, c_{lm}$ are functions of c alone.

When the series in equation 8.40 is substituted for nf_0 in equation 8.39 and the resulting terms are arranged in increasing order of spatial derivatives with an appropriate association of the temporal derivative, the equation begins with the lowest-order derivatives as follows:

$$\frac{1}{3c^2}\frac{d}{dc}\left[c^2\nu_{\mathrm{el}}\left(\frac{\nu^2}{\nu^2+\omega^2}VV'+\overline{C^2}\right)f_0^*\frac{d}{dc}\left(\frac{f_0^*}{f_0^*}\right)\right]n$$

$$+\left(\frac{\nu^2}{\nu^2+\omega^2}\right)\left(\frac{cV'}{3}\right)\left(\frac{d}{dc}f_0^*\right)\left(\frac{\omega}{\nu}\frac{\partial}{\partial x}n+\frac{\partial}{\partial z}n\right)$$

$$+W_xf_0^*\frac{\partial}{\partial x}n+W_zf_0^*\frac{\partial}{\partial z}n$$

$$+\frac{1}{3c^2}\frac{d}{dc}\left\{c^2\nu_{\mathrm{el}}\left[\frac{\nu^2}{\nu^2+\omega^2}\left(\frac{cV}{\nu}\right)f_0^*\left(-\frac{\omega}{\nu}\frac{\partial}{\partial x}n+\frac{\partial}{\partial z}n\right)\right.\right.$$

$$\left.\left.+\left(\frac{\nu^2}{\nu^2+\omega^2}VV'+\overline{C^2}\right)f_0^*\frac{d}{dc}\left(a_1\frac{\partial}{\partial x}n+b_1\frac{\partial}{\partial z}n\right)\right]\right\}$$

$$-f_0^*\frac{d}{dt}n+\left(\frac{\nu^2}{\nu^2+\omega^2}\right)\left(\frac{c^2}{3\nu}\right)f_0^*\left(\frac{\partial^2}{\partial x^2}n+\frac{\partial^2}{\partial z^2}n\right)+\frac{c^2}{3\nu}f_0^*\frac{\partial^2}{\partial y^2}n$$

$$+\left[\left(\frac{\nu^2}{\nu^2+\omega^2}\right)\left(\frac{cV'}{3}\right)\frac{\omega}{\nu}\frac{d}{dc}(a_1f_0^*)+W_xa_1f_0^*\right]\frac{\partial^2}{\partial x^2}n$$

$$+\left[\left(\frac{\nu^2}{\nu^2+\omega^2}\right)\left(\frac{cV'}{3}\right)\frac{d}{dc}(b_1f_0^*)+W_zb_1f_0^*\right]\frac{\partial^2}{\partial z^2}n$$

$$+\frac{1}{3c^2}\frac{d}{dc}\left(c^2\nu_{\mathrm{el}}\left\{\frac{\nu^2}{\nu^2+\omega^2}\left(\frac{cV}{\nu}\right)f_0^*\left(-a_1\frac{\omega}{\nu}\frac{\partial^2}{\partial x^2}n+b_1\frac{\partial^2}{\partial z^2}n\right)\right.\right.$$

$$+\left(\frac{\nu^2}{\nu^2+\omega^2}VV'+\overline{C^2}\right)f_0^*\left[\left(\frac{d}{dc}a_2\right)\frac{\partial^2}{\partial x^2}n+\left(\frac{d}{dc}b_2\right)\frac{\partial^2}{\partial z^2}n+\left(\frac{d}{dc}\beta_2\right)\frac{\partial^2}{\partial y^2}n\right]\Bigg\}\Bigg)$$

$$+\left(\left(\frac{\nu^2}{\nu^2+\omega^2}\right)\left(\frac{cV'}{3}\right)\left[\frac{\omega}{\nu}\frac{d}{dc}(b_1f_0^*)+\frac{d}{dc}(a_1f_0^*)\right]+W_xb_1f_0^*+W_za_1f_0^*\right.$$

$$+\frac{1}{3c^2}\frac{d}{dc}\left\{c^2\nu_{\text{el}}\left[\frac{\nu^2}{\nu^2+\omega^2}\left(\frac{cV}{\nu}\right)f_0^*\left(-b_1\frac{\omega}{\nu}+a_1\right)\right.\right.$$

$$\left.\left.+\left(\frac{\nu^2}{\nu^2+\omega^2}VV'+\overline{C^2}\right)f_0^*\frac{d}{dc}a_{11}\right]\right\}\right)\frac{\partial^2}{\partial x\,\partial z}n+S=0, \quad (8.41)$$

where S is a sequence of terms in $(d/dt)n$ and the mixed and unmixed derivatives of higher orders, all with coefficients that are functions of c involving the $a_k, b_k, \ldots, a_{lm}$.

We remark that the coefficients a_k and so forth are determined through constants of integration to have the property, as did the corresponding coefficients in Chapter 4,

$$4\pi\int_0^\infty (\text{coefficient})f_0^* c^2\,dc=0. \quad (8.42)$$

We now consider in turn the groups of terms shown in equation 8.41. The coefficient of n vanishes identically because $(d/dc)(f_0^*/f_0^*)=0$.

We equate in the next group the coefficients of $(\partial/\partial x)n$ and $(\partial/\partial z)n$ separately to zero and obtain the following pair of equations from which to determine a_1 and b_1:

$$\left(\frac{\nu^2}{\nu^2+\omega^2}\right)\left(\frac{cV'}{3}\right)\frac{\omega}{\nu}\frac{d}{dc}f_0^*+W_xf_0^*+\frac{1}{3c^2}\frac{d}{dc}\left\{c^2\nu_{\text{el}}\left[\frac{\nu^2}{\nu^2+\omega^2}\left(\frac{cV}{\nu}\right)\left(-\frac{\omega}{\nu}\right)f_0^*\right.\right.$$

$$\left.\left.+\left(\frac{\nu^2}{\nu^2+\omega^2}VV'+\overline{C^2}\right)f_0^*\frac{d}{dc}a_1(c)\right]\right\}=0,$$

$$(8.43)$$

$$\left(\frac{\nu^2}{\nu^2+\omega^2}\right)\left(\frac{cV'}{3}\right)\frac{d}{dc}f_0^*+W_zf_0^*+\frac{1}{3c^2}\frac{d}{dc}\left\{c^2\nu_{\text{el}}\left[\frac{\nu^2}{\nu^2+\omega^2}\left(\frac{cV}{\nu}\right)f_0^*\right.\right.$$

$$\left.\left.+\left(\frac{\nu^2}{\nu^2+\omega^2}VV'+\overline{C^2}\right)f_0^*\frac{d}{dc}b_1\right]\right\}=0.$$

The coefficients $a_1(c)$ and $b_1(c)$ are found from these equations, and the constants of integration are chosen so that $a_1(c)$ and $b_1(c)$ satisfy equation 8.42. It is evident that expressions analogous to the one for $b_1(c)$ of equation 4.25 will result. We note that when all terms of these equations are subjected to the operation $4\pi \int_0^\infty (\quad)c^2\,dc$, the first two terms vanish in each equation (see equation 8.38), and also the terms $\frac{1}{3}\{c^2\nu_{el}[\quad]\}|_0^\infty$ vanish at both limits.

As in Chapter 4, we replace $(d/dt)n$ by an expansion in terms of partial spatial derivatives and rearrange terms so that equation 8.41 is an equation in terms of partial spatial derivatives whose coefficients are then equated individually to zero. Since the procedure results in cumbersome algebra, we shall proceed immediately to the final result. We find that

$$-\frac{d}{dt}n + D_\perp \left(\frac{\partial^2}{\partial x^2}n + \frac{\partial^2}{\partial z^2}n\right) + D\frac{\partial^2}{\partial y^2}n$$

$$+\left[\frac{4\pi}{3}\int_0^\infty \left(\frac{\nu^2}{\nu^2+\omega^2}\right)(cV')\frac{\omega}{\nu}\frac{d}{dc}(a_1 f_0^*)c^2\,dc\right]\frac{\partial^2}{\partial x^2}n$$

$$+\left[\frac{4\pi}{3}\int_0^\infty \left(\frac{\nu^2}{\nu^2+\omega^2}\right)(cV')\frac{d}{dc}(b_1 f_0^*)c^2\,dc\right]\frac{\partial^2}{\partial z^2}n$$

$$+\left\{\frac{4\pi}{3}\int_0^\infty \left(\frac{\nu^2}{\nu^2+\omega^2}\right)(cV')\left[\frac{\omega}{\nu}\frac{d}{dc}(b_1 f_0^*) + \frac{d}{dc}(a_1 f_0^*)\right]c^2\,dc\right\}\frac{\partial^2}{\partial x\,\partial z}n$$

$$+\left[\frac{c^2\nu_{el}}{3}\left(\text{terms in }\frac{\partial^2}{\partial x^2}n, \frac{\partial^2}{\partial z^2}n, \frac{\partial^2}{\partial y^2}n, \nabla_{xy}{}^2 n, \frac{\partial^2}{\partial x\,\partial z}n\right)\right]\Bigg|_0^\infty + S'' = 0,$$

where $D_\perp$ is defined in equation 8.31 and D in equation 8.28.

It is found that the differential equation for n in the moving system is

$$-\frac{d}{dt}n + D_{\omega x}\frac{\partial^2}{\partial x^2}n + D_{\omega z}\frac{\partial^2}{\partial z^2}n + D\frac{\partial^2}{\partial y^2}n + D_{\omega xz}\frac{\partial^2}{\partial x\,\partial z}n + S'' = 0, \quad (8.44)$$

where S'' is a sum of terms in the higher partial spatial derivatives and where

$$
\begin{aligned}
D_{\omega x} &= D_\perp + \frac{4\pi}{3} \int_0^\infty \left(\frac{v^2}{v^2+\omega^2} \right)(cV') \frac{\omega}{v} \frac{d}{dc} (a_1 f_0^*) c^2 \, dc \\[2mm]
&= \frac{4\pi}{3} \int_0^\infty \left(\frac{v^2}{v^2+\omega^2} \right)\left[\frac{c^2}{v} f_0^* + \frac{\omega}{v} cV' \frac{d}{dc}(a_1 f_0^*) \right] c^2 \, dc, \\[2mm]
D_{\omega z} &= \frac{4\pi}{3} \int_0^\infty \left(\frac{v^2}{v^2+\omega^2} \right)\left[\frac{c^2}{v} f_0^* + cV' \frac{d}{dc}(b_1 f_0^*) \right] c^2 \, dc, \\[2mm]
D_{\omega xz} &= \frac{4\pi}{3} \int_0^\infty \left(\frac{v^2}{v^2+\omega^2} \right)(cV')\left[\frac{\omega}{v} \frac{d}{dc}(b_1 f_0^*) + \frac{d}{dc}(a_1 f_0^*) \right] c^2 \, dc, \\[2mm]
D &= \frac{4\pi}{3} \int_0^\infty \frac{c^2}{v} f_0^* c^2 \, dc.
\end{aligned}
\tag{8.45}
$$

In the laboratory system of coordinates equation 8.44 becomes

$$
\frac{\partial}{\partial t} n + D_{\omega x} \frac{\partial^2}{\partial x^2} n + D_{\omega z} \frac{\partial^2}{\partial z^2} n + D \frac{\partial^2}{\partial y^2} n + D_{\omega xz} \frac{\partial^2}{\partial x \, \partial z} n
$$
$$
- W_x \frac{\partial}{\partial x} n - W_z \frac{\partial}{\partial z} n + S'' = 0.
\tag{8.46}
$$

In accordance with the discussion in Section 4.5 and the restriction stated in Section 5.1, it will be supposed that equation 8.46 is employed only in circumstances where it may be replaced by its asymptotic form, in which S'' is omitted. We therefore adopt as the working form of equation 8.46

$$
- \frac{\partial}{\partial t} n + D_{\omega x} \frac{\partial^2}{\partial x^2} n + D_{\omega z} \frac{\partial^2}{\partial z^2} n + D \frac{\partial^2}{\partial y^2} n + D_{\omega xz} \frac{\partial^2}{\partial x \, \partial z} n
$$
$$
- W_x \frac{\partial}{\partial x} n - W_z \frac{\partial}{\partial z} n = 0.
\tag{8.47}
$$

A corresponding analysis of the case with $\boldsymbol{\omega}$ directed parallel to $e\mathbf{E}$ yields the following working equation:

$$
- \frac{\partial}{\partial t} n + D_\perp \left(\frac{\partial^2}{\partial x^2} n + \frac{\partial^2}{\partial y^2} n \right) + D_L \frac{\partial^2}{\partial z^2} n - W \frac{\partial}{\partial z} n = 0,
\tag{8.48}
$$

where $D_\perp$ is defined as in equation 8.31, and D_L is the longitudinal coefficient of diffusion (Chapter 4).

8.5. DISTRIBUTION OF ELECTRON NUMBER DENSITY IN A STEADY STREAM FROM A POINT SOURCE IN THE PRESENCE OF CROSSED ELECTRIC AND MAGNETIC FIELDS

Since the behaviour of steady streams of electrons in gases in the presence of crossed electric and magnetic fields has been the subject of a number of experimental investigations, we develop in this section the theory of experiments of this kind. It is supposed that $e\mathbf{E}$ is directed parallel to $+Oz$ and $\boldsymbol{\omega}$ to $+Oy$ and that in consequence we seek to derive expressions for the number density from equation 8.47.

The presence of the term $D_{\omega xz}(\partial^2/\partial x\,\partial z)n$ is a source of complication, but despite the fact that this term can be removed by a simple rotation of the coordinate axes about the Oy axis we take advantage of the fact that in the majority of measurements of the type under consideration the value of $\omega/\bar{\nu}$ is very small compared with unity, where $\bar{\nu}$ is the collision frequency of the electrons whose speeds c lie in the vicinity of c_m, the speed for which $c^2 f_0^*(c)$ is a maximum. In many instances $\omega/\bar{\nu} \cong W_x/W_z$ is of the order of magnitude $1/50$, and it is then justifiable to employ a simpler, if less precise, form of equation 8.47.

We first consider equations 8.45. When $\omega/\nu \ll 1$, it follows that $\nu^2/(\nu^2 + \omega^2)$ may with undetectable error be given the value unity and that, in consequence, $D_{\omega z}$ is indistinguishable from D_L as defined in equation 4.26.

We refer next to equations 8.43 and note that, excluding the terms with $(d/dc)a_1$ and $(d/dc)b_1$ as factors, the terms in the first of these equations are of the order of magnitude ω/ν of the corresponding terms in the second equations. We conclude that $(d/dc)a_1 \cong (\omega/\nu)(d/dc)b_1$, and therefore that the term $(\omega/\nu)cV(d/dc)(a_1 f_0^*)$ in the formula for $D_{\omega x}$ is of the order of magnitude $(\omega/\nu)^2 cV(d/dc)(b_1 f_0^*)$ in the expression for $D_{\omega z}$ and is negligible in comparison with $(c^2/\nu)f_0^*$. Therefore we may with close approximation put

$$D_{\omega x} \cong \frac{4\pi}{3} \int_0^\infty \frac{c^2}{\nu} f_0^* c^2 \, dc = D,$$

(8.49)

$$D_{\omega z} \cong \frac{4\pi}{3} \int_0^\infty \left[\frac{c^2}{\nu} f_0^* + cV' \frac{d}{dc}(b_1 f_0^*) \right] c^2 \, dc = D_L.$$

A less accurate approximation, which has unconsciously been incorporated in most, if not all, previous discussions of the subject, is to ignore the term $D_{\omega xz}(\partial^2/\partial x\,\partial z)n$ in equation 8.47. We note from equation 8.45 that $D_{\omega xz}$ is of the order of magnitude $\omega/\bar{\nu}$ of the other coefficients D. We shall also ignore the term $D_{\omega xz}(\partial^2/\partial x\,\partial z)n$, while bearing in mind that it should occasion no surprise if experiment and theory were found to be in poorer agreement at the larger than at the smaller values of $W_x/W_z\cong\omega/\bar{\nu}$. We therefore develop the theory on the basis of the following approximation to equation 8.47, valid for the case $W_x/W_z\ll1$:

$$-\frac{\partial}{\partial t}n+D\left(\frac{\partial^2}{\partial x^2}n+\frac{\partial^2}{\partial y^2}n\right)+D_L\frac{\partial^2}{\partial z^2}n-W_x\frac{\partial}{\partial x}n-W_z\frac{\partial}{\partial z}n=0. \quad (8.50)$$

Our concern is with steady streams of electrons, for which $(\partial/\partial t)n$ vanishes everywhere. Equation 8.50 with $(\partial/\partial t)n=0$ is then an extension of equation 5.10, which we also transform by use of the variables

$$x'=\left(\frac{D_L}{D}\right)^{1/2}x, \qquad y'=\left(\frac{D_L}{D}\right)^{1/2}y.$$

Equation 8.50 now becomes

$$\frac{\partial^2}{\partial x'^2}n+\frac{\partial^2}{\partial y'^2}n+\frac{\partial^2}{\partial z^2}n=\frac{W_x}{(DD_L)^{1/2}}\frac{\partial}{\partial x'}n+\frac{W_z}{D_L}\frac{\partial}{\partial z}n$$

or

$$\frac{\partial^2}{\partial x'^2}n+\frac{\partial^2}{\partial y'^2}n+\frac{\partial^2}{\partial z^2}n=2\beta\frac{\partial}{\partial x'}n+2\gamma\frac{\partial}{\partial z}n, \quad (8.51)$$

where

$$2\beta=\frac{W_x}{(DD_L)^{1/2}}, \qquad 2\gamma=\frac{W_z}{D_L}.$$

In the more usual form of apparatus employed to investigate the magnetic deflexion of a steady stream of electrons, the receiving electrode is divided by a cut parallel to the $\pm Oy$ axis and the ratio of the currents to the portions of the divided electrode is measured. We therefore adapt equation 8.51 to conform with measurements of this type by eliminating the coordinate y' by integration. If we let

$$p(x',z)=\int_{-\infty}^{\infty}n\,dy'.$$

then, because n and its derivatives vanish at $y'=\pm\infty$, it follows from the integration of equation 8.51 with respect to y' that

$$\frac{\partial^2}{\partial x'^2} p + \frac{\partial^2}{\partial z^2} p = 2\beta \frac{\partial}{\partial x'} p + 2\gamma \frac{\partial}{\partial z} p. \tag{8.52}$$

Let

$$p(x',z) = V \exp(\beta x' + \gamma z). \tag{8.53}$$

It then follows from equation 8.52 that

$$\frac{\partial^2}{\partial x'^2} V + \frac{\partial^2}{\partial z^2} V = \eta^2 V, \tag{8.54}$$

where $\eta^2 = \beta^2 + \gamma^2$. In practice $\eta \cong \gamma$.

Equation 8.54, expressed in terms of cylindrical polar coordinates $r' = (x'^2 + z^2)^{1/2}$, $\theta = \tan^{-1} x'/z$, is

$$\frac{\partial^2}{\partial r'^2} V + \frac{1}{r'} \frac{\partial}{\partial r'} V + \frac{1}{r'^2} \frac{\partial^2}{\partial \theta^2} V - \eta^2 V = 0. \tag{8.55}$$

We put

$$V = R(r') \begin{matrix} \cos \\ \sin \end{matrix} k\theta,$$

where $k = 0, 1, 2, \ldots$; then

$$\frac{\partial^2}{\partial r'^2} R + \frac{1}{r'} \frac{\partial R}{\partial r'} - \left(\frac{k^2}{r'^2} + \eta^2 \right) R = 0,$$

which is of the form assumed by equation 5.16 when in that equation $\alpha = 0$, $\gamma = 1$, $\beta = \eta$, $\nu = k$. Consequently a general solution of equation 8.55 that tends to zero as $r' \to \infty$ is

$$V = \sum_{k=0}^{\infty} A_k K_k(\eta r') \begin{matrix} \cos \\ \sin \end{matrix} k\theta. \tag{8.56}$$

When the stream originates in an isolated pole source, we require the first term only; consequently the function $p(x',z)$, appropriate to an uninterrupted stream, is

$$p(x',z) = \text{const}[\exp(\beta x' + \gamma z)] K_0(\eta r'). \tag{8.57}$$

The following solution is appropriate to a stream that is received by a plane metal electrode in the plane $z = h$, over which n (and therefore p) is zero (see Section 5.1.2),

$$p(x'z) = \text{const}[\exp(\beta x' + \gamma z)][K_0(\eta r') - K_0(\eta r'')] \qquad (8.58)$$

where $r'^2 = x'^2 + z^2$, $r''^2 = x'^2 + (2h - z)^2$.

When the source is a small hole in a plane cathode through which electrons pass into the diffusion chamber, a solution of the two-dimensional asymptotic equation (equation 8.55), which gives $n = 0$ over the plane $z = 0$ (except at the origin) as well as at $z = h$, is the dipole term with its image. With dipole terms the equation for $p(x', z)$ is

$$p(x', z) = -\text{const}[\exp(\beta x' + \gamma z)]\left[\frac{\partial}{\partial z} K_0(\eta r') + \frac{\partial}{\partial z} K_0(\eta r'')\right]$$

$$= A[\exp(\beta x' + \gamma z)]\left[\frac{z}{r'} K_1(\eta r') + \left(\frac{z - 2h}{r''}\right) K_1(\eta r'')\right] \qquad (8.59)$$

since $K_1(s) = -(d/ds)K_0(s)$ and $A = \eta \times \text{const}$. When $z = h$, $r' = r''$; consequently $p(x', h) = 0$ as required.

The current received by a strip of the electrode with width dx whose edges are the lines $x = \text{constant}$ and $x + dx = \text{constant}$ is

$$-eD_L\left[\frac{\partial}{\partial z} p(x', z)\right]_{z = h} dx$$

and the current received by a strip whose edges are the lines $x = a$ and $x = b$ with $b > a$ is

$$i_{a,b} = -eD_L \int_a^b \left[\frac{\partial}{\partial z} p(x', z)\right]_{z = h} dx$$

$$= -(D_L/D)^{1/2} e \int_{a'}^{b'} \left[\frac{\partial}{\partial z} p(x', z)\right]_{z = h} dx', \qquad (8.60)$$

where $a' = (D_L/D)^{1/2} a$ and $b' = (D_L/D)^{1/2} b$.

We require therefore an expression for $(\partial/\partial z)p(x', z)$. It follows from equation 8.59 that

$$\frac{\partial}{\partial z} p(x', z) = \gamma p(x', z)$$

$$+ A[\exp(\beta x' + \gamma z)]\frac{\partial}{\partial z}\left[\frac{z}{r'} K_1(\eta r') + \left(\frac{z - 2h}{r''}\right) K_1(\eta r'')\right].$$

Consider

$$\frac{\partial}{\partial z}\left[\frac{z}{r'}K_1(\eta r')\right] = \frac{1}{r'}K_1(\eta r') + \frac{z^2}{r'}\frac{d}{dr'}\left[\frac{K_1(\eta r')}{r'}\right]$$

$$= \frac{1}{r'}K_1(\eta r') + (z^2\eta^3)\frac{1}{(\eta r')}\frac{d}{d(\eta r')}\left[\frac{K_1(\eta r')}{\eta r'}\right]$$

$$= \frac{1}{r'}K_1(\eta r') + z^2\eta^3(-1)\frac{K_2(\eta r')}{(\eta r')^2} \qquad \text{(from equation 5.22)}.$$

Similarly,

$$\frac{\partial}{\partial z}\left[\frac{z-2h}{r''}K_1(\eta r'')\right] = \frac{K_1(\eta r'')}{r''} - \frac{(z-2h)^2\eta^3 K_2(\eta r'')}{(\eta r'')^2}.$$

Consequently,

$$\frac{\partial}{\partial z}p(x',z) = \gamma p(x',z) + A\left\{\left[\frac{K_1(\eta r')}{r'} + \frac{K_1(\eta r'')}{r''}\right]\right.$$

$$\left. - \eta\left[\left(\frac{z}{r'}\right)^2 K_2(\eta r') + \left(\frac{z-2h}{r''}\right)^2 K_2(\eta r'')\right]\right\}\exp(\beta x' + \gamma z),$$

whence

$$\left[\frac{\partial}{\partial z}p(x',z)\right]_{z=h} = 2A\left[\frac{K_1(\eta r')}{r'} - \eta\left(\frac{h}{r'}\right)^2 K_2(\eta r')\right]_{z=h}\exp(\beta x' + \gamma h).$$

Equation 8.60 therefore assumes the form

$$i_{a,b} = 2e(DD_L)^{1/2}A(\exp\gamma h)\int_{a'}^{b'}\left\{\eta h^2\frac{K_2\left[\eta(h^2+s^2)^{1/2}\right]}{h^2+s^2}\right.$$

$$\left. - \frac{K_1\left[\eta(h^2+s^2)^{1/2}\right]}{(h^2+s^2)^{1/2}}\right\}(\exp\beta s)\,ds \qquad (8.61)$$

where s is a dummy variable. When $\omega = 0$, then $\beta = 0$ and $\eta = \gamma = \lambda_L$ and it is seen that, apart from the constant of proportionality, equation 8.61 in effect reduces to equation 5.34.

We require to express A in terms of the total current $i \equiv i_{-\infty, \infty}$. If we let $a' = -\infty$ and $b' = \infty$, it becomes necessary, in order to calculate $i_{-\infty, \infty}$ from equation 8.61, to evaluate the class of integrals

$$\int_{-\infty}^{\infty} (\exp \beta s) \frac{K_\nu \left[\eta (h^2 + s^2)^{1/2} \right]}{(h^2 + s^2)^{\nu/2}} \, ds = 2 \int_0^{\infty} (\cosh \beta s) \frac{K_\nu \left[\eta (h^2 + s^2)^{1/2} \right]}{(h^2 + s^2)^{\nu/2}} \, ds.$$

Consider the definite integral (Watson, 1944, p. 416)

$$\int_0^{\infty} J_\mu(bs) \frac{K_\nu \left[\eta (h^2 + s^2)^{1/2} \right]}{(h^2 + s^2)^{\nu/2}} s^{\mu+1} \, ds$$

$$= \frac{b^\mu}{\eta^\nu} \left[\frac{(\eta^2 + b^2)^{1/2}}{h} \right]^{\nu - \mu - 1} K_{\nu - \mu - 1} \left[h(\eta^2 + b^2)^{1/2} \right], \quad (8.62)$$

where $\mathrm{Re}(\mu) > -1$, h is positive, and $\mathrm{Re}(\mu)$ is the real part of μ. Let $\mu = -\frac{1}{2}$ and $b = i\beta$, $(i^2 = -1)$; it then follows that

$$\int_0^{\infty} (\cosh \beta s) \frac{K_\nu \left[\eta (h^2 + s^2)^{1/2} \right]}{(h^2 + s^2)^{\nu/2}} \, ds$$

$$= \left(\frac{\pi}{2} \right)^{1/2} \eta^{-\nu} \left[\frac{(\eta^2 - \beta^2)^{1/2}}{h} \right]^{\nu - 1/2} K_{\nu - 1/2} \left[h(\eta^2 - \beta^2)^{1/2} \right]. \quad (8.63)$$

The values of ν that appear in equation 8.61 are $\nu = 1$ and $\nu = 2$, and it follows that, since $\eta^2 - \beta^2 = \gamma^2$,

$$i = 4eA (DD_L)^{1/2}$$

$$\times (\exp \gamma h) \left(\frac{\pi}{2} \right)^{1/2} \left[\left(\frac{h^2}{\eta} \right) \left(\frac{\gamma}{h} \right)^{3/2} K_{3/2}(\gamma h) - \left(\frac{\gamma}{h} \right)^{1/2} \eta^{-1} K_{1/2}(\gamma h) \right]$$

$$= 4eA (DD_L)^{1/2} \frac{(\exp \gamma h)(\pi/2)^{1/2}}{\eta} \left(\frac{\gamma}{h} \right)^{1/2} \left[(\gamma h) K_{3/2}(\gamma h) - K_{1/2}(\gamma h) \right].$$

But, from equations 5.21, $(\gamma h)K_{3/2}(\gamma h) - K_{1/2}(\gamma h) = (\gamma h)K_{1/2}(\gamma h)$; consequently

$$i = \frac{4eA(DD_L)^{1/2}}{\eta}(\exp \gamma h)\left(\frac{\pi}{2}\right)^{1/2}\left(\frac{\gamma}{h}\right)^{1/2}(\gamma h)K_{1/2}(\gamma h)$$

$$= 2\pi e(DD_L)^{1/2}\left(\frac{\gamma}{\eta}\right)A.$$

We are now able to present equation 8.61 in a more useful form:

$$i_{a,b} = i\left(\frac{\eta}{\gamma}\right)\frac{\exp \gamma h}{\pi}\int_{a'}^{b'}\left\{(\eta h^2)\frac{K_2\left[\eta(h^2+s^2)^{1/2}\right]}{h^2+s^2}\right.$$

$$\left. -\frac{K_1\left[\eta(h^2+s^2)^{1/2}\right]}{(h^2+s^2)^{1/2}}\right\}(\exp \beta s)\,ds. \quad (8.64)$$

When $\omega = 0$, then $\beta = 0$, $\eta = \gamma = \lambda_L$, and equation 8.64 reduces to equation 5.34.

It is necessary also to evaluate integrals of the type

$$I_{\pm\beta} = \int_0^\infty \frac{K_\nu\left[\eta(h^2+s^2)^{1/2}\right]}{(h^2+s^2)^{\nu/2}}(\exp \pm \beta s)\,ds.$$

Since it does not appear to be possible to link these integrals immediately with a standard integral such as that given in equation 8.62, we express the integrals as infinite series. We replace $\exp \beta s$ by its series $\sum_{n=0}^\infty (\beta s)^n/n!$ to obtain

$$I_{+\beta} = \sum_{n=0}^\infty \int_0^\infty \frac{K_\nu\left[\eta(h^2+s^2)^{1/2}\right]}{(h^2+s^2)^{\nu/2}}\frac{(\beta s)^n}{n!}\,ds.$$

If we let $n = 2\mu+1$, that is, $\mu = (n-1)/2$, from equation 5.33

$$I_{+\beta} = \sum_{n=0}^\infty 2^{(n-1)/2}\frac{\Gamma[(n+1)/2]}{n!}\left(\frac{\beta}{\eta}\right)^n(\eta h)^{(n-1)/2}h^{1-\nu}K_{\nu-(n+1)/2}(\eta h);$$

$$(8.65)$$

similarly

$$I_{-\beta} = \sum_{n=0}^{\infty} (-1)^n 2^{(n-1)/2} \frac{\Gamma[(n+1)/2]}{n!} \left(\frac{\beta}{\eta}\right)^n (\eta h)^{(n-1)/2} h^{1-\nu} K_{\nu-(n+1)/2}(\eta h).$$

We are also required to consider the integrals

$$I(b') = \int_0^{b'} \frac{K_\nu\left[\eta(h^2+s^2)^{1/2}\right]}{(h^2+s^2)^{\nu/2}} (\exp \pm \beta s) \, ds$$

when $b/h \ll 1$ and $b' = (D_L/D)^{1/2} b$.

The maximum value of $h^2 + s^2$ across the strip of the electrode for which $0 \leqslant s \leqslant b'$ is $h^2 + b'^2$. When, as in practice, $h \cong 10$ cm and $b \cong 0.2$ cm, it is evident that the difference between h and $(h^2 + s^2)^{1/2}$ on this strip is negligibly small. Therefore with very close approximation we may replace $(h^2 + s^2)^{1/2}$ by h in the integrand and write

$$I(b') = \frac{K_\nu(\eta h)}{h^\nu} \int_0^{b'} (\exp \pm \beta s) \, ds = \frac{K_\nu(\eta h)}{\pm \beta h^\nu} [(\exp \pm \beta b') - 1]. \quad (8.66)$$

We now consider the principles of methods for measuring $\beta = W_x/W_z$ or, alternatively, the magnetic deflexion coefficient ψ.

8.6.   PRINCIPLE OF EXPERIMENTAL METHODS

The aim of these methods is to measure β/η, which when $W_x/W_z \ll 1$ is, from equations 8.51, inappreciably different from $\beta/\gamma = (W_x/W_z)$ $(D_L/D)^{1/2} \cong (W_x/W)(D_L/D)^{1/2}$. From the measured value of β/γ we deduce $W_M = (E/B)(W_x/W_z)$ (equation 8.18), and from W_M we obtain ψ, the magnetic deflexion coefficient (equation 8.19), as $\psi = W_M/W$, where W is the true drift velocity measured by time-of-flight methods. From ψ, information can be deduced about the collision cross section $q_m(c)$ by means of equation 8.21 or 8.22. In each method the quantity actually measured is the ratio of the currents received by portions of a divided electrode, but the method of division is different in the two techniques that have been used.

8.6.1.   TOWNSEND'S METHOD (1912). The first investigations of the magnetic deflexion of an electron stream were made by Townsend and Tizard (1912, 1913) and are referred to in Section 1.10.2. The receiving electrode is divided, as shown in Figure 1.1c, by a pair of cuts parallel to the axis Oy, to which ω is also parallel. The central strip is symmetrically located with respect to the centre of the electrode and is connected electrically to one of the two flanking portions of the electrode, which is then in effect divided by a single cut into two unequal portions. With no magnetic field the axis of the stream coincides with the axis Oz, and the current i_3 to the smaller portion of the electrode is less than the current $i_1 + i_2$ to the remainder of the electrode. When a magnetic field $\mathbf{B}$ is applied in the correct sense parallel to the axis $\pm Oy$, the stream is deflected and its axis, which still lies in the xOz plane, meets the electrode at a distance x. Let the centre of the electrode lie at a distance b from the dividing cut. Townsend made the plausible assumption that when the currents i_3 and $i_1 + i_2$ were equal the axis of the stream met the plane of the electrode at the dividing cut. The length of the diffusion chamber is h, and Townsend assumed that $W_x / W_z = b/h$.

The experimental procedure was to find the value of B at which $i_3 = i_1 + i_2$. The velocity W_M was then found as $W_M = (E/B)(b/h)$ and was taken to be the true drift velocity. However, as already remarked, the true drift velocity is $W = \psi^{-1} W_M$. In modern investigations W is measured by time-of-flight methods, and ψ is found as W_M / W.

Townsend's assumption that $W_x / W_z = b/h$, although plausible, would be expected to be in best accord with the facts at small deflexions, but in any event requires justification.

It follows from equation 8.64 that

$$i_3 = i_{b,\infty} = Ci \int_{b'}^{\infty} \phi(\eta, h, s)(\exp \beta s)\, ds$$

where

$$C = \frac{(\eta/\gamma)(\exp \gamma h)}{\pi}$$

and

$$\phi(\eta, h, s) = \eta h^2 \frac{K_2 \left[\eta (h^2 + s^2)^{1/2} \right]}{h^2 + s^2} - \frac{K_1 \left[\eta (h^2 + s^2)^{1/2} \right]}{(h^2 + s^2)^{1/2}}.$$

Similarly,

$$i_1 + i_2 = i_{-\infty, b} = Ci \int_{-\infty}^{b'} \phi(\eta, h, s)(\exp \beta s)\, ds.$$

We require the value of β at which $i_3 = i_1 + i_2$, that is, when $i_{b,\infty} = i_{-\infty, b}$.

Since $i_{b,\infty}=i_{0,\infty}-i_{0,b}$ and $i_{-\infty,b}=i_{-\infty,0}+i_{0,b}$, the condition is equivalent to

$$i_{0,\infty}-i_{-\infty,0}=2i_{0,b}.$$

It follows therefore from equation 8.64 that β must be chosen to satisfy the following transcendental equation:

$$2\int_0^{b'}\phi(\eta,h,s)(\exp\beta s)\,ds=\int_0^{\infty}(\exp\beta s-\exp-\beta s)\phi(\eta,h,s)\,ds$$

$$=\int_0^{\infty}(\exp\beta s-\exp-\beta s)$$

$$\times\left\{\eta h^2\frac{K_2\left[\eta(h^2+s^2)^{1/2}\right]}{h^2+s^2}-\frac{K_1\left[\eta(h^2+s^2)^{1/2}\right]}{(h^2+s^2)^{1/2}}\right\}\,ds.$$

Thus, on making use of equations 8.65, it can be seen that, since $K_{-\nu}=K_{\nu}$,

$$2\int_0^{b'}\phi(\eta,h,s)(\exp\beta s)\,ds=\sum_{m=0}^{\infty}\frac{2^{m+1}(m!)}{(2m+1)!}\left(\frac{\beta}{\eta}\right)^{2m+1}(\eta h)^m$$

$$\times\left[\eta hK_{m-1}(\eta h)-K_m(\eta h)\right].\qquad(8.67)$$

When $b/h\ll1$, the variation of h^2+s^2 across the interval $0\leqslant s\leqslant b$ is negligible. We therefore make the following approximation:

$$\int_0^{b'}\phi(\eta,h,s)(\exp\beta s)\,ds\cong\left[\eta h^2\frac{K_2(\eta h)}{h^2}-\frac{K_1(\eta h)}{h}\right]\int_0^{b'}(\exp\beta s)\,ds$$

$$=\frac{1}{\beta h}\left[(\eta h)K_2(\eta h)-K_1(\eta h)\right]\left[(\exp\beta b')-1\right].$$

If $\beta b'$ is small compared with unity, $(\exp\beta b')-1\cong\beta b'$; consequently, if the first term of the right-hand side of equation 8.67 is adequately approximated by its first term when $\beta/\eta\ll1$, the equation reduces to

$$\frac{b'}{h}\left[(\eta h)K_2(\eta h)-K_1(\eta h)\right]=\frac{\beta}{\eta}\left[\eta hK_1(\eta h)-K_0(\eta h)\right].$$

However, when h is several centimetres, ηh is a large quantity. Consequently the K functions approach their asymptotic form $(\pi/2\eta h)^{1/2}$ $(\exp - \eta h)$, and it follows that $(\beta/\eta) \rightarrow b'/h$, that is to say,

$$\frac{\beta}{\eta} \cong \frac{b'}{h} = \left(\frac{D_L}{D}\right)^{1/2} \frac{b}{h}, \qquad \frac{\beta}{\eta} \cong \frac{\beta}{\gamma}.$$

But, from equation 8.51, $(\beta/\gamma) = (W_x/W_z)(D_L/D)^{1/2}$. Thus, when $\beta b \ll 1$, $\beta/\eta \ll 1$, and $\eta h \gg 1$, as usually occurs, $W_x/W_z \cong b/h$, as assumed by Townsend.

In practice b is of the order 0.2 cm and h is several centimeters, say 8 or 10. Consequently $b/h = W_x/W_z \cong 1/50$, which is sufficiently small to ensure an accurate result. Moreover with $h \cong 10$ cm the quantity ηh is large (of the order 100 to several hundred). The asymptotic formulae then represent the Bessel K functions with negligible error. The division of the electrode into a central strip of width $2b$ with flanking portions as shown in Figure 1.1c permits $\lambda = W/2D$ and therefore D/μ to be measured in the same diffusion chamber as is used to measure magnetic deflexion. The theory of the measurement of D/μ is given in Section 5.3.4.

8.6.2.   METHOD OF HUXLEY AND ZAAZOU (1949). In this method the receiving electrode is divided symmetrically by a cut running parallel to Oy and passing through the centre of the electrode (Figure 1.1b), which is also divided to have a central disk so that measurements of λ, that is, of D/μ, can also be made.

For use with a magnetic field the electrical connections to the various portions are such that the electrode functions as two independent halves. When no magnetic field is present, the currents i_1 and i_2 to the two halves of the electrode are equal; when the magnetic field $\mathbf{B}$ is established parallel to Oy, however, the stream is deflected and one current is increased and the other diminished. Suppose that the current is deflected in the sense of $+Ox$ and that i_1 is increased and i_2 diminished. The quantity that is measured is the ratio $R_{1,2} = i_1/i_2$, for which a formula is required.

It follows from equation 8.64 that

$$i_1 = Ci \int_0^\infty \phi(\eta, h, s)(\exp \beta s)\, ds$$

and   (8.68)

$$i_2 = Ci \int_{-\infty}^0 \phi(\eta, h, s)(\exp \beta s)\, ds = Ci \int_0^\infty \phi(\eta, h, s)(\exp - \beta s)\, ds,$$

where, as before, $C = (\eta/\gamma)(\exp \gamma h)/\pi$ and

$$\phi(\eta,h,s) = \eta h^2 \frac{K_2\left[\eta(h^2+s^2)^{1/2}\right]}{h^2+s^2} - \frac{K_1\left[\eta(h^2+s^2)^{1/2}\right]}{(h^2+s^2)^{1/2}}.$$

The current ratio $R_{1,2}$ is therefore

$$R_{1,2} = \frac{\displaystyle\int_0^\infty \phi(\eta,h,s)(\exp \beta s)\,ds}{\displaystyle\int_0^\infty \phi(\eta,h,s)(\exp - \beta s)\,ds}$$

which, from equations 8.65, is equivalent to

$$R_{1,2} = \frac{\displaystyle\sum_{n=0}^\infty 2^{(n-1)/2}\left[\Gamma\left(\frac{n+1}{2}\right)/n!\right]\left(\frac{\beta}{\eta}\right)^n (\eta h)^{(n-1)/2} \times \left[\eta h K_{2-(n+1)/2}(\eta h) - K_{1-(n+1)/2}(\eta h)\right]}{\displaystyle\sum_{n=0}^\infty (-1)^n 2^{(n-1)/2}\left[\Gamma\left(\frac{n+1}{2}\right)/n!\right]\left(\frac{\beta}{\eta}\right)^n (\eta h)^{(n-1)/2} \times \left[\eta h K_{2-(n+1)/2}(\eta h) - K_{1-(n+1)/2}(\eta h)\right]}$$

$$= \frac{S_e + S_o}{S_e - S_o}, \tag{8.69}$$

where S_e is the sum of the terms in the numerator for which n is even (i.e., of the form $2m$), and S_o is the sum of the terms in which $n = 2m + 1$. It follows that

$$S_e = \sum_{m=0}^\infty 2^{m-1/2} \frac{\Gamma(m+\frac{1}{2})}{(2m)!}\left(\frac{\beta}{\eta}\right)^{2m}(\eta h)^{m-1/2}\left[\eta h K_{m-3/2}(\eta h) - K_{m-1/2}(\eta h)\right]$$

and

$$S_o = \sum_{m=0}^\infty 2^m \frac{\Gamma(m+1)}{(2m+1)!}\left(\frac{\beta}{\eta}\right)^{2m+1}(\eta h)^m\left[\eta h K_{m-1}(\eta h) - K_m(\eta h)\right]$$

$$\tag{8.70}$$

with $K_{-\nu} = K_\nu$.

Equation 8.69 is therefore equivalent to

$$\frac{S_e}{S_o} = \frac{R_{1,2}+1}{R_{1,2}-1}.$$

(8.71)

When ηh is large, as in practice it is, the Bessel K functions in equation 8.70 are each negligibly different from $(\pi/2\eta h)^{1/2}(\exp-\eta h)$ unless m is large. If, however, the series for S_e and S_o converge sufficiently rapidly for their values to be given with adequate accuracy by the sum of their first few terms, the factors $[\eta h K_{m-3/2}(\eta h) - K_{m-1/2}(\eta h)]$ and $[\eta h K_{m-1}(\eta h) - K_m(\eta h)]$ are independent of m and are also in effect equal and may be removed from the ratio S_e/S_o. Consequently

$$\frac{S_e}{S_o} \cong \frac{\displaystyle\sum_{m=0}^{\infty} 2^{m-1/2}[\Gamma(m+1/2)/(2m)!](\beta/\eta)^{2m}(\eta h)^{m-1/2}}{\displaystyle\sum_{m=0}^{\infty} 2^{m}[\Gamma(m+1)/(2m+1)!](\beta/\eta)^{2m+1}(\eta h)^{m}}.$$

Since $\Gamma(m+\tfrac{1}{2})=\pi^{1/2}(2m)!/(2^{2m}m!)$ this expression for S_e/S_o can be given the more convenient form

$$\frac{S_e}{S_o} \cong \left[\frac{1}{2(\beta/\eta)^2(\eta h)}\right]^{1/2}\left[\frac{\exp\left[(\beta/\eta)^2(\eta h)/2\right]}{\displaystyle\sum_{m=0}^{\infty}\left[(\beta/\eta)^2(\eta h)/2\right]^m/[2\Gamma(m+\tfrac{3}{2})]}\right].$$

(8.72)

The terms of the infinite series in the denominator are less than the corresponding terms in the expansion of the exponential function in the numerator; consequently the infinite series is convergent for all values of $(\beta/\eta)^2(\eta h)/2$. However, in laboratory experiments with magnetically deflected streams, the conditions are such that the value of $(\beta/\eta)^2(\eta h)/2$ is much less than unity and equation 8.72 can then be more usefully expressed in the form

$$\frac{S_e}{S_o} \cong \left[\frac{\pi}{2(\beta/\eta)^2(\eta h)}\right]^{1/2}\left[\frac{1+(\beta/\eta)^2(\eta h)/2+\cdots}{1+(\beta/\eta)^2(\eta h)/3+\cdots}\right],$$

that is, to the first order of approximation (with $\eta \cong \gamma$),

$$\frac{\beta}{\eta} \cong \left(\frac{W_x}{W_z} \right) \left(\frac{D_L}{D} \right)^{1/2} \cong \left(\frac{\pi}{2\eta h} \right)^{1/2} \frac{S_o}{S_e}$$

$$= \left(\frac{\pi}{2\eta h} \right)^{1/2} \left(\frac{R_{1,2}-1}{R_{1,2}+1} \right). \tag{8.73}$$

Since, from equations 8.51 and 8.53, $\eta \cong \gamma = W_z/2D_L$, equation 8.73 can be expressed in the form

$$\frac{W_x}{W_z} = \left(\frac{D}{D_L} \right)^{1/2} \left(\frac{\pi D_L}{W_z h} \right)^{1/2} \left(\frac{R_{1,2}-1}{R_{1,2}+1} \right) = \left(\frac{\pi}{2\lambda h} \right)^{1/2} \left(\frac{R_{1,2}-1}{R_{1,2}+1} \right), \tag{8.74}$$

where $2\lambda = W/D \cong W_z/D$.

The quantity λ can be determined with the same apparatus when the receiving electrode is constructed as shown in Figure 1.1b, so that by a change of electrical connections it can be converted into the system of a central disk with a surrounding annulus. Trial experiments were carried out by Huxley and Zaazou (1949) with this method, and more extensive investigations were undertaken by Hall (1955) and by Jory (1965). In all these studies the theoretical formulae were based on the assumption that the source behaves as a pole, whereas in the present analysis a dipole source is postulated. However, as shown in the transition from equation 8.69 to 8.72, the Bessel K functions can be eliminated when ηh is large, with the result that the final formula 8.72 does not depend on which of the assumptions as to the nature of the source is adopted. The present method possesses the advantage that β/η can be derived from any value of $R_{1,2}$ within a suitable range and a check obtained by comparing results found with various values of B. In this way it is possible to check the degree of approximation that is made in employing a simplified form of equation 8.72 such as equation 8.74. The method is therefore more flexible than that of Townsend.

It is evident that other combinations of portions of the electrode could be adopted, but these do not appear to offer advantages over the ones already used.

We note that although the determination of the magnetic deflexion coefficient ψ is usually the aim of these investigations the component W_x itself is equally useful, since it is linked with the distribution function and the collisional cross section $q_m(c)$ through equation 8.15.

Another type of investigation is based on equation 8.48, which relates to the case in which $\boldsymbol{\omega}$ is paralleled to $e\mathbf{E}$. This equation is essentially the same as equation 5.10 with D replaced by $D_\perp$, which is then capable of controlled variation through its dependence on ω.

We therefore arrive at equation 5.11 but with the following interpretation of the symbols: $x' = (D_L/D_\perp)^{1/2}x$, $y' = (D_L/D_\perp)^{1/2}y$, $2\lambda_L = W/D_L$ (as before). Hence the formulae for the current ratios are as given in equation 5.28 but with $d'^2 = h^2 + b'^2$ and $b' = (D_L/D_\perp)^{1/2}b$, where $D_\perp$ is defined in relation to equations 8.31 and 8.28.

Experimental investigations of diffusing streams in a longitudinal magnetic field were made by Bailey (1930), and the theory was also discussed by Huxley (1940).

We proceed in the next chapter to develop the theory of electron motion in gases in the presence of a magnetic field and a high-frequency electric field.

BIBLIOGRAPHY AND NOTES

It was remarked in the bibliography for Chapter 7 that the early formulae for W and D were of the form of "constant free time" equations in which $\nu = 1/T$ is independent of c. These formulae are

$$D = \tfrac{1}{3}lc = \frac{c^2}{3}T \quad \text{and} \quad \mathbf{W} = \frac{e\mathbf{E}}{m}T.$$

These formulae were extended by J. S. Townsend (1915) to cover motion in a constant magnetic field. Townsend found that diffusion normal to the magnetic field took place with a coefficient $D_\perp = D/(1 + \omega^2 T^2) = \nu^2 D/(\nu^2 + \omega^2)$, where D is the isotropic coefficient of diffusion and $\boldsymbol{\omega} = -(e/m)\mathbf{B}$. For the components of the drift velocity, when $\boldsymbol{\omega}$ is perpendicular to $e\mathbf{E}$, he found

$$W_\parallel = \frac{eE}{m}\frac{T}{1 + \omega^2 T^2} \quad \text{and} \quad W_\perp = \frac{eE}{m}\frac{\omega T^2}{1 + \omega^2 T^2},$$

where $W_\parallel$ and $W_\perp$ are the components of $\mathbf{W}$ parallel and normal, respectively, to $e\mathbf{E}$. According to these formulae, the angle of deflexion of an electron stream is θ, where

$$\tan \theta = \frac{W_\perp}{W_\parallel} = \omega T.$$

When $\omega^2 T^2 \ll 1$, that is, $\nu^2 \gg \omega^2$, then $W_\parallel \cong (eE/m)T = W$ and $\tan\theta \cong \omega m W/eE = BW/E$. This is the relation through which Townsend determined $W_M = (E\tan\theta)/B$ in the belief that W_M was the same as W (Sections 8.2.2 and 1.10).

In 1937 he derived more accurate expressions for $W_\parallel$ and $W_\perp$ which showed that in general W_M and W were not equal. The expressions for $W_\parallel$ and $W_\perp$ were derived on the assumption that $\omega T \ll 1$.

Also in 1937, Huxley derived free path formulae in which this restriction was removed. He found

$$W_{\parallel} = \frac{2}{3}\frac{E}{B}\,\overline{\frac{\omega T}{1+\omega^2 T^2}\left(1+\frac{\omega^2 T^2}{1+\omega^2 T^2}\right)} = \frac{eE}{3m}\,\overline{\frac{1}{c^2}\frac{d}{dc}\left(\frac{c^3 v}{v^2+\omega^2}\right)},$$

$$W_{\perp} = -\frac{E}{B}\,\overline{\frac{\omega^2 T^2}{1+\omega^2 T^2}\left(\tfrac{1}{3}+\omega^2 T^2\right)} = -\frac{eE}{3m}\,\overline{\frac{1}{c^2}\frac{d}{dc}\left(\frac{\omega c^3}{v^2+\omega^2}\right)}.$$

The average values in these expressions are taken over all shells (c, dc), and it can be seen that they are equivalent to equations 8.16 and 8.17 when the latter are transformed by partial integration. Huxley (1960) showed that the correct relationship between W and $\tan\theta$ is $W = C(E/B)\tan\theta$, where C is the dimensionless factor

$$C = \frac{\tfrac{1}{3}\left[\,\overline{[c^{-2}(d/dc)(lc^2)]}\,\right]^2}{\left[\,\overline{c^{-2}(d/dc)(l^2 c)}\,\right]} \quad\text{with } l = \frac{1}{Ncq_m}.$$

The magnetic deflexion of an electron stream was also investigated by Frost and Phelps (1962), who wrote $W_M = \psi W$. They termed the coefficient ψ the "magnetic deflexion coefficient." Evidently $\psi = 1/C$. An extensive experimental investigation of magnetic deflexion was undertaken by Jory (1965).

The theory of drift and diffusion in a magnetic field according to the Maxwell-Boltzmann theory was given by Allis and Allen (1937) and by Allis (1956).

W. P. Allis and H. W. Allen, *Phys. Rev.*, **52**, 703, 1937.

W. P. Allis, "Motions of Ions and Electrons," *Handbuch der Physik*, Vol. 21, Springer-Verlag, Berlin, 1956.

V. A. Bailey, *Phil. Mag.*, **9**, 560, 1930; **9**, 625, 1930.

A. G. Engelhardt, A. V. Phelps, and C. G. Risk, *Phys. Rev.*, **135**, A1566, 1964.

L. S. Frost and A. V. Phelps, *Phys. Rev.*, **127**, 1621, 1962.

B. I. H. Hall, *Proc. Phys. Soc. B*, **68**, 334, 1955.

L. G. H. Huxley, *Phil. Mag.*, **23**, 210, 1937.

L. G. H. Huxley, *Phil. Mag.*, **30**, 396, 1940.

L. G. H. Huxley and A. A. Zaazou, *Proc. Roy. Soc. A*, **196**, 402, 1949.

L. G. H. Huxley, *Aust. J. Phys.*, **13**, 718, 1960.

L. G. H. Huxley and R. W. Crompton, "The Motions of Slow Electrons in Gases," *Atomic and Molecular Processes*, (Ed., D. R. Bates) Academic Press, New York, 1962.

R. L. Jory, *Aust. J. Phys.*, **18**, 237, 1965.

J. S. Townsend and H. T. Tizard, *Proc. Roy. Soc. A*, **87**, 357, 1912.

J. S. Townsend and H. T. Tizard, *Proc. Roy. Soc. A*, **88**, 336, 1913.

J. S. Townsend, *Electricity in Gases*, Clarendon Press, Oxford, 1915, pp. 96-102.

J. S. Townsend, *Phil. Mag.*, **23**, 880, 1937.

G. S. Watson, *A Treatise on the Theory of Bessel Functions*, 2nd ed., Cambridge University Press, 1944.

9

MOTION IN A HIGH-FREQUENCY

ELECTRIC FIELD

9.1. INTRODUCTION

In all that has preceded, an electric field is postulated to be uniform and constant. There are many practical circumstances, however, where it is important to understand the motion of electrons under the influence of alternating electric fields, as, for example, in studies of the propagation of electromagnetic waves in slightly ionized gases. In such studies it is essential to possess formulae that provide a quantitative description of the electron motion; consequently the chief concern of this chapter is to derive theoretical formulae that relate to the macroscopic properties of electron motion in gases in alternating electric fields. These fields are assumed to be uniform, and in order to extend the generality of the discussion it is also assumed that a uniform and time-independent magnetic field is present.

At the outset we draw attention to a convenient simplification of the theory of motion in high-frequency fields, as compared with that of motion in static fields, by referring to the scalar equation 3.30 and to Section 3.11. Consider the first term of the group $-(1/3c^2)(\partial/\partial c)\{\quad\}$, which is

$$\frac{1}{3c^2}\frac{\partial}{\partial c}\left[c^3\frac{e\mathbf{E}}{mv}\cdot\mathrm{grad}_r(nf_0)\right].$$

Since $\mathrm{grad}_r(nf_0)$ at a position $\mathbf{r}$ preserves a fixed direction and magnitude during an interval t that contains many cycles of the oscillating field $\mathbf{E}(t)$, it follows that the mean value of $e\mathbf{E}\cdot\mathrm{grad}_r(nf_0)$ is in effect equal to zero and the associated contribution to $\sigma_E(c)$ can be neglected. The disappearance of the term $(1/3c^2)(\partial/\partial c)[c^3\mathbf{V}'\cdot\mathrm{grad}_r(nf_0)]$ greatly simplifies the scalar equation, especially with regard to diffusion which in the presence of a high-frequency field is isotropic provided that no magnetic field is present. However, in the presence of a magnetic field parallel to Oz, the differential equation for n is

$$\frac{\partial}{\partial t} n + D_\perp \left(\frac{\partial^2}{\partial x^2} n + \frac{\partial^2}{\partial y^2} n \right) + D \frac{\partial^2}{\partial z^2} n = 0,$$

where $D_\perp$ and D are defined as in equation 8.28 (but with $D_L \equiv D$). In the formulae for these coefficients the generalized distribution function is replaced by its modified forms in equations 9.55 and 9.56.

In some presentations of the theory of the subject, formulae and equations are developed in the first instance on the assumption that the electric field oscillates at high frequency. It is then assumed that the appropriate formulae for motion in a constant field are obtained simply as special cases by assigning the value zero to the angular frequency in formulae derived for a high-frequency field. This is a spurious generalization, however, since as we have remarked above it is necessary to consider the important term $(1/3c^2)(\partial/\partial c)[c^3 \mathbf{V'} \cdot \text{grad}_r(nf_0)]$ when the field is constant, whereas when the field oscillates at high frequency this term is eliminated. It is this term that is responsible for the difference between the coefficients of diffusion D_L and D. The phenomenon of anisotropic diffusion was therefore overlooked.

We begin the discussion with a recapitulation of what was stated in Sections 3.3 and 5.4.1 about the orders of magnitude of the time constants of decay or adjustment of momentum and mean energy, but we adapt the treatment to motion in high-frequency fields accompanied by a constant magnetic field.

9.2. TIME CONSTANTS OF MOMENTUM AND ENERGY

We consider the behaviour of an isolated group of n_0 electrons. The vector equation for the subclass $n\,d\mathbf{r}(4\pi c^2 f_0 dc)$ of electrons of the group is (equation 3.8)

$$\frac{\partial}{\partial t}(n\mathbf{f}_1) + c\,\text{grad}_r(nf_0) + \frac{\partial}{\partial c}(nf_0)\frac{e\mathbf{E}}{m} - \boldsymbol{\omega} \times (n\mathbf{f}_1) + \nu n\mathbf{f}_1 = 0, \quad (9.1)$$

where $\boldsymbol{\omega} = -e\mathbf{B}/m$.

If we perform the integration $\int_\mathbf{r}(\quad)d\mathbf{r}$ of each term and recall that

$$n_0 = \int_\mathbf{r} n\,d\mathbf{r}, \qquad n_0 f_0^* = \int_\mathbf{r} nf_0 d\mathbf{r}, \qquad \text{and} \qquad n_0 \mathbf{f}_1^* = \int_\mathbf{r} n\mathbf{f}_1\,d\mathbf{r},$$

it follows that

$$\frac{\partial}{\partial t}\mathbf{f}_1^* + \frac{e\mathbf{E}}{m}\frac{\partial}{\partial c}f_0^* - \boldsymbol{\omega}\times\mathbf{f}_1^* + \nu\mathbf{f}_1^* = 0. \qquad (9.2)$$

But the momentum of the electrons of the shell (c, dc) is

$$n_0 m\mathbf{W}(c) = \frac{n_0 mc\mathbf{f}_1^*}{3f_0^*},$$

and the total momentum of the group of n_0 electrons is

$$\mathbf{P} = n_0 m\int_0^\infty 4\pi\mathbf{W}(c)f_0^* c^2\,dc = n_0 m\frac{4\pi}{3}\int_0^\infty c\mathbf{f}_1^* c^2\,dc.$$

We form the integral $n_0 m(4\pi/3)\int_0^\infty (\quad)c^3\,dc$ of equation 9.2 to find

$$\frac{d}{dt}\mathbf{P} + n_0 e\mathbf{E}\frac{4\pi}{3}\int_0^\infty c^3\frac{\partial}{\partial c}f_0^*\,dc - \boldsymbol{\omega}\times\mathbf{P} + n_0 m 4\pi\int_0^\infty \nu\mathbf{W}(c)f_0^* c^2\,dc = 0,$$

that is,

$$\frac{d}{dt}\mathbf{P} + n_0 e\mathbf{E} - \boldsymbol{\omega}\times\mathbf{P} + \nu_P\mathbf{P} = 0, \qquad (9.3)$$

where

$$\nu_P\mathbf{P} = n_0 m 4\pi\int_0^\infty \nu\mathbf{W}(c)f_0^* c^2\,dc$$

and ν_P is an effective collision frequency that accounts for the rate of loss of momentum in encounters. We shall assume that $\nu_P \cong \bar{\nu} = \overline{Nq_m(c)c}$.

When the field is switched off, equation 9.3 becomes

$$\frac{d}{dt}\mathbf{P} - \boldsymbol{\omega}\times\mathbf{P} + \nu_P\mathbf{P} = 0. \qquad (9.4)$$

The rate of change of the magnitude of $\mathbf{P}$ is therefore

$$\frac{d}{dt}P = -\left(\omega^2 + \nu_P^2\right)^{1/2}P,$$

whence

$$P = P_0\left[\exp - \left(\omega^2 + \nu_P^2\right)^{1/2}t\right] = P_0\exp\left(-\frac{t}{\tau_P}\right). \qquad (9.5)$$

The value of $\bar{\nu} \cong \nu_p$ in a gas at a pressure of 1 torr is of the order of magnitude 10^9 to 10^{10} sec^{-1}. Because $\bar{\nu} \propto N$, it follows that the time constant of decay τ_P is very small at this and greater pressures and that in equation 9.3, when E is a function $E(t)$ of time, the momentum P follows in step with the changing field unless $E(t)$ changes by an appreciable proportion of itself in a time comparable with or much less than τ_P. Even in the E region of the ionosphere, where $\tau_P \sim 10^{-5}$ to 10^{-6} sec, this value of the time constant is small compared with the period of audio-frequency modulation of a broadcast transmission. On the other hand, the time constant is comparable with the period of a broadcast carrier wave.

When the time constant τ_ϵ of decay of the mean energy $\frac{1}{2}m\overline{c^2}$ of the group n_0 of electrons is estimated as in Section 5.4.1, it is found to be, for a diatomic gas at a pressure of 1 torr, of the order of magnitude 10^{-7} to 10^{-6} sec, which is greater by a factor of 100 or more than τ_P, the time constant of the momentum. The disparity is even more marked in the monatomic gases. It follows that under the influence of an oscillating electric field with period T comparable with τ_P the momentum can oscillate with appreciable amplitude, with, in general, a difference in phase at the frequency $1/T$, whereas the energy fluctuates inappreciably about the mean value $\frac{1}{2}m\overline{c^2}$, at which it is maintained by the power supplied by the oscillating field.

It is assumed in what follows that the fluctuation of the energy about its mean value is unimportant and that only the time dependence of the drift velocity, which implies that of the momentum, need be considered.

9.3. DRIFT VELOCITY UNDER AN OSCILLATING ELECTRIC FIELD WHOSE PERIOD IS MUCH LESS THAN THE TIME CONSTANT OF DECAY OF THE MEAN ELECTRON ENERGY

Let the Cartesian components of the electric vector be

$$E_k = E_{0_k} \cos(pt + \alpha_k), \qquad k = x, y, \text{or } z. \tag{9.6}$$

In the notation of complex quantities equation 9.6 is equivalent to

$$2E_k = |E_k| \exp ipt + |E_k|^* \exp - ipt, \tag{9.7}$$

where $|E_k| \equiv E_{0_k} \exp i\alpha_k$ is the complex amplitude, and $|E_k|^*$ is its complex conjugate.

It follows from equations 9.6 and 9.7 that

$$E_k = \text{Re}(|E_k| \exp ipt). \tag{9.8}$$

The components of the convective velocity of a shell, which are also sinusoidal, are similarly represented by

$$2W_k(c) = |W_k(c)|\exp ipt + |W_k(c)|^* \exp -ipt,$$

that is,

$$W_k(c) = \mathrm{Re}[|W_k(c)|\exp ipt]. \tag{9.9}$$

We seek formulae that relate the $W_k(c)$ to the $E_k(c)$.

In equation 9.1 (the vector equation) we let the vector $\mathbf{f}_1$ be represented as $\mathbf{f}_1 = \mathbf{f}_E + \mathbf{f}_G$ and separate the equation as follows:

$$\frac{\partial}{\partial t}(n\mathbf{f}_E) + \frac{e\mathbf{E}}{m}\frac{\partial}{\partial c}(nf_0) - \boldsymbol{\omega} \times n\mathbf{f}_E + \nu n\mathbf{f}_E = 0, \tag{9.10}$$

$$\frac{\partial}{\partial t}(n\mathbf{f}_G) + c\,\mathrm{grad}_r(nf_0) - \boldsymbol{\omega} \times n\mathbf{f}_G + \nu n\mathbf{f}_G = 0. \tag{9.11}$$

We now make use of the supposition that the time constant for decay of the mean energy $\frac{1}{2}m\,\overline{c^2}$ is long in comparison with the period $2\pi/p$ of the oscillation. In this circumstance the distribution function f_0 is that appropriate to diffusion in a magnetic field with coefficients uninfluenced by the electric field. The group distribution function f_0^* (Section 9.6) is the generalized function of equations 3.11 and 4.7 or the corresponding function, according to whether or not all encounters are elastic. We proceed as in Section 3.3 by neglecting $(\partial/\partial t)(n\mathbf{f}_G)$ in comparison with the other terms in equation 9.11 because ν is large. In other words, we assume quasi equilibrium conditions for the process of diffusion over a short interval. Equation 9.11 is therefore equivalent to

$$c\,\mathrm{grad}_r(nf_0) - \boldsymbol{\omega} \times n\mathbf{f}_G + \nu n\mathbf{f}_G = 0. \tag{9.12}$$

Moreover the presence of $\mathrm{grad}_r(nf_0)$ requires that f_0 is of the form $f_0 \equiv f_0(c, \mathbf{r}, t)$. We shall, however, assume, unless otherwise stated, that this time and spatial dependence is slight and write $f_0 \equiv f_0(c)$. Since $\mathbf{W}_E(c) = c\mathbf{f}_E/3f_0$, equation 9.10 can be written

$$(nf_0)\frac{\partial}{\partial t}\mathbf{W}_E(c) + \frac{ne\mathbf{E}}{3m}c\frac{d}{dc}f_0 - (nf_0)\boldsymbol{\omega} \times \mathbf{W}_E(c) + (nf_0)\nu\mathbf{W}_E(c) = 0. \tag{9.13}$$

Equation 9.13 is a compact representation of the following set of three equations when $E_k = \mathrm{Re}(|E_k| \exp ipt)$:

$$(v+ip)|W_x(c)| + \omega_z|W_y(c)| - \omega_y|W_z(c)| = -\frac{c}{3f_0}\frac{d}{dc}f_0\frac{e}{m}|E_x|,$$

$$-\omega_z|W_x(c)| + (v+ip)|W_y(c)| + \omega_x|W_z(c)| = -\frac{c}{3f_0}\frac{d}{dc}f_0\frac{e}{m}|E_y|, \quad (9.14)$$

$$\omega_y|W_x(c)| - \omega_x|W_y(c)| + (v+ip)|W_z(c)| = -\frac{c}{3f_0}\frac{d}{dc}f_0\frac{e}{m}|E_z|,$$

which in matrix notation are equivalent to (cf. equation 8.7)

$$|K||[W_E(c)]| = -\frac{c}{3f_0}\frac{d}{dc}f_0\left[\frac{e}{m}|E|\right], \quad (9.15)$$

where [ ] denotes a column vector and $|K|$ is the 3×3 matrix of the coefficients of the $|W_k(c)|$ in equations 9.14. The reciprocal of the matrix $|K|$ is (cf. equation 8.8)

$$|K|^{-1} = \frac{1}{(v+ip)\left[(v+ip)^2 + \omega^2\right]}$$

$$\times \begin{vmatrix} (v+ip)^2 + \omega_x^2 & \omega_x\omega_y - (v+ip)\omega_z & \omega_x\omega_z + (v+ip)\omega_y \\ \omega_y\omega_x + (v+ip)\omega_z & (v+ip)^2 + \omega_y^2 & \omega_y\omega_z - (v+ip)\omega_x \\ \omega_z\omega_x - (v+ip)\omega_y & \omega_z\omega_y + (v+ip)\omega_x & (v+ip)^2 + \omega_z^2 \end{vmatrix}, \quad (9.16)$$

in which $\omega^2 = \omega_x^2 + \omega_y^2 + \omega_z^2$. It follows that

$$[|W_E(c)|] = -\frac{c}{3f_0}\frac{d}{dc}f_0|K|^{-1}\left[\frac{e}{m}|E|\right]. \quad (9.17)$$

Since f_0 is independent of t and $\mathbf{r}$, the drift velocity is the same for a group as a whole as for its elements $n\,d\mathbf{r}$ and is

$$[|W|] = \left[4\pi \int_0^\infty |W_E(c)| f_0 c^2\, dc\right],$$

whence

$$[|W|] = |\mu|[|E|],\tag{9.18}$$

in which the elements of the complex mobility matrix (cf. equations 8.11) are as follows:

$$\mu_{xx,yy,zz} = -\frac{4\pi e}{3m} \int_0^\infty \frac{c^3\left[(\nu+ip)^2 + \omega_{x,y,z}^2\right]}{(\nu+ip)\left[(\nu+ip)^2 + \omega^2\right]} \frac{d}{dc}f_0\, dc,$$

$$\mu_{xy,yx} = -\frac{4\pi e}{3m} \int_0^\infty \frac{c^3\left[\omega_x\omega_y \mp (\nu+ip)\omega_z\right]}{(\nu+ip)\left[(\nu+ip)^2 + \omega^2\right]} \frac{d}{dc}f_0\, dc,$$

$$\mu_{xz,zx} = -\frac{4\pi e}{3m} \int_0^\infty \frac{c^3\left[\omega_x\omega_z \pm (\nu+ip)\omega_y\right]}{(\nu+ip)\left[(\nu+ip)^2 + \omega^2\right]} \frac{d}{dc}f_0\, dc,$$

$$\mu_{yz,zy} = -\frac{4\pi e}{3m} \int_0^\infty \frac{c^3\left[\omega_y\omega_z \mp (\nu+ip)\omega_x\right]}{(\nu+ip)\left[(\nu+ip)^2 + \omega^2\right]} \frac{d}{dc}f_0\, dc.$$

$$\tag{9.19}$$

These expressions evidently reduce to those in equations 8.11 when $p = 0$. The conductivity matrix $|\sigma|$ is defined through

$$[|J|] = ne[|W|] = |\sigma|[|E|],$$

where J is the current density. Consequently

$$|\sigma| = ne|\mu|.$$

It follows from equation 9.18 that

$$\begin{bmatrix} W_x \\ W_y \\ W_z \end{bmatrix} = \mathrm{Re}\left\{ |\mu| \begin{bmatrix} |E_x| \\ |E_y| \\ |E_z| \end{bmatrix} \exp ipt \right\}.\tag{9.20}$$

9.3.1. SPECIAL CASE: $\omega_x = \omega_y = 0$, $\omega_z = \omega$. Although this case appears special, there is in fact no loss of generality. The elements of $|\mu|$ are now

$$\mu_{xz} = \mu_{zx} = \mu_{yz} = \mu_{zy} = 0, \qquad \mu_{xy} = -\mu_{yx}, \qquad \mu_{xx} = \mu_{yy}$$

and $|\mu|$ becomes

$$|\mu| = \begin{vmatrix} \mu_{xx} & \mu_{xy} & 0 \\ -\mu_{xy} & \mu_{xx} & 0 \\ 0 & 0 & \mu_{zz} \end{vmatrix} \tag{9.21}$$

with

$$\left.\begin{aligned}
\mu_{xx} &= -\frac{4\pi}{3}\frac{e}{m}\int_0^\infty \frac{c^3(\nu+ip)}{(\nu+ip)^2+\omega^2}\frac{d}{dc}f_0\,dc \\[2mm]
&= -\frac{4\pi e}{6m}\int_0^\infty \left[\frac{1}{\nu+i(p+\omega)} + \frac{1}{\nu+i(p-\omega)}\right]c^3\frac{d}{dc}f_0\,dc \\[2mm]
&= \frac{e}{m}[I_- + I_+].
\end{aligned}\right.$$

Thus,

$$I_- = \frac{1}{6}\left\{ \overline{c^{-2}\frac{d}{dc}\left[\frac{c^3}{\nu+i(p+\omega)}\right]} \right\}$$

and

$$I_+ = \frac{1}{6}\left\{ \overline{c^{-2}\frac{d}{dc}\left[\frac{c^3}{\nu+i(p-\omega)}\right]} \right\}$$

$$\left.\right\} \tag{9.22}$$

after integration by parts.

Similarly

$$\mu_{xy} = -\mu_{yx} = -\frac{e}{m}i[I_- - I_+], \tag{9.23}$$

whereas

$$\mu_{zz} = \frac{e}{m}I_0,$$

where

$$I_0 = -\frac{4\pi}{3}\int_0^\infty \frac{c^3}{v+ip}\frac{d}{dc}f_0\,dc$$

$$= \frac{1}{3}\left[\overline{c^{-2}\frac{d}{dc}\left(\frac{c^3}{v+ip}\right)}\right]. \tag{9.24}$$

It will be found convenient to represent the matrix $|\mu|$ in equation 9.21 as a sum of simpler matrices as follows:

$$|\mu| = \frac{e}{m}I_-\begin{vmatrix} 1 & -i & 0 \\ i & 1 & 0 \\ 0 & 0 & 0 \end{vmatrix} + \frac{e}{m}I_+\begin{vmatrix} 1 & i & 0 \\ -i & 1 & 0 \\ 0 & 0 & 0 \end{vmatrix} + \frac{e}{m}I_0\begin{vmatrix} 0 & 0 & 0 \\ 0 & 0 & 0 \\ 0 & 0 & 1 \end{vmatrix}.$$

$$\tag{9.25}$$

The expressions for I_-, I_+, and I_0 are equivalent to

$$I_- = |I_-|\exp-i\beta_-, \qquad I_+ = |I_+|\exp-i\beta_+, \qquad I_0 = |I_0|\exp-i\beta_0, \tag{9.26}$$

where

$$\tan\beta_\mp = \frac{(p\pm\omega)\int_0^\infty\left\{c^3/\left[v^2+(p\pm\omega)^2\right]\right\}(d/dc)f_0\,dc}{\int_0^\infty\left\{vc^3/\left[v^2+(p\pm\omega)^2\right]\right\}(d/dc)f_0\,dc}$$

$$\tan\beta_0 = \frac{p\int_0^\infty c^3/(v^2+p^2)(d/dc)f_0\,dc}{\int_0^\infty vc^3/(v^2+p^2)(d/dc)f_0\,dc}. \tag{9.27}$$

We note in passing that these expressions for $\tan\beta_\mp$ and $\tan\beta_0$ are the same as those for $W_x/W_z = \tan\theta$, as given by equations 8.38, when ω is replaced by $(p+\omega)$, $(p-\omega)$, and p, respectively, so that in principle the phase angles $\beta_\mp$ and β_0 can be determined from measurements of W_x and W_z in steady electric fields.

The moduli $|I_-|$, $|I_+|$, and $|I_0|$ are as follows:

$$|I_{\mp}| = \frac{4\pi}{6}\left\{\left[\int_0^\infty \frac{vc^3}{v^2+(p\pm\omega)^2}\frac{d}{dc}f_0\,dc\right]^2\right.$$

$$\left.+\left[\int_0^\infty \frac{(p\pm\omega)c^3}{v^2+(p\pm\omega)^2}\frac{d}{dc}f_0\,dc\right]^2\right\}^{1/2},$$

$$|I_0| = \frac{4\pi}{3}\left[\left(\int_0^\infty \frac{vc^3}{v^2+p^2}\frac{d}{dc}f_0\,dc\right)^2 + \left(\int_0^\infty \frac{pc^3}{v^2+p^2}\frac{d}{dc}f_0\,dc\right)^2\right]^{1/2}.$$

$$\left.\right\} \quad (9.28)$$

It follows from equations 9.20 to 9.25 that when $E_k = |E_k|\exp ipt = E_{0_k}\exp i(pt+\alpha_k)$ then

$$W_x = \frac{e}{m}\mathrm{Re}\left\{\left[I_-(|E_x|-i|E_y|)+I_+(|E_x|+i|E_y|)\right]\exp ipt\right\}$$

$$= \frac{e}{m}\left\{|I_-|\left[E_{0_x}\cos(pt+\alpha_x-\beta_-)+E_{0_y}\cos\left(pt+\alpha_y-\beta_--\frac{\pi}{2}\right)\right]\right.$$

$$\left.+|I_+|\left[E_{0_x}\cos(pt+\alpha_x-\beta_+)+E_{0_y}\cos\left(pt+\alpha_y-\beta_++\frac{\pi}{2}\right)\right]\right\}$$

$$W_y = \frac{e}{m}\mathrm{Re}\left\{\left[I_-(i|E_x|+|E_y|)+I_+(-i|E_x|+|E_y|)\right]\exp ipt\right\}$$

$$= \frac{e}{m}\left\{|I_-|\left[E_{0_x}\cos\left(pt+\alpha_x-\beta_-+\frac{\pi}{2}\right)+E_{0_y}\cos(pt+\alpha_y-\beta_-)\right]\right.$$

$$\left.+|I_+|\left[E_{0_x}\cos\left(pt+\alpha_x-\beta_+-\frac{\pi}{2}\right)+E_{0_y}\cos(pt+\alpha_y-\beta_+)\right]\right\}$$

$$W_z = \frac{e}{m}|I_0|E_{0_z}\cos(pt+\alpha_z-\beta_0).$$

$$\left.\right\} \quad (9.29)$$

Special cases of the response to the general oscillating field are covered by equations 9.29.

9.3.2. **E** NORMAL TO **ω**. Consider, for instance, the special oscillating field for which $E_{0_x} = E_{0_z} = 0$, $E_{0_y} = E_0$, $\alpha_y = 0$. Equations 9.29 then give:

$$W_x = \frac{e}{m} E_0 [|I_-| \cos(pt - \beta_-) + |I_+| \cos(pt - \beta_+)]$$

$$W_y = \frac{e}{m} E_0 \left[|I_-| \cos\left(pt - \beta_- + \frac{\pi}{2}\right) + |I_+| \cos\left(pt - \beta_+ - \frac{\pi}{2}\right) \right] \quad (9.30)$$

$$W_z = 0.$$

9.3.3. ROTATING ELECTRIC VECTOR. Let the vector $\mathbf{E}_+$ possess amplitude $E_{0_{R+}}$, and suppose it to rotate in the xOy plane in an anticlockwise sense; that is, its angular velocity **p** is directed along $+Oz$. Let its phase constant be α_+. The components of the field are then

$$E_x = E_{0_{R+}} \cos(pt + \alpha_+),$$

$$E_y = E_{0_{R+}} \sin(pt + \alpha_+) = E_{0_{R+}} \cos\left(pt + \alpha_+ - \frac{\pi}{2}\right),$$

$$E_z = 0.$$

Consequently from equations 9.29, with $\alpha_x = \alpha_+$, $\alpha_y = \alpha_+ - \pi/2$,

$$W_{x_+} = \frac{e}{m} 2E_{0_{R+}} |I_+| \cos(pt + \alpha_+ - \beta_+),$$

$$W_{y_+} = \frac{e}{m} 2E_{0_{R+}} |I_+| \cos\left(pt + \alpha_+ - \beta_+ - \frac{\pi}{2}\right) \quad (9.31)$$

$$= \frac{e}{m} 2E_{0_{R+}} |I_+| \sin(pt + \alpha_+ - \beta_+).$$

Thus $\mathbf{W}_+$ is also a vector rotating with angular velocity $+\mathbf{p}$ and with a phase retardation β_+ with respect to $\mathbf{E}_+$.

Next consider the response to an electric vector $\mathbf{E}_-$ that rotates with angular velocity $-\mathbf{p}$ (i.e., in a clockwise sense) in the xOy plane and with a phase constant $-\alpha_-$. The components of this vector are

$$E_x = E_{0_{R-}} \cos(pt + \alpha_-),$$

$$E_y = -E_{0_{R-}} \sin(pt + \alpha_-) = E_{0_{R-}} \cos\left(pt + \alpha_- + \frac{\pi}{2}\right),$$

$$E_z = 0.$$

Thus $\alpha_x = \alpha_-$, $\alpha_y = \alpha_- + \pi/2$.

Equations 9.29 now give

$$W_{x_-} = \frac{e}{m} 2E_{0_{R-}} |I_-| \cos(pt + \alpha_- - \beta_-),$$

$$W_{y_-} = \frac{e}{m} 2E_{0_{R-}} |I_-| \cos\left(pt + \alpha_- - \beta_- + \frac{\pi}{2}\right) \qquad (9.32)$$

$$= -\frac{e}{m} 2E_{0_{R-}} |I_-| \sin(pt + \alpha_- - \beta_-).$$

Thus W_- is also a vector that rotates at angular velocity $-p$ and follows E_- with a phase lag of β_-.

The superposition of oppositely rotating electric fields with $E_{0_{R+}} = E_{0_{R-}} = E_{0_R}$ and $\alpha_+ = \alpha_- = 0$ gives resultant components

$$E_x = 2E_{0_R} \cos pt = E_0 \cos pt, \qquad E_y = 0,$$

that is, a simple harmonic oscillation along the axis Ox. Superposition of the components of W_+ and W_- as given in equations 9.31 and 9.32 recovers equation 9.30.

The significant feature of equations 9.31 and 9.32 is the association of $(|I_+|, \beta_+)$ with $+p$ and $(|I_-|, \beta_-)$ with $-p$.

9.3.4.   LIMITING FORMS OF I_- AND I_+. We suppose that $\nu^2 \ll (p \pm \omega)^2$, as is usually the case in studies of radio wave propagation in the ionosphere. Equations 9.22 then simplify to:

$$I_{\mp} \cong -\frac{4\pi}{6} \int_0^\infty \frac{[\nu - i(p \pm \omega)]c^3}{(p \pm \omega)^2} \frac{d}{dc} f_0 \, dc$$

$$\left. = -\frac{1}{2(p \pm \omega)^2} \left[\frac{4\pi}{3} \int_0^\infty \nu c^3 \frac{d}{dc} f_0 \, dc - i(p \pm \omega) \frac{4\pi}{3} \int_0^\infty c^3 \frac{d}{dc} f_0 \, dc \right] \right\} (9.33)$$

and

$$\left. \tan\beta_{\mp} \cong \frac{(p \pm \omega) \int_0^\infty c^3 (d/dc) f_0 \, dc}{\int_0^\infty \nu c^3 (d/dc) f_0 \, dc}. \right.$$

When q_m is independent of c, these expressions, after partial integration, reduce to

$$I_{\mp} \cong \frac{1}{2(p \pm \omega)} \left[\tfrac{4}{3}\bar{\nu} + i(p \pm \omega) \right] \qquad (9.34)$$

and

$$\tan\beta_{\mp} = \frac{p \pm \omega}{\tfrac{4}{3}\bar{\nu}},$$

where $\bar{\nu} = Nq_m\bar{c}$. Thus, as $\bar{\nu}/(p \pm \omega) \to 0$, the resistive component of the conductivity approaches zero and the medium becomes entirely reactive. The effective collision frequency to be used in the magneto-ionic theory of Appleton and Hartree (Ratcliffe, 1959) is

$$\nu_{\text{eff}} = \tfrac{4}{3}\bar{\nu} = \tfrac{4}{3}Nq_m\bar{c},$$

which is the value obtained from general expressions for the conductivity (Huxley, 1960) for the special case of q_m independent of c.

9.3.5. GYROMAGNETIC RESONANCE. Suppose that $\nu \gg (p - \omega)$. It follows from equations 9.22 that in these circumstances

$$I_{-} = -\frac{4\pi}{6} \int_0^\infty \frac{\nu - i(p + \omega)}{\nu^2 + (p + \omega)^2} c^3 \frac{d}{dc} f_0 \, dc,$$

$$ \qquad (9.35)$$

$$I_{+} \cong -\frac{4\pi}{6} \int_0^\infty \frac{c^3}{\nu} \frac{d}{dc} f_0 \, dc.$$

Consequently at and near resonance ($p \cong \omega$) the conductivity associated with I_{+} appropriate to a positively rotating vector normal to $\boldsymbol{\omega}$ is resistive and greatly exceeds that associated with I_{-}, appropriate to a rotating vector with an angular velocity $-\mathbf{p}$ (equations 9.31 and 9.32).

It follows from equations 9.35 that when q_m is independent of c

$$I_{+} \cong \frac{1}{Nq_m} \frac{4\pi}{3} \int_0^\infty c f_0 \, dc = \frac{\overline{c^{-1}}}{3Nq_m} = \frac{\overline{c^{-1}}\bar{c}}{3\bar{\nu}},$$

where as before $\bar{\nu} = Nq_m\bar{c}$.

With Maxwell's distribution function

$$\overline{c^{-1}}\bar{c} = \frac{4}{\pi} = 1.27,$$

but with that of Druyvesteyn

$$\overline{c^{-1}}\,\bar{c} = \frac{\pi^{1/2}}{\left[\Gamma(\tfrac{3}{4})\right]^2} = 1.18.$$

Thus at gyromagnetic resonance the effective collision frequency is, for the positively rotating component of the field,

$$\nu_{\text{eff}} = \frac{3\bar{\nu}}{\overline{c^{-1}}\,\bar{c}} = 2.36\bar{\nu} \qquad \text{(Maxwell)}$$

$$= 2.54\bar{\nu} \qquad \text{(Druyvesteyn)}. \tag{9.36}$$

We consider next the mean rate at which an electron receives energy from an oscillating field.

9.4. MEAN VALUE OVER A CYCLE OF THE PRODUCT OF TWO QUANTITIES OSCILLATING SINUSOIDALLY AT THE SAME FREQUENCY

Let $x = A\cos(pt + \alpha_x)$ and $y = B\cos(pt + \alpha_y)$ be the two quantities, that is,

$$2x = |A|\exp ipt + |A|^*\exp - ipt,$$

$$2y = |B|\exp ipt + |B|^*\exp - ipt.$$

Consider

$$xy = AB\cos(pt + \alpha_x)\cos(pt + \alpha_y)$$

$$= \frac{AB}{2}\left[\cos(2pt + \alpha_x + \alpha_y) + \cos(\alpha_x - \alpha_y)\right].$$

The mean value of xy taken over an interval equal to the periodic time $T = 2\pi/p$ is:

$$\overline{xy} = \frac{AB}{2T}\int_{t_0}^{t_0+T}\left[\cos(2pt + \alpha_x + \alpha_y) + \cos(\alpha_x - \alpha_y)\right]dt$$

$$= \frac{AB}{2}\cos(\alpha_x - \alpha_y) = \tfrac{1}{2}\text{Re}\{|A||B|^*\}.$$

Thus

$$\overline{xy} = \tfrac{1}{2}\text{Re}\{|A||B|^*\} = \tfrac{1}{2}\text{Re}\{|A|^*|B|\}. \tag{9.37}$$

9.5.   THE MEAN POWER COMMUNICATED TO AN ELECTRON

The mean power communicated to an electron of a group n_0 of electrons is most easily derived in a general manner by making use of equation 9.37. We assume as before that ω is parallel to $+Oz$.

The energy supplied in time dt to a group of n_0 drifting electrons is $n_0 e \mathbf{E} \cdot \mathbf{W} \, dt$; consequently the mean power communicated to an electron is

$$w = e \frac{1}{T} \int_0^T (E_x W_x + E_y W_y + E_z W_z) \, dt, \tag{9.38}$$

in which

$$E_x = E_{0_x} \cos(pt + \alpha_x), \qquad E_y = E_{0_y} \cos(pt + \alpha_y), \qquad E_z = E_{0_z} \cos(pt + \alpha_z).$$

Although w may be calculated directly from equation 9.38, it is more instructive in the first instance to use equivalent rotating vector fields. The components E_x and E_y can be expressed as

$$E_x = E_{0_x} \cos(pt + \alpha_x) = E_{0_x} \cos \alpha_x \cos pt - E_{0_x} \sin \alpha_x \sin pt,$$

$$E_y = E_{0_y} \cos(pt + \alpha_y) = E_{0_y} \cos \alpha_y \cos pt - E_{0_y} \sin \alpha_y \sin pt.$$

Consider Figure 9.1a. The vectors OA and OB rotate with angular velocity $\mathbf{p}$ parallel to ω, whereas OD and OC rotate with $-\mathbf{p}$. The magnitudes of the vectors are as follows: $OA = OC = (E_{0_x}/2)\cos \alpha_x$, $OD = OB = (E_{0_x}/2)\sin \alpha_x$, and the sum of their projections on Ox is given by $E_{0_x}\cos(pt + \alpha_x)$.

Similarly $E_{0_y}\cos(pt + \alpha_y)$ resolves into the two pairs of oppositely rotating vectors (Figure 9.1b): $OF = OH = (E_{0_y}/2)\cos \alpha_y$, $OG = OK = (E_{0_y}/2)\sin \alpha_y$.

The vectors of Figures 9.1a and 9.1b can be combined to give the system depicted in Figure 9.1c, which shows

$$OR = OA - OG = \tfrac{1}{2}(E_{0_x}\cos \alpha_x - E_{0_y}\sin \alpha_y),$$

$$OS = OB + OF = \tfrac{1}{2}(E_{0_x}\sin \alpha_x + E_{0_y}\cos \alpha_y),$$

$$OT = OC + OK = \tfrac{1}{2}(E_{0_x}\cos \alpha_x + E_{0_y}\sin \alpha_y),$$

$$OU = OD - OH = \tfrac{1}{2}(E_{0_x}\sin \alpha_x - E_{0_y}\cos \alpha_y).$$

The final resultants are the vectors OP and OQ, rotating at angular velocities $+\mathbf{p}$ and $-\mathbf{p}$, respectively.

The magnitudes of OP and OQ are

$$|OP| = (OR^2 + OS^2)^{1/2} = \tfrac{1}{2}\Big[E_{0_x}{}^2 + E_{0_y}{}^2 + 2E_{0_x}E_{0_y}\sin(\alpha_x - \alpha_y)\Big]^{1/2} = E_{0_+},$$

(9.39)

$$|OQ| = (OT^2 + OU^2)^{1/2} = \tfrac{1}{2}\Big[E_{0_x}{}^2 + E_{0_y}{}^2 - 2E_{0_x}E_{0_y}\sin(\alpha_x - \alpha_y)\Big]^{1/2} = E_{0_-}.$$

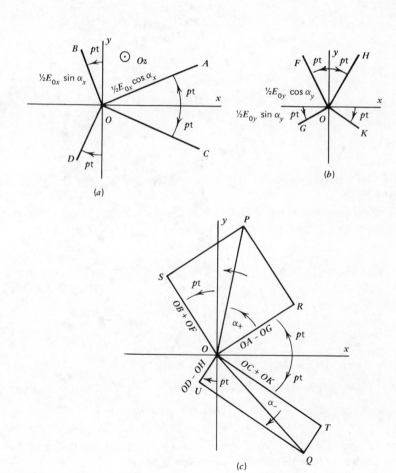

FIG. 9.1

The phase constant of the rotating vector OP is the angle $POR = \alpha_+$; that of OQ, the angle $QOT = \alpha_-$. Thus

$$
\left.
\begin{aligned}
\cos\alpha_+ &= \frac{OR}{OP} = \frac{E_{0_x}\cos\alpha_x - E_{0_y}\sin\alpha_y}{2E_{0_+}}, \\[2mm]
\sin\alpha_+ &= \frac{OS}{OP} = \frac{E_{0_x}\sin\alpha_x + E_{0_y}\cos\alpha_y}{2E_{0_+}}, \\[2mm]
\tan\alpha_+ &= \frac{\sin\alpha_+}{\cos\alpha_+};
\end{aligned}
\right\} \quad (9.40)
$$

and

$$
\left.
\begin{aligned}
\cos\alpha_- &= \frac{E_{0_x}\cos\alpha_x + E_{0_y}\sin\alpha_y}{2E_{0_-}}, \\[2mm]
\sin\alpha_- &= \frac{E_{0_x}\sin\alpha_x - E_{0_y}\cos\alpha_y}{2E_{0_-}}, \\[2mm]
\tan\alpha_- &= \frac{\sin\alpha_-}{\cos\alpha_-}.
\end{aligned}
\right\} \quad (9.41)
$$

This system is equivalent to the initial system for E_x and E_y given in equation 9.38.

It follows from equations 9.31 and 9.32 that the rotating vectors $\mathbf{W}_+$ and $\mathbf{W}_-$ driven by E_{0_+} and E_{0_-} are, respectively,

$$
\begin{aligned}
W_{x_+} &= \frac{e}{m} 2E_{0_+}|I_+|\cos(pt + \alpha_+ - \beta_+), \\[2mm]
W_{y_+} &= \frac{e}{m} 2E_{0_+}|I_+|\cos\left(pt + \alpha_+ - \beta_+ - \frac{\pi}{2}\right),
\end{aligned}
\quad (9.42)
$$

and

$$
\begin{aligned}
W_{x_-} &= \frac{e}{m} 2E_{0_-}|I_-|\cos(pt + \alpha_- - \beta_-), \\[2mm]
W_{y_-} &= \frac{e}{m} 2E_{0_-}|I_-|\cos\left(pt + \alpha_- - \beta_- + \frac{\pi}{2}\right),
\end{aligned}
\quad (9.43)
$$

where E_{0_+} and E_{0_-} are defined in equations 9.39. Equations 9.42 and 9.43 can be shown to be equivalent to equations 9.29.

We return to equation 9.38, which because of equation 9.37 can be transformed to

$$w = \tfrac{1}{2} e \, \mathrm{Re}(|E_x|^* |W_x| + |E_y|^* |W_y| + |E_z|^* |W_z|).$$

We relate $\mathbf{W}$ to $\mathbf{E}$ through equations 9.20 and 9.25; consequently

$$\frac{mw}{e^2} = \frac{\mathrm{Re}}{2} \underset{\text{row matrix}}{(|E_x|^*, |E_y|^*, |E_z|^*)}$$

$$\times \left(I_- \begin{vmatrix} 1 & -i & 0 \\ i & 1 & 0 \\ 0 & 0 & 0 \end{vmatrix} + I_+ \begin{vmatrix} 1 & i & 0 \\ -i & 1 & 0 \\ 0 & 0 & 0 \end{vmatrix} + I_0 \begin{vmatrix} 0 & 0 & 0 \\ 0 & 0 & 0 \\ 0 & 0 & 1 \end{vmatrix} \right) \begin{bmatrix} |E_x| \\ |E_y| \\ |E_z| \end{bmatrix},$$

$$(9.44)$$

where

$$|E_x| = E_{0_x} \exp i\alpha_x, \qquad |E_y| = E_{0_y} \exp i\alpha_y, \qquad |E_z| = E_{0_z} \exp i\alpha_z,$$

$$I_{\mp} = |I_{\mp}| \exp - i\beta_{\mp}, \qquad I_0 = |I_0| \exp - i\beta_0.$$

Thus

$$\frac{mw}{e^2} = \frac{1}{2} Re \begin{bmatrix} (|E_x|^* + i|E_y|^*) I_- + (|E_x|^* - i|E_y|^*) I_+ \\ (-i|E_x|^* + |E_y|^*) I_- + (i|E_x|^* + |E_y|^*) I_+ \\ |E_z|^* I_0 \end{bmatrix}^{Tr} \begin{bmatrix} |E_x| \\ |E_y| \\ |E_z| \end{bmatrix}$$

$$= \tfrac{1}{2} Re \left\{ \left[E_{0_x}^{\,2} + iE_{0_x} E_{0_y} \exp i(\alpha_x - \alpha_y) - iE_{0_x} E_{0_y} \exp - i(\alpha_x - \alpha_y) + E_{0_y}^{\,2} \right] I_- \right.$$

$$+ \left[E_{0_x}^{\,2} - iE_{0_x} E_{0_y} \exp i(\alpha_x - \alpha_y) + iE_{0_x} E_{0_y} \exp - i(\alpha_x - \alpha_y) + E_{0_y}^{\,2} \right] I_+$$

$$\left. + E_{0_z}^{\,2} I_0 \right\}$$

$$= \tfrac{1}{2} \left\{ \left[E_{0_x}^{\,2} - 2 E_{0_x} E_{0_y} \sin(\alpha_x - \alpha_y) + E_{0_y}^{\,2} \right] Re I_- \right.$$

$$+ \left[E_{0_x}^{\,2} + 2 E_{0_x} E_{0_y} \sin(\alpha_x - \alpha_y) + E_{0_y}^{\,2} \right] Re I_+$$

$$\left. + E_{0_z}^{\,2} Re I_0 \right\}. \qquad (9.45)$$

The real parts of $I_\mp$ and I_0 are to be found from equations 9.22 and 9.24; thus

$$\mathrm{Re}I_\mp = -\frac{4\pi}{6}\int_0^\infty \frac{\nu c^3}{\nu^2+(p\pm\omega)^2}\frac{d}{dc}f_0\,dc,$$

$$\mathrm{Re}I_0 = -\frac{4\pi}{3}\int_0^\infty \frac{\nu c^3}{\nu^2+p^2}\frac{d}{dc}f_0\,dc.$$

(9.46)

Consequently, by use of equation 9.39 the expression for w can be reduced to

$$w = -\frac{4\pi}{3}\frac{e^2}{m}\int_0^\infty \nu c^3\left[\frac{E_{0_+}{}^2}{\nu^2+(p-\omega)^2}+\frac{E_{0_-}{}^2}{\nu^2+(p+\omega)^2}+\frac{E_{0_z}{}^2}{2(\nu^2+p^2)}\right]\frac{d}{dc}f_0\,dc.$$

(9.47)

When $p\to\omega$, the predominant term is the one dependent on $E_{0_+}{}^2$.

9.6.   THE DISTRIBUTION FUNCTION $f_0(c)$

It was seen that the vector $\mathbf{f}_1(c)$, which satisfies equation 9.1, can be separated into components $\mathbf{f}_1=\mathbf{f}_E+\mathbf{f}_G$, which individually satisfy equations 9.10 and 9.11.

The scalar equation is

$$\frac{\partial}{\partial t}(nf_0)+\frac{c}{3}\,\mathrm{div}_r(n\mathbf{f}_1)+\frac{1}{4\pi c^2}\frac{\partial}{\partial c}[\sigma_E(c)-\sigma_{\mathrm{coll}}(c)]=0,$$

where

$$\sigma_E(c)=\frac{4\pi}{3}c^2\frac{e\mathbf{E}}{m}\cdot n\mathbf{f}_1=\frac{4\pi n}{3}c^2\frac{e\mathbf{E}}{m}\cdot(\mathbf{f}_E+\mathbf{f}_G).$$

The field E is assumed to oscillate with a frequency such that the time constant of the mean energy $\frac{1}{2}mc^2$ is long relative to the periodic time T of the oscillation (Section 9.2); consequently f_0 can change only by an inappreciable proportion of itself in the time T. Similarly $\mathbf{f}_G$ alters little in this time, and the mean value of $\mathbf{E}\cdot\mathbf{f}_G$ taken over several cycles is zero.

We seek the mean value $\overline{\sigma_E(c)}$ of $\sigma_E(c)$ taken over a cycle. Since the mean value of $\mathbf{E}\cdot\mathbf{f}_G$ is zero, we are in effect required to evaluate the mean value of $\mathbf{E} \cdot \mathbf{f}_E$, which according to equation 9.37 is

$$\overline{\mathbf{E} \cdot \mathbf{f}_E} = \tfrac{1}{2} \mathrm{Re} \left(|E_x|^* |f_{E_x}| + |E_y|^* |f_{E_y}| + |E_z|^* |f_{E_z}| \right).$$

Since $\mathbf{f}_E = (3f_0/c) \mathbf{W}_E(c)$, it follows from equation 9.17 that

$$[|f_E|] = \left(-\frac{d}{dc} f_0 \right) |K|^{-1} \left[\frac{e}{m} |E| \right], \tag{9.48}$$

where $|K|^{-1}$ is the matrix displayed in equation 9.16. Without loss of generality the direction of $+Oz$ is taken to be that of $\boldsymbol{\omega}$ so that, in $|K|^{-1}$, $\omega_x = \omega_y = 0$, $\omega_z = \omega$. Equation 9.48 then becomes

$$\begin{bmatrix} |f_{E_x}| \\ |f_{E_y}| \\ |f_{E_z}| \end{bmatrix} = \begin{bmatrix} \phi_{xx} & \phi_{xy} & 0 \\ \phi_{yx} & \phi_{yy} & 0 \\ 0 & 0 & \phi_{zz} \end{bmatrix} \begin{bmatrix} |E_x| \\ |E_y| \\ |E_z| \end{bmatrix}, \tag{9.49}$$

where

$$\left. \begin{aligned} \phi_{xx} = \phi_{yy} &= \left(-\frac{e}{m} \frac{d}{dc} f_0 \right) \frac{v + ip}{(v + ip)^2 + \omega^2} \\ &= \left(-\frac{e}{m} \frac{d}{dc} f_0 \right) (s_- + s_+), \\ \phi_{xy} = -\phi_{yx} &= \left(-\frac{e}{m} \frac{d}{dc} f_0 \right) [-i(s_- - s_+)], \\ \phi_{zz} &= \left(-\frac{e}{m} \frac{d}{dc} f_0 \right) s_0, \end{aligned} \right\} \tag{9.50}$$

and

$$s_- = \frac{1}{2[v + i(p + \omega)]}, \qquad s_+ = \frac{1}{2[v + i(p - \omega)]}, \qquad s_0 = \frac{1}{v + ip}.$$

If we replace $|\phi|$ by a sum of matrices as follows:

$$|\phi| = \left(-\frac{e}{m} \frac{d}{dc} f_0 \right) \left(s_- \begin{vmatrix} 1 & -i & 0 \\ i & 1 & 0 \\ 0 & 0 & 0 \end{vmatrix} + s_+ \begin{vmatrix} 1 & i & 0 \\ -i & 1 & 0 \\ 0 & 0 & 0 \end{vmatrix} + s_0 \begin{vmatrix} 0 & 0 & 0 \\ 0 & 0 & 0 \\ 0 & 0 & 1 \end{vmatrix} \right),$$

it follows that, as in equation 9.44,

$$\frac{\overline{\mathbf{E}\cdot\mathbf{f}_E}}{-(e/m)(d/dc)f_0} = \tfrac{1}{2}\mathrm{Re}\,(\underbrace{|E_x|^*|E_y|^*|E_z|^*}_{\text{row matrix}})$$

$$\times\left(s_-\begin{vmatrix}1 & -i & 0 \\ i & 1 & 0 \\ 0 & 0 & 0\end{vmatrix}+s_+\begin{vmatrix}1 & i & 0 \\ -i & 1 & 0 \\ 0 & 0 & 0\end{vmatrix}+s_0\begin{vmatrix}0 & 0 & 0 \\ 0 & 0 & 0 \\ 0 & 0 & 1\end{vmatrix}\right)\begin{bmatrix}|E_x| \\ |E_y| \\ |E_z|\end{bmatrix}$$

$$=\tfrac{1}{2}\Big\{\Big[E_{0_x}^{\,2}-2E_{0_x}E_{0_y}\sin(\alpha_x-\alpha_y)+E_{0_y}^{\,2}\Big]\mathrm{Res}_-$$

$$+\Big[E_{0_x}^{\,2}+2E_{0_x}E_{0_y}\sin(\alpha_x-\alpha_y)+E_{0_y}^{\,2}\Big]\mathrm{Res}_+$$

$$+E_{0_z}^{\,2}\,\mathrm{Res}_0\Big\}$$

$$=\frac{\nu E_{0_-}^{\,2}}{\nu^2+(p+\omega)^2}+\frac{\nu E_{0_+}^{\,2}}{\nu^2+(p-\omega)^2}+\frac{\nu E_{0_z}^{\,2}}{2(\nu^2+p^2)}$$

from equations 9.39 and 9.50. Thus

$$\overline{\sigma_E(c)}=-\frac{4\pi}{3}nc^2\Big(\frac{e}{m}\Big)^2\left[\frac{\nu E_{0_-}^{\,2}}{\nu^2+(p+\omega)^2}+\frac{\nu E_{0_+}^{\,2}}{\nu^2+(p-\omega)^2}+\frac{\nu E_{0_z}^{\,2}}{2(\nu^2+p^2)}\right]\frac{d}{dc}f_0.$$

$$(9.51)$$

We compare this expression with the formula for $\sigma_E(c)$ for a steady field E, which is

$$\sigma_E(c)=-\frac{4\pi}{3}nc^2\nu_{\mathrm{el}}VV'\frac{d}{dc}f_0,$$

where $V'=eE/m\nu$, $\nu=Ncq_m(c)$, $V=eE/m\nu_{\mathrm{el}}$, and $\nu_{\mathrm{el}}=Ncq_{m_{\mathrm{el}}}(c)$. We can therefore give $\overline{\sigma_E(c)}$ this form since equation 9.51 can be written as

$$\overline{\sigma_E(c)}=-\frac{4\pi}{3}nc^2\nu_{\mathrm{el}}V_{\mathrm{eq}}V'_{\mathrm{eq}}\frac{d}{dc}f_0, \qquad (9.52)$$

where

$$V_{\mathrm{eq}}V'_{\mathrm{eq}}=\left\{\frac{[(e/m)E_{0_-}]^2}{\nu^2+(p+\omega)^2}+\frac{[(e/m)E_{0_+}]^2}{\nu^2+(p-\omega)^2}+\frac{[(e/m)E_{0_z}]^2}{2(\nu^2+p^2)}\right\}\frac{\nu}{\nu_{\mathrm{el}}}. \qquad (9.53)$$

Thus $V_{eq}V'_{eq}$ is the equivalent for high-frequency fields of VV' for static fields.

The scalar equation with high-frequency electric fields is therefore

$$\frac{\partial}{\partial t}(nf_0) + \frac{c}{3}\,\text{div}_r(n\mathbf{f}_1) + \frac{1}{4\pi c^2}\frac{\partial}{\partial c}\Big[\,\overline{\sigma_E(c)} - \sigma_{\text{coll}}(c)\Big] = 0. \quad (9.54)$$

Since the gradient term has vanished from $\overline{\sigma_E(c)}$ in equation 9.54, there is no interaction between $\text{grad}_r(nf_0)$ and $e\mathbf{E}$. We shall regard the influence of the presence of $(\partial/\partial t)n$ upon $f_0(c)$, as discussed in relation to equation 4.11, as negligible.

When all encounters are elastic, the distribution function for an isolated group of electrons, or in a steady stream when the spatial gradients are zero or negligible, is to be derived from the relation $\overline{\sigma_E(c)} = \sigma_{\text{coll}}(c)$, that is to say,

$$\frac{V_{eq}^2}{3}\frac{d}{dc}f_0 + \frac{mc}{M}f_0 + \frac{\overline{C^2}}{3}\frac{d}{dc}f_0 = 0,$$

where V_{eq}^2 is equal to the right-hand side of equation 9.53 when $\nu = \nu_{el}$. Whence

$$f_0 = A\exp\left(-\frac{3m}{M}\int_0^c \frac{c\,dc}{V_{eq}^2 + \overline{C^2}}\right). \quad (9.55)$$

When inelastic encounters are important, then, as shown in Chapter 6, the flux $\sigma_{\text{coll}}(c)$ is increased by additional terms which are functions of c. Consequently we replace $3mc/M$ by $(3mc/M) + \phi(c)$. Equation 9.55 is then replaced by

$$f_0 = A\exp\left[-\int_0^c \frac{(3mc/M) + \phi(c)}{V_{eq}V'_{eq} + \overline{C^2}}\,dc\right]. \quad (9.56)$$

BIBLIOGRAPHY AND NOTES

The majority of theoretical investigations of drift in high-frequency electric fields have been stimulated by experimental investigations of the ionosphere and by laboratory studies employing microwaves. It was early pointed out (Huxley 1940) that the Appleton-Hartree equations for the propagation of electromagnetic waves in the ionosphere were based on a model in which the electrons behaved as charged particles lacking random motion but forced into oscillating motion in a resistive medium. It was remarked that in a more faithful representation it would be necessary to derive expressions for the conductivity tensor of the medium considered as a slightly ionized gas containing free electrons in a magnetic field. Expressions for the components of the drift velocity and the complex conductivity were derived by Huxley in several papers. They are related to the corresponding equations found by Maxwell-Boltzmann methods through the usual transformation involving integration by parts.

An early and important investigation of the high-frequency conductivity due to electrons in a slightly ionized gas was that of Margenau (1946), who employed Maxwell-Boltzmann methods. In Margenau's investigations the magnetic field is absent. Margenau derived an expression for the velocity distribution function that is equivalent to equation 9.55 but with $\omega = 0$. It will suffice to give references to typical publications in this field.

FREE PATH METHODS

L. G. H. Huxley, *Phil. Mag.*, **23**, 442, 1937.

L. G. H. Huxley, *Phil. Mag.*, **29**, 313, 1940.

L. G. H. Huxley, *J. Atmos. Terr. Phys.*, **16**, 46, 1959.

L. G. H. Huxley, *Aust. J. Phys.*, **13**, 718, 1960.

MAXWELL-BOLTZMANN METHODS

H. Margenau, *Phys. Rev.*, **69**, 508, 1946; **73**, 297, 1948.

W. P. Allis, "Motions of Ions and Electrons," *Handbuch der Physik*, Vol. 21, Springer-Verlag, Berlin, 1956.

V. L. Ginsburg and A. V. Gurevič, "Nichtlineare Erscheinungen in einem Plasma, das sich in einem veränderlichen elektromagnetischen Feld befindet," *Fortschr. Phys.*, **8**, 97, 1960.

H. K. Sen, "The Generalized Magneto-ionic Theory," Guenter Loeser Memorial Lecture, Air Force Cambridge Research Laboratories, Bedford, Mass., 1966.

R. C. Whitten and I. G. Poppoff, *Physics of the Lower Ionosphere*, Prentice-Hall, Englewood Cliffs, N. J., 1965, Chap. 7.

MAGNETO-IONIC THEORY

J. A. Ratcliffe, *The Magneto-Ionic Theory and its Application to the Ionosphere*, Cambridge University Press, London, 1959.

Some investigators, in order to adapt the form of the distribution function to gases in which inelastic encounters are important, replace $V_{eq}^2(c)$ in equation 9.55 by $[(2m/M)/G(c)] V_{eq}^2(c)$, where $G(c)\frac{1}{2}mc^2$ is the energy lost on the average by an electron with speed c in an encounter. Thus for elastic encounters $G = 2m/M$, but in general G exceeds $2m/M$. It is obvious that this procedure is not consistent with equation 9.56, which has been correctly derived. The chief effect of the presence of inelastic encounters is to increase $\sigma_{coll}(c)$; there is a less important influence on $q_m(c)$ in V_{eq}. Thus the change is to the numerator rather than to the denominator of the integrand in equation 9.56.

APPENDIX A

MAXWELL'S VECTOR REPRESENTATION

OF AN ENCOUNTER

In Figure A.1a, masses m and M are located at vector positions $\mathbf{s}$ and $\mathbf{S}$, respectively, in configuration space, with O as origin. The vector position of their centroid is

$$\mathbf{X} = \frac{m\mathbf{s} + M\mathbf{S}}{M + m}.$$

Let the velocities of m and M be $\mathbf{c} = d\mathbf{s}/dt$ and $\mathbf{C} = d\mathbf{S}/dt$, and the velocity

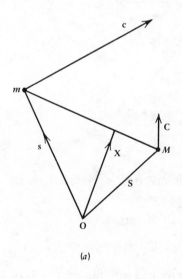

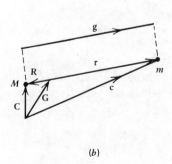

(a)   (b)

FIG. A.1

of the centroid be $\mathbf{G} = d\mathbf{X}/dt$. It follows that

$$(m + M)\mathbf{G} = m\mathbf{c} + M\mathbf{C}$$

and

$$\mathbf{G} = \frac{m\mathbf{c} + M\mathbf{C}}{M + m}.$$

The velocity of m relative to M is

$$\mathbf{g} = \mathbf{c} - \mathbf{C}$$

and that relative to the centroid is

$$\mathbf{r} = \mathbf{c} - \mathbf{G} = \frac{(M + m)\mathbf{c} - (m\mathbf{c} + M\mathbf{C})}{M + m}$$

$$= \frac{M}{M + m}(\mathbf{c} - \mathbf{C}) = \frac{M}{M + m}\mathbf{g}.$$

The velocity of M relative to the centroid is

$$\mathbf{R} = \mathbf{C} - \mathbf{G} = \frac{(M + m)\mathbf{C} - (m\mathbf{c} + M\mathbf{C})}{M + m}$$

$$= \frac{m}{M + m}(\mathbf{C} - \mathbf{c}) = -\frac{m}{M + m}\mathbf{g},$$

whence

$$m\mathbf{r} + M\mathbf{R} = 0.$$

Also

$$r + R = \frac{M + m}{M + m}g = g.$$

These relations between the velocities are summarized in Figure A.1b, in which the velocity vectors are shown in velocity space.

Momentum. The total momentum of the pair of particles is

$$\mathbf{P} = m\mathbf{c} + M\mathbf{C} = (m + M)\mathbf{G}.$$

The momentum in a system of coordinates travelling with the velocity $\mathbf{G}$ of the centroid is

$$m\mathbf{r} + M\mathbf{R} = 0.$$

Kinetic energy. The total kinetic energy of translation is

$$\tfrac{1}{2}(mc^2 + MC^2) = \tfrac{1}{2}\left[m(\mathbf{G+r})^2 + M(\mathbf{G+R})^2\right]$$

$$= \tfrac{1}{2}\left[(m+M)G^2 + mr^2 + MR^2 + 2\mathbf{G}\cdot(m\mathbf{r}+M\mathbf{R})\right]$$

$$= \tfrac{1}{2}(m+M)G^2 + \tfrac{1}{2}mr^2 + \tfrac{1}{2}MR^2.$$

Encounter. Let $\mathbf{s}$, $\mathbf{S}$, $\mathbf{c}$, and $\mathbf{C}$ be such that in due course m and M approach sufficiently closely to each other, and therefore to their centroid, that their velocities are changed through mutual forces of interaction between them. If these forces are short range, the time of effective interaction is small, and after the encounter m and M move with their velocities changed from $\mathbf{c}$ to $\mathbf{c}'$ and $\mathbf{C}$ to $\mathbf{C}'$.

The relative velocities become

$$\mathbf{g}' = \mathbf{c}' - \mathbf{C}', \qquad \mathbf{r}' = \frac{M}{M+m}\mathbf{g}', \qquad \mathbf{R}' = -\frac{m}{M+m}\mathbf{g}'.$$

If $\mathbf{G}'$ is the new velocity of the centroid, the new momentum is

$$\mathbf{P}' = m\mathbf{r}' + M\mathbf{R}' = (m+M)\mathbf{G}'.$$

But momentum is conserved; consequently

$$\mathbf{P}' = \mathbf{P}, \qquad (m+M)\mathbf{G}' = (m+M)\mathbf{G},$$

that is,

$$\mathbf{G}' = \mathbf{G}.$$

The motion of the centroid is unchanged by the encounter.

The velocities before and after the encounter are therefore given by the vector representation shown in Figure A.2, which was first presented by Maxwell.*

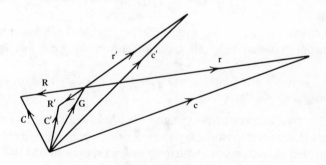

FIG. A.2.   Maxwell's velocity diagram of an encounter.

*J. C. Maxwell, *Collected Papers*, Vol. I, Cambridge University Press, 1890, p. 379.

APPENDIX B

TRANSFERENCE OF

MOMENTUM AND ENERGY FROM

ELECTRONS TO MOLECULES

B.1. DYNAMICS OF ENCOUNTERS

B.1.1. MEAN MOMENTUM LOST BY AN ELECTRON IN AN ENCOUNTER.

From equation 2.3 the total number of encounters in $d\mathbf{r}$ in time dt is

$$n_s = (dt\, d\mathbf{r})nNgq_0(g),$$

and it follows from equation 2.8 that the average momentum imparted by an electron to a molecule is

$$\frac{\Delta\mathbf{P}}{n_s} = \frac{q_m(g)}{q_0(g)}\left(\frac{mM}{m+M}\right)\mathbf{g}. \tag{B.1}$$

Equation 2.8 may be expressed in the form

$$\Delta\mathbf{P} = (dt\, d\mathbf{r})nv\left(\frac{mM}{m+M}\right)\mathbf{g} = (dt\, d\mathbf{r})nvm\mathbf{r}, \tag{B.2}$$

in which $v \equiv Ngq_m(g)$ is an equivalent (aggregate) collision frequency for momentum transfer, as if all the momentum $m\mathbf{r}$ were lost in each encounter.

When scattering is isotropic in encounters of every type, $q_{1_{el}}(g) = q_{1_{in}}(g) = 0$ and $q_m(g) = q_0(g)$.

B.1.2. TRANSFERENCE OF ENERGY.

It is useful also to consider the rate at which electrons impart energy to molecules within an element of volume $d\mathbf{r}$ immersed in both beams, which we again suppose to traverse each other obliquely.

276

The velocities before encounter of the electrons and the molecules are, respectively (Figure A.2),

$$\mathbf{c} = \mathbf{G} + \mathbf{r} \qquad \text{and} \qquad \mathbf{C} = \mathbf{G} + \mathbf{R}$$

and after encounter the velocities are

$$\mathbf{c}' = \mathbf{G} + \mathbf{r}' \qquad \text{and} \qquad \mathbf{C}' = \mathbf{G} + \mathbf{R}'.$$

It follows that

$$c^2 - c'^2 = 2\mathbf{G} \cdot (\mathbf{r} - \mathbf{r}') + (r^2 - r'^2).$$

But

$$\mathbf{r} = \frac{M}{M+m}\mathbf{g}, \qquad \mathbf{r}' = \frac{M}{M+m}\mathbf{g}', \qquad \mathbf{G} = \mathbf{C} - \mathbf{R} = \mathbf{C} + \frac{m}{M+m}\mathbf{g}.$$

Consequently

$$c^2 - c'^2 = \frac{2M}{M+m}\left(\frac{m}{M+m}g^2 + \mathbf{C} \cdot \mathbf{g} - \frac{m}{m+M}\mathbf{g} \cdot \mathbf{g}' - \mathbf{C} \cdot \mathbf{g}'\right) + (r^2 - r'^2).$$

Let (γ, ψ) be the direction of $\mathbf{r}'$ and $\mathbf{g}'$ in a system of coordinates in which the direction of $\mathbf{r}$ and $\mathbf{g}$ is the polar axis (Figure B.1; the azimuth angle ψ is not shown), and let β be the angle between $\mathbf{C}$ and $\mathbf{g}$, with OX chosen to lie in the plane of $\mathbf{C}$ and $\mathbf{g}$. Then

$$\mathbf{g} \cdot \mathbf{g}' = gg'\cos\gamma, \qquad \mathbf{C} \cdot \mathbf{g} = Cg\cos\beta,$$

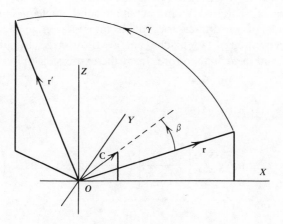

FIG. B.1

and

$$\mathbf{C} \cdot \mathbf{g}' = Cg'(\cos\beta\cos\gamma + \sin\beta\sin\gamma\cos\psi).$$

The expression for $c^2 - c'^2$ may therefore be written as

$$c^2 - c'^2 = \frac{2M}{M+m}\left[\left(\frac{m}{M+m}g^2 + Cg\cos\beta\right)\left(1 - \frac{g'}{g}\cos\gamma\right)\right.$$

$$\left. - Cg'\sin\beta\sin\gamma\cos\psi\right] + (r^2 - r'^2). \qquad (B.3)$$

The kinetic energy lost by the particle m in this encounter is therefore

$$\Delta Q = \tfrac{1}{2}mc^2 - \tfrac{1}{2}mc'^2$$

$$= \frac{mM}{M+m}\left[\left(\frac{m}{M+m}g^2 + Cg\cos\beta\right)\left(1 - \frac{g'}{g}\cos\gamma\right)\right.$$

$$\left. - Cg'\sin\beta\sin\gamma\cos\psi\right] + \tfrac{1}{2}m(r^2 - r'^2). \qquad (B.4)$$

Consider the energy transferred to the particles M in $d\mathbf{r}$, in time dt, by the particles m scattered within a small solid angle $d\omega = \sin\gamma\,d\gamma\,d\psi$ in specific kinds of encounter for which the differential scattering cross sections are written $p_k(\gamma, g)$. The energy transferred by particles m to particles M within $d\mathbf{r}$ is therefore, from the expression preceding equation 2.3,

$$(dt\,d\mathbf{r})nNg\sum_k \Delta Q_k p_k(\gamma, g)\,d\omega$$

$$= (dt\,d\mathbf{r})nNg\sum_k \left\{\frac{mM}{M+m}\left[\left(\frac{m}{M+m}g^2 + Cg\cos\beta\right)\left(1 - \frac{g'_k}{g}\cos\gamma\right)\right.\right.$$

$$\left.\left. - Cg'_k\sin\beta\sin\gamma\cos\psi\right] + \tfrac{1}{2}m(r^2 - r_k'^2)\right\}p_k(\gamma, g)\sin\gamma\,d\gamma\,d\psi.$$

When the particles m are electrons and the particles M are molecules, the magnitudes of the scattered velocities r'_k are constant and independent of γ and ψ.

Next, integrate for all values of γ and ψ, keeping β, c, and C fixed, that is to say, g is fixed, as it must be for the pair of beams. The total energy transferred in time dt within $d\mathbf{r}$ is therefore

$$\Delta Q = (dt\, d\mathbf{r}) nNg \sum_k \left[\frac{mM}{M+m} \left(\frac{m}{M+m} g^2 + Cg\cos\beta \right) q_{mk}(g) \right.$$

$$\left. + q_{0k}(g)\tfrac{1}{2}m(r^2 - r_k'^2) \right], \tag{B.5}$$

in which $q_{mk}(g)$ is the momentum transfer cross section for the kth process and is defined in equation 2.9, and $q_{0k}(g)$ is the associated total scattering cross section. Elastic scattering is included in equation B.5, but for it $r'_k = r$ and $q_{0k}(g)(m/2)(r^2 - r_k'^2)$ vanishes. Let $k = 0$ refer to elastic encounters, and write $q_{0_{el}} \equiv q_{00}$.

In the case of inelastic scattering $g' \neq g$, so that

$$\tfrac{1}{2}m(r^2 - r_k'^2) = \frac{m}{2}\left(\frac{M}{M+m} \right)^2 (g^2 - g_k'^2),$$

which, according to equation 2.6, is $[M/(M+m)]\epsilon_k$. Equation B.5 is thus equivalent to

$$\Delta Q = (dt\, d\mathbf{r})(nNg)\left\{ \frac{mM}{M+m}\left(\frac{m}{M+m} g^2 + Cg\cos\beta \right) q_{m_{el}}(g) \right.$$

$$\left. + \sum_{k=1} \left[\frac{mM}{M+m}\left(\frac{m}{M+m} g^2 + Cg\cos\beta \right) q_{mk}(g) + \frac{M}{M+m}\epsilon_k q_{0k}(g) \right] \right\}.$$

$$\tag{B.6}$$

B.2.   DERIVATION OF EXPRESSION FOR $\sigma_{\text{coll}}(c)$ WHEN ALL ENCOUNTERS ARE ELASTIC

We seek an expression for the net increase, brought about by encounters in time dt, of the population of the class $(n\, d\mathbf{r} \cdot 4\pi f_0 c^2\, dc)$ of electrons whose velocity points lie in the shell (c, dc) of velocity space.

First, we consider the spherical surface in velocity space whose centre is the origin and whose radius is c, which for convenience we call "the sphere c." In general, when the motion of an electron is changed by an encounter with a molecule, the speed c_2 of the electron after an encounter is not the same as its speed c_1 before the encounter. If the speed C of the molecule before an encounter is zero, c_2 is always less than c_1; when C is not zero, however, c_2 is less than c_1 in some encounters but greater in others, according to the magnitudes and directions of the velocities $\mathbf{C}$, $\mathbf{c_1}$, and $\mathbf{c_2}$.

Consider the sphere c (Figure B.2a and B.2b). In some encounters in which c_1 exceeds c, the speed is diminished by $\Delta c = c_1 - c_2$; if, therefore, c_2 is less than c, the velocity point of the electron is carried by the encounter across the surface c and comes to rest within the sphere. In other encounters, although $c_2 < c_1$, the reduction in speed Δc is insufficient to make $c_2 < c$, in which event the velocity point comes to rest outside the sphere. In still other encounters in which $c_1 < c$ but c_2 exceeds c_1, it is possible that $-\Delta c = c_2 - c_1$ is sufficient to make $c_2 > c$. In this event the velocity point is transferred from within the sphere c across its surface to come to rest outside it. Again, if $c_1 < c$, it is possible that $c_2 < c$ even when $c_2 > c_1$ and the point remains within the sphere.

Thus a flux of points inwards and outwards is brought about by encounters across the surface of any sphere c. These inward and outward fluxes are equal when electrons move in thermal equilibrium with the gas in the absence of an electric field; consequently the net inward flux is zero and the distribution function $f(c)$ is that of Maxwell. When an electric

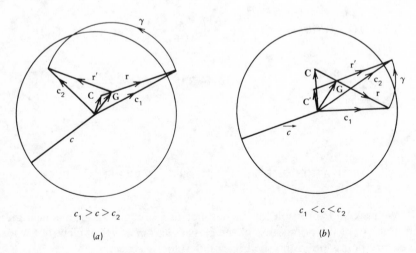

FIG. B.2.   Velocity diagram. In (b) the change in $\mathbf{C}$ is also shown.

force eE acts upon each electron, however, the equilibrium state of motion is one in which the net inward flux from encounters is not zero, but positive, and balances a net outward flux of points across the surface c due to the action of the electric field (see equation 2.19).

We let

$$(dt \, d\mathbf{r}) \sigma_{coll}(c) \equiv (dt \, d\mathbf{r}) \sigma_{coll}$$

be the net number of points that cross the surface c inwards in time dt as a result of the displacements Δc due to encounters. Then the increase in time dt of the point population of the shell (c, dc) is

$$(dt \, d\mathbf{r} \, dc) \frac{\partial}{\partial c} \sigma_{coll}.$$

It is necessary therefore to derive an expression for $\sigma_{coll}(c)$.

When all encounters are elastic, the expression is relatively simple but ceases to be so when encounters other than elastic ones are important. In this appendix therefore we shall restrict the discussion to elastic encounters. Inelastic encounters are considered in Chapter 6.

It is evident that the first step is to derive an expression for $\Delta c = c_1 - c_2$, and to this end we refer to equation B.3. If we write $c = c_1$, $c' = c_2$, and consider all encounters to be elastic ($g = g'$, $r = r'$), equation B.3 reads

$$c_1^2 - c_2^2 = \frac{2M}{M+m} g \left[\left(\frac{m}{M+m} g + C \cos \beta \right) (1 - \cos \gamma) - C \sin \beta \sin \gamma \cos \psi \right].$$

It is more convenient, however, in the present application, to replace g by r through the relation $g = [(M+m)/M]r$; thus

$$c_1^2 - c_2^2 = 2r \left[\left(\frac{m}{M} r + C \cos \beta \right) (1 - \cos \gamma) - C \sin \beta \sin \gamma \cos \psi \right],$$

in which the angles β, γ, ψ have the meaning already given them. Since C is the velocity of a molecule and $m/M \cong 10^{-4}$ or less, the right-hand side of this equation is very small compared with either c_1 or c_2, so that we may write $c_1 = c_2 + \Delta c$, with $\Delta c / c_2 \ll 1$.

Consequently

$$c_1^2 - c_2^2 = (c_1 + c_2)(c_1 - c_2) \cong 2c_2 \cdot \Delta c,$$

whence

$$\Delta c \cong \frac{r}{c_2}\left[\left(\frac{m}{M}r + C\cos\beta\right)(1 - \cos\gamma) - C\sin\beta\sin\gamma\cos\psi\right].$$

We consider next the geometrical interpretation of this expression for Δc (Figure B.3). The velocity of the centroid is $\mathbf{G} = (m/M)\mathbf{r} + \mathbf{C}$, where $\mathbf{r}$ is the velocity of m relative to the centroid. The velocity of m relative to the centroid after the encounter is $\mathbf{r}'$, and the velocity in the system at rest, to which $\mathbf{C}$ and $\mathbf{c}_1$ are referred, is $\mathbf{c}_2$. In a system of coordinates in which $\mathbf{r}$ is the polar axis, the direction of $\mathbf{C}$ is $(\beta, 0)$ and that of $\mathbf{r}'$ is (γ, ψ) (Figure B.1); then

$$\cos\beta = \frac{\mathbf{C}\cdot\mathbf{r}}{Cr}, \qquad \cos\gamma = \frac{\mathbf{r}'\cdot\mathbf{r}}{r'r},$$

and the cosine of the angle between the directions of $\mathbf{C}$ and $\mathbf{r}'$ is

$$\frac{\mathbf{C}\cdot\mathbf{r}'}{Cr'} = \cos\beta\cos\gamma + \sin\beta\sin\gamma\cos\psi.$$

Since C, G, $(m/M)r$ are each very small compared with c_1, c_2, and $r = r'$, it can be seen from Figure B.3 that

$$c_1 \cong r + \mathbf{G}\cdot\frac{\mathbf{r}}{r} = r + \left(\mathbf{C} + \frac{m}{M}\mathbf{r}\right)\cdot\frac{\mathbf{r}}{r} = r + \frac{m}{M}r + C\cos\beta,$$

$$c_2 \cong r' + \mathbf{G}\cdot\frac{\mathbf{r}'}{r'} = r + \left(\mathbf{C} + \frac{m}{M}\mathbf{r}\right)\cdot\frac{\mathbf{r}'}{r'}$$

$$= r + \frac{m}{M}r\cos\gamma + C(\cos\beta\cos\gamma + \sin\beta\sin\gamma\cos\psi),$$

whence

$$\Delta c = c_1 - c_2 \cong \left(\frac{m}{M}r + C\cos\beta\right)(1 - \cos\gamma) - C\sin\beta\sin\gamma\cos\psi,$$

which agrees with the analytical expression for Δc given above when, in the latter, r/c_2 is considered to be unity.

Let us consider the sphere with radius c and write $c = c_2$. We seek the net increase in the point content of the sphere c in time dt. In Figure B.4, the velocity $\mathbf{c}_2$ is represented by $\mathbf{OB}$ with $c_2 = c$. We seek $\mathbf{c}_1$ such that $\mathbf{r}$ after a change of direction γ carries the velocity point from A to B on the surface of the sphere where $\mathbf{OA} = \mathbf{c}_1$. The vector $\mathbf{G} = \mathbf{C} + (m/M)\mathbf{r}$, where $r \cong c$. We draw $\mathbf{r}$ to make an angle γ with $\mathbf{r}'$; then $\mathbf{r} = \mathbf{XA}$, and $\mathbf{c}_1 = \mathbf{OA}$ is an initial

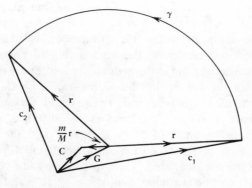

FIG. B.3

velocity such that the velocity point is carried from A to B on the surface of the sphere when γ is the angle of scattering. The interval $AD = c_1 - c_2$ is Δc, whose value for the angle γ is given by the expression deduced above.

In the case shown in the diagram in which $\mathbf{G} \cdot \mathbf{r}$ exceeds $\mathbf{G} \cdot \mathbf{r}'$, A is external to the sphere c. But $XA = XB$, and $XD < XB$; consequently the point D on the surface of the sphere, when scattered through the angle γ to

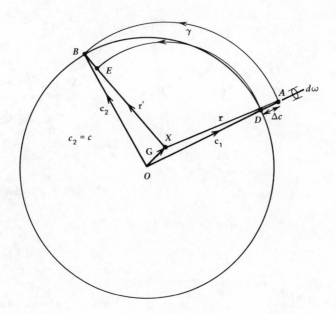

FIG. B.4

reside on $\mathbf{r}'$, comes to rest at E within the sphere. Thus all points on the interval AD, initially outside the sphere c, after scattering through γ come to rest within the depth BE below the surface of the sphere c. Such a scattering of points contributes to the inward flux of points across the surface c. Points scattered from A through angles less than $\gamma = AXB$ do not come to rest within the sphere, but do so for angles greater than AXB. Points on OA extended, lying beyond A, can enter the sphere only at angles greater than AXB. In the limit, as the distance beyond A is increased, the scattering angle of entry reaches π and the point enters diametrically opposite to D. At greater distances from A points do not enter the sphere, however scattered. Thus, in summary, all velocity points on the interval AD can be scattered through an angle γ and come to rest within the sphere within a distance $EB = AD$ of the surface.

Figure B.5 illustrates a case in which Δc is negative. Here $c_2 = c$, and $\Delta c = DA$. If $OA = c_1$, then A lies within the sphere and points scattered from A through γ onto $\mathbf{r}'$ reach the surface of the sphere at B. Points similarly scattered from D reach E, and points from the interval AD come to rest outside the sphere. Such scattering contributes to the outward flux across the sphere c. Points starting from within the interval XA do not leave the sphere, and points scattered from beyond D do not cross the surface.

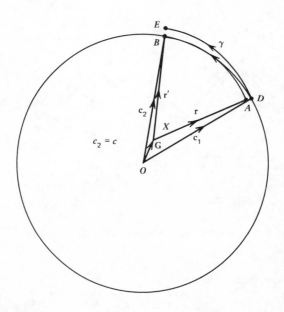

FIG. B.5

We now proceed to derive an expression for the net transport of velocity points across the surface of the sphere c in time dt. In Figure B.4, $d\omega$ is an elementary solid angle about the direction of $\mathbf{c}_1$. The element of volume whose length is the interval AD and which is bounded by $d\omega$ is $c^2 d\omega\, AD = c^2 d\omega\, \Delta c = d\mathbf{c}$. The number of electrons scattered through the interaction in $d\mathbf{r}$ of the elementary crossed beams with strengths $nf d\mathbf{c}$ and $NF d\mathbf{C}$ into the elementary solid angle $d\omega_\gamma = \sin\gamma\, d\gamma\, d\psi$ is, in time dt,

$$(d\mathbf{r}\, dt)(nf d\mathbf{c})(NF d\mathbf{C})g p_{el}(\gamma, g)\, d\omega_\gamma \cong (nNc)(fF d\mathbf{C})$$

$$\times (c^2 d\omega\, \Delta c) p_{el}(\gamma, c)\, d\omega_\gamma\, (dt\, d\mathbf{r})$$

where $f(\mathbf{c}) = f_0(c) + \sum_1^\infty f_k(c) P_k(\cos\theta)$ and $\mathbf{r}'$ is the axis of $d\omega_\gamma$. Although Δc is small in comparison with c, its existence is responsible for the flux. The appropriate value of f to take is the mean value for the whole interval Δc, namely,

$$f(\mathbf{c}) + \frac{\partial f}{\partial c}\frac{\Delta c}{2}.$$

When, in the expression for Δc, r is replaced by c, since $r \cong c$, the expression for the number becomes

$$(dt\, d\mathbf{r})(nNF d\mathbf{C})\left[f(\mathbf{c}) + \frac{\Delta c}{2}\frac{\partial f}{\partial c} \right] c^3 \Delta c\, p_{el}(\gamma, c)\, d\omega_\gamma\, d\omega$$

with

$$\Delta c = \left(\frac{m}{M}c + C\cos\beta \right)(1 - \cos\gamma) - C\sin\beta \sin\gamma \cos\psi$$

$$d\omega_\gamma = \sin\gamma\, d\gamma\, d\psi, \qquad d\omega = \sin\theta\, d\theta\, d\phi,$$

in which (θ, ϕ) is the direction of $\mathbf{c}_1$ with respect to a fixed system of axes with the direction of $\mathbf{f}_1$ as polar axis.

In order to find the number of points scattered from $d\mathbf{c}$ into the sphere, it is necessary to integrate over all directions of scattering (γ, ψ) and over all elementary beams $NF d\mathbf{C}$.

The factors that depend on directions are

$$p_{el}(\gamma, c)\, \Delta c f(\mathbf{c})\sin\gamma\, d\gamma\, d\psi$$

and

$$\frac{(\Delta c)^2}{2}\frac{\partial f}{\partial c}p_{el}(\gamma, c)\sin\gamma\, d\gamma\, d\psi.$$

We are thus required to evaluate

$$f(\mathbf{c}) \int_0^\pi \int_0^{2\pi} \left[\left(\frac{m}{M} c + C \cos\beta \right)(1 - \cos\gamma) - C \sin\beta \sin\gamma \cos\psi \right]$$

$$\times p_{el}(\gamma, c) \sin\gamma \, d\gamma \, d\psi$$

$$= f(\mathbf{c}) \left(\frac{m}{M} c + C \cos\beta \right) q_{m_{el}}(c)$$

and

$$\frac{1}{2} \frac{\partial f}{\partial c} \int_0^\pi \int_0^{2\pi} \left[\left(\frac{m}{M} c + C \cos\beta \right)^2 (1 - \cos\gamma)^2 - 2 \left(\frac{mc}{M} + C \cos\beta \right)(1 - \cos\gamma) \right.$$

$$\left. \times C \sin\beta \sin\gamma \cos\psi + C^2 \sin^2\beta \sin^2\gamma \cos^2\psi \right] p_{el}(\gamma, c) \sin\gamma \, d\gamma \, d\psi$$

$$= \frac{1}{2} \frac{\partial f}{\partial c} 2\pi \int_0^\pi \left[\left(\frac{m}{M} c + C \cos\beta \right)^2 (1 - \cos\gamma)^2 + \frac{C^2 \sin^2\beta \sin^2\gamma}{2} \right]$$

$$\times p_{el}(\gamma, c) \sin\gamma \, d\gamma.$$

Next we sum for all elementary beams $NF d\mathbf{C} = NFC^2 dC \sin\beta \, d\beta \, d\phi'$ (where β is the polar angle of $\mathbf{C}$ as appears in equation B.3 and ϕ' is the associated angle of azimuth) with C held constant to find

$$\left[4\pi f(\mathbf{c}) \left(\frac{m}{M} c \right) q_{m_{el}}(c) \right] NFC^2 dC$$

and

$$4\pi NFC^2 dC \frac{\partial f}{\partial c} \left[2\pi \int_0^\pi \frac{C^2}{3} (1 - \cos\gamma) p_{el}(\gamma, c) \sin\gamma \, d\gamma \right.$$

$$\left. + \pi \int_0^\pi \left(\frac{mc}{M} \right)^2 (1 - \cos\gamma)^2 p_{el}(\gamma, c) \sin\gamma \, d\gamma \right]$$

$$= 4\pi NFC^2 dC \frac{\partial f}{\partial c} \frac{C^2}{3} q_{m_{el}}(c) + 4\pi NFC^2 dC \frac{\partial f}{\partial c} \left(\frac{mc}{M} \right)^2$$

$$\times \pi \int_0^\pi (1 - \cos\gamma)^2 p_{el}(\gamma, c) \sin\gamma \, d\gamma.$$

Next we integrate for all C and note that

$$4\pi \int_0^\infty FC^2 \, dC = 1, \qquad 4\pi \int_0^\infty FC^4 \, dC = \overline{C^2}.$$

The factors become

$$Nf(\mathbf{c}) \frac{mc}{M} q_{m_{el}}(c)$$

and

$$\frac{\partial f}{\partial c} N \left[\frac{\overline{C^2}}{3} q_{m_{el}}(c) + \left(\frac{mc}{M} \right)^2 \pi \int_0^\pi (1 - \cos\gamma)^2 p_{el}(\gamma, c) \sin\gamma \, d\gamma \right].$$

Thus, in effect, the net number of points that enter the sphere from $d\omega$ in time dt is (Figure B.4)

$$d\omega \, (dt \, d\mathbf{r}) \left[Ncq_{m_{el}}(c) \right] nc^2$$

$$\times \left\{ \frac{m}{M} cf + \frac{\partial f}{\partial c} \left[\frac{\overline{C^2}}{3} + \left(\frac{mc}{M} \right)^2 \frac{\pi}{q_{m_{el}}(c)} \int_0^\pi (1 - \cos\gamma)^2 p_{el}(\gamma, c) \sin\gamma \, d\gamma \right] \right\}.$$

The result sought, after integration over all elementary solid angles $d\omega = \sin\theta \, d\theta \, d\phi$, is [since $f(\mathbf{c}) = f_0(c) + \sum_1^\infty f_k(c) P_k(\cos\theta)$]

$$(dt \, d\mathbf{r}) \left[Ncq_{m_{el}}(c) \right] (4\pi nc^2)$$

$$\times \left\{ \frac{m}{M} cf_0 + \frac{\partial f_0}{\partial c} \left[\frac{\overline{C^2}}{3} + \left(\frac{mc}{M} \right)^2 \frac{\pi}{q_{m_{el}}(c)} \int_0^\pi (1 - \cos\gamma)^2 p_{el}(\gamma, c) \sin\gamma \, d\gamma \right] \right\}.$$

The last term is of the order of magnitude $(\partial f_0 / \partial c)(mc/M)^2$, and we therefore compare the magnitudes of $(mc/M)^2$ and $\overline{C^2}$.

When electrons move in thermal equilibrium with a gas, $m\overline{c^2} / M\overline{C^2}$ $= 1$, that is, $(m/M)^2(\overline{c^2} / \overline{C^2}) = m/M \cong 10^{-4}$ or less. We conclude that the term $(mc/M)^2(\partial f_0 / \partial c)$ may be neglected as unimportant.

The expression therefore reduces to

$$(dt \, d\mathbf{r}) \left[Ncq_{m_{el}}(c) \right] 4\pi nc^2 \left(\frac{m}{M} cf_0 + \frac{\overline{C^2}}{3} \frac{\partial f_0}{\partial c} \right)$$

$$= (dt \, d\mathbf{r}) 4\pi nc^2 v_{el} \left(\frac{m}{M} cf_0 + \frac{\overline{C^2}}{3} \frac{\partial f_0}{\partial c} \right)$$

where the collision frequency for momentum transfer $\nu_{el} = Ncq_{m_{el}}(c)$. It follows that the expression sought for $\sigma_{coll}(c)$ is (elastic encounters)

$$\sigma_{coll}(c) = 4\pi nc^2\nu_{el}\left(\frac{mc}{M}f_0 + \frac{\overline{C^2}}{3}\frac{\partial f_0}{\partial c}\right).$$

This formula can also be derived from the collision integral. The analysis is much simpler for the case in which the molecules are at rest ($\mathbf{C}=0$).

Although this formula for $\sigma_{coll}(c)$ is correct, the analysis has been simplified by the omission of terms that, although small, might have been significant in the final result. Such terms appear individually in the final expression for the coefficient of $f_0(c)$, the terms being of order C^2/c and thus of comparable magnitude with the term $(m/M)c$. However, it is found that two such terms arise that are of equal magnitude but opposite sign and hence contribute nothing to the final result. The correction terms to the coefficient of $\partial f_0/\partial c$ are found to integrate to zero individually and thus do not affect the final result to the second order of approximation.

We consider first the approximations made in the derivation of the expression for Δc. The expression is

$$c_1^2 - c^2 = 2r\left\{\frac{m}{M}r(1-\cos\gamma) + C[(1-\cos\gamma)\cos\beta - \sin\beta\sin\gamma\cos\psi]\right\}.$$

Since $m \ll M$, we may replace r by $(c + \Delta c - C\cos\beta)$, since $r \cong g$, and $(m/M)r$ by $(m/M)c$, since the corrections to the final result from this last substitution are trivial. Then, if we define δc as

$$\delta c = \frac{mc}{M}(1-\cos\gamma) + C[(1-\cos\gamma)\cos\beta - \sin\beta\sin\gamma\cos\psi],$$

that is, the approximate expression for Δc used in the derivation above, it follows that

$$c_1^2 - c^2 = (2c + \Delta c)\Delta c = 2(c - C\cos\beta + \Delta c)\delta c,$$

that is,

$$\Delta c = \left(1 - \frac{C}{c}\cos\beta + \frac{\Delta c}{c}\right)\delta c - \frac{(\Delta c)^2}{2c}.$$

Since $(C/c) \ll 1$ and $(\Delta c/c) \ll 1$, it follows that, to a first approximation, $\Delta c = \delta c$, but to a second approximation

$$\Delta c = \left(1 - \frac{C}{c}\cos\beta + \frac{\delta c}{2c}\right)\delta c.$$

The other approximations that lead to the neglect of terms of the order of magnitude of those now under consideration are as follows:

1. That the collision frequency for scattering into the solid angle $d\omega_\gamma$ was taken to be $Ncp_{el}(\gamma,c)d\omega_\gamma$, whereas the *average* value of the *relative* speed $r - \Delta c/2$ appropriate to the range of speeds Δc should have been used rather than c.

2. That the elementary volume of velocity space from which the prescribed scattering of points takes place was taken to be the volume element $c^2 d\omega \Delta c$ (where $\Delta c = AD$ in Figure B.4). However, more correctly, the volume element is specified by considering the solid angle subtended at the centre of mass X rather than the angle subtended at 0, and its volume found by using the relative speed $r - \Delta c/2$ appropriate to the centre of the range of speeds Δc rather than the speed c as used previously.

It follows that a more accurate expression for the number of points scattered into $d\omega_\gamma$ in time dt, which also cross the surface of the sphere c, is found from the original expression

$$dt\,d\mathbf{r}(nNF(C)\,d\mathbf{C})\left[f(\mathbf{c}) + \frac{\Delta c}{2}\frac{\partial}{\partial c}f(\mathbf{c})\right]c^3\Delta c\,p_{el}(\gamma,c)\,d\omega_\gamma\,d\omega$$

by replacing c by

$$r - \frac{\Delta c}{2} \cong \left(c - C\cos\beta + \frac{\Delta c}{2}\right) = c\left(1 - \frac{C}{c}\cos\beta + \frac{\Delta c}{2c}\right)$$

and Δc by

$$\left(1 - \frac{C}{c} + \frac{\delta c}{2c}\right)\delta c.$$

The new expression, which is correct to the second order of approximation, thus is

$$(dt\,d\mathbf{r})(nNF(C)\,C^2\,dC)\left[f(\mathbf{c}) + \frac{\delta c}{2}\frac{\partial}{\partial c}f(\mathbf{c})\right]c^3\delta c\left(1 - 4\frac{C}{c}\cos\beta + 2\frac{\delta c}{c}\right)$$

$$\times p_{el}(\gamma,c)\,d\omega_\gamma\,d\Omega\,d\omega,$$

since $d\mathbf{C} = C^2\,dC\,d\Omega$, where $d\Omega$ is an elementary solid angle.

In the factor $(1 - 4(C/c)\cos\beta + 2\delta c/c)$ the first term (unity) evidently leads to the expression given above for $\sigma_{coll}(c)$, whereas the correction terms $-4(C/c)\cos\beta$ and $2\delta c/c$ give equal and opposite contributions to the coefficient of f_0 when integration is taken over all elementary solid angles $d\omega_\gamma$, $d\Omega$, and $d\omega$. The contributions of these terms to the coefficient of $(\partial/\partial c)f_0$ vanish individually. It follows that the more detailed derivation leaves the formula found for $\sigma_{coll}(c)$ unchanged.

APPENDIX C

EXPANSION OF $f(\mathbf{c}, \mathbf{r}, t)$ IN SPHERICAL HARMONICS WHEN $\mathbf{E}$ AND $\mathrm{grad}_r(n)$ ARE NOT PARALLEL

When $\mathrm{grad}_r(nf)$ and $e\mathbf{E}$ are not parallel, it is impossible to choose a direction for the polar axis such that $f(\mathbf{c})$ is independent of azimuth ϕ, as supposed in the representation

$$f = \sum_{k=0}^{\infty} f_k P_k(\cos\theta).$$

We take an arbitrary direction for the polar axis, and assume that when both $\mathrm{grad}_r(nf)$ and $e\mathbf{E}$ are present each "polarizes" the distribution of velocity point density in the shell (c, dc) with an axis of symmetry about its own direction. We let the direction of $\mathrm{grad}_r(nf)$ be (θ_G, ϕ_G) and of $e\mathbf{E}$ be (θ_E, ϕ_E) in the reference system, and γ_G and γ_E be the angles between an arbitrary direction (θ, ϕ) and the directions (θ_G, ϕ_G) and (θ_E, ϕ_E), respectively. We suppose that

$$f = f_0 + \sum_{k=1}^{\infty} f_{k_G} P_k(\cos\gamma_G) + \sum_{k=1}^{\infty} f_{k_E} P_k(\cos\gamma_E), \qquad \text{(C.1)}$$

where

$$\cos\gamma_G = \cos\theta\cos\theta_G + \sin\theta\sin\theta_G\cos(\phi - \phi_G)$$

and

$$\cos\gamma_E = \cos\theta\cos\theta_E + \sin\theta\sin\theta_E\cos(\phi - \phi_E).$$

But, from the addition theorem for spherical harmonics,

$$P_k(\cos\gamma_G) = P_k(\cos\theta)P_k(\cos\theta_G)$$

$$+2\sum_{m=1}^{\infty}\frac{(k-m)!}{(k+m)!}\cos m\,(\phi-\phi_G)\cdot P_k{}^m(\cos\theta)P_k{}^m(\cos\theta_G)$$

and

$$P_k(\cos\gamma_E) = P_k(\cos\theta)P_k(\cos\theta_E)$$

$$+2\sum_{m=1}^{\infty}\frac{(k-m)!}{(k+m)!}\cos m\,(\phi-\phi_E)\cdot P_k{}^m(\cos\theta)P_k{}^m(\cos\theta_E),$$

where

$$P_k{}^m(\mu) = (1-\mu^2)^{m/2}\frac{d^m}{d\mu^m}P_k(\mu).$$

We let the polar axis $\theta=0$ lie in the plane containing the directions (θ_G,ϕ_G) and (θ_E,ϕ_E); that is to say, the polar points on the sphere lie on the same line of longitude, which we may take as $\phi=0$.

It then follows that

$$f=f_0+\sum_{k=1}^{\infty}\left\{\left[f_{k_G}P_k(\cos\theta_G)+f_{k_E}P_k(\cos\theta_E)\right]P_k(\cos\theta)\right.$$

$$\left.+2\sum_{m=1}^{\infty}\left[f_{k_G}P_k{}^m(\cos\theta_G)+f_{k_E}P_k{}^m(\cos\theta_E)\right]\frac{(k-m)!}{(k+m)!}P_k{}^m(\cos\theta)\cos m\phi\right\}.$$

$$(C.2)$$

Since $P_k{}^m(1)=0$, the value of f when $\theta=0$ (at the pole of reference) is

$$f_{\theta=0}=f_0+\sum_{k=1}^{\infty}\left[f_{k_G}P_k(\cos\theta_G)+f_{k_E}P_k(\cos\theta_E)\right]$$

in which the lowest-order term dependent on θ is $(k=1)$

$$f_{1_G}\cos\theta_G+f_{1_E}\cos\theta_E.\qquad\qquad(C.3)$$

Let the direction of the axis $\theta = 0$, which lies in the plane containing the directions $(\theta_G, 0)$, and $(\theta_E, 0)$, be chosen to give $f_{1_G} \cos \theta_G + f_{1_E} \cos \theta_E$ its maximum value. Since $\theta_G - \theta_E = $ constant, it follows that $d\theta_G / d\theta = d\theta_E / d\theta$ and that the condition for a maximum is therefore

$$f_{1_G} \sin \theta_G + f_{1_E} \sin \theta_E = 0. \tag{C.4}$$

Define vectors $\mathbf{f}_{1_G}$ and $\mathbf{f}_{1_E}$, which have the magnitude and directions $[f_{1_G}, (\theta_G, 0)]$ and $[f_{1_E}, (\theta_E, 0)]$, respectively. Then the vector $\mathbf{f}_1 = \mathbf{f}_{1_G} + \mathbf{f}_{1_E}$ has the direction of the axis $\theta = 0$, since the projection of $(\mathbf{f}_{1_G} + \mathbf{f}_{1_E})$ on the plane normal to $\mathbf{f}_1$ is $(f_{1_G} \sin \theta_G + f_{1_E} \sin \theta_E)$ and is zero.

It is this choice of axis that is adopted in the derivation of the basic equations in Chapter 2, that is to say, the direction of $\mathbf{f}_1$ is adopted as the polar axis of reference. Equation 2.24, which is based on considerations of conservation of momentum, exhibits $\mathbf{f}_1$ as the vector sum of $\mathbf{f}_{1_G} = -(c/\nu) \mathrm{grad}_r (n f_0)$ and $\mathbf{f}_{1_E} = -V(\partial/\partial c)(n f_0)$.

We return to equation C.1, in which the polar axis $\theta = 0$ is arbitrarily directed with respect to the axes of polarization. Equation C.1 is equivalent to

$$f = f_0 + \sum_{k=1}^{\infty} f_k P_k(\cos \theta)$$

$$+ 2 \sum_{k=1}^{\infty} \sum_{m=1}^{\infty} A_{km} \frac{(k-m)!}{(k+m)!} [\cos m (\phi - \phi_G) + \cos m (\phi - \phi_E)] P_k^{\,m}(\cos \theta),$$

$$\tag{C.5}$$

where

$$f_k = f_{k_G} P_k(\cos \theta_G) + f_{k_E} P_k(\cos \theta_E)$$

and

$$A_{km} = f_{k_G} P_k^{\,m}(\cos \theta_G) + f_{k_E} P_k^{\,m}(\cos \theta_E).$$

In particular,

$$f_1 = f_{1_G} \cos \theta_G + f_{1_E} \cos \theta_E$$

and because f_{1_G} and f_{1_E} are the magnitudes of $\mathbf{f}_{1_G}$ and $\mathbf{f}_{1_E}$, the coefficient f_1, when the reference axis and the axes of polarization are arbitrarily directed, becomes the sum of projections of $\mathbf{f}_{1_G}$ and $\mathbf{f}_{1_E}$ on the reference axis $\theta = 0$. The mean value of f around a parallel of latitude $\theta = $ constant is $(1/2\pi) \int_0^{2\pi} f \, d\phi$, and in the process of integration terms containing $\cos m (\phi - \phi_G)$ and $\cos m (\phi - \phi_E)$ are eliminated.

The right-hand side of equation C.5 is thereby reduced to

$$f_0 + \sum_{k=1}^{\infty} f_k P_k(\cos\theta),$$

which is the form adopted for f in Section 2.6.1, where, however, in addition the direction of the axis $\theta = 0$ is taken as that of $\mathbf{f}_1$. Thus the representation of f in a form independent of ϕ is, in fact, the replacement of f by its mean value around a parallel of latitude. However, had ϕ been retained at this stage, it would have been eliminated in the course of integration over the whole sphere. It is therefore convenient to eliminate the dependence on ϕ at the outset.

PART 2

10

EXPERIMENTAL METHODS

FOR MEASURING

ELECTRON DRIFT VELOCITIES

10.1. INTRODUCTION

Of the parameters that describe the macroscopic properties of electron swarms in an electric field, the drift velocity $\mathbf{W}$ has possibly been the subject of the greatest number of experimental investigations. In this chapter we do not attempt to give a complete survey of all these investigations since comprehensive accounts are to be found elsewhere,[1,2] but instead restrict our account to a description of experimental methods that have already been developed as techniques for precise measurement or that appear capable of being developed in this way. As we shall see, the definition of "precise" is somewhat flexible in this context since it depends to some extent on whether other phenomena accompany the straightforward processes of drift and diffusion.

We begin our account with descriptions of experiments that have been successfully applied to measurements at lower values of E/N, where ionization is negligible. The experiments are discussed in four sections, each section covering experiments that are essentially similar in principle. After this general summary we describe in considerably more detail experiments that have achieved greatest success in this regime. The next section of the chapter deals with measurements of W in the presence of ionization and attachment, where the experiments become more difficult and some loss of precision is to be expected. The final section describes experiments that have been made using abnormally large gas number densities and that have revealed some previously unobserved phenomena.

As was pointed out in Chapter 8, the first experiments designed to measure the electron drift velocity $\mathbf{W}$ did, in fact, measure not $\mathbf{W}$ but rather a related quantity, the magnetic drift velocity $\mathbf{W}_M$. The basic difference between the two transport coefficients is not always understood. As a consequence it is not unusual to find, even in recent publications, values for W_M either compared directly with values of W or quoted in place of them when no other data are available. With modern experimental techniques the difference between the two quantities is usually well in excess of experimental error; in some circumstances the difference can be so large as to make comparison or substitution valueless. For this reason there will be no further discussion in this chapter of measurements based on the deflexion of a stream by a transverse magnetic field, the measurement of W_M being discussed as a separate topic in Chapter 12.

10.2.   THE PULSED TOWNSEND DISCHARGE TECHNIQUE

The time-of-flight technique that is conceptually the most straightforward and that is based on the simplest form of drift tube is the so-called pulsed Townsend discharge. Developed in the first instance by Hornbeck[3] for the measurement of both electron and ion mobilities, the technique was subsequently used by Bowe[4] to measure accurately electron drift velocities

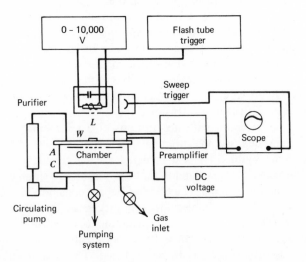

FIG. 10.1.   Schematic diagram of the apparatus used by Bowe.[4] Photoelectrons are released from the cathode C by UV light from a quartz flash tube L, which passes through a quartz window W and perforations in the anode A.

in the monatomic gases and to study the influence of molecular impurities on these measurements.

The principle of the experiment can be understood with the aid of Figure 10.1. The drift tube consists essentially of plane parallel electrodes, the anode being perforated to allow the incidence of light on the cathode from a flash tube external to the chamber. A light pulse from the flash tube, having a duration of the order of 0.5 μsec and a repetition rate of the order of 100 Hz, releases from the cathode photoelectrons which drift to the anode in the uniform field established between the electrodes. If the time constant of the anode circuit is sufficiently small, a constant potential difference is rapidly established across a load resistance in the circuit. The potential difference is caused by the displacement current which flows in the circuit as a consequence of the transit of the electron group between cathode and anode; the current ceases when the group arrives at the anode. A voltage pulse is therefore produced, typical examples of which are shown in Figure 10.2.[4] The finite width of the light pulse and the time constant of the anode circuit reduce the gradient of the leading edge of the pulse; diffusion of the group in transit reduces still further the (negative) gradient of the trailing edge. The transit time t_m of the group can be estimated from the time interval between the mid-points of the leading and trailing edges (see Figure 10.2).

In some experiments in neon and argon at the lowest values of E/N, Hornbeck[3] found a modification to the typical current wave forms in the

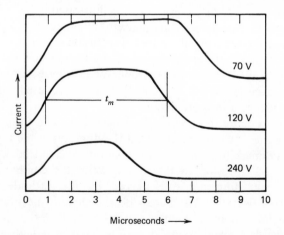

FIG. 10.2. Typical current pulses obtained by Bowe[4] in helium with $p \sim 100$ torr. The increase of W with cathode-anode voltage, and the time resolution of the method are illustrated.

form of a spike at the beginning of the pulse. He reported, "Increasing the voltage increases the amplitude of the pulse and rapidly eclipses the spike effect. Thereafter, the pulses behave like helium." These observations are consistent with an explanation[5] based on back diffusion to the cathode (see p. 303). The current in the gap is given by $I(t) = N'(t)eW/h$, where $N'(t)$ is the total number of electrons in the gap at time t and h is the electrode separation; consequently the current decreases rapidly initially, when the loss of electrons by back diffusion to the cathode is large, but approaches an asymptotic value as the distance between the group and the cathode increases. The effects are seen in neon and argon, but not in helium, because the ratio D/W is much larger in the heavier gases for the same experimental conditions. The disappearance of the spike with increasing voltage corresponds to the reduction of D/W with increasing E.

The most accurate electron drift velocity measurements based on this method are those of Bowe.[4] The electrodes of his apparatus were approximately 30 cm in diameter and spaced 2.54 cm apart on Kel-F stand-off insulators. The Pyrex plate-glass cathode was tin-coated to provide a photosensitive surface. Bowe estimated the error in determining t_m to be $\pm 2\frac{1}{2}\%$ and the overall error associated with the calibration of the sweep speed and the determination of E and p to be $\pm 5\%$.

A major aim of Bowe's work was to investigate the effect of molecular impurity on measurements in the monatomic gases. With this in mind he designed his apparatus to incorporate a recirculating gas purification system consisting of a pump and a tube containing calcium turnings heated to 450°C. The marked effect of the removal of impurity by the purifier, particularly in the case of argon, is illustrated in his original paper.

The majority of Bowe's measurements were made with high gas pressures, and as a consequence the effect of diffusion on the interpretation of his results was small. Nevertheless, as shown by Lowke,[5] the straightforward deduction of W from the oscillographic records, using $W = h/t_m$, may not be accurate when low gas pressures are used, and we now examine this problem in more detail.

In order to examine the time dependence of the current density in the gap, we make the usual assumption that the duration of the initiatory light flash is infinitesimal, but we now also take into account back diffusion to the cathode and diffusion to the anode. With the commonly accepted assumptions regarding the rôle played by the cathode and anode boundaries (see Section 5.1.2) it can be shown that the number density within an electron group released at a metal cathode in the plane $z = 0$ and travelling towards a metal anode at $z = h$ is given by (equation 5.5)

$$n(z,t) = \frac{N_0}{(4\pi D_L)^{1/2} W t^{3/2}} \left\{ z \exp\left[-\frac{(z-Wt)^2}{4D_L t} \right] \right.$$

$$\left. + (z-2h)\left(\exp\frac{hW}{D_L} \right)\exp\left[\frac{(z-2h-Wt)^2}{4D_L t} \right] \right\}, \qquad (10.1)$$

where it is assumed that the group originates at time $t=0$ as a uniform plane highly concentrated distribution at the cathode ($z=0$). The constant N_0 represents the number of electrons which ultimately escape from unit area of the cathode.†

If we denote by $N(t)$ the number of electrons in a column of unit cross-sectional area extending from cathode to anode, it can be shown that the current density in the gap is[6]

$$i(t) = \frac{N(t)eW}{h}. \qquad (10.2)$$

Consequently

$$i(t) = \frac{N_0 e}{\pi^{1/2} h t} \left\{ \int_0^h \frac{z}{(4D_L t)^{1/2}} \exp\left[-\frac{(z-Wt)^2}{4D_L t} \right] dz \right.$$

$$\left. + \int_0^h \frac{z-2h}{(4D_L t)^{1/2}} \exp\left[\frac{hW}{D_L} - \frac{(z-2h-Wt)^2}{4D_L t} \right] dz \right\}.$$

In the first integral let $u=(z-Wt)/(4D_L t)^{1/2}$ and in the second let $v=(z-2h-Wt)/(4D_L t)^{1/2}$. It follows that

$$i(t) = \frac{N_0 e}{\pi^{1/2} h t} \left\{ \int_{-Wt/(4D_L t)^{1/2}}^{(h-Wt)/(4D_L t)^{1/2}} \left[Wt + (4D_L t)^{1/2}u \right] \exp(-u^2)\, du \right.$$

$$\left. + \int_{-(2h+Wt)/(4D_L t)^{1/2}}^{-(h+Wt)/(4D_L t)^{1/2}} \left[Wt + (4D_L t)^{1/2}v \right] \exp\left(\frac{hW}{D_L} - v^2 \right) dv \right\}.$$

†This result can be verified by considering the case when the anode is at infinity and forming the integral $\int_0^\infty n(z,t)\,dz$, when it will be found that $N(t) \to N_0$ as $t \to \infty$.

Regrouping terms, we have

$$
i(t) = \frac{N_0 e W}{\pi^{1/2} h} \left[\int_{-Wt/(4D_L t)^{1/2}}^{(h-Wt)/(4D_L t)^{1/2}} \exp(-u^2)\, du \right.
$$

$$
\left. + \int_{-(2h+Wt)/(4D_L t)^{1/2}}^{-(h+Wt)/(4D_L t)^{1/2}} \exp\left(\frac{hW}{D_L} - v^2\right) dv \right]
$$

$$
+ \frac{2N_0 e}{h}\left(\frac{D_L}{\pi t}\right)^{1/2}\left[\int_{-Wt/(4D_L t)^{1/2}}^{(h-Wt)/(4D_L t)^{1/2}} u\exp(-u^2)\, du \right.
$$

$$
\left. + \int_{-(2h+Wt)/(4D_L t)^{1/2}}^{-(h+Wt)/(4D_L t)^{1/2}} v\exp\left(\frac{hW}{D_L} - v^2\right) dv \right]
$$

$$
= \frac{N_0 e W}{\pi^{1/2} h} \left\{ \left[\mathrm{Erf}\,\frac{Wt}{(4D_L t)^{1/2}} + \mathrm{Erf}\,\frac{h-Wt}{(4D_L t)^{1/2}} \right] \right.
$$

$$
\left. + \left[\mathrm{Erf}\,\frac{2h+Wt}{(4D_L t)^{1/2}} - \mathrm{Erf}\,\frac{h+Wt}{(4D_L t)^{1/2}} \right] \exp\frac{hW}{D_L} \right\}
$$

$$
+ \frac{2N_0 e}{h}\left(\frac{D_L}{\pi t}\right)^{1/2}\left[\int_{-Wt/(4D_L t)^{1/2}}^{(h-Wt)/(4D_L t)^{1/2}} u\exp(-u^2)\, du \right.
$$

$$
\left. + \int_{-(2h+Wt)/(4D_L t)^{1/2}}^{-(h+Wt)/(4D_L t)^{1/2}} v\exp\left(\frac{hW}{D_L} - v^2\right) dv \right]. \quad (10.3)
$$

A graph of the function $i(t)$ for a typical set of experimental conditions is shown in Figure 10.3. The figure confirms the explanation given earlier of the spike which Hornbeck observed in his oscillographic records.

When the electron group is well clear of the cathode, that is, when there is no further loss of electrons by back diffusion, the current attains an asymptotic value i_0 (see Figure 10.3). Let t_m denote the time at which the current falls to $i_0/2$. It has been normally assumed that negligible error is

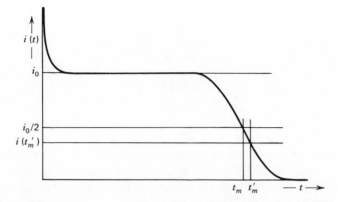

FIG. 10.3. Theoretical curve of current as a function of time in a pulsed Townsend discharge experiment.

made in calculating the drift velocity from the simple formula $W = h/t_m.$†
Using equation 10.3, we now examine this approximation in some detail in order to estimate the magnitude of the error that results from its use.

Let $t'_m = h/W$. Then it follows from the application of Newton's rule that, to a good approximation,

$$t_m \cong t'_m - \left[\frac{i(t'_m) - i_0/2}{(di/dt)_{t=t'_m}} \right]$$

since $t_m \cong t'_m$.

It is shown in Appendix 10.1 that

$$i(t'_m) - \frac{i_0}{2} \cong - \frac{3N_0eW}{2\pi^{1/2}h(hW/D_L)^{1/2}},$$

while Lowke[5] has shown that

$$\left(\frac{di}{dt} \right)_{t=t'_m} \cong - \frac{N_0eW^2}{h(4\pi D_L t'_m)^{1/2}}.$$

Consequently

$$t'_m - t_m \cong \frac{3D_L}{W^2},$$

†In practice t_m is determined from an oscillogram such as that shown in Figure 10.2, in which both the leading and the trailing edges of the pulse exhibit a slope largely contributed to by the finite width of the UV light pulse. In this analysis it is assumed that the finite width introduces no additional error in determining t_m.

that is,

$$t_m \cong \frac{h}{W}\left(1 - \frac{3D_L}{hW}\right)$$

or

$$W \cong \frac{h}{t_m}(1 - 3\beta), \tag{10.4}$$

where $\beta = D_L/hW = (D_L/\mu)/V$, and $V = Eh$ is the potential difference between the cathode and the anode of the drift tube. As an example, we may take the case of a measurement in helium at $E/N = 3$ Td and $p = 50$ torr in an apparatus with a 2-cm electrode separation. Then $D_L/\mu \cong 1$ V[7] and $V \cong 100$ V. Consequently the correction factor 3β is approximately 3%.

The experimental conditions used by Bowe were apparently such that the errors resulting from the omission of the correction factor were usually considerably less than in the example above. Nevertheless the correction cannot be ignored when high accuracy is to be achieved.

10.3.   ELECTRICAL SHUTTER TECHNIQUES

The original interpretation of time-of-flight experiments based on the use of electrical shutters was extremely simple. The first shutter was seen simply as a gate which admitted a well defined electron group into the drift space at a known time, while it was supposed that the second shutter could be used to sample the time variation of the electron density at a plane a known distance from the first shutter without disturbing the number density at any time. With these assumptions, the transit time of the group was taken as the time interval between the opening of the shutters that corresponded to maximum transmission of the second shutter. As will be shown subsequently, the interpretation of the results of such experiments in terms of this simple model requires modification; nevertheless the model is adequate for a qualitative description of the experiments.

10.3.1.   ION MOBILITY EXPERIMENTS OF VAN DE GRAAFF AND OF TYN-DALL ET AL. The first experiments of this kind appear to be those of Van de Graaff[8] and Tyndall, Starr, and Powell,[9] who independently devised very similar techniques for the measurements of ion mobilities. In these experiments an electrical shutter was used to form the groups as well as to determine their arrival times at a given plane. The apparatus of Tyndall et al. is shown schematically in Figure 10.4. Each of the electrical shutters S_1 and S_2 consists of two closely spaced plane grids, giving rise to the name "four-gauze" method for the technique. The electrodes $G_1 \cdots G_4$ are provided in order to establish a uniform field throughout the drift space between S_1 and S_2.

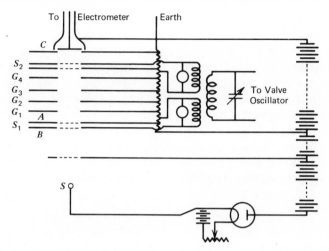

FIG. 10.4. Schematic diagram of the apparatus used by Tyndall et al. (From A. M. Tyndall, *The Mobility of Positive Ions in Gases*, Cambridge University Press, 1938.)

The connections to the voltage divider chain are such that, in the absence of the alternating potential difference applied between the elements of each shutter, the electric field within each shutter is directed oppositely to the main drift field. Consequently with the application of an AC signal to A and B a steady stream of ions incident on S_1 from a source S is intercepted except during a small interval of time once per cycle, when the field within S_1 reverses direction and its strength is approximately equal to that of the drift field. In this way a regular succession of ion groups is introduced at known times into the drift space.

The velocity of the groups is measured by sampling the ion density at S_2, a known distance (in this instance 2 cm) from S_1. Since the action of S_2 is identical with that of S_1, and since the AC signals applied to both shutters are exactly in phase, the ion current transmitted by S_2 is taken to be proportional to the number density of the group arriving at S_2 which was formed at the first shutter τ seconds earlier, τ being the period of the AC signal. On the simple model, therefore, the drift velocity of the ions is calculated from $W = f_m d$, d being the effective drift distance between S_1 and S_2, and $f_m = 1/\tau_m$ the frequency at which maximum current is recorded at C.

Van de Graaff's apparatus was identical in principle with that described above, differing in detail mainly in the absence of the upper element of the second shutter. The interrupted reverse field used to sample the ion groups was established in this instance between a single grid at S_2 and the collecting electrode.

A disadvantage of this form of shutter apparatus is the uncertainty in the effective drift distance, which arises from the difficulty in determining the exact positions of the plane at which the ion group may be said to originate and of the plane at which it is sampled.* Tyndall and Powell[10] sought to overcome this difficulty by performing an absolute determination which could be used as a calibrating standard. To this end they constructed an apparatus in which the distance S_1S_2 could be varied. By keeping the drift field and the frequency of the AC shutter signal constant and finding successive positions of the movable shutter at which the transmitted current was a maximum, they were able to determine, free of end effects, the distance travelled by the ion groups in the period of the AC signal.

10.3.2. APPLICATION OF THE "FOUR-GAUZE" METHOD TO ELECTRON DRIFT VELOCITY MEASUREMENTS. Although little or no use of the "four-gauze" method for the determination of *electron* drift velocities appears to have been made until recently, an application of the method can now be found in the work of Blevin and Hasan.[11] These authors made measurements in hydrogen and nitrogen at very high values of E/N, where difficulty has been experienced in applying the more commonly used Bradbury-Nielsen shutter apparatus because of the reduced efficiency of the electrical shutters.

The achievement of high accuracy in experiments based on electrical shutter techniques depends primarily on the ability to determine accurately the frequency of the AC signal at which maximum electron (or ion) current is transmitted. This in turn depends on the effectiveness of the shutters in collecting all the electrons at all times other than during the "open" time interval (an interval that must be small compared with the drift time) and on minimizing the dispersion of the electron groups by diffusion. The effects of diffusion are reduced by the use of high gas number densities, but there is an attendant difficulty when measurements are to be made at high values of E/N in that large values of the product Nd (d being the drift distance) may lead to electrical breakdown within the apparatus.

Blevin and Hasan attempted to overcome these difficulties in two ways. In the first place, they used the same form of shutter as Tyndall and others since they found it to be more effective than the Bradbury-Nielsen shutter at high values of E/N. Secondly, they reduced the effectiveness of the secondary process leading to electrical breakdown by introducing an electrode with a small central aperture midway between the two shutters.

---

*In the original apparatus, the distance AB was 0.5 cm and the drift distance S_1S_2 was 2 cm. The uncertainty in the drift distance was less in later experiments, where AB was considerably reduced and S_1S_2 increased.

With this arrangement they found that the values of Nd could be greatly increased before breakdown occurred.

These authors were able to determine values of f_m with sufficient accuracy for values of E/N up to 300 Td or more. Their published data, however, do not extend beyond $E/N = 150$ for hydrogen and $E/N = 240$ for nitrogen since they were unsure of the correct procedure to adopt to take proper account of the effects of primary and secondary ionization.

The accuracy claimed by the authors for their data is about $\pm 2\%$ at the lower end of the range and ± 3 or 4% at the upper end, the limitations being set primarily by insufficient stability of the electron source and the difficulty of measuring accurately pressures of the order of 1 torr or less. Support for their claim is afforded by the good agreement with Lowke's[12] data for hydrogen and nitrogen in the region of overlap.

10.3.3.   EXPERIMENTS OF BRADBURY AND NIELSEN. Perhaps the most important development of the electrical shutter apparatus, leading to the high accuracy now obtainable with it, is that due to Bradbury and Nielsen.[13]* These authors replaced the earlier form of electrical shutter by the coplanar arrangement shown schematically in Figure 10.5. This form of grid, in which the two sets of interconnected wires are mounted on a mica, glass, or ceramic former and are therefore insulated from each other, was proposed by Loeb and used in the first instance by Cravath[14] as an electron filter. In Cravath's apparatus the two sections of the grid were connected to the output of a high-frequency oscillator. Since the mobility of the electrons is very much greater than that of the ions, the majority of the electrons were swept to the wires by the alternating field, whereas the motion of the negative ions was only slightly perturbed.

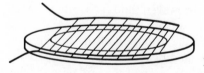

FIG. 10.5.   Sketch of a typical electrical shutter.

*Historically, credit must again go to Van de Graaff for devising the coplanar grid and applying it to drift velocity measurements. In a brief letter to *Nature* (**10**, 124, 1929) he refers to a description in his doctoral thesis of May 1928 of his experiments with the new form of shutter. His letter was prompted by A. M. Cravath's[14] paper in *Physical Review* describing the use of a similar grid as an electron filter. However, Van de Graaff did not publish any electron or ion drift velocity data measured with an apparatus incorporating his new device. Although Bradbury and Nielsen were aware of Van de Graaff's earlier (1928) work, they were apparently not aware of the later contribution, as no reference was made to it.

For drift velocity measurements, the principal advantage of this form of shutter is the accuracy with which it is possible to specify the positions of the planes at which the electron groups are formed and subsequently sampled. This is particularly important in *electron* drift velocity measurements, where high diffusion rates would tend to invalidate the procedure used by Tyndall and Powell to calibrate ion mobility drift tubes of fixed length. Experimental evidence will be given in Section 10.6 to show that, with proper attention to the design and construction of the coplanar shutters, the uncertainty regarding the positions of the planes may be made less than the thickness of the grid wires.

In their first experiments, Bradbury and Nielsen employed an apparatus in which the shutters were made from 0.08-mm wire spaced 1.0 mm apart on mica formers. The shutters were used in conjunction with a set of guard electrodes similar to those shown in Figure 10.4 to give a drift space approximately 6 cm in length. The construction of their apparatus was such that it could be evacuated to less than 10^{-7} torr before each set of experiments. Electrons were released photoelectrically from a zinc plate.

Sinusoidal voltages were applied to the electrical shutters, but difficulty was experienced in these early experiments in maintaining a constant amplitude as the frequency of the signal was varied. Furthermore, the transmission of each shutter was found to be strongly dependent on the amplitude of the signal (see Section 10.6.5). For this reason, it was recognized that a technique for determining f_m in which the frequency itself was varied would lead to significant errors. The problem was largely overcome by holding the amplitude and frequency of the shutter signals constant and varying the drift field until maximum current was recorded. Even with this technique it was necessary to make an allowance for a shift of the current maximum from its true position, the shift now arising from an inherent dependence of the transmitted current on the strength of the drift field. The correction was small, however, and subsequent work verified the validity of the procedure that was used to make it.

In a second paper, Nielsen[13] described the extension of the earlier work in H_2 to N_2, He, Ne, and Ar. Some modifications were made to the apparatus, the most significant being the reduction of the grid wire spacing to 0.5 mm in order to increase the efficiency of the grids. The photoelectric electron source was replaced by a plane grid of tungsten filaments. A thermocouple mounted between the shutters was used to check that there were no significant heating effects from the filaments. Modifications to the oscillator providing the shutter signals achieved a voltage output that was largely independent of the frequency, thus enabling these and later experiments to be made with a constant drift field.

In the third and final paper, Nielsen and Bradbury[13] reported the application of their technique to the measurement of electron drift velocities in the presence of attachment and to the measurement of the mobility of the negative ions formed. In order to ensure that some free electrons survived, it was necessary to perform the experiments at pressures considerably lower than those used previously. Therefore these experiments in oxygen, air, nitrous oxide, and ammonia were restricted to the pressure range 2 to 10 torr.

It is difficult to assess the probable error in these experiments, particularly those arising from diffusive effects (see Section 10.3.6), because insufficient detail was given concerning the pressures used for the various sets of measurements. For the experiments in H_2 it was stated that pressures of 1 to 50 torr were used. It can therefore be estimated that, over the full range of E/N, the results taken at the highest pressure at each value were probably subject to errors from diffusive effects of less than 1%; in most cases the error was possibly considerably less. At the lowest pressures for which measurements were made at each value of E/N, however, it is likely that diffusive effects were responsible for errors of several per cent. Insufficiently accurate pressure measurement, particularly at the lower pressures, was probably responsible for the fact that no systematic dependence of the results on gas pressure was observed.

Although there have been many later measurements of electron drift velocities for the same gases and over the same range of values E/N as those studied by Bradbury and Nielsen, it is a tribute to the work of these authors that, in the 35 years since it was completed, few investigations have shown a comparable accuracy. The most significant errors in their work appear to have arisen from their difficulty in obtaining gases of adequate purity, notably neon and argon. This difficulty is scarcely surprising in the light of recent work described in Section 10.6, which shows that the presence in the heavier monatomic gases of molecular impurities at levels of a few parts per million can change the drift velocity significantly.

10.3.4. LATER EXPERIMENTS USING THE BRADBURY–NIELSEN TECHNIQUE. Despite the inherent simplicity of the technique and its demonstrated effectiveness it does not appear to have been applied without modification except in the work described in Section 10.6. There seems to have been some reluctance to use the technique because of reservations about the end effects caused by distortion of the drift field in the vicinity of the shutters. However, a comparison of the results obtained using several different techniques, including those described in this chapter, shows that, where it has been possible to apply the electrical shutter method, the accuracy achieved with it is greater than that so far attained with other methods. The reasons for this will be discussed in Section 10.6.

10.3.5.   EXPERIMENTS OF PHELPS AND OTHERS. In a series of experiments begun about 1960, Phelps and his coworkers at the Research Laboratories of the Westinghouse Electric Corporation made an extensive set of drift velocity measurements in a number of gases over a wide range of the experimental parameters E/N and T. Although they set out to construct an apparatus and to use procedures identical in principle with those of Bradbury and Nielsen, the presence of end effects caused them to modify the experimental procedures in the way described below.

The apparatus designed by Phelps, Pack, and Frost,[15] shown schematically in Figure 10.6, differed in important details from the apparatuses of Bradbury and Nielsen. The shutters were made of 0.08-mm-diameter wire spaced 0.8 mm apart and mounted on fired lava formers, which were shielded from the electron stream by metal facings. The shutter wires and facings were gold plated to reduce stray fields introduced by contact potential differences. The guard electrodes were so constructed (see Figure 10.6) that the drift space was shielded electrically from surfaces charges on the glass vacuum wall and to a large extent optically from the UV light source. The design of the apparatus and the materials used in its construction conformed to ultrahigh-vacuum specifications, so that base pressures of 10^{-9} torr and low outgassing rates could be achieved.

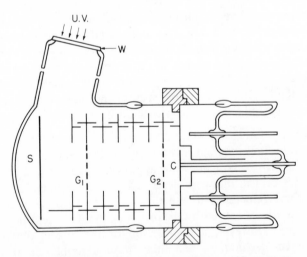

FIG. 10.6.   Diagram of the drift tube used by Phelps et al.[15] Light from a UV source enters through the quartz window W and falls on the photocathode S. G_1 and G_2 are the electrical shutters, and C is the collector.

The sine wave gating signals used by Bradbury and Nielsen to operate the electrical shutters were replaced by rectangular pulses. The disadvantage of sine wave signals arises from the need in some circumstances to use large amplitudes to achieve adequate resolution. This may be understood by referring to Figure 10.7a, where it is seen that if V_B is the voltage between the grid wires necessary to absorb, say, 99% of the electrons falling on the grid, a peak-to-peak voltage of V_{pp} is required to achieve an open time of δt.

In practice the effective open time is somewhat less than δt because the transmission of the shutter is already greatly reduced for voltages less than V_B. Nevertheless peak-to-peak voltages many times greater than V_B are required to obtain adequate performance, with the possibility of significant field distortion in the vicinity of the shutters.

In their experiments Phelps et al. reduced the effects of field distortion near the shutters by applying a DC bias voltage V_B between the wires, together with the signals comprised of rectangular pulses of duration δt and amplitude $V_B/2$ (see Figure 10.7b). Field distortion caused by the bias voltage was reduced to a minimum by maintaining the mean potential of the grid wires at the level appropriate to the position of the shutter plane in the drift field. The curves in Figure 10.7b therefore represent the potential of each half of the shutter with respect to the mean potential. It can be seen from the figure that a comparable performance could be obtained with rectangular pulses of much smaller amplitude with a consequent reduction in field distortion. The question as to whether the distortion is, in fact, necessarily a significant source of error is discussed in Section 10.6.

Notwithstanding this new mode of operation, the results at low E/N were found to depend on the drift distance, differences of 10 to 20% being

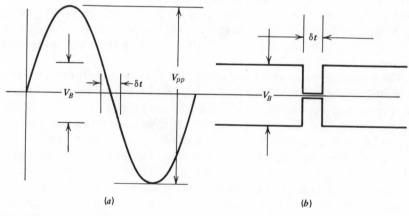

(a)   (b)

FIG. 10.7

obtained between the results recorded using drift distances of 2.54 and 6.35 cm. Since the variations could not be ascribed to contact potential differences,[15] it was assumed that some unknown end effect caused the effective drift distance to depend slightly on the experimental conditions. To eliminate these end effects the final results were obtained by measuring the transit times in drift tubes having the drift distances already quoted and finding the ratio of the difference between the drift distances to the difference between the transit times. This was accomplished by making a complete set of meaurements with the drift tube in one configuration, dismantling the tube to change its dimensions, and repeating the measurements with the new drift distance.

This somewhat cumbersome procedure was eliminated by Pack and Phelps,[15] whose principal interest lay in the study of electron swarms having very low mean energies. The drift tube was redesigned with the specific objects of minimizing end effects and making measurements at low gas temperatures. In the new tube, the electron groups were formed by the action of a pulsed UV source on the photocathode rather than by the first shutter. As a consequence both grids could serve as the sampling grid, allowing two drift distances, in this case 2.54 and 5.08 cm, to be used. As shown in Section 10.3.6, the determination of W from the difference between the two transit times should eliminate, at least to first order, errors from end effects.

Even with this modification, however, there was still some evidence of end effects. These may have been caused by the somewhat wide spacing of the shutter wires (~3.5 mm). The wide spacing would have increased the transmission to the second shutter when the first was quiescent, but would also have led to increased field distortion in the vicinity of the shutters. In neon and argon, for example, it was necessary to use values of V_B that were comparable with the potential drop across the drift space. The use of such large voltages, coupled with the wide wire spacing, considerably enlarged the region adjacent to the shutters in which there was severe field distortion.

The continued presence of end effects led Pack and Phelps to develop a modified procedure which they called the "zero-bias" or "rejection" mode. In this method the two sections of the grid are held at the same potential except during the short time interval in which a rectangular pulse is applied to reduce its transmission. As a result minimum current is received by the collector when the application of the pulse coincides with the arrival of the electron group from the photocathode. The operation of the shutter in this way results in the field in the drift space being maintained free of distortion except during the short time interval in which the flux through the shutter is sampled. The systematic errors from end effects in the results obtained by this method were less than the random errors.

Reference should be made to the original papers for details of the precautions taken to obtain accurate measurements of the experimental parameters and to ensure adequate gas purity.

10.3.6. INTERPRETATION OF RESULTS FROM ELECTRICAL SHUTTER EXPERIMENTS. In early time-of-flight experiments in which electrical shutters were used the choice of the experimental parameters and the magnitude of the experimental errors were such that no detectable errors arose from the use of the simple formula $W = h/t_m$ to calculate the drift velocity from the time t_m at which maximum current was transmitted by the sampling shutter. In these experiments the influence of end effects at the shutters was masked by other experimental errors. The possible importance of such effects was recognized by Duncan,[16] Crompton, Hall, and Macklin,[17] and Phelps, Pack, and Frost.[15] As we have already seen, Phelps and his collaborators designed their experiments to eliminate as far as possible errors caused by end effects but introduced other sources of error in so doing. On the other hand, in the work of Crompton and others described later in this chapter (Section 10.6), the principal stress was laid on the reduction of all sources of random and systematic error, although no specific attempt was made to reduce end effects by a differencing technique. As a consequence the form of the systematic errors introduced by the use of the simple formula became clearly evident and an attempt was made to develop a satisfactory theoretical explanation for them. We now summarize the theory, taking as the starting point the theory developed in Chapter 5.

The effect of diffusive processes in time-of-flight experiments was first analyzed by Duncan.[16] In his analysis of the Bradbury-Nielsen experiment he assumed (a) that the source of electrons was an isolated point source, (b) that the number of electrons received by the collecting electrode was proportional to the axial flux across a geometrical plane located in the plane of the second shutter, and (c) that the open time of the shutters was a constant fraction of the period of the gating signal applied to them, that is, that a sinusoidal signal or a square wave signal with a constant duty cycle was used. He also considered the consequence of overlap from successive groups within the drift space, although his analysis showed that this effect could be disregarded except in circumstances that would lead to very low accuracy because of the poor definition of the maxima in the curves of transmitted current versus frequency (see Appendix 10.2).

The equation used by Duncan to represent the travelling groups was the same as equation 5.1a since the same assumptions were used in each case. However, since Duncan's work antedated the realization of the existence of anisotropic diffusion, he assumed that $D_L = D$. From equation 5.1a it follows that, with this assumption, the electron number density on the z axis ($\rho = 0$) is given by

$$n(0,z,t) = \frac{n_0}{(4\pi Dt)^{3/2}} \exp\left[-\frac{(z-Wt)^2}{4Dt}\right],\tag{10.5}$$

where n_0 electrons are released at the origin by the first shutter at time $t = 0$, the shutter being assumed to have an infinitesimally small aperture.

Equation 10.5 may now be used to determine the axial number density at the second shutter (situated in the plane $z = h$) at the instant the shutter opens. Let the frequency of the AC gating signal be f; the second shutter then opens at a time $t = \tau = 1/2f$ after the release of the electron group by the first shutter at time $t = 0$, and consequently the axial number density at $z = h$ arising from the group is, at time $t = \tau$,

$$n(0,h,\tau) = \frac{n_0}{(4\pi D\tau)^{3/2}} \exp\left[-\frac{(h-W\tau)^2}{4D\tau}\right].$$

However, the third assumption requires that n_0 be directly proportional to τ, so that we may write $n_0 = n'_0\tau$, where n'_0 is a constant. Thus

$$n(0,h,\tau) = \frac{n'_0}{(4\pi D)^{3/2}\tau^{1/2}} \exp\left[-\frac{(h-W\tau)^2}{4D\tau}\right].\tag{10.6}$$

Since this equation is formally the same as equation 5.1b, and since the assumption made in Section 5.2.2b is also made here, that is, that the flux to the collector is proportional to the flux $(nW - D\,\partial n/\partial z)_{z=h}$ across a geometrical plane at $z = h$, it follows that the drift speed W is related to t_m through the equation

$$W = \frac{h}{t_m}(1 - 2\beta),$$

where $\beta = D/Wh$.

It is interesting to note that the assumption made by Duncan (that the number of electrons collected is proportional to the *axial* flux density rather than to the *total* flux across the plane $z = h$) would lead to a correction factor of 4β if the populations of the electron groups were assumed to be independent of the gating frequency. However, the increase in the factor from 2β to 4β is exactly offset by the factor introduced by the particular mode of operation of the shutters in the original Bradbury-

Nielsen method, that is, by the substitution of $n'_0\tau$ for n_0 in equation 10.5. Consequently Duncan's analysis gives the same result as that derived in Section 5.2.2b, although the two cases are in fact different, the earlier result being based on the assumption that the collector receives *all* the electrons transmitted by the second shutter, and that the populations of the groups are independent of τ.

The predictions of Duncan's analysis were investigated experimentally by Crompton, Hall, and Macklin,[17] who used electrical shutter drift tubes of three different lengths, 3, 6, and 10 cm, and gas pressures low enough to ensure measurable effects. Although their results were in qualitative agreement with those of the analysis, the variation of the values with pressure and drift distance was considerably greater than that predicted theoretically. In the light of subsequent evidence, Lowke's[5] suggestion that the enhanced effects were caused by a decreasing amplitude of the gating signal with increasing frequency seems likely to have been correct.

A considerably more detailed analysis of the problem was made by Lowke,[5] who took account of diffusion to both shutters in addition to the factors included in Duncan's analysis. Although differing in detail, the analysis of Lowke is similar to that given in Section 5.2.1 apart from the replacement, where appropriate, of the constant n_0 by $n'_0\tau$, as in Duncan's analysis. We summarize the conclusions from this analysis as it applies to several types of experiment.

(a) Bradbury-Nielsen Experiment

With the assumption that the number of electrons in each group admitted by the first shutter is inversely proportional to the frequency of the AC gating signal, the expression for the integrated number density across a plane normal to the z axis at time τ becomes (see equation 5.5)

$$q = \text{const}\,\tau^{-1/2}\left\{z\exp\left[-\frac{(z-W\tau)^2}{4D_L\tau}\right] + (z-2h)\left(\exp\frac{hW}{D_L}\right)\exp\left[-\frac{(z-2h-W\tau)^2}{4D_L\tau}\right]\right\},$$

$$(10.7)$$

from which it may be shown, by following the procedure of section 5.2.1 in which β was defined as $\beta = D_L/Wh$, that

$$W = \frac{h}{t_m}(1-3\beta).$$
$$(10.8)$$

If the collecting electrode is assumed to be sufficiently small that it samples the axial flux density rather than the total flux, the correction factor is increased to 5β.

Burch[18] has suggested that the analysis requires modification to take account of the fact that some of the electrons admitted to the space beyond the second shutter, and therefore collected by the receiving electrode, are lost by back diffusion to the shutter when it closes; the higher the gating frequency, that is, the smaller the open time, the fewer will be the electrons that survive.* Therefore the effect reduces the factor 3β and, according to Burch, may in some circumstances eliminate it.

Experimental results that confirm the general validity of the analysis outlined here will be given in Section 10.6.

(b) Experiments with Square Wave Pulses of Constant Duration

As discussed in the preceding section, there may be some advantage in operating the electrical shutters with rectangular pulses of constant duration rather than with a sinusoidal signal. In this case the analysis of Section 5.2.1 applies without modification, and a correction factor of the order of 5β should be applied to the simple formula.

(c) Experiments Using Variable Drift Distances

In order to overcome end effects that were attributed to field distortion in the vicinity of the electrical shutter, Phelps et al. used techniques in which two different drift distances were employed, the drift velocity being calculated from the ratio of the difference between the drift distances to the difference between the transit times. It is not difficult to see that such techniques also eliminate errors from diffusion to the shutters apart from terms in β^2, which arise through distortion of the shape of the group from the Gaussian form.[15] Thus it has already been shown (Section 5.2.2) that the position of maximum density within the group is moved forward by an amount $2D_L/W$ by back diffusion to the first shutter (or photoelectric cathode). Similarly, as a consequence of the diffusion to the shutter and the decay of the pulse, the flux is a maximum at the second shutter at a time when the position of maximum density within an isolated travelling group would fall short of the shutter by $3D_L/W$. To first order the effects of these displacements cancel when the differencing technique is used.

10.4. EXPERIMENTS BASED ON DETECTION OF PHOTONS FROM DRIFTING ELECTRON GROUPS

A novel technique that appears in principle to be capable of giving accurate results at higher values of E/N was devised by Breare and von

---

*This effect is not to be confused with the more straightforward effect of reduced sampling time with increased frequency, which is compensated for by an increase in the number of groups arriving in unit time.[5,16]

Engel.[19] Their method is based on a technique for locating the position of a travelling electron group from the UV light emitted by molecules excited by the electron swarm. The authors suggest that a principal advantage of this method is that it eliminates uncertainty in the calculation of electron transit times from the displacement currents that are recorded in the usual forms of time-of-flight experiment. However, they are not specific as to the cause of this uncertainty and its significance is not obvious. On the other hand, an obvious advantage is the possibility of using a very simple electrode structure, such as those employed in pulsed Townsend discharge techniques, to form the drift chamber. Although such a structure was not employed in the original experiments, its use would enable measurements to be made at very high values of E/N with less likelihood of extraneous electrical breakdown.

A schematic diagram of the apparatus used for the initial experiments at low pressures (~2 torr) is shown in Figure 10.8. Electrons were produced by a continuous discharge in the cavity behind S_2 and drifted into the region between S_3 and G under the action of an electric field. Holes were positioned in S_2 and S_3 in such a way as to allow electrons to drift and diffuse into S_3G but to prevent light from the discharge from entering the drift space between C and A. Electrodes S_3 and G formed two elements of a shutter similar to those used in the "four-gauze" method. A gating signal of approximately 10^{-7}-sec duration and 10^{-8}-sec rise time was applied to G to form the electron groups, which then entered the drift space through the grid C. The distance from C to the side arm containing the UV light detector was varied by sliding the whole cathode assembly $S_1 \cdots C$ along the rods R.

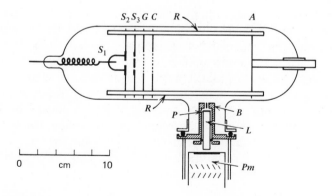

FIG. 10.8. Simplified diagram of the discharge vessel and light detector used by Breare and von Engel.[19]

The light detector consisted of a photomultiplier P_m, placed behind a collimating aperture B, and a glass rod L, which acted as a light guide. A thin layer of sodium salicylate P was deposited on the front face of the glass rod to convert the UV radiation to visible light; in this way radiation of approximately 1000 Å from electronic excitation of molecular hydrogen was converted to 4150 Å radiation suitable for detection by the photomultiplier. The diameter of the collimator was such that only a small region of the drift space, approximately 3 mm in diameter, was exposed to the photomultiplier.

Under the conditions of the low-pressure experiments it was predicted theoretically, and later confirmed experimentally, that about two photons would be recorded with the passage of each group past the detector. The trace shown in Figure 10.9a therefore represents a typical oscillogram of the output of the photomultiplier as a function of time. From the data for a large number of events, a histogram such as that shown in Figure 10.9b was obtained, representing the probability distribution of the photons generated at the line of sight of the detector as a function of delay time. Because the lifetime of the excited states is small compared with the drift time, this distribution also characterizes the spectrum of arrival times of electrons at this position. The drift velocity was then calculated by estimating from the histogram the arrival time of the centre of mass of the group.

At higher pressures a modified technique was used. A more efficient electron source was made by discharging a capacitor across a gap formed by the cathode and an electrode placed a short distance behind it, the discharge being completed in 10^{-7} sec. The number of electrons per discharge was controlled by the size of the capacitor.

The photomultiplier signals were considerably larger as a consequence of the greater number of excitation collisions per unit length in the direction of drift, as well as improvements made to the detection system. Consequently the drift velocity of each group could be determined by measuring the time interval between pulses received from two spaced detectors, since a sufficient number of photons was recorded from the transit of each group to produce a pulse representative of the histogram shown in Figure 10.9b. It was found that the time of transit of the centre of mass of a group past each detector could no longer be inferred accurately from the shape of the output pulses; consequently the time interval was taken as that between the leading edges of the pulses. The drift times were therefore falsified by diffusion, and drift velocities that were a function of the gas pressure were obtained. A typical set of results is shown in Table 10.1.

For a number of reasons the results for drift velocities and (longitudinal) diffusion coefficients in hydrogen obtained by Breare and von Engel are

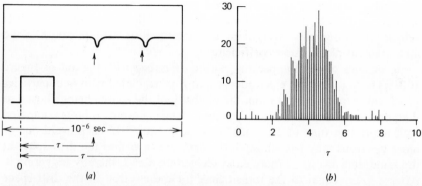

FIG. 10.9.   (a) Typical oscillogram obtained by Breare and von Engel,[19] using the apparatus shown in Figure 10.8: upper trace, photomultiplier output; lower trace, gating pulse applied to G. (b) Typical histogram showing the number of light pulses observed at a time τ after opening the shutter. Abscissa in units of 10^{-7} sec.

not as accurate as the values obtained by other methods. Nevertheless their method is included here because it is fundamentally different from the other ones described in this chapter and may well find important applications.

10.5.   METHODS DEPENDENT ON MEASUREMENT OF ARRIVAL TIME DISTRIBUTIONS

A somewhat different approach to time-of-flight measurements was taken by the group at Oak Ridge National Laboratory (ORNL) under the initial direction of G. S. Hurst. The technique developed by this group, which in its most recent and highly developed form enables measurements of both drift velocities and diffusion coefficients to be made simul-

TABLE 10.1.   The variation with pressure of values of W calculated from transit times estimated from the leading edges of the photomultiplier pulses

E/p ($V\,cm^{-1}$ $torr^{-1}$)	$W(10^6\ cm\ sec^{-1})$ at a pressure (torr) of			
	20	40	80	90
15	5.76	5.55	5.19	5.03
17	6.30	6.08	5.74	5.75
19	7.16	6.85	6.70	6.70

taneously, was based on the original suggestion of Stevenson.[20] In this section we follow the development of the method as applied to drift velocity measurements, postponing to a later section its application to the measurement of diffusion coefficients.

Stevenson's original apparatus can be described with the aid of Figure 10.10. The gap between electrodes A and B is irradiated with beta particles which cause volume ionization, the intensity of the source being chosen so that, on the average, the interval between ionizing events is large compared with the transit time of electrons between B and C. In each event one or more electrons may pass through the aperture in B into the drift space at the same time as one or more enter Geiger counter 2, thereby triggering it. After a delay equal to the transit time for the electron in the drift space BC, counter 1 is triggered. From a succession of such events, the drift velocity can be calculated as the drift distance BC divided by the average delay between the triggering of the two counters.

In its original form Stevenson's technique had neither wide application nor the capability of high accuracy. In the first place, the density of ionization produced by the beta particles was sufficiently low that it was necessary to operate the counters in the Geiger mode to obtain sufficient sensitivity to detect the small number of electrons entering them. Hence the number of gases and the range of gas pressures that could be used were restricted. Second, the use of an uncollimated source resulted in a possible lack of simultaneity of events at electrode B which could be as much as 20% of the shortest transit times.

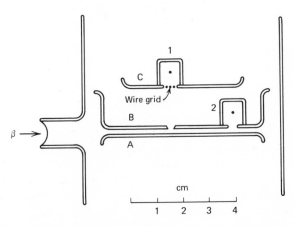

FIG. 10.10. Schematic diagram of the coincidence counting arrangement used by Stevenson[20] to measure electron drift velocities.

These difficulties were overcome by Bortner, Hurst, and Stone,[21]who replaced the beta source by a highly collimated alpha source, thus ensuring that all ionizing events took place in a plane parallel to B. Furthermore the number of electrons entering the counters at each ionizing event was then sufficient that adequate sensitivity could be obtained by operating the counters in the proportional mode in conjunction with high-gain amplifiers. The operating restrictions referred to above were therefore considerably reduced. Finally, the measuring accuracy was significantly improved by using a technique whereby the pulses from the two counters were brought into coincidence by a calibrated variable delay line. In this way it was no longer necessary to rely on a cathode ray oscilloscope display as the primary device for measuring time intervals since its sole function now was to determine the coincidence of the pulses.

Despite these major improvements there appear to be no advantages of this technique over the time-of-flight methods already described which use some kind of electrical shutter. There are, however, some disadvantages, notably the restriction of the technique to high gas pressures to obtain adequate counter efficiency, and the use of more elaborate procedures for signal detection and interval timing. The only data available from the method are contained in the original paper by Bortner et al. and are restricted to some gases and gas mixtures commonly used in ionization chambers and proportional counters. The range of values of E/N that were investigated was also restricted because of the limited range of pressures that could be used.

Although this technique was not used extensively, it is important in that it was the precursor to further experiments by the group at ORNL. For reasons that will become apparent, Hurst et al. reverted, in these later experiments, to techniques whereby single electrons were detected. Initially a counter operating in the G-M mode was used as the detector,[22, 23] as in Stevenson's original work, and as a consequence measurements were again restricted to gases and gas pressures suitable for the operation of such detectors. In later work[24] the apparatus was modified so that a particle detector could be used, thus removing both restrictions.

The theory underlying the experiments, as given by Hurst and his colleagues, is as follows. Figure 10.11 shows a schematic representation of the apparatus. A uniform distribution of electrons is established momentarily over the plane $z=0$ at $t=0$, and the electron concentration is sampled subsequently at the point P as a function of time. A uniform electric field is maintained at all times between the planes $z=0$ and $z=d$. The one-dimensional time-dependent equation that relates the electron number density $n(z,t)$ to the drift speed W and longitudinal diffusion

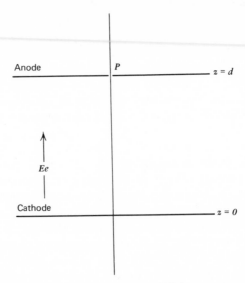

FIG. 10.11.   Diagram of the geometry of the time-of-flight apparatus used by Hurst and others.

coefficient D_L is (equation 4.30)

$$\frac{\partial n(z,t)}{\partial t} = D_L \frac{\partial^2 n(z,t)}{\partial z^2} - W \frac{\partial n(z,t)}{\partial z}.†$$

In this equation it is assumed that there are no sources or sinks of electrons and that there is neither ionization nor attachment. If in seeking a solution for this equation it is further assumed that diffusion to the electrodes at $z=0$ and $z=d$ can be ignored, it may be shown that the solution which represents the travelling group is (Section 5.2).

$$n(z,t) = N_0 (4\pi D_L t)^{-1/2} \exp\left[-\frac{(z-Wt)^2}{4D_L t} \right],$$

where N_0 is the number of electrons per unit area at the cathode at time $t=0$.

†In the original development of Hurst et al. the isotropic diffusion coefficient D was used in place of D_L since their work came before the general recognition of anisotropic diffusion. Their work, in fact, played a major part in the discovery of this phenomenon in electron diffusion and drift (see Chapter 4).

In order to determine the number of electrons $E(t)\Delta t$ arriving at the detector P between t and $t+\Delta t$ it was assumed[22] that the flux due to diffusion was negligibly small. With this assumption, the rate of arrival of electrons at P at time t is simply the instantaneous flux $Wn(d,t)$ multiplied by the area a of the sampling hole, and it follows that

$$E(t)\Delta t = N_0 aW(4\pi D_L t)^{-1/2}\left[\exp-\frac{(d-Wt)^2}{4D_L t}\right]\Delta t. \qquad (10.9)$$

It is clear that the drift velocity can be determined to a first approximation by dividing the drift distance d by the time t_m at which a maximum in $E(t)$ is recorded. The value so obtained is inexact, however, since the time at which $E(t)$ is a maximum does not coincide with the transit of the centre of mass of the group through the plane $z=d$ because of the decay of the group by diffusion. A more exact relation between W and t_m is found as follows.

The maximum value of $E(t)$ occurs when $\partial E(t)/\partial t=0$, that is, when

$$-\frac{E(t)}{2t}+\left[\frac{2W(d-Wt)}{4D_L t}+\frac{(d-Wt)^2}{4D_L t^2}\right]E(t)=0,$$

that is,

$$W^2 t_m{}^2 + 2D_L t_m - d^2 = 0,$$

or

$$d-Wt_m = \frac{2D_L t_m}{d+Wt_m}.$$

But, to first order, $d \cong Wt_m$; consequently

$$d-Wt_m \cong \frac{D_L}{W}$$

that is,

$$W \cong \frac{d}{t_m}\left(1-\frac{D_L}{Wd}\right)$$

$$= \frac{d}{t_m}(1-\beta), \qquad (10.10)$$

where $\beta = D_L/Wd = (D_L/\mu)/V$, and $V = Ed$ is the potential difference between the cathode and the anode of the drift chamber.

This result may now be compared with that of the more exact analysis of Section 5.2, where it was shown that, when account is taken of diffusion to the cathode and the anode, the relation between W and the time t_m at which $E(t)$ is a maximum is

$$W = \frac{d}{t_m}(1 - 5\beta).$$

Thus, if the experimental conditions are such that it is necessary to allow for the correction factor β in equation 10.10, which takes account only of the decay of the group (see, e.g., ref. 24), it is also necessary to allow in the theory for other diffusive effects, that is, the change in the electron density distribution caused by diffusion to cathode and anode and the flux through the hole resulting from diffusion. The foregoing analysis suggests that the correction factor is thereby increased by a factor of 5.

In addition to the correction factor already discussed, several other corrections arise for which allowance was made.[24] Since these are instrumental in origin, details of the experiment itself must first be described.

Figure 10.12 shows a schematic diagram of the apparatus developed by Wagner, Davis, and Hurst.[24] The main vacuum chamber is divided into three sections connected only by small apertures. With this arrangement and the use of high-speed pumps to evacuate the transition and detection regions, pressures from 1 to 25 torr (and in some cases up to 200 torr) could be used in the drift region while maintaining a pressure of the order

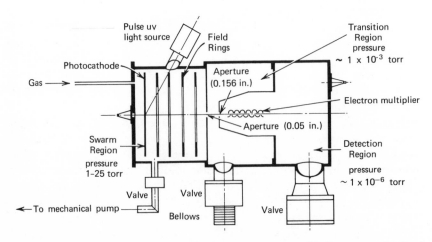

FIG. 10.12. Time-of-flight apparatus developed by Wagner et al.[24] The vacuum chamber is divided into three regions, operating in the pressure ranges shown.

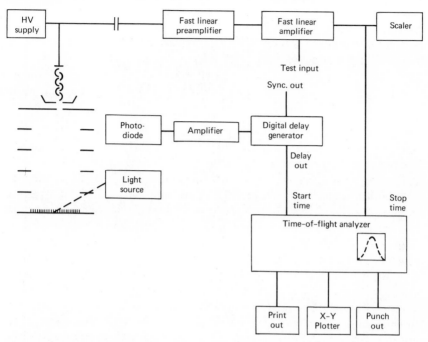

FIG. 10.13.   Block diagram of the electronic logic used in a typical time-of-flight experiment. (From ref. 24.)

of 10^{-6} torr in the detection region, thereby enabling an electron multiplier to be employed as the detector.

Large-diameter guard electrodes were used to form a drift chamber of 24.5-cm length, in which a uniform drift field could be established. Electrons were produced at the cathode by pulses of UV light incident on the cathode through a side arm terminated in a quartz window. Pulse durations of the order of 250 nsec and repetition rates of 300 pulses per second were used.

Figure 10.13 shows a block diagram of the electronic logic. The signal from a photodiode, caused by a flash from the pulsed UV light source, serves as the starting signal for the time-of-flight spectrometer, which consists of a time-to-voltage-amplitude converter and a multichannel analyzer. The stop signal is derived from an output pulse from the electron multiplier caused by an electron arriving at the collector, the delay between the arrival at the anode of the drift chamber and the output pulse being negligible compared with the drift time. Each event is recorded in the appropriate channel of the analyzer, and the experiment proceeds until

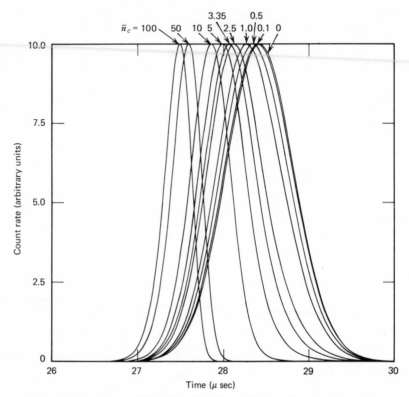

FIG. 10.14. Calculated curves illustrating the effect of Poisson distortion for various values of $\bar{n}_c$ as listed at the top of the drawing. (From ref. 23.)

adequate statistics are obtained. Since the width of the channels of the analyzer can be varied from approximately 30 nsec to 60 μsec, a wide range of experimental conditions can be used.

The use of a relation for $E(t)$ such as that given by equation 10.9 to derive the transport coefficients from the arrival time spectrum assumes (1) that the initial distribution of electrons at the cathode can be approximated by a delta function, (2) that the effect on the distribution of electronic fluctuations is negligible, and (3) that no distortion in the spectrum results from the dead-time of the detection system. In practice the effect of the finite width of the initial distribution and electronic fluctuations is simply to broaden the distribution. On the other hand, the effect of the dead-time is to displace as well as to broaden the spectrum (the so-called Poisson distortion[22]). Its influence on the determination of W must therefore be discussed.

The analyzer output records $E(t)$ accurately, provided that every electron entering the collector from every pulse is recorded. However, the dead-time of the analyzer or detector is usually long compared with the width of the arrival time spectrum so that, if the source intensity is such that two or more electrons enter the collector from a significant number of pulses, electrons in the tail of the pulse are discriminated against, since they fail to be recorded. The distribution is therefore biased towards electrons in the leading part of the pulse, and an interpretation of the spectrum that ignores this discrimination will lead to measured values of the drift velocity that are too large.

Figure 10.14 shows calculated spectra for various values of $\bar{n}_c$, where $\bar{n}_c$ is the average number of electrons entering the detector per pulse. When $\bar{n}_c \ll 1$, the probability of two or more electrons per group entering the collector becomes vanishingly small and, as expected, the distortion of the spectrum becomes negligible.

Using a procedure similar to that described in Section 11.6.1, Wagner et al.[24] calculated a correction factor for Poisson distortion as a function of the ratios C/F and $\delta t/2t_m$, where C is the total number of counts recorded from F pulses, $\delta t = |t_1 - t_m|$, and t_1 is a value of t such that $E(t_1) = E(t_m)/e$.

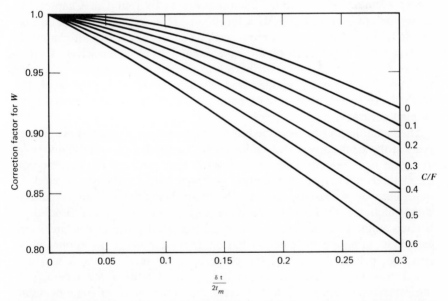

FIG. 10.15. Correction factor for W due to Poisson distortion and the use of the equation $W = d/t_m$, which neglects the decay of the electron groups. C/F is the number of electrons counted per light flash. (From ref. 24.)

The relationship between C/F and $\bar{n}_c$ is discussed in Section 11.6.1, where it is shown that $C/F \to \bar{n}_c$ as $\bar{n}_c \to 0$. The fact that the error is a function of $\delta t / 2t_m$ is not unexpected since, from the argument above, no error in W would arise from this cause if all electrons arrived simultaneously at the anode.

Figure 10.15 shows the correction factor calculated by Wagner et al., which accounts for the combined effects of the decay of the group and Poisson distortion. It may be noted that the variable $\delta t / 2t_m$ is approximately equal to $\beta^{1/2} = (D_L / Wd)^{1/2}$ (see Section 11.6.1). Since the curves were computed on the basis of equation 10.9, the correction for diffusive effects appears to be underestimated. Nevertheless the approximate magnitude of the corrections and their dependence on the experimental conditions can be seen from the figure.

10.6.   HIGH-PRECISION MEASUREMENTS USING THE BRADBURY-NIELSEN TECHNIQUE

The results that have been published from the experiments already described in this chapter, as well as from other experiments relying on the application of similar techniques, cover many gases and a wide range of values of E/N and T (see Chapter 14), but for some applications results of higher accuracy are required. Examples are to be found in Chapter 13, where the application of the data for W and D/μ to the determination of elastic and inelastic collision cross sections is described. In that chapter it is shown that inaccuracy of the experimental results themselves is often one of the largest factors contributing to the lack of uniqueness or to the inaccuracy of the cross sections, and that the maximum capability of this method can be achieved only by basing the analysis on experimental data of high accuracy.

The techniques to be described in this section were developed over some years at the University of Adelaide and the Australian National University to obtain drift velocity data of a precision that matched the accuracy of the methods of cross-section analysis that were being developed in parallel. For a number of reasons, the most important being the ease and accuracy with which the effective transit time can be measured, the Bradbury-Nielsen method was selected as having the greatest promise as a precision technique, although the method has some limitations to its applicability (see, e.g., p.306). As part of the programme of development it was necessary to study in some detail the factors that determine the accuracy of the experiment. Some of these factors are common to all time-of-flight experiments; others are peculiar to the Bradbury-Nielsen method. It was hoped that as a by-product of this work a clearer understanding would emerge of the causes of the poor agreement often found in previously published data from different sources. As was to be expected, the study led not only to a better appreciation of the importance of many experimental difficulties but also to a more searching analysis of electron transport theory, in particular, the theory as applied to time-of-flight experiments. This work was reviewed in Part 1, and its application to drift tube experiments has been described in this chapter.

In this section the general principles that govern the design of high-precision experiments of this type are discussed, and the apparatus and experimental techniques that have been developed with these considerations in mind are described.

10.6.1.   DESIGN CRITERIA. The drift velocity in a given gas is a function of the ratio E/N and the gas temperature T and is determined from a transit time measurement, using a formula of the form

$$W = \frac{h}{t_m}(1 - C\beta), \tag{10.11}$$

where $\beta = D_L / Wh$ (see Section 10.3.6). It follows that seven factors, some of them interrelated, must be considered in the design of the experiment.

1. The establishment and measurement of a uniform electric field E.
2. The determination of the gas number density N.
3. The determination of the drift distance h.
4. The measurement of an effective transit time t_m.
5. The establishment of uniform gas temperature in the drift tube and the measurement of the temperature T.
6. The production of pure gas and the maintenance of its purity.
7. The choice of operating conditions to ensure the required accuracy in the calculation of W from equation 10.11.

We now examine each of these factors in turn.

10.6.2.   THE ELECTRIC FIELD E. By considering the total transit time of an electron group moving between two specified planes, it can be shown that nonuniformity of the electric field has only a second-order effect on the measurement of the drift velocity. Consequently, unless the field is severely distorted, it is necessary to know only the average field, as calculated from the potential difference between the planes and their separation, to relate W correctly to E/N. This has been confirmed by an experiment in which one third of the electrodes of a 6-cm drift tube were connected together in order to introduce gross distortion of the field.[12] The error thus introduced was found to be unexpectedly small. However, despite this insensitivity, the possibility of significant distortion from two sources cannot be overlooked when high accuracy is to be achieved. First, appreciable distortion may be caused by the proximity of the vacuum envelope, whether it be an equipotential metal surface or a glass wall which may carry surface charge. Second, the method of mounting the shutter grids may introduce distortion, because it is not easy to accommodate the insulating formers carrying the grids in such a way that the field in their vicinity remains undisturbed.

The apparatus shown in Figure 10.16 was designed to minimize these effects. The drift chamber, 100 mm in length, was formed from from a set of thick copper annuli, of inner and outer diameter 90 and 114 mm, respectively, and thickness 16.1 mm, separated by glass spacers 0.5 mm thick. A cross section through one of the shutters is shown in Figure 10.16b. Each consisted of a hollow annulus housing a glass former, which held the wires of the grid, the thickness of the annulus and its inner diameter matching those of the guard electrodes. The grid consisted of about 200 nichrome wires of 0.08-mm diameter with their centres accurately spaced 0.4 mm apart and held in position by Pyroceram* glass frit between two plate-glass annuli. The grid assembly was so positioned within the hollow annular ring that the plane of the grid coincided with the mid-plane of the ring.

As will be shown in Chapter 11, the electric field established by applying appropriate potentials to a guard electrode structure of this form is highly uniform over a volume extending to within a short distance of the electrodes, and it is of more than adequate uniformity in the central region to which the electrons are confined. Moreover the field contains equipotential plane surfaces located at the mid-planes of the electrodes and at the planes passing through the centres of the gaps separating the electrodes. The location of the grids is such, therefore, that the field distribution within the chamber is undisturbed by them. An additional advantage of this design is that there is negligible penetration of the field external to the guard rings through the narrow gaps between them.

Two other factors affect the uniformity of the electric field, namely, the AC potentials applied to the grids and contact potential differences between various parts of the apparatus.

Field distortion in the vicinity of the grids was minimized by ensuring that the grid wires were coplanar and that the wires were closely and uniformly spaced. The first condition was met in the manufacture of the shutter by ensuring that the wires were laid down on a plane surface (either plate glass, as already described, or an accurately machined ceramic surface in the apparatus to be described subsequently) and that the bonding material did not come between the surface and the wires. This, as

*Corning Glass Company.

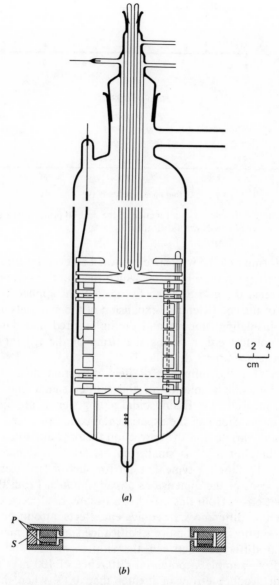

0 2 4
cm

(a)

P
S

(b)

FIG. 10.16. Electrical shutter apparatus designed for maximum uniformity of the elecric field in the drift space. (a) Section through the apparatus, (b) Section through one of guard modules containing an electrical shutter: S, cylindrical glass space; P, plate glass annuli supporting the grid wires. (From ref. 26.)

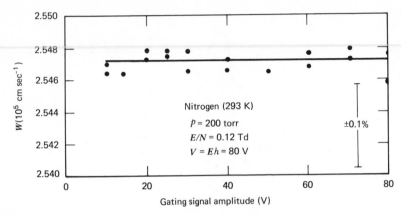

FIG. 10.17. Measured drift velocity as a function of the peak-to-peak voltage applied to the electrical shutters of a Bradbury-Nielsen drift tube.

well as the uniformity of the wire spacing, was achieved by first winding a grid of the correct pitch on a jig, after which the glass or ceramic annulus was pressed against the grid and the Pyroceram frit applied from above. Correct choice of the coefficients of expansion of the materials used for the grid wires, the insulating support, and the jig ensured that the tension of the wires was not lost either during the firing of the frit (at 450°C) or subsequently at temperatures down to 77 K.

A critical test for the influence of field distortion near the shutters consists in varying the amplitude of the shutter signal and noting the variation of t_m so produced. If sinusoidal gating signals are applied, care must be taken in the choice of the experimental conditions used for the test to ensure that any variation in the correction factor β due to change in the open time of the shutter[5,18] is small, that is, $(D_L/\mu)/V$ must be made small. Figure 10.17 shows a typical result for such a test, using a 10-cm drift tube with grids of the dimensions already quoted. From the figure it can be seen that errors from this effect can usually be disregarded.

Contact potential differences, or equivalent effects introduced by surface charges on the shutters, are of more significance since they may affect the actual potential difference between the shutter planes and hence the average field. For example, a potential difference of 100 mV, which may well occur, will produce errors of greater than 0.1% when the potential difference between the shutter planes is less than 100 V. Only contact potential differences between the two shutters are important, potential differences between the shutters and the guard electrodes having no more than a second-order effect.

Errors from contact potential differences are difficult to distinguish from

TABLE 10.2. Comparison of the results obtained in helium at 293 K with different drift tubes.

The results in the column headed CEJ were obtained with the tube designed to have negligible field distortion (Figure 10.16),[26] while those under CER were obtained with a tube having an electrode structure similar to the tube shown in Figure 10.18 but a 10 cm drift length.[29] The results under CEMcI were obtained with the apparatus shown in Figure 10.18.[27] Gas pressures were chosen to make errors from diffusive effects less than 0.1%.

	200 torr		500 torr			500 torr	
	CEJ	CER	CEJ	CER		CEJ	CEMcI
E/N (Td)	$W(10^5$ cm sec$^{-1})$				E/N (Td)	$W(10^5$ cm sec$^{-1})$	
0.607	1.054	1.055	1.054	1.053	0.0303	0.638	0.638
0.1517	1.842	1.840			0.1214	1.622	1.619
0.2124	2.215	2.213					
0.303	2.671	2.669					

those introduced by diffusive effects. For example, if the correction factor for diffusive effects is taken as $3\beta = 3(D_L/\mu)/V$, the error introduced into the measurement of W for thermal electrons at room temperature ($D_L/\mu = \kappa T/e = 0.025$ V) is of the same magnitude and has the same functional dependence on V as that resulting from a potential difference of 75 mV between the shutters, the first shutter being negative with respect to the second.

Contact potential differences were minimized by coating the shutter wires with gold by vacuum deposition, the two shutters being treated in a single operation. In some of the drift tubes the guard electrodes were also coated, although this was not essential.

The apparatus shown in Figure 10.16 was designed to be as free as possible from field distortion and to provide, therefore, experimental data against which the results from other drift tubes could be standardized. However, its construction, particularly that of the shutters, made it unsuitable for outgassing at high temperatures and for measurements at low temperatures, because of the risk of damage through differential expansion of its components.

On the other hand, it has already been shown that field distortion introduced by the electrode configuration is unlikely to be a serious factor since considerable nonuniformity of the electric field can be tolerated. An apparatus, shown in Figure 10.18,[27] was therefore designed specifically for use over a large temperature range, but with some field distortion unavoidably introduced. By checking the results from this apparatus against those from the "standard" drift tube the significance of the errors caused by the field distortion could be determined. The results in Table 10.2 show that, as was expected, the effect of the less uniform field in the second drift tube is barely significant. Further details of the construction of the apparatus are given in the original papers.

10.6.3. THE GAS NUMBER DENSITY N. Provided that there are no temperature gradients within the drift tube and that the temperature is accurately known (see Section 10.6.6), the accuracy of the determination of N depends on the accuracy with which the gas pressure can be determined. In the more recent experiments pressures were measured with quartz spiral manometers* calibrated against a primary pressure standard.[32] The resolution of the gauges and the accuracy of the pressure standard were such that pressures in excess of 5 torr could be measured with errors of less than 0.1%.

10.6.4. THE DRIFT DISTANCE h. The drift distance h must be determined more accurately than any of the other experimental parameters since both the transit time t_m and the electric field E (for a known value of V) depend linearly on it. Thus, if an error of $x\%$ is made in the determination of h, an error of $2x\%$ will result in the measurement of W at a given value of E/N if W is proportional to E/N. It is therefore essential to ensure the following:

1. That the planes of the shutter grids are well defined, that is, the wires of the grid are coplanar.
2. That the shutter planes are parallel.
3. That the electrode structure is designed to have long-term mechanical stability to withstand cycling over a large temperature range.

For these reasons, as well as for those already given, the guard electrodes of the drift tubes shown in Figures 10.16 and 10.18 were made of thick material. They were lapped and polished to a high degree of flatness and surface finish, and their dimensions were kept within close tolerances, especially in the vicinity of the glass spacers that separate them. The whole structure was held together under moderate tension by tie rods that ran the length of the apparatus, relief for differential expansion being provided by crimped washers between the end electrodes and the tie-rod nuts. The shutters themselves were held in position by spring clips, care being taken to ensure an accurate register between the insulating former and the supporting surface of the electrode upon which it was mounted.

*Texas Instruments, Precision Pressure Gauge model 145.

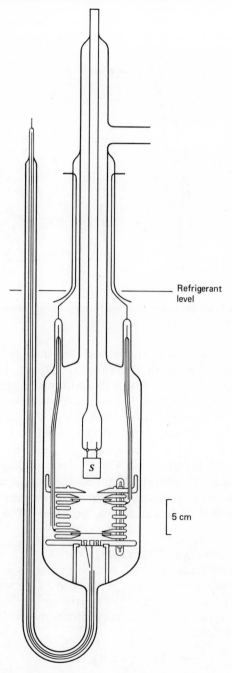

Refrigerant
level

5 cm

FIG. 10.18. Drift tube designed for low-temperature measurement. The electron source S consists of an ^{241}Am foil mounted within a hollow cylinder. (From ref. 27.)

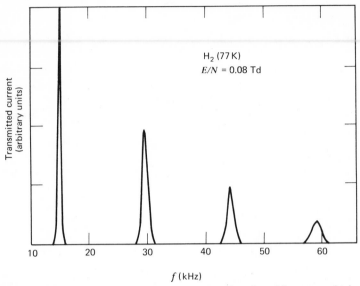

FIG. 10.19. Typical curve of transmitted current as a function of frequency obtained from a Bradbury-Nielsen drift tube.

The measurement of h is difficult because the fragility of the grids prevents the application of the more conventional methods. The technique that was employed relies on the use of a fine wire probe attached to the carriage of a vernier height gauge to locate a single wire in each shutter, contact with a wire being determined electrically. The height of each shutter plane relative to a reference plane was measured by transferring the setting of the gauge to a height micrometer.* In this way the distance between the wires could be measured with an error of less than 4 μm. When account was taken of anomalies in the location of the wires relative to a plane through the mean position, the error in the determination of h was estimated to be less than 40 μm, that is, less than 0.05% for the drift distance of 100 mm.

10.6.5. MEASUREMENT OF THE EFFECTIVE TRANSIT TIME t_m. In Bradbury-Nielsen experiments in which sine wave signals are applied to the shutters, the effective transit time t_m is equal to $1/2f_1$, where f_1 is the frequency of the signal at which the first maximum in the transmitted current is recorded. Equation 10.11, from which the drift velocity is calculated, therefore becomes

$$W = 2hf_1(1 - C\beta).$$ (10.12)

*Mitutoya Height Master.

In practice, f_1 may be determined from higher-order maxima, as well as from the first. Figure 10.19 shows a typical curve of transmitted current as a function of the frequency of the AC signal, from which it is clear that in this instance f_1 may be determined from three or four of the maxima, using the relation $f_1 = f_n/n$, where n is the order of the peak. The average value of f_1 is then used in equation 10.12. Furthermore a comparison of the values of f_1 serves as a useful overall check on the experiment.

A number of factors determine the accuracy with which f_1 can be measured. These factors are most easily discussed by reference to a specific example. By means of the procedure suggested by Duncan,[16] the following expression for the transmitted current as a function of the frequency of the AC signal can be derived from equation 10.7 (see Appendix 10.2):

$$i \cong \text{const} \cdot \sum_{m=1}^{\infty} \frac{f^{3/2}}{m^{5/2}} \exp\left[-\left(\frac{hW}{4D_L}\right)\frac{f}{mf_1}\left(1 - \frac{mf_1}{f}\right)^2 \right], \quad (10.13)$$

where $f_1 = W/2h$. In deriving this expression small terms that arise from the residual asymmetry of the travelling group as it approaches the second shutter have been neglected, the initial asymmetry being caused by back diffusion to the cathode. This approximation is equivalent to neglecting terms in β^2 in the derivation of equation 5.6.

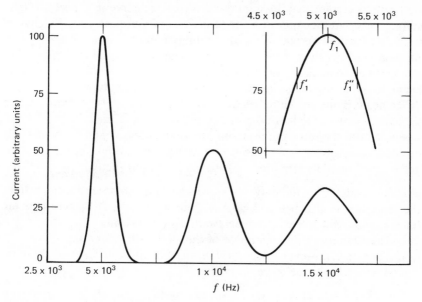

FIG. 10.20. Theoretical current-frequency curve for a Bradbury-Nielsen apparatus.

Figure 10.20 shows a theoretical curve plotted for the experimental conditions $h = 10$ cm, $W = 10^5$ cm sec^{-1}, and $D_L = 0.333 \times 10^4$ cm^2 sec^{-1}, using equation 10.13. Experimental curves taken under the same conditions may show maxima that are slightly broader than those in the figure and also a less rapid decrease in the heights of the maxima as the frequency increases. These characteristics indicate that the groups originating at the first shutter are insufficiently narrow to be well represented by equation 10.1 at small times. However, as shown in section 11.6.2, it is usually not difficult to operate the shutters with negligible open times, in which case the experimental curves closely resemble the theoretical curve.

The experimental parameters that have been chosen for this example give a value of $\beta = (D_L/\mu)/V$ of $1/300$, that is, the correction to be applied for diffusive effects is about 1% if the coefficient C is approximately 3. In the majority of high-precision experiments, [25-31] the experimental parameters are such that the values of β are considerably smaller than this value and, as a consequence, the current maxima are somewhat narrower. Nevertheless, even in the present case, the transmitted current changes rapidly with frequency at frequencies somewhat removed from the maximum, as can be seen from the enlarged portion of the first maximum. One method of determining f_1 accurately, is to find the average value of two frequencies somewhat removed from f_1, such as f'_1 and f''_1 in the figure, for which equal currents are transmitted. In this way, by making determinations of f where di/df is large, errors due to fluctuations in the transmitted current may be minimized. From the curve in Figure 10.20, it can be seen that a change in f'_1 or f''_1 of 0.1% produces about a 1% change in the transmitted current. Since the stability and resolution of the apparatus are usually such that the two currents can be set equal to within 0.5%, the average frequency can be determined to within 0.1% even in this example; it is not uncommon to obtain three such averages, taken at different positions on the maximum, agreeing to within 0.05%, with similar agreement between values obtained from higher-order maxima.[29]

Apart from the error in measuring the frequencies f'_1 and f''_1, which can be made insignificant with the use of standard counting techniques, the accuracy of the determination of f_1 depends on four factors:

1. The stability of the current source and other components that affect the stability of the transmitted current, for example, the DC power supplies for the main field, the electrometer, and associated leads.

2. The stability of the amplitude of the AC shutter signals.

3. The phase difference between the signals applied to the two shutters.

4. Asymmetry in the current maximum.

The overall stability required of the system can be seen from the example already given, that is, the transmitted current must remain constant to within about 1% over the time required to set f'_1 and f''_1 if the average is to have an error of less than 0.1%. Such stability could not always be readily obtained with thermionic electron sources,[12,25-27] despite the use of highly stabilized power supplies. The radioactive sources used in later experiments[28-31] were greatly superior in this respect.

The need to have shutter signals whose amplitudes are independent of frequency may also be understood from this example. For a given frequency the transmitted current is strongly dependent on the amplitude of the signal. The number in each electron group is approximately proportional to the open time of the first shutter, and the open time is inversely proportional to the amplitude of the signal for large amplitudes. Similarly the number of electrons transmitted by the second shutter is inversely proportional to the amplitude; consequently the transmitted current varies as $1/(\text{amplitude})^2$. It follows from our example that if the amplitude of the signal changes by 1% for a 10% change in frequency the average of f'_1 and f''_1 will be in error by 0.1%. This effect obviously becomes accentuated as the width of the maximum increases; for example, if the correction for diffusive effects is 6%, the determination of f_1 by a similar procedure (i.e., by determining f'_1 and f''_1 at 80% of maximum current), using an AC gating signal with the same characteristics, will lead to an error of about 1%. For this reason some care is required in the design and construction of the amplifier for feeding the shutters since it may be required to produce a signal of the order of 50 V rms into a capacitive load of up to 100 pF. Alternatively the amplitude of the signal must be carefully monitored and adjusted during each determination of f_1.

These difficulties can be avoided by the use of square wave gating signals since the open time of the shutter is then largely independent of amplitude. However, if a signal with a constant duty cycle is to be used in order to reduce the factor C (see Section 10.3.6) and to preserve a higher order of symmetry in the current-frequency maxima, a number of technical problems are introduced that are more formidable than producing an AC signal of adequate amplitude stability. For this reason, and also because it has been demonstrated that the end effects introduced by the use of relatively large amplitude sine wave signals are negligibly small (see Section 10.6.2), sine wave signals were preferred for these experiments.

Zero phase difference between the signals applied to the two shutters is an obvious requirement. It can be met without a great deal of difficulty, however, by a suitable choice of components in the resistance-capacitance network feeding the shutters.[12]

Finally, it is necessary to show that asymmetry of the current maxima introduces negligible error in the determination of f_1 by the averaging procedure. Inspection of equation 10.13 shows that the largest asymmetry is to be expected when the width of the maximum is greatest. Since, as has already been stated, the curve shown in Figure 10.20 represents a comparatively broad maximum, the error in the determination of f_1 arising from asymmetry may be regarded as being larger than normal. Nevertheless, even in this example $(f'_1 + f''_1)/2$, as determined at 80% of the current maximum, differs from f_1 by only about 0.15%.

Asymmetry of the maxima may occur for reasons other than diffusive effects, for example, as a result of a background current from negative ions formed by attachment. Since the background current increases with increasing shutter frequency,[12] its effect is to give a negative slope to the line passing through values of $(f'_1 + f''_1)/2$ taken at different values of transmitted current, that is, to enhance the asymmetry due to diffusion. Since the latter effect is usually negligibly small, asymmetry of the maxima is a useful indicator of the presence of negative ions.

10.6.6.   THE GAS TEMPERATURE T. The temperature of the gas affects the results in two ways. In the first place, if measurements are being made at low values of E/N, where the electrons are approaching the condition of thermal equilibrium, the energy distribution function is largely governed by the gas temperature, and hence the drift velocity depends on both E/N and T. For example, for a gas in which q_m is independent of c, equation 3.13 shows that W is proportional to $T^{-1/2}$ if E/N is held constant. However, at high values of E/N, where the electron energy is essentially independent of T,[15,27] this effect ceases to be important. Of more universal importance is the need for a uniform and accurately known temperature in order to calculate the gas number density from the pressure.

The establishment and measurement of uniform temperature in the drift tube presents few problems at room temperature but is more difficult at low temperature, where heat transfer by conduction and radiation from portions of the apparatus at room temperature must be reduced to a minimum. The apparatus shown in Figure 10.18 was designed to overcome this problem. Since the Kovar-to-glass lead-throughs can withstand immersion in liquid nitrogen, the liquid level could be maintained above the seals, thus ensuring that there was no heat transfer by the electrical leads. The bore of the tube, which served as the pump tube and access for the electron source, was made comparatively small so that only a small fraction of the area of the surface surrounding the electrode structure was at room temperature. Heat influx by radiation was thereby minimized. The replacement of thermionic electron sources by radioactive ones[28] removed the last significant source of heat, and no temperature gradients were observed within drift tubes having this form of construction.

Although it is reasonable to assume that the electrode structure and the gas eventually came into thermal equilibrium with the liquid nitrogen bath, thermocouples were attached to the electrodes adjacent to the shutters to check this assumption and to detect the presence of thermal gradients. The copper-constantan thermocouples were made of 0.15-mm-diameter wire and were calibrated before the assembly of the apparatus. In operation the emf's of the thermocouples were measured to within 1 μV, thus allowing the temperatures to be determined to within 0.1 K. These measurements confirmed that equilibrium was established and made it possible to infer the temperature of the gas in later experiments from the purity of the liquid nitrogen and the barometric pressure.

10.6.7. GAS PURITY. As is well known, no general statement can be made about the effects of impurities in the gas samples used in the experiments because the magnitude of such effects depends not only on the gas and the impurity but also on a number of experimental parameters. Thus a few parts per million of O_2 can prevent accurate measurement of W in many gases at low values of E/N and at low temperatures because of the large negative-ion background that results from the three-body reaction to form O_2^-. On the other hand, the presence of O_2 in the molecular gases at high values of E/N is much less important, while 100 ppm of H_2 or N_2 in another molecular gas would have negligible effect on the results. Water vapour can seriously affect the accuracy of the results in many gases at low values of E/N because of its very large low-energy elastic and inelastic cross sections.

The effect of small traces of the molecular gases in the heavy monatomic gases was recognized by Townsend and his collaborators. Also, it was demonstrated clearly in Bowe's[4] experiments with argon when the operation of his recirculating purification system lowered the drift velocity by as much as a factor of 3 by removing impurity from gas that was initially 99.99% Ar.

Because of the extreme sensitivity of the results to impurity in some instances and because of the static conditions in which the experiments are carried out, it is essential to use clean vacuum and gas-handling techniques in order to achieve low outgassing rates. As a general rule the construction of the drift tubes and associated vacuum systems used in the experiments described in this section was based on the technology developed for small, static UHV systems; that is, the materials employed in the construction of the drift tubes were either metal, glass, or ceramic, while no elastomers were used either in the systems themselves or in the gas-handling plants associated with them.

In the earlier experiments[12,26] where semipermanent seals were employed in the glass envelope of the drift tube itself, outgassing rates were measured before the experiments and the effects of likely impurity determined by experiment. The results so obtained showed that no significant error was introduced by the presence of impurity.

In later work[29-31] directed primarily towards the heavy monatomic gases, permanently sealed envelopes were used and a systematic study was made of the more important and difficult problem raised by the presence of minute traces of molecular impurity. The results in argon, for example, showed that a nitrogen concentration of 5 ppm could cause errors considerably larger than the other errors discussed in this section. Consequently further investigations were made of the importance of the outgassing of the apparatus and of ways in which the gas might be purified and analyzed.[33]

Tables 10.3 and 10.4 show the results of experiments in which known amounts of H_2 and N_2 were added to neon and argon that initially contained less than 2 ppm of each. In each case it can be seen that hydrogen has less effect than nitrogen and that, in the case of nitrogen, the effects become very large at higher electron energies, a fact that has been attributed to the onset of relatively large energy losses through vibrational excitation.[30,31] This may be understood by comparing the energy loss when an electron collides elastically with an argon atom with that which occurs when it vibrationally excites a nitrogen molecule; for an electron with 0.5-eV energy, for example, the average energy lost in elastic collisions is only about 1.5×10^{-5} eV, whereas 0.3 eV is required to excite the least energetic vibrational transition.

When examined in relation to the other sources of experimental error

TABLE 10.3. The change in W at 293 K resulting from the addition of known quantities of H_2 and N_2 to neon

E/N	Change in W (%)	
(Td)	20 ppm H_2	20 ppmN_2
0.03	0.4	
0.06	0.4	0.7
0.15	0.5	0.5
0.21		2.0
0.30	0.5	4.2
0.45	0.4	4.6
0.60		5.2

TABLE 10.4. The change in W at 293 K resulting from the addition of known quantities of H_2 and N_2 to argon

E/N (Td)	Change in W (%)	
	20 ppm H_2	20 ppmN_2
0.08	1.3	1.3
0.10	1.3	0.6
0.20	1.4	4.3
0.30	1.2	8.5
0.40	1.0	11.0
0.50	1.1	11.6
0.60		11.3
0.70		10.5

discussed in this section, the results of the experiments with argon demonstrate that inadequate gas purity can be the most serious source of experimental error unless adequate precautions are taken. By recording the change in drift velocity with time when a sample of gas was stored in the apparatus for a lengthy period, it was shown that contamination caused by long-term outgassing contributes negligibly to the impurity of the gas. This was the case whether or not the apparatus had been baked. It was also shown by analyzing samples of gas taken from the drift tube itself that negligible contamination results from the displacement of adsorbed gas on the surfaces of the apparatus by the newly admitted gas. Although both processes undoubtedly take place, the use of high pressures in experiments of this kind prevents either from being significant. Thus a partial pressure of impurity of 10^{-4} torr is required to give a level of contamination of 1 ppm in gas at 100 torr. These figures contrast strongly with those for other kinds of scattering experiments, where background pressures of 10^{-9} torr may be required to obtain equally pure samples of gas. It should be pointed out, however, that this requirement is rarely, if ever, necessary because of the different nature of the experiments.

Although contamination within the drift tube and its associated vacuum system is seldom a problem, it is more difficult to obtain gas of adequate purity and to introduce it into the system without contamination. For example, great care must be taken in the use of commercially available gas regulators because, to date, all of them contain some kind of elastomer material. In the experiments in question, their use was avoided by isolating the gas cylinder from the system with a UHV leak valve* that not only was capable of withstanding the maximum pressure of the cylinder but also gave adequate control of the gas flow. The procedure for obtaining gas of adequate purity depended on the gas itself and the impurity. Thus diffusion through heated palladium was used to obtain hydrogen of high purity, fractional distillation to remove traces of oxygen from CO_2, purification by passing the gas over heated titanium to eliminate molecular impurities from the monatomic gases, and so on. The reader is referred to the original papers for more detailed descriptions of the techniques.

10.6.8.   THE CORRECTION FACTOR $C\beta$. Notwithstanding the simplifications made in its development, the analysis given in Sections 5.2 and 10.3.6 is important for several reasons. First, it shows that it may be necessary in some circumstances to allow for diffusive effects in calculating W from the effective transit time t_m. Second, it draws attention to the fact that the magnitude of the correction depends on the type of time-of-flight experiment or even on the particular mode of operation of a specified type, for example, the use of sine wave or square wave gating signals in a Bradbury-Nielsen experiment. Third, the analysis shows that the correction factor is inversely proportional to N so that, where it is possible to increase the pressure without limit, the correction factor may be made as small as desired. However, the analysis represents a further oversimplification in the case of electrical shutter experiments in that it ignores such factors as the perturbation to the electron energy distribution caused by the strong electric fields in the vicinity of the shutters, the loss of electrons by back diffusion to the second shutter,[18] the effect of the finite open time of the shutters,[5, 18] and the fact that the variation in the energy distribution throughout the travelling group results in the efficiency of the shutters being dependent on the position of the group in relation to the shutter at the time it opens.[5] In view of the smallness of the correction factor in most circumstances and the complexity of the theory that would account for these additional effects, the development of such a theory has not so far been warranted. The procedure has been, therefore, to determine C experimentally from the hyperbolic dependence of $W' = h/t_m = 2hf_1$ on the pressure p in order to apply equation 10.12.

Figure 10.21 shows typical results for W' measured at different pressures. One set of data was taken with an apparatus having a drift distance h of 6 cm,[5] and the other with $h = 10$ cm. In this example the points lie on a single curve which can be fitted by equation 10.11 to within the experimental error, using a value of $C = 4.5$ for the data taken with $h = 6$ cm and $C = 6$ for the other. In general the data taken with larger values of h show

*Granville-Phillips series 203.

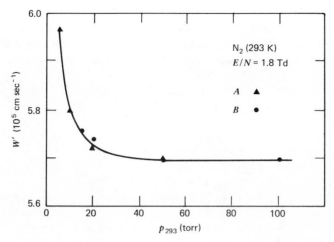

FIG. 10.21. Typical experimental results for W' as a function of gas pressure; points A were taken in an apparatus for which $h = 6$ cm, and points B with a longer apparatus ($h = 10$ cm).

less pressure dependence, as would be expected if C were in fact a true constant.[30] However, the results shown in the figure illustrate the necessity to determine C empirically, since effects other than the diffusive effects described by the theory used to derive equation 10.11 can contribute to the difference between the values of C required to fit the data in each case. Such effects include those referred to above as well as those arising from contact potential differences and other surface phenomena.

In the majority of experiments that have been performed using the techniques described in this section, the value of the correction factor is of the same order as the total random error. Even at the lowest pressures at which the measurements are made the correction factor is often no more than several times the random error. In some circumstances, particularly at low or high values of E/N, one or more factors (e.g., the onset of significant errors from contact potential differences or of electrical break-down) may limit the range of pressures that can be investigated. For a number of different reasons, therefore, the accurate determination of C is nearly always difficult. On the other hand, since the corrections to be made to the results taken at the highest pressures are very often considerably less than 0.25%, an error of 50% in the estimation of C is of little consequence. An exception is to be found in the data for argon,[31] where the corrections are of the order of 1% even for the results at the highest pressures. In this gas, therefore, errors arising from end effects contribute significantly to the total error, and considerable modification to the apparatus and techniques would be required to reduce them significantly.

TABLE 10.5. Electron drift velocities in neon at 76.8 K, obtained with a 10-cm drift tube. The best estimate values were obtained by combining these data, with the data obtained with a 5-cm drift distance (see ref. 30)

E/N (Td)	W (10^5 cm sec^{-1}) at pressure (torr) of						Best Estimate of W (10^5 cm sec^{-1})
	700	600	500	200	150	100	
1.594×10^{-3}	0.424	0.424					0.424
3.187	0.571	0.571					0.572
4.781	0.671	0.670	0.670				0.671
6.374	0.750	0.750	0.750				0.751
7.968	0.816	0.817	0.817				0.817
1.594×10^{-2}	1.060	1.061	1.060	1.063	1.063		1.062
2.390	1.234	1.234	1.234	1.237	1.239	1.239	1.235
3.984		1.497	1.498	1.500	1.503	1.504	1.500
5.578		1.704	1.705	1.708	1.712	1.711	1.707
7.171		1.881	1.883	1.886	1.888	1.889	1.885
1.195×10^{-1}				2.318	2.323	2.324	2.320
1.992				2.878	2.882	2.886	2.883
2.390				3.119	3.122	3.127	3.125
3.984					3.918	3.917	3.92
5.259						4.430	4.44

10.6.9. RESULTS. As an illustration of the application of the techniques that have been described in this section a sample of the results published for neon is given in Table 10.5. As has already been implied, errors from some of the sources will be larger in some gases and smaller in others, and it is suggested that the reader examine similar tables in the references cited to obtain an overall picture of the accuracy of the data that are available. Neon has been chosen because it is representative. Results for some of the molecular gases such as hydrogen and nitrogen can be obtained with even greater precision because diffusive effects are less important, while the data for argon and oxygen are in general less precise for a variety of reasons. Overall, however, it may be stated that the measurement of electron drift velocities has attained a level of accuracy that, with few exceptions, matches the uses to which the data are put.

10.7. MEASUREMENT OF W IN THE PRESENCE OF IONIZATION AND ATTACHMENT

Apart from the work of Blevin and Hasan[11] and Breare and von Engel,[19] the experiments described so far were undertaken to provide electron drift velocity data for situations in which the effects of ionization and attachment are insignificant. Where these processes play a significant rôle, however, additional experimental difficulties are encountered. In the case of attachment the problems arise simply through the loss of electrons between the cathode and the anode of the drift tube. On the other hand, the onset of ionization brings a number of problems, not the least of which is the design of a drift tube which can operate in this regime without electrical breakdown. Before describing how these problems have been overcome in a number of laboratories it is useful to discuss them briefly in broad outline.

As we have already seen, the parameter that determines the error in the measurement of the effective transit time t_m in time-of-flight experiments is the ratio D_L/Wh, which may be expressed alternatively as (pD_L/W) $(1/ph)$. For a given value of E/N, the ratio pD_L/W is fixed; consequently the accuracy of the experiments increases as the product ph increases. When electron attachment occurs, of an initial number n_0 of electrons that leave the cathode (or first shutter) of a drift tube, the number of electrons that survive the drift distance h is $n_0 \exp - \alpha_{at}h = n_0 \exp - (\alpha_{at}/p)ph$, where α_{at} is the attachment coefficient. The quantity α_{at}/p is independent of pressure for a two-body attachment process; it follows, therefore, that an increase in the product ph to improve the accuracy of the determination of t_m is accompanied by a reduction in the number of electrons that survive the drift distance. As this means that there are fewer electrons to detect against a larger negative-ion background, it is clear that a compromise has to be found between a satisfactory signal-to-noise ratio, on the one hand, and adequate resolution for the determination of t_m, on the other. When the attachment coefficient is high, and more particularly when the attachment process is a three-body process so that α_{at} is proportional to p^2, it may be difficult or impossible to find such a compromise.

A somewhat similar situation applies with respect to ionization. As is well known, the similarity principle can be applied to determine the condition of electrical breakdown for electrode structures of similar design but different size. To a first approximation it is found that the breakdown potential is determined by the product ph. Thus, when static potentials are applied to the drift tube, it is again necessary to compromise, that is, to find a value of ph large enough to enable t_m to be determined accurately yet small enough to ensure that electrical breakdown does not occur in the apparatus at the value of E/N under investigation. An alternative solution to the problem is to apply short-duration impulse voltages to the tube, as discussed in Section 10.7.2.

There are other difficulties at high values of E/N where ionization is important, particularly those associated with the efficiency of the shutters in an electrical shutter apparatus and with the accurate determination of the transit times, which may be considerably less than 100 nsec. For these reasons it is not to be expected that electron drift velocities can be measured with the same accuracy in this region as has been achieved with the experiments described in the preceding sections. Nevertheless the data now available are of great importance, particularly to the study of electrical breakdown in gases.

10.7.1.   ATTACHMENT BUT NO IONIZATION. The limitations to the experimental conditions already discussed are exemplified in the case of oxygen, where, despite the importance of measurements of W for electrons near thermal equilibrium, no data are to be found for $E/N < 0.6$ Td. Below $E/N \cong 3$ Td the attachment process is predominantly three-body nondissociative attachment to form O_2^- with α_{at}/N^2 varying inversely with E/N.[34] This, coupled with the fact that it is usually necessary to use higher gas number densities at lower values of E/N if experimental errors from other sources are not to increase, means that in practice α_{at} increases rapidly as E/N is reduced. Consequently the attenuation factor $e^{-\alpha_{at}h}$ increases extremely rapidly, and the situation is such that when $E/N \cong 0.5$ Td only a few per cent of the electrons can traverse the drift distance without attachment if the experimental conditions are such that t_m can be determined with reasonable accuracy. Thus we find Nielsen and Bradbury[13] writing that "the background current can be reduced sufficiently to permit accurate measurement of the electron current maxima (only) by restricting the pressure to values between 2 and 10 mm." As a result the cut-off point for their results in oxygen is $E/N \cong 1.5$ Td, compared with a lower limit of less than one fifth of this value for nonattaching gases. Doehring,[35] using a "four-gauze" method, apparently experienced similar limitations, although he was able to employ somewhat higher pressures.

Obviously, measurements at $E/N \cong 0.1$ Td, which are required to examine the transport properties of near-thermal electrons, are out of the question when conventional techniques are used. One possible way of surmounting these difficulties would be to differentiate between the electrons and ions by exciting the electrons in the collecting region of the drift tube by applying a high-amplitude radio-frequency field and detecting them through the excitation radiation so produced, as in Cavalleri's experiments (see Chapter 11).

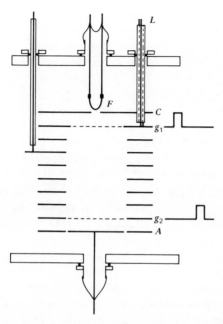

FIG. 10.22. Schematic diagram of the drift tube used by Prasad and Smeaton.[36] Overall length CA is 9 cm; drift distance g_1g_2 is 7 cm; F, platinum wire filament; g_1,g_2, shutters; A, collector; L, coaxial leads.

10.7.2. IONIZATION BUT NO ATTACHMENT.

(a) Electrical Shutter Experiments

Using a drift tube having coplanar shutters but specially designed to withstand the application of high voltages without breakdown, Prasad and Smeaton[36] succeeded in extending the measurements in nitrogen to E/N = 120 Td. The design of this drift tube enabled these authors to use gas pressures of the order of 25 torr and applied voltages up to 10 kV, and thus to eliminate inaccuracy due to diffusive processes. Their experimental results should therefore be superior to those taken with pressures of a few torr, the pressures more normally used for this range of E/N.

The success of their experiments may be attributed largely to the design of the leads to the various electrodes and the circuitry for applying fast gating signals to the shutters. Although the design of the electrode structure followed normal practice, the leads were introduced into the vacuum chamber through O-ring seals in the base plate, the lead wires being encased in alumina tubes which terminated adjacent to the appropriate electrode (see Figure 10.22). Connection to the electrodes and shutter wires was made through spring contacts. This arrangement ensured that the leads were totally enclosed by insulating material throughout their length, thus eliminating the most common cause of failure under these experimental conditions, that is, corona breakdown initiated at the small-diameter lead wires.

The shutter gating pulses were 100 nsec wide and were produced by two identical pulse generators triggered by a delay unit that could be continuously varied in the range 0 to 10 μsec. The delay time between the pulses was measured with a fast oscilloscope.

Despite the use of grids with closely spaced wires (about 0.3 mm) and square pulses of large amplitude (up to 60 V), Prasad and Smeaton experienced the same difficulty as did Blevin and Hasan in obtaining an adequate reduction in the transmission of the shutters when closed. Nevertheless they reported that the ratio of the maximum transmitted current to the background was of the order of 1.2 even under the worst conditions and that the transit time could be determined to within 2%. The agreement between their results and those of Lowke for $E/N < 60$ Td is within the combined experimental error, suggesting that the data of Prasad and Smeaton are also reliable outside the region of overlap. The use of Araldite in the construction of the shutters, a material which can be a significant source of water vapour, and of cylinder gas containing 100 ppm of unspecified impurity means that the question of gas purity cannot be entirely dismissed, although it seems likely to have contributed little to the overall error.

(b) Pulsed Discharge Experiments

The methods developed by Raether and his colleagues at the University of Hamburg have been fully reviewed by Raether[37] and will therefore be described only briefly here. They are based essentially on the technique described in Section 10.2 but with modifications that enable measurements to be made under conditions approaching those of electrical breakdown, that is, where there are large electron multiplication factors as well as a short electron transit times. The pulsed Townsend discharge technique is an obvious choice for this regime since the simplicity of the electrode structure means that the apparatus can be designed to withstand applied voltages that are limited only by the breakdown potential of the gap itself.

Two procedures for measuring drift velocities have been developed, the so-called optical and electrical methods.[37] The optical method developed by Wagner and Raether[38] is based on the use of a high-gain image converter to photograph the electron avalanche and thus to determine its position as a function of the time. The ionizing process responsible for electron multiplication and the growth of the avalanche is accompanied by

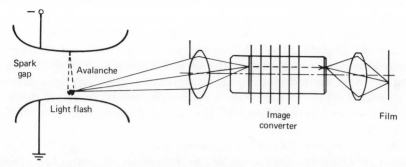

FIG. 10.23. Diagram to illustrate the principle of the optical method for measuring the transit time of electron avalanches as used by Raether and others. (From ref. 38.)

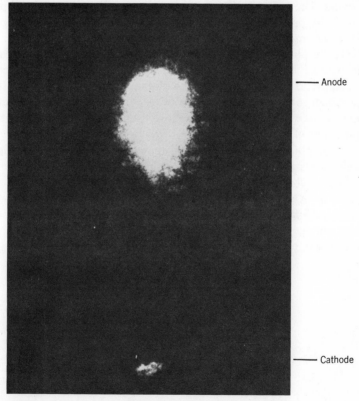

FIG. 10.24. Typical photograph of an electron avalanche recorded with the apparatus shown diagrammatically in Figure 10.23. (From ref. 38.)

a large number of excitation collisions. Since the mean lifetime of the excited states is of the order of 10^{-8} sec, the distribution of excited atoms, which can be photographed by the image converter, marks the position of the avalanche at a given time with reasonable precision provided that the transit time is not too short. In these experiments the transit times were of the order of 50 to 100 nsec. Figure 10.23 shows the experimental arrangement schematically, and Figure 10.24 a typical photograph from which the drift velocity was estimated.

An advantage of the optical method is that it can be used for experiments in which the application of the voltage to the gap would, given sufficient time, produce breakdown. In these circumstances the voltage is applied to the gap as a rectangular pulse having sufficient duration to enable the transit time to be measured but terminating before breakdown occurs. The large displacement current surge caused by the application of the pulse prevents this technique from being used in conjunction with the electrical method.

The application of the electrical method to drift velocity measurements is typified by the work of Frommhold.[39] A spark light source specially designed to produce light pulses of extremely small duration (~ 10 nsec) and high intensity[37] was used to release of the order of 10^4 electrons from the cathode. Experimental conditions were chosen to give amplifications of the order of 10^2.

A simplified circuit diagram of the apparatus is shown in Figure 10.25; full details of the circuits used for supplying steady and transient voltages to the discharge gap and for detecting the transient currents in the gap are given in Raether's book[37] and the original papers. The drift velocities of the electrons and ions are found from a study of the time dependence of the voltage drop $V_R(t)$, which is amplified and displayed on a fast oscilloscope. We give now a brief resumé of the relevant theory.

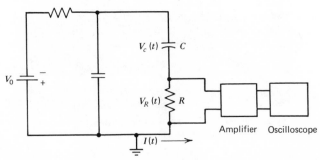

FIG. 10.25.   Circuit diagram of the electrical method for recording electron avalanches.

The current I flowing through R is the sum of the currents I_- and I_+ induced by the electrons and positive ions moving in the gap and the displacement current caused by any reduction in the voltage drop across the gap (i.e., the capacity C) due to this transient current flow, that is,

$$I = I_- + I_+ + C\frac{dV_C}{dt}.$$

The interpretation of the voltage wave form recorded by the oscilloscope would be straightforward if the displacement current could be neglected, in which case $V_R(t) = R[I_-(t) + I_+(t)]$. However, the circuit parameters cannot, in general, be chosen to satisfy this condition, and the oscillograms must therefore be interpreted in the following way.

Let us assume that the circuit elements are chosen so that (a) the total voltage drop $V_C(t) + V_R(t)$ is V_0 throughout the measurement, and (b) the time constant $RC \gg T_+$, where T_+ is the transit time of the ions formed adjacent to the anode (i.e. $T_+ = h/W_I$ where W_I is the drift velocity of the ions). The second assumption implies that the charge that flows through R for time intervals $t < T_+$ is negligible compared with the charge carried by the electrons and ions.

From the first assumption it follows that

$$\frac{dV_R}{dt} = -\frac{dV_C}{dt}$$

$$= \frac{1}{C}(I_- + I_+ - I),$$

while from the second it follows that, provided $t < T_+$, it is legitimate to neglect $\int_0^t I\,dt$ when integrating the right-hand side of the equation above. Consequently

$$V_R(t) = \frac{1}{C}\int_0^t [I_-(t) + I_+(t)]\,dt.$$

If it is reasonable to assume that there is no dispersion of the electron group by diffusion, the number of electrons in the gap at time $t \leqslant t_m = h/W$ is given by

$$N = N_0 \exp \alpha_i Wt,$$

where N_0 electrons are produced at the cathode by the initial light flash, and α_i is the primary ionization coefficient. Thus it follows from equation 10.2 that

$$I_-(t) = \frac{N_0 eW}{h} \exp \alpha_i Wt, \qquad t \leqslant t_m,$$

and that the instantaneous voltage drop $V_{R_-}(t)$ due to the electron component of the current is

$$V_{R_-}(t) = \frac{N_0 e W}{Ch} \int_0^t (\exp \alpha_i W t)\, dt$$

$$= \frac{1}{\alpha_i h} \frac{N_0 e}{C} \{ [\exp \alpha_i W t] - 1 \} \qquad \text{for} \quad t \leqslant t_m. \qquad (10.14)$$

Similarly, if it is reasonable to assume that the majority of the positive ions are formed adjacent to the anode,[37] an adequate approximate expression for the voltage drop due to the positive-ion component is

$$V_{R_+}(t) \cong \left(\frac{N_0 e}{C} \exp \alpha_i h \right) \frac{t}{T_+}, \qquad t_m < t < T_+. \qquad (10.15)$$

Now from equations 10.14 and 10.15 it can be seen that $V_{R_-}(t)$ reaches a maximum value of $(1/\alpha_i h)(N_0 e/C)\exp \alpha_i h$ at $t = t_m$ (since $1 \ll \exp \alpha_i h$), and $V_{R_+}(t)$ attains its maximum of $(N_0 e/C)\exp \alpha_i h$ at $t = T_+$. Since $t_m \ll T_+$, it follows that the contribution to $V_R(t)$ from the positive-ion component is small at $t = t_m$, even though the maximum of $V_{R_-}(t)$ is only $1/\alpha_i h$ of the maximum of $V_{R_+}(t)$. The voltage pulse therefore has the form shown in Figure 10.26, where, in accordance with our original assumption, the time constant RC of the circuit is chosen to be so large that the collapse of $V_R(t)$ due to current flow through R is negligible in the time scale of the observation.

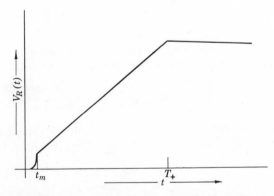

FIG. 10.26.   Voltage pulse $V_R(t)$ produced by an avalanche when $RC \gg T_+$.

From this approximate analysis and from Figure 10.26 it would appear that the transit time t_m could be determined accurately from the sharp discontinuity in $V_R(t)$ at t_m. However, in practice the discontinuity is not so well defined since $I_-(t)$ does not fall abruptly to zero at $t = t_m$, owing to the finite width of the avalanche in the field direction caused by factors such as the finite duration of the initiatory light pulse. Figure 10.27a shows typical voltage pulses from the time of the initial light flash to $t \approx t_m$, the rest of the pulses contributed by the positive-ion component not being shown. The oscillograms are usually analyzed by plotting $V_R(t)$ against t on a semilogarithmic graph, as shown in Figure 10.27b; in this way the

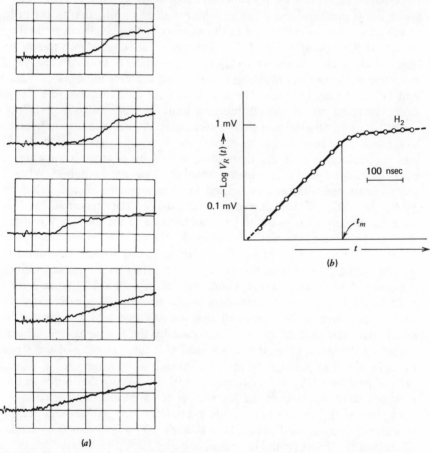

(a)

(b)

FIG. 10.27.   (a) Typical voltage pulses obtained in experiments to observe the transit times of electron avalanches. (From ref. 39.) (b) Semilogarithmic plot of a typical voltage pulse.

discontinuity corresponding to the arrival of the avalanche at the anode is more clearly seen.

Although obviously less accurate than the experiments at lower values of E/N, where the experimental difficulties are far less formidable, these techniques have supplied valuable data for a number of gases for values of E/N as high as 300 Td.

10.7.3.   SIMULTANEOUS IONIZATION, ATTACHMENT, AND DETACHMENT. When both ionization and attachment occur simultaneously, their combined effect is to enlarge the range of experimental conditions to which measurements are restricted when either process acts singly. On the one hand, the problem of electron loss by attachment is partly or fully offset by the electron multiplication process that accompanies ionization. The factor which limits measurement when attachment alone occurs, that is, the ratio of the number of negative ions to the number of free electrons, is thereby decreased. On the other hand, the presence of attachment postpones the onset of electrical breakdown to higher values of E/N because it causes a reduction in the net amplification factor. As an example the work of Naidu and Prasad[40] may be cited. These authors were able to measure electron drift velocities in dichlorodifluoromethane in the range $330 < E/N < 620$ Td, using an electrical shutter technique. An important modification to the technique used earlier by Prasad and Smeaton[36] (see Section 10.7.2) was the replacement of the thermionic source by photoemission from a back-illuminated gold film, which enabled the source to be pulsed. When a continuously operating source is used for measurements at high values of E/N, the ratio of the current transmitted at the maximum to the background current is found to be small because of the inefficiency of the shutters. In these experiments, for example, the residual transmission with the shutters permanently closed (i.e., with no gating signal) was found to be 10% of the current with the shutters open. Thus with a pulse width of the order of 100 nsec and a repetition rate of, say, 50 kHz, that is, a duty cycle of 0.5%, the current maximum would be only 5% greater than the background current if it assumed that no negative ions are formed by attachment and that all electrons admitted by the first shutter are transmitted by the second (i.e., if it is assumed that the spreading of the group by diffusion is negligible). Naidu and Prasad, in fact, reported a typical ratio of peak to background currents of 1.03. By pulsing the source so that electrons were incident on the top shutter for a period that included the open time of the shutter but slightly exceeded it, the background current was greatly reduced and, typically, a ratio of 1.30 was obtained. A similar advantage is offered by the technique used by Pack and Phelps.[15]

Finally, it is interesting to note the new problems that arise when both these phenomena are accompanied by detachment, as occurs, for example, at higher values of E/N in oxygen. The binding energy of the O^- ions formed by the dissociative attachment of O_2 is approximately 1 eV. Hence at higher values of E/N (>120 Td), where the O^- ions become sufficiently energetic, collisions of the ions with neutral molecules can result in detachment to form a free electron and either O_3 or $O_2 + O$. Since this detachment process can take place only at values of E/N at which ionization also takes place, each detached electron initiates a new electron avalanche. In pulsed discharge experiments, therefore, a new component is added to the transient currents produced by the main electron avalanche and the positive ions that accompany its formation. The relatively simple transient *current* wave form of Figure 10.28a, which corresponds to the fast transit of the avalanche and the slower transit of the positive ions, is replaced by the transient wave form of Figure 10.28b. Here the peak current caused by the electron avalanche no longer decays rapidly almost to zero for $t > t_m$ but is followed by an aftercurrent that initially may be almost as large as the peak. The aftercurrent is caused by the avalanches initiated by the detached electrons, and it persists after the main avalanche current has ceased because the detached electrons have spent part of their transit time as negative ions of greatly reduced mobility.

The theory of avalanche formation in the presence of attachment and detachment has been developed in detail (see, e.g., the work of Frommhold[41]), but it will not be reproduced here. It is sufficient to say that the presence of detachment adds a further complicating factor to the interpretation of the oscillograms obtained in this type of experiment and may lead

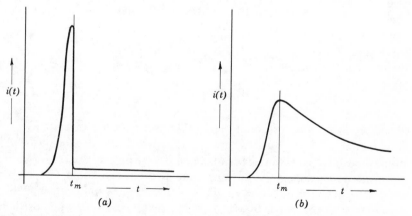

FIG. 10.28.   When electron detachment occurs, the current transient is transformed from the shape shown in (a) to that shown in (b), the tail corresponding to the transit of detached electrons.

to loss of accuracy in the determination of W because of the rounding of the main avalanche peak from which the transit time is determined.

10.8.   MEASUREMENTS OF W AT ABNORMALLY LARGE GAS NUMBER DENSITIES

In theoretical derivations of the formula for W in terms of the experimental parameters E and N and the cross sections for the collision processes between the electrons and gas molecules, it is normally assumed that all encounters between electrons and molecules are binary and that the interaction time is negligibly small compared with the free time between collisions. In terms of the mathematical formulation given in Chapter 3, this assumption implies that each encounter is described by the instantaneous transfer of a representative point from one location in velocity space to another. With this assumption the drift velocity is predicted to be independent of N provided that E/N is held constant, a result that has been confirmed in a large number of experiments when proper allowance is made for end effects such as those caused by diffusion.

The first experiments to reveal an unexplained dependence of W on gas number density which was clearly greater than experimental error were those of Lowke.[12] In the course of his experiments with nitrogen at 77 K, Lowke found that the measured values of W decreased by as much as 3% as the pressure was increased from 50 to 500 torr. End effects could be dismissed as a possible explanation for two reasons. First, the variation was approximately an order of magnitude larger than could be accounted for in terms of known effects. Second, the variation of the measured value W' was found to be accurately described by an equation of the form

$$W' = W(1 - \alpha_f N)$$ (10.16)

rather than by the equation

$$W' = W\left(1 + \frac{\alpha_e}{N}\right),$$ (10.17)

which describes the variation due to known end effects (see Section 10.6.8); that is, the measured values showed a linear dependence on N rather than the hyperbolic dependence shown in Figure 10.21.

A density dependence in nitrogen and hydrogen at room temperature was later found by Grünberg[42] in a series of measurements in which pressures as large as 31,000 torr were used. The apparatus used by Grünberg was similar to that described in Section 10.7.2b. Because the measurements were made at low values of E/N, where ionization was negligible, and because very large gas number densities were used, Grün-

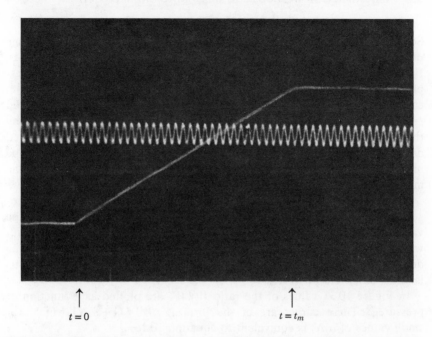

FIG. 10.29. A typical oscillogram of the cathode voltage $V_R(t)$ obtained in Grünberg's experiments.[42] A 50-MHz signal is superimposed to calibrate the time scale.

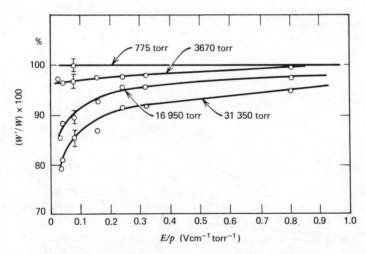

FIG. 10.30. The pressure dependence of W' in nitrogen at 293 K observed by Grünberg.[42] The values of W' at a pressure of 775 torr were taken as a reference set, and the values measured at higher pressures expressed as a percentage of these values.

berg's experiments were free of the experimental difficulties described in Section 10.7.2*b* and the values of W could be determined with the same order of accuracy as that obtainable in electrical shutter experiments. Figure 10.29 shows a typical oscillogram of $V_R(t)$ obtained at the highest pressure. The 50-MHz signal was superimposed on the record in order to calibrate the sweep speed. The absence of rounding in the trace at $t=0$ and $t=t_m$ indicates that the initial spread of the electron group produced by the 6-nsec light pulse was negligible and that negligible diffusion occurred during the transit of the group. The transit time could therefore be measured with high accuracy. Error limits of $\pm 1\%$ were placed on the majority of the measured values W'.

Figure 10.30 shows the density dependence that was observed in nitrogen at room temperature for $0.04 \leqslant E/p_{293}$ (V cm^{-1} torr^{-1}) $\leqslant 0.8$, the values recorded at a pressure of 775 torr being taken as a standard set with which the high-pressure data were compared. The curves show that the density dependence is largest at low values of E/N, where variations as large as 20% were observed.

In Figure 10.31 values of the ratio W/W' are plotted as a function of pressure.[43] These curves are of the form $W/W'=(1+\alpha_f N)$ which, for small values of $\alpha_f N$, is equivalent to equation 10.16.

Further experiments with low temperatures and moderately large number densities were carried out in hydrogen and deuterium by Cromp-

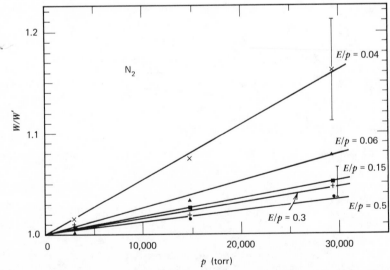

FIG. 10.31. Curves showing data similar to those plotted in Figure 10.30, replotted to demonstrate the linear dependence of W' on N. (From ref. 43.)

ton, Elford, and McIntosh[27] and in parahydrogen by Crompton and McIntosh.[28] In each case small variations of W' with N were observed when the gas pressure was varied in the range 50 to 500 torr.

The first attempt to explain these results was made by Frommhold,[44] who suggested that the lower drift velocities at high gas number densities were due to the electrons spending part of their transit time as temporary negative ions. Let us suppose that ν_I is the collision frequency for temporary negative-ion formation, and that the complex autoionizes after a mean lifetime of τ_I. It can then be shown[41,45] that the measured drift velocity W' is related to the drift velocity W at zero density by the equation

$$W' \cong \frac{W}{1 + \nu_I \tau_I}$$

$$\cong W(1 - \nu_I \tau_I) \qquad (10.18)$$

provided that $\nu_I \tau_I \ll 1$. A simple proof, which is adequate for our purpose, is as follows.

Let the total transit time of the group be t_m, of which a total time τ is spent, on the average, by each electron as a free electron. During this time the electrons drift an average distance of $W\tau$. Since the collision frequency for the capture process is ν_I, the total number of collisions which result in the formation of temporary negative ions is $\nu_I \tau$, and the total lifetime of the complexes is $\nu_I \tau \tau_I$, during which time they drift a distance $\nu_I \tau \tau_I W_I$, where W_I is the drift velocity of the negative ions. It follows that

$$\tau(1 + \nu_I \tau_I) = t_m \qquad (10.19)$$

and

$$\tau \left[1 + \left(\frac{W_I}{W} \right) \nu_I \tau_I \right] = \frac{h}{W}. \qquad (10.20)$$

Since $W_I \ll W$ and $\nu_I \tau_I \ll 1$, it follows from equation 10.20 that $\tau \cong h/W$, and hence

$$W' = \frac{h}{t_m} = \frac{h}{\tau(1 + \nu_I \tau_I)}$$

$$\cong \frac{W}{1 + \nu_I \tau_I},$$

in agreement with equation 10.18.

Although an explanation based on temporary electron capture leads to the correct form for the variation of W' with N, it is not obviously applicable in the present instance because there are no known negative-ion states in hydrogen and nitrogen with energies comparable to the mean energies of the swarms in the experiments where the density dependence is most marked. Such an explanation requires a capture process for electrons having energies of 0.1 eV or less. Frommhold therefore proposed that the formation of the temporary negative-ion states is due to a Feshbach type of resonance associated with an inelastic transition. In the case of hydrogen it was suggested that rotational excitation is the relevant process, and that the resonances therefore occur at energies somewhat below the thresholds of the rotational transitions, that is, at a few hundredths of an electron volt.

This suggestion was taken up by Crompton and Robertson,[46] who made comparative measurements at 77 K in normal hydrogen, parahydrogen,

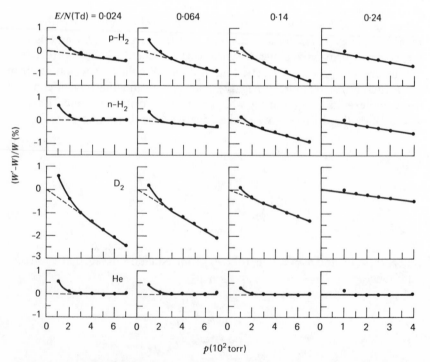

FIG. 10.32. Fractional change in drift velocity $(W' - W)/W$ in normal hydrogen, parahydrogen, deuterium, and helium at 77 K, plotted as a function of the pressure p for four values of E/N. (From ref. 46.)

deuterium, and helium. On the basis of Frommhold's hypothesis four predictions were made.

1. If resonance capture is associated with the $J = 0 \to 2$ transition only, the values of $\alpha_f = (\nu_I / N)\tau_I$ (equations 10.16 and 10.18) would be expected to be four times as large in parahydrogen as in normal hydrogen because of the different populations of the ground state in the two forms (see Table 13.2).

2. Alternatively, if a resonance is associated only with the $J = 1 \to 3$ transition, a density dependence would be expected in normal hydrogen but not in parahydrogen.

3. Because of the low rotational thresholds in deuterium, measurable pressure dependence might be expected even at the lowest value of E/N if the resonance was associated with the $J = 0 \to 2$ transition.

4. No density dependence would be expected in helium.

Figure 10.32 shows a typical example of the results that were obtained. The density dependence is clearly observable at all values of E/N in normal hydrogen, parahydrogen, and deuterium (except in normal hydrogen at $E/N = 0.024$ Td), but no density dependence is seen in the results for helium. Figure 10.33 shows α_f plotted as a function of eD/μ; the

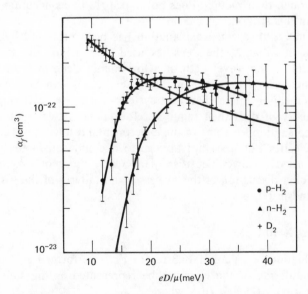

FIG. 10.33. Density-dependence coefficient α_f for normal hydrogen, parahydrogen, and deuterium at 77 K, plotted as a function of the characteristic energy eD/μ. (From ref. 46.)

difference in the variation of α_f with eD/μ in normal hydrogen and parahydrogen is compatible with Frommhold's theory provided that resonance capture is associated with both rotational transitions. The results for deuterium and helium are also in qualitative agreement with the theory.

Despite this apparently convincing support for the resonance capture theory, attempts to predict the temporary negative-ion states theoretically have met with varying success.[47,48] At the present time their existence has not been confirmed either by theory or by other experiments.

There have been a number of experimental investigations at very high pressures following Frommhold's paper, some of them providing supporting evidence for his theory and others giving results which cannot be explained on a hypothesis of this kind. Huber[49] observed marked density dependence in ethane and propane. Lehning[50] found a very large density dependence in carbon dioxide, although the effect was nonlinear. At pressures in excess of 20,000 torr the drift velocity was less than 1% of the corresponding value at 500 torr. These results were confirmed by Allen and Prew.[51] In methane, Lehning[50] found that W' *increased* with increasing N. Grünberg[52] reported further measurements in hydrogen, nitrogen, helium, and argon. In argon no density dependence was observed; in helium the density dependence was less than the experimental error for $p < 8000$ torr but was measurable for pressures above this value. De Munari, Mambriani, and Giusiano[53] observed a density dependence in xenon, although their method does not appear to be as accurate as others described in this section.

An alternative theoretical explanation has been proposed by Legler,[54] who points out that at the densities used in some of the experiments $(N \geqslant 10^{20}\ cm^{-3})$ the wavelength of a low-energy electron is of the same order as the mean free path. In this case the gas can be treated as a continuous medium rather than as a system of discrete scattering centres, and the theory of drift then requires reformulation. Legler's theory provides an explanation for some of the experimental results at extremely high number densities but does not appear to account satisfactorily for the results recorded at lower densities, particularly those of Crompton and others. A fuller discussion of the successes and failures of the two theories will be found in ref. 46.

APPENDIX 10.1

The evaluation of equation 10.3 for $t = t'_m$ is performed as follows. Let the right-hand side of the equation be represented as the sum of four terms, $i_1(t)$, $i_2(t)$, $i_3(t)$, and $i_4(t)$, where

$$i_1(t) = \frac{N_0 e W}{\pi^{1/2} h} \left[\operatorname{Erf} \frac{Wt}{(4D_L t)^{1/2}} + \operatorname{Erf} \frac{h - Wt}{(4D_L t)^{1/2}} \right],$$

$$i_2(t) = \frac{N_0 e W}{\pi^{1/2} h} \left[\operatorname{Erf} \frac{2h + Wt}{(4D_L t)^{1/2}} - \operatorname{Erf} \frac{h + Wt}{(4D_L t)^{1/2}} \right] \exp \frac{hW}{D_L},$$

$$i_3(t) = \frac{2N_0 e}{h} \left(\frac{D_L}{\pi t} \right)^{1/2} \int_{-Wt/(4D_L t)^{1/2}}^{(h - Wt)/(4D_L t)^{1/2}} u \exp(-u^2)\, du,$$

$$i_4(t) = \frac{2N_0 e}{h} \left(\frac{D_L}{\pi t} \right)^{1/2} \int_{-(2h + Wt)/(4D_L t)^{1/2}}^{-(h + Wt)/(4D_L t)^{1/2}} v \exp\left(\frac{hW}{D_L} - v^2 \right) dv.$$

Then it follows that for $t = t'_m = h/W$

$$i_1(t'_m) = \frac{N_0 e W}{\pi^{1/2} h} \operatorname{Erf} \frac{W t_m'^{1/2}}{(4D_L)^{1/2}}$$

$$= \frac{N_0 e W}{\pi^{1/2} h} \operatorname{Erf} \left(\frac{hW}{4D_L} \right)^{1/2}$$

$$\cong \frac{N_0 e W}{\pi^{1/2} h} \left[\frac{\pi^{1/2}}{2} - \left(\frac{hW}{D_L} \right)^{-1/2} \exp\left(-\frac{hW}{4D_L} \right) \right]$$

since

$$\operatorname{Erf} x = \frac{\pi^{1/2}}{2} - \frac{e^{-x^2}}{2x} \left[1 - \frac{1}{2x^2} + \frac{3}{(2x^2)^2} - \cdots \right]$$

$$\cong \frac{\pi^{1/2}}{2} - \frac{e^{-x^2}}{2x} \qquad \text{for large } x;$$

$$i_2(t'_m) = \frac{N_0 e W}{\pi^{1/2} h} \left(\exp \frac{hW}{D_L} \right) \left[\operatorname{Erf} \left(\frac{9}{4} \frac{hW}{D_L} \right)^{1/2} - \operatorname{Erf} \left(\frac{hW}{D_L} \right)^{1/2} \right]$$

$$\cong \frac{N_0 eW}{\pi^{1/2} h} \left(\exp \frac{hW}{D_L} \right) \left[\frac{\pi^{1/2}}{2} - \frac{1}{3} \left(\frac{hW}{D_L} \right)^{-1/2} \exp \left(-\frac{9}{4} \frac{hW}{D_L} \right) \right.$$

$$\left. - \frac{\pi^{1/2}}{2} + \frac{1}{2} \left(\frac{hW}{D_L} \right)^{-1/2} \exp \left(-\frac{hW}{D_L} \right) \right]$$

$$= \frac{N_0 eW}{\pi^{1/2} h} \frac{1}{2(hW/D_L)^{1/2}} \left[1 - \tfrac{2}{3} \exp \left(-\frac{5}{4} \frac{hW}{D_L} \right) \right]$$

$$\cong \frac{N_0 eW}{\pi^{1/2} h} \frac{1}{2(hW/D_L)^{1/2}} ;$$

$$i_3(t'_m) = \frac{2N_0 e}{\pi^{1/2} h} \frac{D_L^{1/2}}{t'^{1/2}_m} \int_{-(hW/4D_L)^{1/2}}^{0} u e^{-u^2} du$$

$$= - \frac{N_0 e}{\pi^{1/2} h} \frac{D_L^{1/2}}{t'^{1/2}_m} e^{-u^2} \Big|_{-(hW/4D_L)^{1/2}}^{0}$$

$$= \frac{N_0 eW}{\pi^{1/2} h} \left(\frac{hW}{D_L} \right)^{-1/2} \left[\exp \left(-\frac{hW}{4D_L} \right) - 1 \right] ;$$

$$i_4(t'_m) = \frac{2N_0 e}{\pi^{1/2} h} \frac{D_L^{1/2}}{t'^{1/2}_m} \left(\exp \frac{hW}{D_L} \right) \int_{-(9/4 \cdot hW/D_L)^{1/2}}^{-(hW/D_L)^{1/2}} v e^{-v^2} dv$$

$$= - \frac{N_0 eW}{\pi^{1/2} h} \left(\frac{hW}{D_L} \right)^{-1/2} \left(\exp \frac{hW}{D_L} \right) e^{-v^2} \Big|_{-(9/4 \cdot hW/D_L)^{1/2}}^{-(hW/D_L)^{1/2}}$$

$$= - \frac{N_0 eW}{\pi^{1/2} h} \left(\frac{hW}{D_L} \right)^{-1/2} \left[1 - \exp \left(-\frac{5}{4} \frac{hW}{D_L} \right) \right]$$

$$\cong - \frac{N_0 eW}{\pi^{1/2} h} \left(\frac{hW}{D_L} \right)^{-1/2} .$$

Consequently

$$i(t'_m) = i_1(t'_m) + i_2(t'_m) + i_3(t'_m) + i_4(t'_m)$$

$$\cong \frac{N_0 e W}{\pi^{1/2} h} \left[\frac{\pi^{1/2}}{2} - \frac{3}{2(hW/D_L)^{1/2}} \right],$$

and

$$i(t'_m) - \frac{i_0}{2} \cong - \frac{3 N_0 e W}{2\pi^{1/2} h (hW/D_L)^{1/2}}, \quad \text{since} \quad i_0 = \frac{N_0 e W}{h}.$$

APPENDIX 10.2

When an alternating gating signal of frequency f is applied to the shutters of a Bradbury-Nielsen drift tube, the first shutter releases electron groups into the drift space at time intervals of $1/2f$. A succession of groups therefore travels down the tube, each group travelling with velocity W and being separated from its neighbours by $W/2f$. Figure A10.1 shows, as an example, the distribution of number density within a drift tube, at the time

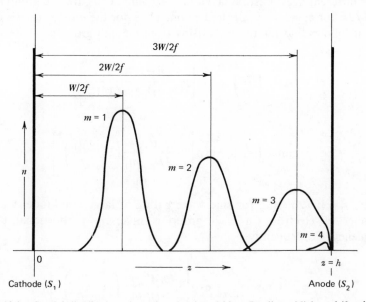

FIG. A10.1.   Spatial distribution of number density within a Bradbury-Nielsen drift tube.

at which the shutters open, when the frequency is chosen to be such that $3W/2f$ is somewhat less than h. It can be seen that the leading edge of the group denoted by $m=3$ has arrived at the second shutter S_2, and that only a small portion of the group $m=4$ remains within the drift space. The contributions from groups with $m \geqslant 5$ are negligible. Both shutters S_1 and S_2 are regarded as metal planes at which $n=0$ except during the infinitesimally short time in which they are open.

When shutter S_2 opens, the total instantaneous flux through the shutter resulting from a group that was released t sec earlier from S_1 is (see Section 5.2)

$$\phi = \text{const} \cdot t^{-3/2} \left(\frac{h^2}{D_L t} - 2 \right) \exp \left[- \frac{(h - Wt)^2}{4D_L t} \right]$$

$$\cong \text{const} \cdot t^{-5/2} \exp \left[- \frac{(h - Wt)^2}{4D_L t} \right]$$

since $2 \ll h^2/D_L t$ for most experiments. In deriving this equation it is assumed for the moment that the initial number in each group is independent of the gating frequency f.

In determining the contributions to the flux from each of the groups their different ages must be taken into account. Thus, for the group $m=1$, $t=1/2f$; for $m=2$, $t=2/2f$; and so on. Thus, for the mth group, $t=m/2f$, and it follows that the flux ϕ_m arising from the mth group is

$$\phi_m = \text{const} \cdot \left(\frac{f}{m} \right)^{5/2} \exp \left[- \frac{(h - mW/2f)^2}{4D_L(m/2f)} \right]$$

$$= \text{const} \cdot \left(\frac{f}{m} \right)^{5/2} \exp \left[- \frac{hW}{4D_L} \frac{f}{mf_1} \left(1 - \frac{mf_1}{f} \right)^2 \right],$$

where $f_1 = W/2h$. But, for sine wave gating of large amplitude, the total number of electrons in each group is inversely proportional to the frequency. Consequently

$$\phi_m = \text{const} \cdot \frac{f^{3/2}}{m^{5/2}} \exp \left[- \frac{hW}{4D_L} \frac{f}{mf_1} \left(1 - \frac{mf_1}{f} \right)^2 \right],$$

and the total flux ϕ is given by

$$\phi = \text{const} \cdot \sum_{m=1}^{\infty} \frac{f^{3/2}}{m^{5/2}} \exp\left[-\frac{hW}{4D_L} \frac{f}{mf_1}\left(1 - \frac{mf_1}{f}\right)^2 \right].$$

Since the decreasing open time of the second shutter as the frequency increases is exactly compensated for by the increase in the number of pulses in unit time transmitted to the collecting electrode, it follows that the transmitted current i is directly proportional to ϕ. Equation 10.13 then follows immediately.

It may be noted that, in most cases of practical interest, the contributions to the nth-order maximum of the current-frequency curve from terms in the series other than the term $m = n$ are negligible, provided that n is small.

REFERENCES

1. L. B. Loeb, *Basic Processes of Gaseous Electronics*, University of California Press, Berkeley, 1955.

2. E. W. McDaniel, *Collision Phenomena in Ionized Gases*, John Wiley & Sons, New York, 1964.

3. J. A. Hornbeck, *Phys. Rev.*, **73**, 570, 1948.

4. J. C. Bowe, *Phys. Rev.*, **117**, 1411, 1960; *Bull. Am. Phys. Soc.*, **12**, 232, 1967.

5. J. J. Lowke, *Aust. J. Phys.*, **15**, 39, 1962.

6. See, for example, J. D. Cobine, *Gaseous Conductors*, McGraw-Hill, New York, 1941.

7. J. H. Parker and J. J. Lowke, *Phys. Rev.*, **181**, 290, 1969; J. J. Lowke and J. H. Parker, *Phys. Rev.*, **181**, 302, 1969.

8. R. J. van de Graaff, *Phil. Mag.*, **6**, 210, 1928.

9. A. M. Tyndall, L. H. Starr, and C. F. Powell, *Proc. Roy. Soc. A*, **121**, 172, 1928.

10. A. M. Tyndall and C. F. Powell, *Proc. Roy. Soc. A*, **134**, 125, 1931.

11. H. A. Blevin and N. Z. Hasan, *Aust. J. Phys.*, **20**, 735, 1967; **20**, 741, 1967.

12. J. J. Lowke, *Aust. J. Phys.*, **16**, 115, 1963.

13. N. E. Bradbury and R. A. Nielsen, *Phys. Rev.*, **49**, 388, 1936; R. A. Nielsen, *Phys. Rev.*, **50**, 950, 1936; R. A. Nielsen and N. E. Bradbury, *Phys. Rev.*, **51**, 69, 1937.

14. A. M. Cravath, *Phys. Rev.*, **33**, 605, 1929.

15. A. V. Phelps, J. L. Pack, and L. S. Frost, *Phys. Rev.*, **117**, 470, 1960; J. L. Pack and A. V. Phelps, *Phys. Rev.*, **121**, 798, 1961.

16. R. A. Duncan, *Aust. J. Phys.*, **10**, 54, 1957.

17. R. W. Crompton, B. I. H. Hall, and W. C. Macklin, *Aust. J. Phys.*, **10**, 366, 1957.

18. D. S. Burch, personal communication.

19. J. N. Breare and A. von Engel, *Proc. Roy. Soc. A*, **282**, 390, 1964.

20. A. Stevenson, *Rev. Sci. Instrum.*, **23**, 93, 1952.

21. T. E. Bortner, G. S. Hurst, and W. G. Stone, *Rev. Sci. Instrum.*, **28**, 103, 1957.

22. G. S. Hurst, L. B. O'Kelly, E. B. Wagner, and J. A. Stockdale, *J. Chem. Phys.*, **39**, 1341, 1963.

23. G. S. Hurst and J. E. Parks, *J. Chem. Phys.*, **45**, 282, 1966.

24. E. B. Wagner, F. J. Davis, and G. S. Hurst, *J. Chem. Phys.*, **47**, 3138, 1967.

25. M. T. Elford, *Aust. J. Phys.*, **19**, 629, 1966.

26. R. W. Crompton, M. T. Elford, and R. L. Jory, *Aust. J. Phys.*, **20**, 369, 1967.

27. R. W. Crompton, M. T. Elford, and A. I. McIntosh, *Aust. J. Phys.*, **21**, 43, 1968.

28. R. W. Crompton and A. I. McIntosh, *Aust. J. Phys.*, **21**, 637, 1968.

29. R. W. Crompton, M. T. Elford, and A. G. Robertson, *Aust. J. Phys.*, **23**, 667, 1970.

30. A. G. Robertson, *J. Phys. B: At. Mol. Phys.*, **5**, 648, 1972.

31. A. G. Robertson, Ph.D thesis, Australian National University, 1970.

32. J. Gascoigne, *Vacuum*, **21**, 21, 1971.

33. M. T. Elford and H. B. Milloy, *J. Vac. Sci. Technol.*, **9**, 1084, 1972.

34. L. M. Chanin, A. V. Phelps, and M. A. Biondi, *Phys. Rev.*, **128**, 219, 1962.

35. A. Doehring, *Z. Naturforsch.*, **7a**, 253, 1952.

36. A. N. Prasad and G. P. Smeaton, *Brit. J. Appl. Phys.*, **18**, 371, 1967.

37. H. Raether, *Electron Avalanches and Breakdown in Gases*, Butterworths, London, 1964.

38. K. H. Wagner and H. Raether, *Z. Phys.*, **170**, 540, 1962.

39. L. Frommhold, *Z. Phys.*, **160**, 554, 1960.

40. M. S. Naidu and A. N. Prasad, *Brit. J. Appl. Phys. (J. Phys. D)*, **2**, 1431, 1969.

41. L. Frommhold, *Fortschr. Phys.*, **12**, 597, 1964.

42. R. Grünberg, *Z. Phys.*, **204**, 12, 1967.

43. R. Grünberg, *Z. Naturforsch.*, **23a**, 1994, 1968.

44. L. Frommhold, *Phys. Rev.*, **172**, 118, 1968.

45. R. H. Ritchie and J. E. Turner, *Z. Phys.*, **200**, 259, 1967.

46. R. W. Crompton and A. G. Robertson, *Bull. Am. Phys. Soc.*, **14**, 259, 1969; *Aust. J. Phys.*, **24**, 543, 1971.

47. D. J. Kouri, W. N. Sams, and L. Frommhold, *Phys. Rev.*, **184**, 252, 1969, and references therein.

48. R. J. W. Henry and N. F. Lane, *Phys. Rev.*, **183**, 221, 1969.

49. B. Huber, *Z. Naturforsch.*, **23a**, 1228, 1968; **24a**, 578, 1969.

50. H. Lehning, *Phys. Lett.*, **28A**, 103, 1968; **29A**, 719, 1969.

51. N. L. Allen and B. A. Prew, *J. Phys. B: At. Mol. Phys.*, **3**, 1113, 1970.

52. R. Grünberg, *Z. Naturforsch.*, **24a**, 1838, 1969.

53. G. N. de Munari, G. Mambriani, and F. Giusiano, *Lett. Nuovo Cimento*, **3**, 849, 1970.

54. W. Legler, *Phys. Lett.*, **31A**, 129, 1970.

11

EXPERIMENTAL METHODS

FOR DETERMINING

ELECTRON DIFFUSION COEFFICIENTS

11.1. INTRODUCTION

In Chapter 10 we surveyed the techniques that have been used for the determination of electron drift velocities, restricting detailed discussion to methods which have led to the most accurate measurements. In this chapter we undertake a similar discussion in relation to the measurement of electron diffusion coefficients. Although there are fewer experimental techniques to be described, the interpretation of the experiments is less straightforward and has been the subject of considerable controversy. As a consequence we have found it necessary to devote rather more space to these measurements than to those described in the preceding chapter.

Until recently only one basic method, based on the measurement of the lateral diffusion of an electron stream, had been developed,[1] although there have been significant variations in the application of this technique in attempts either to improve the precision of the measurements or to adapt it to different situations, as, for example, to diffusion in the presence of ionization and attachment.[2] In the last 5 years, however, this situation has changed with the introduction of two new methods, the first based on measurements of arrival time distributions in time-of-flight experiments[3] and the second on the decay of electron number density within a diffusion cell of known dimensions.[4]

The results obtained with the first of the new methods caused confusion initially because they differed greatly from those yielded by the older method. However, it was soon realized that these differences arose from the anisotropic nature of diffusion in an electric field, and the significance of this phenomenon in a number of electron diffusion experiments then became apparent. It is now recognized that the study of arrival time distributions yields data for a new transport coefficient, the longitudinal diffusion coefficient D_L, rather than confirmatory or supplementary data for the lateral diffusion coefficient D.

The chief advantage of the second of the newly developed methods, whereby D is measured directly rather than determined through measurements of the ratio D/W as in the original method, lies in the ease with which it can be applied to otherwise unfavourable experimental situations, that is, to the measurement of D when ionization and attachment accompany diffusion. A further advantage of the method is that only one diffusion process controls the distribution of number density within the apparatus.

Doubts have sometimes been expressed about the validity of the results derived from the original lateral diffusion method because of the influence of the boundaries on the distribution of electron number density within the diffusion chamber. This doubt is in fact largely unjustified because it has been shown that, for the experimental conditions that have been used for the majority of the measurements, the results are scarcely affected by phenomena that occur near the electrodes. Nevertheless, reinforced by the sometimes poor agreement between the results from different experimental groups, the doubt was one of the factors responsible for attempts to measure D by other methods. The discovery of anisotropic diffusion, which was a direct consequence of the search for alternative methods, resulted in further doubt being cast on the results of lateral diffusion experiments since it then became apparent that the analysis of all experiments up to that time had been based on the incorrect assumption of isotropic diffusion.

In order to clarify the situation, we commence this chapter by summarizing the history of the theoretical background of these experiments, describing the developments that have taken place and pointing out their significance in relation to the interpretation of the experiments. Our aim in Section 11.2 is, therefore, not only to trace the history but also to explain why so much of the earlier data do not require reanalysis despite the increasing sophistication of the theory.

In later sections we discuss the work of several groups who have applied the lateral diffusion technique to the study of diffusion in both the presence and the absence of the additional processes of electron attachment and ionization. Finally, we describe the method of measuring D directly by electron density sampling and the determination of D_L by time-of-flight techniques.

11.2.   A SUMMARY OF THE PROGRESS TOWARDS AN ADEQUATE THEORY OF THE TOWNSEND-HUXLEY LATERAL DIFFUSION EXPERIMENT

The lateral diffusion experiment for measuring the ratio D/W was devised by Townsend[5] in 1908 and applied by him in the first instance to the measurement of this ratio for positive ions. A typical form of the apparatus for studying electron diffusion is shown in Figure 11.1.

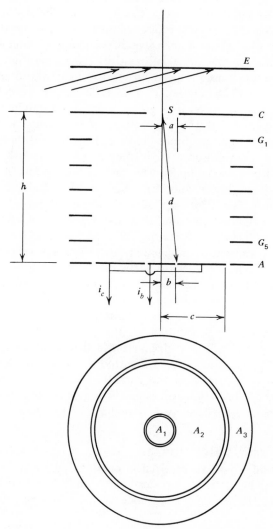

FIG. 11.1.   Diagram to illustrate the geometry of a typical lateral diffusion apparatus.

Electrons, released photoelectrically from the metal plate E, drift and diffuse towards the anode A in a uniform electric field established throughout the cylindrical volume of the apparatus by the guard electrodes $G_1, G_2, \ldots$, to which the appropriate potentials are applied. The cathode C, which forms one boundary of the diffusion chamber, contains a central hole S of radius a which serves as the electron source. It is supposed that the initial conditions at the cathode are well defined, that is, that the electrons enter the diffusion chamber having an energy distribution which is the same as that applying between C and A, that the distribution of number density over the source aperture is known, and that $n=0$ over the cathode plane for all r greater than a. The first condition is met as closely as possible by ensuring that the electrons drift for an adequate distance before arriving at the source hole in an electric field whose strength is equal to that of the field between C and A. When the source hole is reasonably large, that is, when a is greater than a few millimetres, it is customary to assume that the electron number density is constant everywhere within the aperture. Since the area of the photocathode E from which the electrons are released is large compared with the area of S, this assumption would be reasonable if it were not for the the loss of electrons to the edges of the aperture and the radial density gradients that are thereby established. The final supposition, that $n=0$ for $r>a$ at the cathode, rests on the assumption that the reflection coefficient for electrons at the metal surfaces is zero.

The electron stream whose initial radial distribution is assumed to be defined in this way proceeds towards the anode A, which forms the other boundary of the diffusion chamber. The anode is divided with axial symmetry into the segments A_1, A_2, and A_3 as shown in the figure. The distance h and the radii b and c are accurately known. The experiment consists in determining the distribution of current at the anode by measuring the currents i_b and i_c for a given electric field strength E and gas number density N. The ratio W/D is then found through the appropriate theory relating it to the ratio of the currents and the geometry of the apparatus. We now follow, as nearly as possible in chronological order, the theories which have been developed to analyze the results of this experiment.

11.2.1. TOWNSEND'S SOLUTION. Townsend's original solution[6] was based on the assumption that the electron number density $n(x,y,z)$ within the diffusion chamber satisfies the steady-state differential equation that applies when the electric field is parallel to the z axis and diffusion is isotropic, that is,

$$D\nabla^2 n = W\frac{\partial n}{\partial z}. \tag{11.1}$$

The assumption of isotropic diffusion was, of course, to remain unchallenged for almost 60 years. For the boundary conditions necessary to solve this equation Townsend assumed the cathode boundary condition already described and the further condition that n vanishes at a cylindrical surface of diameter equal to the inner diameter c (see Figure 11.1) of the guard electrodes. Because an insignificant number of electrons reach the radial boundary under the majority of experimental conditions, negligible error is introduced by this simplification of a somewhat complex situation. With these boundary conditions the solution of equation 11.1, expressed in cylindrical coordinates (ρ, z, ϕ), takes the form

$$n(\rho, z) = \sum_k \frac{2n_0 a}{c^2} \left[\frac{J_1(\xi_k a) J_0(\xi_k \rho)}{\xi_k J_1^2(\xi_k c)} e^{-\theta_k z} \right], \qquad (11.2)$$

where J_0 and J_1 are Bessel functions of the first kind, $\xi_k = x_k / c$ [x_k being the kth root of $J_0(x) = 0$], and $\theta_k = (\lambda^2 + \xi_k^2)^{1/2} - \lambda$. As defined previously, $\lambda = W/2D$.

In order to calculate the ratio of the currents received by the portions of the divided anode Townsend made two assumptions:

1. The number density within the stream is unaffected by the presence of the anode.
2. The current di to an element dS of the anode arises solely from drift, that is, $di = di_W = eWn \, dS$, the diffusive current $di_D = -eD(\partial n / \partial z) dS$ being negligibly small by comparison.

The fraction of the total current collected by the central disk, of radius b (see Figure 11.1), is then given by

$$R = \frac{i_b}{i_{\text{total}}} = \frac{2\pi eW \int_0^b n\rho \, d\rho}{2\pi eW \int_0^c n\rho \, d\rho}$$

$$= \frac{b \sum_k J_1(\xi_k a) J_1(\xi_k b) e^{-\theta_k z} / x_k^2 J_1^2(\xi_k c)}{c \sum_k J_1(\xi_k a) e^{-\theta_k z} / x_k^2 J_1(\xi_k c)}. \qquad (11.3)$$

Because the series converges slowly, considerable computation is involved in calculating R from equation 11.3.

For the sake of subsequent comparison it is interesting to note the formula which is obtained for R when the radius of the source aperture a

approaches zero, that is, the source becomes a point source, and $c \to \infty$. It can then be shown by an alternative analysis that the formula for R becomes (see Section 5.3)

$$R = 1 - \left(\frac{h}{d} \right) \exp - \lambda (d - h), \qquad (11.4)$$

where $d = (h^2 + b^2)^{1/2}$.

Although applying to a somewhat different form of the apparatus, in which the circular source aperture was replaced by a slit and the anode was divided into strips parallel to the slit source, the solution developed by Pidduck[7] is worthy of note because it appears to have been the first to make use of the boundary condition $n = 0$ at the anode as well as the other boundary conditions. Having solved equation 11.1 for the correct boundary conditions, Pidduck computed the current to an element dS of the anode by considering the diffusive component only, that is, $di_D = - eD(\partial n / \partial z) dS$. A step towards a more accurate treatment of the problem had been taken earlier by Mackie,[8] who included the diffusive term as well as the drift term in calculating the current across a geometrical plane representing the anode.

A simple formula for the two-dimensional case of diffusion from a narrow slit was developed by Townsend and Tizard[9] and used subsequently by Bailey[10] as the basis for the theory of his method of studying electron attachment. It is based on the assumption that longitudinal diffusion can be ignored in determining the lateral spread of the stream. We will comment further on this approximation in Section 11.2.8, where it will be applied to a stream that originates from a point source.

In spite of the shortcomings of most of the foregoing analyses, the majority of the results derived from experiments using one or another of them was not significantly in error as a consequence.[11] This is true mainly because the experimental conditions were chosen to ensure that the lateral spread of the stream at the anode was little affected by the process of longitudinal diffusion in free space and by diffusion to the cathode and anode boundaries. It follows that, since longitudinal diffusive effects were small in most of the experiments, the results were scarcely affected by the alternative assumptions used to derive them, these assumptions being that longitudinal diffusion can be ignored, or that the longitudinal diffusion coefficient is the same as that for lateral diffusion.

11.2.2. HUXLEY'S SOLUTION. As we have seen, formulae relating the current ratio R to $\lambda = W/2D$ that were developed before 1940 all suffered from one or more of the following disadvantages:

1. They were based on the assumption that n was constant over the

finite-sized source aperture, an assumption of questionable validity in many circumstances.

2. The calculation of R was cumbersome because the formulae required the evaluation of many terms in slowly converging infinite series.

3. The boundary condition at the anode was not satisfied.

Huxley[12] now proposed a solution that overcame the first two of these difficulties and sought to overcome the third. However, in developing it he made an error that, as we shall see, had remarkable consequences. The essence of Huxley's treatment lay in the reduction of the diameter of the source aperture to the degree where it could be regarded as a point source. He also took advantage of the fact that the gas number densities and electric field strengths can usually be adjusted without difficulty to ensure that vanishingly small numbers of electrons reach the guard electrodes even in an apparatus of modest dimensions. He was therefore able to assume with negligible error that the radius of the diffusion chamber was infinite. Following Pidduck,[7] Huxley assumed that the influence of the cathode and anode boundaries could be simulated by postulating that $n = 0$ at the metal surfaces and that the perturbation to the steady-state energy distribution caused by diffusion to the boundaries could be neglected. Implicit in his solution is the further assumption that the energy distribution is spatially invariant. (See also Section 11.2.5. These assumptions are also made in the analyses described in Sections 11.2.3 and 11.2.4.)

Assuming the boundary condition described above and that the aperture can be regarded as a point source, Huxley developed a solution of equation 11.1 which consisted of the contributions from an infinite number of dipole-like terms of the form

$$n_k(\rho, z) = -A(\exp \lambda z)\left[\frac{\partial}{\partial z}\left(\frac{\exp - \lambda r_k}{r_k}\right)\right], \qquad k = 0, \pm 1, \pm 2, \ldots, \quad (11.5)$$

where $r_k = [\rho^2 + (2kh - z)^2]^{1/2}$ is the distance of the point P from the kth dipole (see Figure 11.2). The first dipole ($k = 0$), situated at the origin and representing the source itself, satisfies the condition that n is zero over the cathode, except at the origin; the second dipole ($k = 1$), placed at $z = 2h$, neutralizes the number density (but not the flux) at the anode from the source dipole, thus satisfying the boundary condition $n = 0$ at this electrode also; the third dipole ($k = -1$) placed at $z = -2h$, neutralizes the small perturbation to n at the cathode caused by the second dipole; and so on. In practice, apart from exceptional experimental circumstances, contributions from terms other than those corresponding to $k = 0$ and 1 are negligible and will be disregarded in what follows. The number density within the diffusion chamber is therefore represented by

$$n(\rho,z) = -A(\exp\lambda z)\left[\frac{\partial}{\partial z}\left(\frac{\exp-\lambda r_0}{r_0}\right) + \frac{\partial}{\partial z}\left(\frac{\exp-\lambda r_1}{r_1}\right)\right] \quad (11.6)$$

and the current to the anode now arises simply from the diffusive flux since the flux due to drift is zero. The formula corresponding to the limiting conditions used to derive equation 11.4, that is, $a \to 0$, $c \to \infty$, is then

$$R = 1 - \left(\frac{h}{d} - \frac{1}{\lambda h} + \frac{h}{d^2\lambda}\right)\left(\frac{h}{d}\right)\exp-\lambda(d-h). \quad (11.7)$$

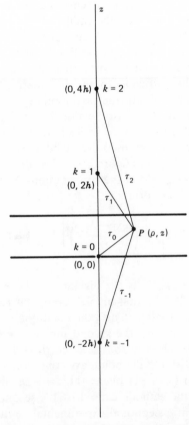

FIG. 11.2. Diagram to illustrate the disposition of the dipole images required to satisfy the condition $n=0$ at the cathode and anode.

Unfortunately an algebraic error was made in the original derivation of the ratio formula based on equation 11.6, which resulted in the following formula being obtained:

$$R = 1 - \left(\frac{h}{d}\right) \exp - \lambda(d - h). \tag{11.8}$$

This formula was used for the analysis of an extensive series of experiments,[13] including some that were made using conditions for which the anode current distribution was significantly affected by longitudinal diffusion. So successful was the formula that no error in it was suspected. With the discovery of the error, therefore, came the disconcerting result that the apparently soundly based formula of equation 11.7 did not lead to a satisfactory interpretation of the experiments, whereas the "incorrect" formula of equation 11.8 did.[14]

The success of the so-called empirical or Huxley formula led to attempts to find some justification for it. It was first shown that the formula results from the use of a pole-like source term and an image to make $n = 0$ over the anode. The formula for n is then

$$n(\rho, z) = A(\exp \lambda z)\left(\frac{\exp - \lambda r_0}{r_0} - \frac{\exp - \lambda r_1}{r_1}\right), \tag{11.9}$$

where, as already defined, r_0 and r_1 are the distances of the source and its image from P (see Figure 11.2). However, although the assumed anode boundary condition was met, the cathode boundary condition was not. Since the reflection coefficient for low-energy electrons from metal surfaces is known to be high in some circumstances, the suggestion of a breakdown of the electrode boundary conditions was not unreasonable, although a breakdown at the cathode and not at the anode was entirely speculative. In this context it is worth noting that the same formula can be obtained by satisfying the boundary condition at the cathode but not that at the anode (see equation 11.4).

A rigorous test of the empirical formula was subsequently made,[15] in which results obtained in the region of significant disagreement between equations 11.7 and 11.8 were tested against results taken under conditions where the two ratio formulae converged, that is, using a diffusion chamber for which h/d was approximately unity and large values of λh. The second condition corresponds to large values of the ratio $V/(D/\mu)$, where V is the voltage between cathode and anode. Figure 11.3, in which the discrepancy between the values of λ calculated from the two formulae is plotted as a function of R over the useful range of this experimental variable, shows that measurements made with $b = 0.5$ and $h = 10$ cm ($h/d = 0.999$) yield

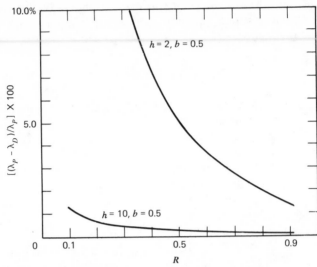

FIG. 11.3.   Curves showing the discrepancy (expressed as a percentage) between the values of λ_P calculated from equation 11.8 and the values of λ_D calculated from equation 11.7 for the same values of R and different geometries. (From ref. 15.)

values of λ that differ by less than 1% when analyzed using the two formulae, provided that $0.2 < R < 0.9$; for $R > 0.3$ the error is usually less than other experimental errors. This result is indicative of the small influence of longitudinal diffusive effects (see also Section 11.2.7) since, in terms of the derivations of equations 11.7 and 11.8, it means that the spread of the stream at the anode is almost the same whether or not the cathode boundary condition is satisfied. On the other hand, when $b = 0.5$ and $h = 2$ cm, the two formulae yield significantly different results, as can be seen from the upper curve in the figure, showing that longitudinal diffusive effects are now important.

Table 11.1 shows some of the results of a set of measurements in hydrogen with $b = 0.5$ and $h = 2.0$ cm.[15] In each case the upper entry is the value calculated by using equation 11.7, the lower being calculated with equation 11.8. The "standard" values, shown for comparison, were determined from measurements with $h = 10$ cm. As has already been stated, these values are insensitive to the choice of ratio formula (see equation 11.7 or 11.8) used to derive them from the current ratios. As will be shown subsequently (Section 11.2.7), these values would also have been obtained if a theory based on anisotropic diffusion (or, in fact, negligible longitudinal diffusion) had been used to derive the current ratio formula since at this distance from the source the lateral spread is little affected by

TABLE 11.1. The results of experiments in hydrogen using the alternative solutions (upper entries from equation 11.7, lower entries from equation 11.8)

E/N (Td)	Pressure (torr)			"Standard" Value
	5	10	20	
0.909		3.70	3.61	3.56
		3.57	3.56	
1.212		4.58	4.52	4.45
		4.45	4.45	
1.818	6.62	6.39	6.31	6.22
	6.23	6.21	6.22	
3.03	9.87	9.51		9.32
	9.34	9.29		
6.06	15.55	15.17		14.95
	14.93	14.86		

longitudinal diffusion. It can be seen that the superiority of equation 11.8 to analyze the data from these measurements in hydrogen was established beyond doubt.

This surprising result prompted further examination of the theory,* in particular a detailed analysis of the problem by Hurst and Liley.[18] Although not strictly the next in chronological order, their work will now be summarized.

11.2.3.   THE SOLUTION OF HURST AND LILEY. These authors set out to analyze the problem, paying particular attention to the inclusion of boundary conditions that might explain the anomalous results described in the preceding section. They argued that, since the physical conditions at

---

*Equation 11.8 was obtained by Lawson and Lucas,[54] although erroneously. The starting point of their analysis was the time-dependent equation $\partial n/\partial t = D\nabla^2 n - W\partial n/\partial z$, a solution of which was obtained[16] as the sum of a number of terms representing the contributions to the number density from a set of drifting and diffusing electron groups released at time $t = 0$. This approach should have led to equation 11.7 rather than 11.8, since the initial strengths and placements of the travelling groups were chosen to give $n = 0$ at cathode and anode (see also the subsequent paper by Huxley[17]). The anomaly arises from a fallacy in the derivation[16] whereby some of the travelling groups were given a velocity of $-W$. Thus, although the assumed boundary conditions were satisfied, the continuity equation was not (D. S. Burch, personal communication).

the boundaries were unknown, the correct procedure to adopt was to formulate a general solution containing adjustable parameters to account for the boundary conditions, and then to determine the parameters by matching the solution to the results of the experiment itself. It was their aim to include also the effect of secondary emission from the electrodes due to both photon and positive-ion bombardment in order to analyze the results of experiments that had been made in which current ratios were markedly affected by secondary processes.[19] In this section we restrict the discussion to that part of their work which deals with the boundary condition problem in the absence of secondary processes, leaving a description of the rest until Section 11.4.3.

A general solution to equation 11.1 in cylindrical coordinates, that is,

$$\frac{1}{\rho}\frac{\partial}{\partial\rho}\left(\rho\frac{\partial n}{\partial\rho}\right) + \frac{\partial^2 n}{\partial z^2} - \frac{W}{D}\frac{\partial n}{\partial z} = 0, \qquad (11.10)$$

which applies to an apparatus in which electrons enter the diffusion chamber through a small hole in the cathode, is

$$n(\rho,z) = \int_0^\infty J_0(\xi\rho)[A(\xi)\exp\beta_1 z + B(\xi)\exp\beta_2 z]\xi\,d\xi, \qquad (11.11)$$

where ξ^2 = constant of separation in equation 11.10, $\beta_1 = \lambda - (\xi^2 + \lambda^2)^{1/2}$, $\beta_2 = \lambda + (\xi^2 + \lambda^2)^{1/2}$, and $\lambda = W/2D$.

It follows that the z component of the flux density is

$$j_z(\rho,z) = nW - D\frac{\partial n}{\partial z}$$

$$= D\int_0^\infty J_0(\xi\rho)[\beta_2 A(\xi)\exp\beta_1 z + \beta_1 B(\xi)\exp\beta_2 z]\xi\,d\xi. \qquad (11.12)$$

The coefficients $A(\xi)$ and $B(\xi)$ that appear in equation 11.11 are determined by the boundary conditions, and the problem therefore reduces to deciding upon appropriate boundary conditions and finding values of A and B that are consistent with them. In effect, Hurst and Liley first wrote down generalized boundary conditions and then used the results of experiment to decide upon the specific boundary conditions appropriate to the particular experiment. They were then able to determine the values of A and B. We now summarize their procedure, omitting the mathematical detail.

Using equation 11.12, it can be shown that the current ratio R, defined previously as

$$R = \frac{\int_0^b j_z(\rho,h)\rho\,d\rho}{\int_0^\infty j_z(\rho,h)\rho\,d\rho},$$

is given by

$$R = \int_0^\infty J_1(\xi b)\frac{\psi(\xi,h)}{\psi(0,h)}d(\xi b), \tag{11.13}$$

where

$$\psi(\xi,h) = \beta_2 A(\xi)\exp\beta_1 h + \beta_1 B(\xi)\exp\beta_2 h. \tag{11.14}$$

We require this result subsequently to compare it with the ratio formula (equation 11.8), which is known to fit the experimental data.

We now seek an appropriate generalization of the boundary conditions. Since the analysis was developed to include specifically secondary emission at the electrodes, the boundary conditions are written in terms of particle flux densities rather than number densities. The z component of the electron flux density may be written as

$$j_z = j_{z^+} + j_{z^-}, \tag{11.15}$$

where j_{z^+} refers to the flux of particles moving towards the anode, and j_{z^-} to those moving towards the cathode, and

$$j_{z^+} = \frac{n\bar{c}}{4} + \frac{j_z}{2},$$

$$j_{z^-} = -\frac{n\bar{c}}{4} + \frac{j_z}{2}. \tag{11.16}$$

Let the axial component of the flux density from the small hole in the cathode of Huxley's version of the lateral diffusion apparatus (see Figure 11.2) be $p(\rho,0) = p'(\rho,0)\delta(\rho)$, where

$$2\pi\int_0^\infty p'(\rho,0)\delta(\rho)\rho\,d\rho = P, \tag{11.17}$$

P being the axial component of the total primary flux at the source hole.

Then, if ζ_c is the reflection coefficient at the cathode, it follows that

$$j_{z+}(\rho,0) = -\zeta_c j_{z-}(\rho,0) + p(\rho,0). \tag{11.18}$$

Consequently, from equations 11.15 and 11.16,

$$j_z(\rho,0) = -\left[\frac{n(\rho,0)\bar{c}a_c}{2}\right] + \frac{2p(\rho,0)}{1+\zeta_c}, \tag{11.19}$$

where

$$a_c = \frac{1-\zeta_c}{1+\zeta_c}.$$

Similarly,

$$j_z(\rho,h) = \frac{n(\rho,h)\bar{c}a_a}{2}, \tag{11.20}$$

where $a_a = (1-\zeta_a)/(1+\zeta_a)$, and ζ_a is the reflection coefficient at the anode.

Using equations 11.11 and 11.12, we can now express equations 11.19 and 11.20, representing the boundary conditions, in terms of the coefficients $A(\xi)$ and $B(\xi)$, thereby obtaining two equations for these coefficients in terms of the reflection coefficients and the source function. The equations are

$$\int_0^\infty J_0(\xi\rho)[A(\xi)(\beta_2 + a_c\lambda_T) + B(\xi)(\beta_1 + a_c\lambda_T)]\xi\, d\xi = \frac{2p(\rho,0)}{D(1+\zeta_c)}$$

$$\tag{11.21}$$

and

$$\int_0^\infty J_0(\xi\rho)[A(\xi)(\beta_2 - a_a\lambda_T)\exp\beta_1 h + B(\xi)(\beta_1 - a_a\lambda_T)\exp\beta_2 h]\xi\, d\xi = 0,$$

$$\tag{11.22}$$

where $\lambda_T = \bar{c}/2D$.

In principle these equations can be solved for $A(\xi)$ and $B(\xi)$ provided that $p(\rho,0)$, ζ_c, and ζ_a are known. In practice, however, since ζ_c and ζ_a are not known and cannot be easily determined under the conditions of the experiments, the empirical relation between R and λ (equation 11.8) is now used to determine ζ_c and ζ_a in the following way.

It can be shown that equation 11.8 is equivalent to

$$R = \int_0^\infty J_1(\xi b) \exp\left\{ \left[\lambda - (\xi^2 + \lambda^2)^{1/2} \right] h \right\} d(\xi b)$$

$$= \int_0^\infty J_1(\xi b)(\exp \beta_1 h) \, d(\xi b). \tag{11.23}$$

A comparison of equations 11.23 and 11.13 then shows that

$$\frac{\psi(\xi, h)}{\psi(0, h)} = \exp \beta_1 h. \tag{11.24}$$

But the left-hand side of equation 11.24 can be evaluated in terms of ζ_c and ζ_a, using equation 11.14 and the values of $A(\xi)$ and $B(\xi)$ found by solving equations 11.21 and 11.22. The solution of the equation which results is

$$\zeta_c = \frac{\bar{c} + W}{\bar{c} - W},$$

$$\zeta_a = -1, \tag{11.25}$$

and the corresponding boundary conditions are

$$j_z(\rho, 0) = \frac{n(\rho, 0) W}{2} + \left(1 - \frac{W}{\bar{c}} \right) p(\rho, 0),$$

$$n(\rho, h) = 0. \tag{11.26}$$

Finally, having determined the reflection coefficients by this empirical method, we may insert them in equations 11.21 and 11.22 and then solve the equations to find A and B explicitly, from which it follows, using equation 11.11, that

$$n = \left(1 - \frac{W}{\bar{c}} \right) \frac{P}{2\pi D} \int_0^\infty J_0(\xi \rho) \frac{2 \exp \beta_1 z}{\beta_2 - \beta_1} \frac{1 - \exp[(\beta_1 - \beta_2)(h - z)]}{1 + \exp[(\beta_1 - \beta_2)h]} \xi \, d\xi,$$

$$\tag{11.27}$$

which may be expressed in series form as

$$n = \left(1 - \frac{W}{\bar{c}} \right) \frac{P}{2\pi D} e^{\lambda z} \sum_{-\infty}^{\infty} (-1)^k \frac{\exp - \lambda r_k}{r_k}, \tag{11.28}$$

where

$$r_k^2 = \rho^2 + (z + 2kh)^2.$$

The terms corresponding to $k = 0$ and -1 lead to equation 11.9. Thus, starting with a general solution of equation 11.1 for isotropic diffusion and with generalized boundary conditions at the anode and cathode, Hurst and Liley found that their solution could be made to conform with the experimental results of Crompton and Jory provided that the cathode was almost perfectly reflecting ($\zeta_c \cong 1$), in which case the pole-like solutions of equations 11.9 and 11.28 apply. However, they pointed out that the reflection coefficients used in their theory were not necessarily to be taken as true reflection coefficients because, to paraphrase the relevant comments in their paper:

1. They must account for the true physical reflecting properties of the electrodes.

2. In general, in the vicinity of a boundary it is to be expected that the electron velocity distribution will be anisotropic, while relatively large density gradients could occur. Such effects could lead to nonuniform and nonscalar electron temperatures and associated transport coefficients. This means that, within a few mean free paths of a boundary, equations 11.10, 11.12, and the two-term expansion for the velocity distribution function (equation 2.15 with $k = 1$) will inadequately describe the behaviour of the electrons. Therefore, if a solution of these particular equations is to be used to determine n, j_z, R, and so on, it follows that the boundary conditions, that is, the "reflection" coefficients, must be of an artificial nature.

3. The ζ's have been defined only in terms of the axial components of flux. Since, at a boundary, particles may also be introduced as a pure radial component, the ζ's must be capable of accounting for this.

4. In association with items 2 and 3, but still an independent possibility, the source of axial flux could be inadequately defined, the true nature and distribution of this source being accounted for only by the ζ's.

Subsequent analysis of the experimental results of Crompton and Jory, using the so-called diffusion equation based on anisotropic diffusion, shows that theory can in fact be matched to the results of experiment when the normally accepted boundary conditions at both the cathode and the anode are employed. Nevertheless the value of Hurst's and Liley's analysis, which could be developed for the case of anisotropic diffusion, lies in its applicability to situations in which the boundary conditions have to be determined through the experiments themselves and in which these conditions are complicated by secondary emission. We will return to this problem in Section 11.4.3.

11.2.4. THE SOLUTION OF WARREN AND PARKER. The apparatus used by these authors[20] (see Section 11.3.3) had a source aperture considerably larger than the one in the apparatus of Crompton and Jory[15] ($a = 0.125$ compared with $a = 0.05$ cm), while the ratio of guard ring inner radius to diffusion chamber length in their longer apparatus was considerably less (approximately 0.2,* compared with approximately 0.4). It was therefore necessary to allow in these experiments for the possible influences of the finite sizes of the source aperture and of the cylindrical boundary of the apparatus, whereas no such allowance had been necessary in the analysis of the experiments of Crompton and Jory.

Warren and Parker took as the basis of their solution the dipole-like solution of equation 11.5, and, initially following Huxley, they constructed a solution for a point aperture and infinite-radius boundary, using an infinite set of such terms corresponding to dipoles placed at $z = 0$, $\pm 2h$, $\pm 4h$, and so forth. To this point, therefore, their solution was identical with that described in Section 11.2.2. Furthermore they found that, for practically the whole range of their experimental parameters, an adequate representation was obtained by taking only the source term at the origin and its image at $z = 2h$. Thus, although not explicitly expressed, the formula for the current ratio obtained by Parker and Warren with these restrictions ($a \to 0$, $c \to \infty$) was that given in equation 11.7.

Two approaches were then used to include the effect of a finite-sized source aperture. In the first, Townsend's solution (Section 11.2.1) was modified to include the effect of the anode boundary condition, resulting in the following expression for $n(\rho, z)$:

$$n(\rho, z) = Ae^{\lambda z} \sum_k \frac{J_1(\xi_k a) J_0(\xi_k \rho) \sinh\left[(\lambda^2 + \xi_k{}^2)^{1/2}(h - z)\right]}{\xi_k J_1{}^2(\xi_k c) \sinh\left[(\lambda^2 + \xi_k{}^2)^{1/2} h\right]}, \quad (11.29)$$

where, as before, $\xi_k = x_k/c$ and x_k is the kth root of $J_0(x) = 0$. Since this solution also accounts for the influence of a cylindrical boundary of radius c, it represents a complete solution to the problem, provided that n is assumed to be constant over the aperture. However, it suffers from the disadvantage that for large values of $\lambda = W/2D$ (where the stream is least divergent, and consequently the finite size of the source aperture the most significant) the evaluation of the current ratio formula based on equation 11.29 is difficult because of the slow convergence of the series.

Warren and Parker therefore adopted an alternative approach in which the finite-sized source was represented by a uniform distribution of di-

---

*The inner radius of the guard electrodes is incorrectly stated in the original paper as 1.5 in. instead of 0.75 in. (Parker, personal communication).

poles, obtaining a solution for $n(\rho, z)$ in terms of modified Bessel functions. The formula obtained by them is complex but has the advantage of more rapid convergence.* Since a solution built up in this way from the simple dipole solution cannot take account of the effect of a cylindrical boundary of finite radius, Parker and Warren could not use the ratio formula based on such a solution to calculate the current ratios for their apparatus at small values of λ. Nevertheless a comparison of the current ratios from this solution with those calculated using the point dipole ratio formula (equation 11.7) made it possible to calculate the effect of the finite size of the aperture for *all* values of λ. It was found that the differences amounted to less than 5% everywhere; moreover, as pointed out by the authors, the calculations overestimate the error since the effective diameter of the source aperture is less than its actual diameter because of the reduction of the number density towards the edge of the aperture caused by diffusion. Because of the small and uncertain size of the correction Warren and Parker did not employ this solution; nor, in fact, were they able to use any of the solutions developed here for the reasons discussed in Section 11.3.3.

11.2.5. PARKER'S SOLUTION ALLOWING FOR THE SPATIAL VARIATION OF THE ENERGY DISTRIBUTION FUNCTION. Common to all the theories that had been developed up to this time is the assumption of a spatially invariant energy distribution function with, as a consequence, spatially invariant transport coefficients. The first attempt to investigate the consequences of this assumption was made by Parker,[21] who obtained an exact solution to the Boltzmann equation for the case of electrons originating at a point source and drifting in an electric field in unbounded space. The results of this analysis were then used to infer the magnitude of the errors in the values of D/μ calculated from the current ratios by means of theories based on the invariance of the energy distribution function. A detailed description of Parker's analysis is beyond the scope of this summary, and we therefore give only an outline of his approach and some conclusions from it. For simplicity we restrict this description to the case of constant collision frequency, although in Parker's paper the solution for constant cross section is also developed.

As was shown in Chapter 3, it has been customary to find simple solutions to the time-independent Boltzmann equation by assuming that the terms depending on the gradients of the electron number density have a negligible influence on f_0, that is, that the function nf_0 can be expressed as the product of two functions, one function, $n(\mathbf{r})$, representing the spatial dependence of the electron number density, and the other, $f_0(c)$, representing a spatially independent velocity distribution. Moreover, in this simpli-

---

*As expected, the ratio formula derived in this way goes over to equation 11.7 as $a \rightarrow 0$.

fied solution, it is assumed that $f_0(c)$ is given by equation 3.11 or alternatively, in terms of ϵ, by eqution 6.51.* Before proceeding to the formal solution, Parker used this simplified approach to establish a criterion whereby the importance of the gradient terms can be estimated.

For the case of constant collision frequency, equation 6.45 becomes, with the omission of the time-dependent term and the term accounting for inelastic collisions,

$$\frac{2mv}{M\epsilon^{1/2}} \frac{\partial}{\partial\epsilon} \left(\epsilon^{3/2} \left\{ (nf_0) + \left[\kappa T + \frac{M}{3} \left(\frac{eE}{mv} \right)^2 \right] \frac{\partial}{\partial\epsilon} (nf_0) \right. \right.$$

$$\left. \left. + \tfrac{1}{3} MeE \left(\frac{1}{mv} \right)^2 \frac{\partial}{\partial z} (nf_0) \right\} \right)$$

$$+ \frac{2eE\epsilon}{3mv} \frac{\partial^2}{\partial\epsilon\,\partial z} (nf_0) + \frac{2\epsilon}{3mv} \nabla^2 (nf_0) = 0, \qquad (11.30)$$

where it is assumed that eE is parallel to $+Oz$. As shown in Chapter 3, if equation 11.30 is multiplied by $\epsilon^{1/2} d\epsilon$ and integrated over all energies, the group of terms in the large parentheses vanishes, leaving only the last two terms of the equation. Thus

$$\frac{\partial}{\partial z} \left[\frac{2eE}{3mv} \int_0^\infty \epsilon^{3/2} \frac{\partial}{\partial\epsilon} (nf_0)\, d\epsilon \right] + \nabla^2 \left[\frac{2}{3mv} \int_0^\infty \epsilon^{3/2} (nf_0)\, d\epsilon \right] = 0,$$

which, after integration of the first term by parts, may be expressed in the form†

$$\nabla^2 (Dn) - W \frac{\partial n}{\partial z} = 0,$$

where

$$D = \frac{2}{3mv} \int_0^\infty \epsilon^{3/2} f_0\, d\epsilon. \qquad\qquad (11.31)$$

If it is now assumed that f_0 is independent of position and given by

---

*From here on, we use ϵ rather than c as an independent variable in order to facilitate comparison with Parker's work.

†In the general case where $v = v(\epsilon)$, a similar form of the differential equation for n may be obtained, but both W and D are then spatially dependent coefficients.[22]

equation 6.51 (with $\nu = $ constant), that is,

$$f_0(\epsilon) = \text{const} \cdot \exp\left[-\frac{\epsilon}{\kappa T + (M/3)(eE/m\nu)^2} \right]$$

$$= \text{const} \cdot \exp\left[-\frac{\epsilon}{\kappa T + 1/B} \right], \qquad (11.32)$$

where $B = (3/M)(m\nu/eE)^2$, equation 11.31 reduces to the familiar form of the steady-state differential equation for n

$$D\nabla^2 n - W\frac{\partial n}{\partial z} = 0. \qquad (11.33)$$

Let it be supposed that an accurate estimate of the relative magnitude of the gradient terms can be obtained by assuming that the presence of density gradients perturbs $f(\epsilon)$ negligibly from the spatially independent form given by equation 11.32. Then equation 11.33 is valid and may be used in equation 11.30 to eliminate the last two terms on the left-hand side. In the absence of density gradients, the first term of the simplified equation, which represents the energy loss in elastic collisions, balances the sum of the remaining terms, which represent, respectively, the energy gain from inverse collisions and from the field. It follows that, in the presence of gradients, a comparison of the gradient term that remains, after the last two terms of equation 11.30 have been eliminated as above, with either the collision loss term or the sum of the other two terms enables its importance to be determined. The ratio R of the terms can be shown to be

$$R = \left(1 - \frac{\epsilon_T}{\epsilon_{av}} \right) \left[\frac{D(\partial n/\partial z)}{nW} \right], \qquad (11.34)$$

where ϵ_T/ϵ_{av} is the ratio of the average energy of the gas molecules to the average energy of the electrons. Thus near thermal equilibrium $(E/N \rightarrow 0)$ the gradient term becomes unimportant, but when $\epsilon_{av} \gg \epsilon_T$ the terms can be neglected only when the diffusion current in the field direction is small compared with the drift current. When applied to the case of a point source of electrons [in the high-field limit for which $(M/3)(eE/m\nu)^2 \gg \kappa T$], equation 11.34 predicts that the ratio of the terms at the point (ρ, z) is

$$R = \frac{1 - \cos\theta - (4\epsilon_{av}\cos^2\theta)/3eEz}{2},$$

where $\tan\theta = \rho/z$. Thus, for points located near the axis ($\cos\theta \approx 1$) and sufficiently removed from the source to ensure that $(4\epsilon_{av}\cos^2\theta)/3eEz$ is negligible, the gradient terms may be neglected and the usual solutions which disregard these terms are adequate approximations.

A formal solution of equation 11.30 (expressed in cylindrical coordinates) that takes account of the spatial variation of f_0 was obtained by Parker, who transformed the equation into separable form through a change of the variables ϵ, z, and ρ to new independent variables, one of which was a function of both ϵ and z. With this transformation a solution for nf_0 was determined which was the sum of contributions from an infinite number of modes. However, since the highest modes decay exponentially with distance from the source, it is in general necessary to consider only the fundamental mode.

In the high-field limit the expression for nf_0 is straightforward and reads

$$nf_0 = A \frac{e^{-B\epsilon}}{z[1-(\epsilon-\epsilon_0)/eEz]} \exp\left[-\frac{eEB\rho^2}{4z}\left(1 - \frac{\epsilon-\epsilon_0}{eEz}\right)\right] \quad \text{when } \epsilon < \epsilon_0 + eEz,$$

$$= 0 \quad \text{when} \quad \epsilon > \epsilon_0 + eEz, \tag{11.35}$$

where ϵ_0 is the initial energy of the electrons released at an isolated point source, and A is a constant. This expression for nf_0 is to be compared with the product nf_0 that is found when the gradient terms are ignored, in which case n is given by the first term of equation 11.9 and f_0 by equation 11.32, that is,[*]

$$(nf_0)_{\text{approx}} = \frac{Ae^{-B\epsilon}}{(\rho^2+z^2)^{1/2}} \exp\left\{ -\frac{eEB\left[(\rho^2+z^2)^{1/2}-z\right]}{2}\right\}. \tag{11.36}$$

We note immediately that equation 11.35, unlike equation 11.36, cannot be written as the product of two functions, one depending only on position and the other only on energy.

Figure 11.4 shows the ratio of the two solutions for the case of $\epsilon_0 = 0$ and the ratio $\epsilon_{av}/eEz = 0.1$. The ratio of the two functions for given values of the parameter ρ/z is shown plotted as a function of ϵ/eEz. For points on the axis ($\rho = 0$) it can be seen that, for all energies, $(nf_0)_{\text{exact}}$ is greater than

---

[*]Since $f_0(\epsilon) = \text{const} \cdot e^{-B\epsilon}$, the electron energy distribution is Maxwellian with mean energy $= \frac{3}{2}(De/\mu) = 3/2B$. It follows that $\lambda = W/2D = eEB/2$.

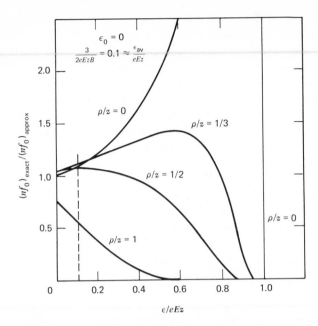

FIG. 11.4.   The ratio $(nf_0)_{exact}/(nf_0)_{approx}$ versus ϵ/eEz for constant collision frequency. The vertical dashed line indicates the approximate position of ϵ_{av}/eEz on the abscissa. (From ref. 21.)

$(nf_0)_{approx}$. Thus both the number density of the electrons near the axis and their average energy are greater than the corresponding values calculated from the approximate theory. Conversely, for points far removed from the axis $(nf_0)_{exact}$ is less than $(nf_0)_{approx}$ for all energies, and both the mean energy and the number density are less than the values derived from the approximate theory.

As noted by Parker, a qualitative explanation can be given of the variation of the average energy with distance from the axis. On the average, electrons within a small volume element at a given point gain energy from the field because more electrons move in the direction of the electric force $e\mathbf{E}$ than against it. In the absence of density gradients, the energy gained from the field as a consequence of the drift motion just balances the energy lost in collisions when $f_0 = \text{const} \cdot e^{-B\epsilon}$, and the average energy is $(3/2)(eD/\mu) = 3/2B$. However, near the axis the density gradient results in an increased flow in the direction of drift, more energy is received from the field, and the average energy is higher; conversely, at points sufficiently far distant from the axis the density gradient in the z direction is in the opposite sense, and the average energy is therefore lower.

Although these arguments, as well as the general discussion of the results shown in Figure 11.4, make it clear that theories based on a spatially invariant energy distribution function may lead to significant errors in the interpretation of the results of lateral diffusion experiments, it is not easy to estimate from the figure the magnitude of such errors, since they depend on whether or not there is a significant number of electrons in the classes for which $(nf_0)_{exact}/(nf_0)_{approx}$ differs significantly from unity. While there is quite clearly a significant depletion of electrons far from the axis, the error incurred by using the approximate theory is likely to be small in most experiments because the stream is sampled close to the axis and because few electrons have an energy which exceeds the average value by a factor of more than 2 or 3. Furthermore Figure 11.4 refers to a somewhat extreme case. If, for example, ϵ_{av} is taken to be 0.2 eV (about 5 times greater than thermal energy at 300 K), the potential difference between the source and the sampling plane will be only 2 V (since $\epsilon_{av}/eEz = 0.1$), whereas in typical experiments the voltage drop across the diffusion chamber is usually not less than 10 times (very often it is more than 100 times) larger than this value. We therefore proceed to obtain a semi-quantitative estimate of the error that would follow from the use of the approximate solution.

When, as is usually the case, $eEz \gg \epsilon_{av}$ [i.e., the point (ρ, z) is far from the source in the axial direction], and to simplify the analysis we put $\epsilon_0 = 0$, the terms $(1 - \epsilon/eEz)^{-1}$ in equation 11.35 can be replaced to first order by $(1 + \epsilon/eEz)$, and the equation then reads

$$nf_0 = Az^{-1}\left(1 + \frac{\epsilon}{eEz}\right)\exp\left\{-B\epsilon\left[1 + \left(\frac{\rho}{2z}\right)^2\right]\right\}\exp\left(-\frac{eEB\rho^2}{4z}\right). \quad (11.37)$$

It follows immediately that when both ϵ/eEz and $(\rho/z)^2$ can be neglected in comparison with unity, that is, for points far distant from the source but close to the axis, the expression for nf_0 reduces to the approximate expression of equation 11.36 (cf. Figure 11.4).

Equation 11.37 can be used to obtain the following approximate expressions for the average energy and the number density:[21]

$$\epsilon_{av} \approx \frac{3(1 + 1/eEBz)}{2B\left[1 + (\rho/2z)^2\right]}, \quad (11.38a)$$

$$n \approx \frac{A'(1 + 3/2eEBz)}{z\left[1 + (\rho/2z)^2\right]^{3/2}}\exp\left[-\frac{eEBz(\rho/z)^2}{4}\right], \quad (11.38b)$$

where A' is a constant.

From equation 11.38a the following facts emerge, illustrating further some of the points discussed in relation to Figure 11.4:

1. The average energy of electrons on the axis is larger than the value $3/2B$ predicted from the approximate theory.

2. For large values of z the average energy approaches the value $3/2B$.

3. For points far removed from the axis the average energy is significantly less than $3/2B$. Thus for large values of z (i.e., $1/eEBz \ll 1$) and $\rho/z = 1$ the average energy is 25% less than predicted by the approximate theory.

Equation 11.38b can be used as a basis for estimating the errors in the values of D/μ that are determined from a lateral diffusion experiment on the basis of the approximate theory. In making the estimate it is assumed for simplicity that the ratio determined experimentally is the ratio of the number density at a radial distance ρ to the number density on the axis. By using equation 11.36 it can be readily shown that this ratio, which will be denoted by R_N, is given by the approximate theory as

$$(R_N)_{\text{approx}} = \left[1 + \left(\frac{\rho}{z}\right)^2\right]^{-1/2} \exp\left(\frac{-eEBz\left\{\left[1+(\rho/z)^2\right]^{1/2}-1\right\}}{2}\right).$$

$$(11.39a)$$

On the other hand, the more exact theory, resulting in equation 11.38b, yields the following expression for R_N:

$$(R_N)_{\text{exact}} = \left[1 + \frac{(\rho/z)^2}{4}\right]^{-3/2} \exp\left[\frac{-eEBz(\rho/z)^2}{4}\right]. \quad (11.39b)$$

Equations 11.39a and 11.39b can be expressed in more useful forms by expanding in powers of $(\rho/z)^2$ and retaining only the significant terms, in which case for small values of $(\rho/z)^2$ they become:

$$(R_N)_{\text{approx}} \approx \left[\frac{1-(\rho/z)^2}{2} + \frac{eEBz(\rho/z)^4}{16}\right] \exp\left[\frac{-eEBz(\rho/z)^2}{4}\right],$$

$$(11.40a)$$

$$(R_N)_{\text{exact}} \approx \left[\frac{1-3(\rho/z)^2}{8}\right] \exp\left[\frac{-eEBz(\rho/z)^2}{4}\right]. \quad (11.40b)$$

TABLE 11.2. Maximum fractional differences in D/μ, estimated using the values of ρ/z for Warren's and Parker's apparatus

$\mu Ez/4D^a$	$(\rho/z)^2$	$\delta(D/\mu)/D/\mu$ (%)
50–500 (long tube)	0.005	0.1
	0.02	0.5
2–10 (short tube)	0.24	5
	0.9	22

^{a}The heading of the first column of Table 1 of Parker's paper should read $\mu Ez/4D$. The estimated error for $(\rho/z)^2 = 0.24$ in the short tube has been revised from 8 to 5% (Parker, personal communication).

These equations can be solved to give "approximate" and "exact" values of D/μ for a given value of R_N. Assuming that $(\rho/z)^2$ is sufficiently small that $\ln[1 - (\rho/z)^2/2 + eEBz(\rho/z)^4/16]$ and $\ln[1 - 3(\rho/z)^2/8]$ can be replaced by $[-(\rho/z)^2/2 + eEBz(\rho/z)^4/16]$ and $[-3(\rho/z)^2/8]$, respectively, and replacing $\ln(1/R_N)$ by $(\mu Ez/4D)(\rho/z)^2$, it can be shown that

$$\frac{(D/\mu)_{approx} - (D/\mu)_{exact}}{(D/\mu)_{approx}} \approx \frac{(\rho/z)^2/8}{\ln(1/R_N)} - \frac{(\rho/z)^2}{4}$$

$$\approx \frac{D}{2\mu Ez} - \frac{(\rho/z)^2}{4}. \qquad (11.41)$$

Table 11.2 shows values of the maximum fractional error in D/μ calculated from equation 11.41 for the values of ρ/z and for the range of values of the parameter $\mu Ez/4D$ used in the experiments of Warren and Parker.[20] From the table it can be seen that the estimated errors in the values of D/μ obtained with the long tube are usually negligible but that significant errors are predicted in the results from the short tube. It must be stressed, however, that these predictions are to be taken as indicating only very approximately the magnitude of the errors and that the actual errors may be larger or smaller than these estimates. Thus, in deriving the expression for n (equation 11.38b), it was not possible to take account of the effect of the boundary conditions in the actual experiment. Furthermore, in estimating the error in D/μ, it was assumed that the measured ratio was equal to the ratio of the number density at a distance ρ from the axis to the axial density, rather than to the ratio of the fluxes falling within, and outside, a circular area of radius ρ. Finally, these estimates refer only to the case of constant collision frequency in a gas in which no inelastic collisions occur.

Notwithstanding these restrictions on the validity of the error analysis and its application, Parker's theoretical treatment of the diffusing stream is important for several reasons. In the first place, it underlines the inadequacy of the assumption that the energy distribution function is uniform throughout the diffusion chamber, since it has been shown that at points close to the source, and at all points for which ρ/z is large, the function differs significantly from that obtained with the usual approximation. Second, it illustrates the difficulty of interpreting accurately the results of lateral diffusion experiments in circumstances where the electric field is large but the ratio $Ez/(D/\mu)$ is small. Third, it predicts that there may be large errors in the values of D/μ calculated on the basis of the approximate theory when $Ez/(D/\mu)$ is small and ρ/z is large. On the other hand, the theory shows that for large values of $Ez/(D/\mu)$ and small values of ρ/z, precisely the conditions that have been used for a large number of experimental measurements both old and new, negligible error has resulted from the usual interpretation in terms of spatially invariant energy distribution functions. A qualitative explanation of this fact will be given in Section 11.2.7.

11.2.6.   OTHER SOLUTIONS BASED ON A SPATIALLY DEPENDENT ENERGY DISTRIBUTION FUNCTION. An analysis similar to that of Parker and based in part on his solution has been given by Francey.[22] Again two special cases, those of constant collision frequency and constant cross section, were treated. In the first case, Francey follows Parker to the point where a solution for nf_0 is found corresponding to the lowest mode of a series of solutions. From there on, however, Francey proceeds to examine the far distant solution, that is, the solution for which it is assumed that $eEz \gg \epsilon - \epsilon_0$, without first making the approximation $B^{-1} \gg \kappa T$. His solution should therefore be generally applicable, that is, not restricted to the high-field case only. With this assumption the following formulae for W and D were obtained:

$$W = \frac{a}{\nu},$$

$$(11.42)$$

$$D = \frac{2\alpha(1+\alpha)}{Bm\nu(1+2\alpha)},$$

where $a = eEm$, $B = 3v^2/a^2M$ (as in Parker's analysis), and $\alpha = \kappa TB$. It follows that

$$\left.\begin{aligned}
\frac{D}{W} &= \frac{2\kappa T(1+\alpha)}{ma(1+2\alpha)}, \\[2mm]
\frac{D}{\mu} &= \frac{2(1+\alpha)\kappa T}{(1+2\alpha)e} \\[2mm]
&= \frac{\kappa T}{e} \quad \text{for } \alpha \gg 1 \\[2mm]
&= \frac{2\kappa T}{e} \quad \text{for } \alpha \ll 1.
\end{aligned}\right\} \qquad (11.43)$$

This last result is surprising since it means that the maximum mean energy is only twice the mean energy of the gas molecules regardless of the value of E/N. An explanation for this paradoxical result has not yet been given.

Francey's treatment of the case of constant cross section differs from Parker's but his conclusion is the same, that is, that no general solution can be found, and that useful solutions appropriate to the lateral diffusion experiment can be obtained only by solving approximate equations.

A somewhat different approach was taken by Desloge and Mitchell,[23] who based their analysis on an approximate solution of the velocity distribution function developed earlier by Desloge,[24] namely,

$$f(\mathbf{r},c) = f_0(\mathbf{r},c)\left[1 + \sum_i 2\xi \bar{c}_i c_i + \frac{8\xi^3}{5m}\sum_i q_i c_i(c^2 - 5/2\xi)\right], \quad (11.44)$$

where

$$f_0(\mathbf{r},c) = n\left(\frac{\xi}{\pi}\right)^{3/2} \exp - \xi c^2,$$

$$\xi = \frac{m}{2\kappa\tau},$$

$\bar{c}_i$ is the ith component of the local mean velocity, and τ and q_i are parameters that are approximately equal to the electron temperature and the ith component of the heat flux vector per electron, respectively.

Using this distribution function, Desloge and Mitchell found a set of moment equations containing n, τ, $\bar{c}_i$, and q_i which could in principle be solved to determine these quantities. In order to proceed, however, a further simplified set of equations was obtained on the assumption that spatial gradient terms could be considered as negligibly small compared with other terms. Desloge and Mitchell showed that the criterion which may be used to decide whether or not the gradient terms are negligibly small is

$$\left| \frac{\mu E}{D} \right| \gg \left| \sum_i \frac{1}{n} \frac{\partial n}{\partial x_i} \frac{E_i}{E} \right|, \tag{11.45}$$

where, as before, the subscript i refers to the ith component. When the field is parallel to the z axis, this criterion is equivalent to that stated by Parker (see quation 11.34 and the following paragraph).

As one might expect, the set of equations that follows from the neglect of the gradient terms is the set on which the usual analyses of lateral diffusion experiments are based (see Sections 11.2.1 to 11.2.4), and the analysis of Desloge and Mitchell is therefore straightforward from this point. The ratio formula obtained by them corresponds to the fraction of the total electron flux from an isolated source of negligible dimensions which crosses a geometrical plane normal to the field and a distance h from the source within a radius b from the axis of symmetry, that is,

$$R = \frac{\displaystyle\int_0^b j_z(\rho, h)\rho \, d\rho}{\displaystyle\int_0^\infty j_z(\rho, h)\rho \, d\rho}, \tag{11.46}$$

where

$$j_z(\rho, h) = n(\rho, h)W - D\left(\frac{\partial n}{\partial z}\right)_{z=h} \tag{11.47}$$

and $n(\rho, z)$ is given by the first term of equation 11.9, that is,

$$n(\rho, z) = A e^{\lambda z} \frac{\exp - \lambda r}{r}, \tag{11.48}$$

with $r = (\rho^2 + z^2)^{1/2}$. It follows that

$$R = 1 - \left(\frac{1 + h/d}{2}\right) \exp - \lambda(d - h), \tag{11.49}$$

which is equation 41 of the original paper. When h/d is small, this formula approaches equations 11.7 and 11.8.

11.2.7. SOLUTIONS BASED ON THE STEADY-STATE DIFFERENTIAL EQUATION FOR n ASSUMING ANISOTROPIC DIFFUSION. Before the work of Lowke[25] and subsequently of Huxley,[17] the ratio formula which had been shown to account satisfactorily for many sets of data from lateral diffusion experiments had been justified only on semiempirical grounds, as described in the preceding sections. Although Parker[21] and others[22,23] had suspected that the anomalies between theory and experiment were a consequence of the assumption of a spatially invariant energy distribution function, and although their work had shown that such an assumption could, in certain circumstances, give rise to serious errors, they were not able to find a formal justification of the empirical formula.

A tractable solution to the problem allowing for a spatially dependent energy distribution function was given by Lowke,[25] whose analysis was based on the assumption that equation 11.1, the differential equation for n based on isotropic diffusion, could be replaced by an equation that accounts for anisotropic diffusion, since the diffusive process had been shown to be anisotropic when an electron swarm drifts in the presence of an electric field. Thus, if D_L and D are the coefficients that characterize diffusion parallel and normal to the electric field $\mathbf{E}$, the equation for n becomes (see Chapter 4)

$$D_L \frac{\partial^2 n}{\partial z^2} + D\left(\frac{\partial^2 n}{\partial x^2} + \frac{\partial^2 n}{\partial y^2}\right) = W \frac{\partial n}{\partial z} \tag{11.50}$$

when $\mathbf{E}$ is parallel to the z axis.

Equation 11.50 was assumed by Lowke on the grounds that the experiments of Wagner, Davis, and Hurst[3] and the theoretical analysis of Parker and Lowke[26] of diffusion and drift in an electric field had demonstrated that the electron motion may be characterized by the three transport coefficients, W, D, and D_L, although it was recognized that longitudinal diffusion is not a straightforward diffusive process but arises from a combination of normal diffusion and differential drifts (see Chapter 4).

When equation 11.50 is solved for n, with the boundary conditions $n = 0$ over the cathode and anode except at the hole in the cathode, which acts as the point source, the following formula for the current ratio may be obtained (see Chapter 5)*

$$R = 1 - \left(\frac{h}{d'} - \frac{1}{\lambda_L h} + \frac{h}{\lambda_L d'^2} \right)\left(\frac{h}{d'} \right)\exp - \lambda_L(d' - h), \quad (11.51)$$

where $\lambda_L = W/2D_L, d'^2 = h^2 + b'^2$, and $b' = (D_L/D)^{1/2}b$.

At first sight it appears that the current ratio is determined primarily by longitudinal diffusion since λ_L appears explicitly in the formula. However, the lateral diffusion coefficient enters implicitly through d', and it can be shown without difficulty that equation 11.51 can be rewritten as

$$R = 1 - \left(\frac{h}{d'} - \frac{1}{\lambda_L h} + \frac{h}{\lambda_L d'^2} \right)\left(\frac{h}{d'} \right)\exp\left(-\frac{\lambda b^2}{d' + h} \right), \quad (11.52)$$

which is to be compared with equation 11.8 written in the same form, namely,

$$R = 1 - \left(\frac{h}{d} \right)\exp\left(-\frac{\lambda b^2}{d + h} \right). \quad (11.53)$$

Since $d' \approx d \approx h$, it can be seen that the ratio formulae converge, provided that h is sufficiently large and b/h sufficiently small, and therefore, as expected, that D rather than D_L primarily determines the spreading of the stream. Moreover, provided that b/h is small and λb is not too large ($\lesssim 100$), equation 11.51 or 11.52 can be replaced with negligible error by

$$R = 1 - \left[1 + \left(\frac{1}{2} - \frac{D_L}{D} \right)\left(\frac{b}{d} \right)^2 \right]\left(\frac{h}{d} \right)\exp - \lambda(d - h) \quad (11.54)$$

(see Chapter 5).

*Equation 11.51 is obtained by considering only the contributions from the dipole at the source and the image which makes $n = 0$ at the anode, since these are the only significant contributions from the infinite hierarchy of dipoles except in exceptional circumstances. Lowke[25] gives the complete expression as

$$R = 1 - \frac{\sum\limits_{k=1}^{\infty} (1/d_k'^3)\left[h^2 d_k'(2k-1)^2 - b'^2/\lambda_L \right]\exp - \lambda_L d_k'}{\sum\limits_{k=1}^{\infty} \exp - (2k-1)\lambda_L h}$$

where $d_k'^2 = (2k-1)^2 h^2 + (D_L/D)b^2$. When only the first term in the series is considered, which is equivalent to considering the source dipole and its image, this expression for R reduces to equation 11.51.

Equation 11.54 apparently explains the success of the semiempirical Huxley formula in explaining Crompton's and Jory's results for hydrogen. In this gas, over the range of electron energies corresponding to the values of E/N for which the measurements were made, the momentum transfer cross-section q_m is approximately constant. It follows[26,17] that the ratio $D_L/D \cong 0.5$ (see Chapter 4) and that equation 11.54 simplifies to the Huxley formula. Although it has not been put to the same test, the empirical formula would be expected to yield equally consistent data in helium since q_m is practically constant in this gas also over a large energy range. On the other hand, in argon one would expect anomalous results to follow from the application of the empirical formula to current ratios determined with a short apparatus ($h \approx 2$ cm) since D_L/D varies widely from 0.5.[26]

It follows also from equation 11.54 that the ratio formula is insensitive to the value of D_L/D when b/d is small, that is, for large values of h, and that in this circumstance the formula approaches the empirical formula regardless of the value of D_L/D. Thus, when a sufficiently long apparatus is used, accurate values of the lateral diffusion coefficient can be calculated from the current ratios without knowledge of the values of D_L.

Figure 11.5 illustrates these points. With the exception of the curves for D_L/D equal to 4 and 10, which require for their calculation the inclusion of the higher-order image terms in the expression for n, the curves were calculated from equation 11.51 without further approximation. The insensitivity of the curves for $h = 10$ cm to the value of the ratio D_L/D and, more particularly, the agreement between the new ratio curves and the Huxley ratio curves when $D_L/D = 0.5$ for $h = 2$ and 10 cm can be seen from the figure.

Thus, in summary, there are two important consequences of this new formulation of the theory of the lateral diffusion experiment. In the first place the theory appears to account satisfactorily for the long-lived paradox that was associated with the success of the empirical formula. Second, and of equal importance, it confirms the validity of the asymptotic form of all the ratio formulae used in the past and therefore of the results of the majority of earlier investigations, at least with regard to the analysis of the current ratios to determine lateral diffusion coefficients. It remains to give a simple physical interpretation of the lack of sensitivity of the current ratios to the longitudinal diffusive processes when the current distribution is sampled sufficiently far from the source.

· 11.2.8. LATERAL SPREAD OF AN ISOLATED TRAVELLING GROUP. In order to gain further physical insight into this problem, we conclude this section with an argument, based on the properties of an isolated drifting

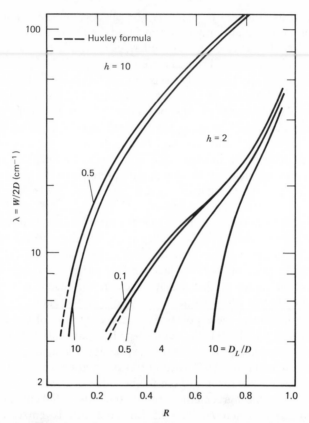

FIG. 11.5. Curves showing the current ratio R, as calculated from equation 11.52, for an apparatus with $b = 0.5$ and $h = 2$ and 10 cm. With $h = 10$ cm the curves for $D_L/D = 0.5$ and 10 are not greatly displaced relative to each other; the curve for $D_L/D = 0.5$ coincides with the empirical ratio curve (dashed curve). The curves with $h = 2$ cm show a strong dependence on the ratio D_L/D, but again the curve for $D_L/D = 0.5$ coincides with the empirical curve (dashed curve). (From ref. 25.)

and diffusing electron group, which indicates why the current ratio R is independent of D_L for large values of $\lambda_L h$ and why, with this condition, equations 11.51 and 11.8 converge provided that b/h is small. We first recall that equation 11.8 can be expressed in the form of equation 11.53 and note that, if b remains fixed, the ratio formula approaches the form

$$R_{\text{lim}} = 1 - \exp\left(-\frac{\lambda b^2}{2h}\right)$$

as $h \to \infty$.

The time-dependent equation for n which allows for anisotropic diffusion is

$$\frac{\partial n}{\partial t} = D_L \frac{\partial^2 n}{\partial z^2} + D\left(\frac{\partial^2 n}{\partial x^2} + \frac{\partial^2 n}{\partial y^2}\right) - W\frac{\partial n}{\partial z} \qquad (11.55)$$

when $e\mathbf{E}$ is parallel to $+Oz$. The solution of this equation, which corresponds to the release of n_0 electrons at the origin of coordinates at time $t = 0$, is

$$n = \frac{n_0}{4\pi Dt(4\pi D_L t)^{1/2}} \exp\left(\frac{-\rho^2}{4Dt}\right) \exp\left[-\frac{(z-Wt)^2}{4D_L t}\right]$$

$$= n_0 p_1(\rho, t)p_2(z, t),$$

where

$$p_1(\rho, t) = \frac{1}{4\pi Dt}\exp\left(-\frac{\rho^2}{4Dt}\right), \qquad (11.56)$$

$$p_2(z, t) = \frac{1}{(4\pi D_L t)^{1/2}}\exp\left[-\frac{(z-Wt)^2}{4D_L t}\right],$$

and $\rho^2 = x^2 + y^2$.

We note that

$$2\pi \int_0^\infty p_1(\rho, t)\rho\, d\rho = \int_{-\infty}^\infty p_2(z, t)\, dz = 1,$$

that is, that p_1 and p_2 are probability functions such that, at time t, $2\pi p_1(\rho, t)\rho\, d\rho$ is the probability of finding an electron in the cylindrical volume bounded by ρ and $\rho + d\rho$ and that $p_2(z, t)dz$ is the probability of finding an electron between the planes z and $z + dz$.

Let us assume that $D_L = 0$. Then $p_2(z, t)$ becomes the delta function $\delta(z - Wt)$ and equation 11.56 becomes

$$n = \frac{n_0}{4\pi Dt}\exp\left(-\frac{\rho^2}{4Dt}\right)\delta(z - Wt), \qquad (11.57)$$

which describes a two-dimensional distribution of charge confined to the plane $z = Wt$, that is, an axially symmetric, infinitesimally thin distribution moving parallel to the z axis with velocity $\mathbf{W}$. When such a distribution arrives at the plane $z = h$, it may easily be shown (by evaluating equation 11.57 at $t = h/W$) that the ratio of the number of electrons incident on the circular area of radius b centred on the axis to the total number of electrons is*

$$R = 1 - \exp\left(-\frac{\lambda b^2}{2h}\right) = R_{\text{lim}}. \tag{11.58}$$

Equation 11.58 also follows immediately from putting $D_L = 0$ in equation 11.51.

The distribution at the receiving plane represented by equation 11.58 is that which occurs when every electron released from the source arrives at the plane at the same time. It follows that even when $D_L \neq 0$ we would expect to observe a distribution approaching R_{lim}, provided that the spread of arrival times is small compared with $t = h/W$. Now it may be shown without difficulty (see, e.g., Section 11.6.1) that if δt is the full width at half maximum of the distribution of arrival times, which is approximately Gaussian, $\delta t/t \propto 1/(\lambda_L h)^{1/2}$; that is, for fixed values of E/N and N, $\delta t/t \to 0$ as $h \to \infty$. Furthermore, since the effects of the cathode and anode boundaries are localized, their influence on the radial distribution of the travelling group also diminishes as h increases. Consequently, it follows that, provided h is sufficiently large, the current ratio should depend only on the lateral diffusion coefficient. That this is also predicted by the formal theory can be shown by letting $h \to \infty$ in equation 11.51 (see Chapter 5).

11.3. EXPERIMENTAL DETERMINATION OF D/μ IN THE ABSENCE OF IONIZATION AND ATTACHMENT

11.3.1. INTRODUCTION: EARLY MEASUREMENTS. Experiments to measure the ratio D/μ fall conveniently into two groups, those which were completed before 1940 and which were carried out either in Oxford under Townsend or in Sydney under V. A. Bailey, and those which have been carried out since 1950, usually with a somewhat different emphasis. The early measurements have been summarized in detail elsewhere, particularly by Healey and Reed[11] and by Loeb,[27] and descriptions of these experiments will not be repeated here. The more recent work began with the experiments of Huxley and his students[28] in Birmingham in the late 1940's and continued with Crompton, Sutton, Elford, and others at the University

---

*A formula for R analogous to equation 11.58 was first derived by Townsend and Tizard[9] and forms the basis of the theory developed by Bailey[10] to analyze his experiments on electron attachment.

of Adelaide and the Australian National University, Canberra. Later, similar work was undertaken at ORNL, Oak Ridge,[29] at the Westinghouse Laboratories in Pittsburgh,[20] and more recently still at the University of Edinburgh.[30]

In the following subsections we describe first the experiments of the last-named groups and then conclude the section with a more detailed account of the investigations of Huxley and his collaborators, their work having resolved some of the problems that were unresolved in other studies.

11.3.2. EXPERIMENTS OF COCHRAN AND FORESTER. These experiments[29] were carried out primarily as part of the programme of the Health Physics Laboratory at ORNL to determine the properties of low-energy electrons in the hydrocarbons. The apparatus was similar in general design to that used by Crompton and Sutton[13] (see Section 11.3.5), and the results were analyzed using the Huxley formula. For some of the experiments the hot-filament electron source was replaced by the corona discharge source used by Huxley and Zaazou.[28] Teflon and fluorothene were used in the construction of the apparatus and may have been responsible for the somewhat high outgassing rate of 0.3 millitorr hr^{-1}. Nevertheless the gas samples, which ranged in purity from 99.99 to 99.0%, were probably not contaminated sufficiently during the experiments to affect the results seriously since all the gases were molecular. There was no evidence of contamination by electronegative gases.

In the absence of details of the ranges of gas number density used and the agreement obtained between the values of D/μ recorded at the same value of E/N but with different N, it is difficult to assess the reliability of the data of Cochran and Forester. Examination of the final data for hydrogen and nitrogen does not help to clarify the situation, since the data for nitrogen are in satisfactory agreement with accepted values, whereas the hydrogen data differ by as much as 20%. Moreover the values of D/μ at low values of E/N in CO_2 do not appear to approach $\frac{3}{2}\kappa T$ as E/N approaches zero, as is evident in the results of later experiments.[20,31] One must conclude, therefore, that the results for the other gases, CH_4, C_2H_4, and C_3H_6, are to be regarded with considerable reservation.

11.3.3. EXPERIMENTS OF WARREN AND PARKER. An extensive set of measurements of D/μ in both monatomic and diatomic gases was made by these authors[20] to complement the drift tube studies of Phelps and his coworkers described in Chapter 10. Since the aim of the programme of which these investigations formed a part was to determine elastic and inelastic cross sections for very-low-energy electrons, provision was made for making measurements at low temperatures and particular attention was

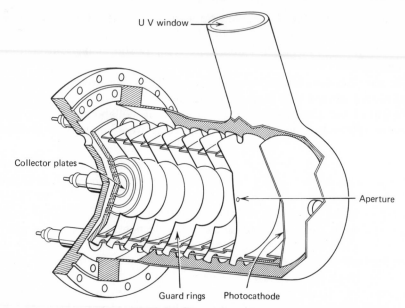

FIG. 11.6.   The apparatus for measuring D/μ used by Warren and Parker.[20] The apparatus is shown here in the longer configuration.

paid to the problems associated with measurement at very low values of E/N.

The geometry of the apparatus used by Warren and Parker was essentially that proposed by Huxley, although the diameter of the source hole was such that it could not be assumed automatically to be small enough to act as a point source. However, the design of the apparatus, which is shown in Figure 11.6, incorporated several new features and was designed to meet UHV specifications. The outer vacuum envelope was constructed of nonmagnetic stainless steel and was provided with a side tube capped with a quartz window. The electrodes were made of Advance metal, both the shields attached to the guard electrodes and the segments of the lower electrode overlapping to prevent the penetration of unknown fields into the drift region. Electrons were released photoelectrically from the gold-coated photocathode by ultraviolet light admitted to the apparatus through the quartz window. All the electrode surfaces, except the photocathode, were coated with colloidal graphite to minimize the scattering of UV light within the diffusion chamber and to reduce contact potential differences. In a separate experiment[32] the uniformity of the surface potential of graphite-coated surfaces was determined using a vibrating capacitor technique. From these experiments it was concluded that the contact potential differences within the apparatus were reduced to about 10 mV. As will be described in Section 11.3.5, contact potential differences of this magnitude, if they had occurred between the segments of the receiving electrode, could have seriously affected the measurements when the cathode-to-anode voltage was small. However, the actual distribution of potential over the collector was not determined and, in any case, would have been altered by the rigorous outgassing procedures that were followed. "Irregular results" were observed at the lowest voltages, which were of the order of 1 V, and were attributed to this cause.

The effects of space charge repulsion were looked for but found to be absent with the small collector currents that were used, ranging from 10^{-11} to 10^{-13} A. Some difficulty was experienced with background currents caused by electrons emitted from the electrode surfaces by scattered light, but these currents, amounting to less than 5% of the total currents, were measured and allowed for in the calculation of the current ratios.

The experimental difficulties encountered by Warren and Parker are best described by discussing the results of their experiments with helium at 77 K. These results were obtained using the longer of their two diffusion chambers ($h\approx9$ cm, $b/h\approx0.07$ and 0.14). With these geometries the current ratios can be analyzed with the aid of equation 11.8 with negligible error for the reasons given in Section 11.2.4. A more serious error arises from the neglect of the finite size of the source aperture, which in this instance was 0.25 cm in diameter (but see Section 11.2.4).

The spread of the data (see Figure 11.7), which for most values of E/N were taken over a wide range of values of N, was disappointingly large (of the order of 30%) and was several times greater than the estimated sum of the experimental errors. Moreover it was observed that the errors were apparently systematic rather than random and that they would have been considerably reduced if a slightly different ratio curve had been used to derive the values of D/μ from the current ratios. Warren and Parker surmised that the systematic errors in their results were probably due to field distortion within the apparatus. They therefore constructed another apparatus with a symmetrically placed transverse cut in the anode in order to obtain an estimate of the distortion by measuring the inequality between the currents falling on either side of the cut. Typically an inequality of 10% was recorded. Since the apparatus had been constructed with great care, these results indicated the extreme sensitivity of the current ratios to an unknown source of inhomogeneity in the electric field. Moreover it was concluded that contact potential differences were not the primary cause of the inhomogeneity since large errors still occurred even when the cathod-to-anode voltage V was large. Similarly it did not seem possible to attribute the results either to insulating surface layers, magnetic fields, or incorrect

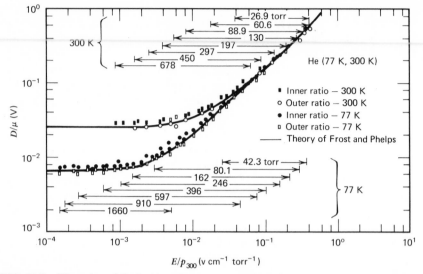

FIG. 11.7. Warren's and Parker's results for D/μ in helium at 77 and 300 K, based on an analysis of the current ratios using equation 11.7. The range of E/p_{300} that could be investigated at each normalized pressure $p_{300} = 300\,p/T$ is shown. The full curve was calculated by Frost and Phelps from an analysis of drift data. (From ref. 20.)

temperature measurements, or to a slight misplacement of the transverse cut in the specially constructed apparatus. It was concluded that the distortion had arisen from some unknown geometric factor (a conclusion borne out by the work described in Section 11.3.5) and that the only solution to the difficulty was to find an empirical ratio curve for use in analyzing the results of subsequent experiments.

The graphical procedure for finding the curve, which is described in the original paper, rested on three assumptions:

1. That each current ratio is a function only of $V/(D/\mu)$.
2. That D/μ is a function only of E/N and T.
3. That $D/\mu \to \kappa T/e$ as $E/N \to 0$.

The success of the empirical ratio curve may be seen from Figure 11.8, in which the values of D/μ have been rederived from the original current ratios. Values calculated from the cross section that is compatible with the drift data from the same laboratory are shown for comparison. It can be seen that the spread of the data has been greatly reduced and that the values are in good agreement with the calculated curve. The same procedure was equally successful when applied to the current ratios measured with the short tube; moreover, the results obtained with all combinations

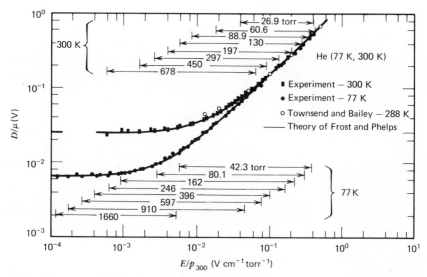

FIG. 11.8.   The data shown in Figure 11.7, reanalyzed using the empirical ratio curve. (From ref. 20.)

of b and h were found to be in satisfactory agreement after reduction in this way. The empirical ratio curve was therefore used to analyze the data from all subsequent experiments.

Warren and Parker used their apparatus to measure D/μ in He, Ar, N_2, H_2, D_2, CO, and CO_2 at both room and low temperatures for values of E/N in the range $3 \times 10^{-4} \lesssim E/N(\text{Td}) \lesssim 10$. Some of their results are presented in Chapter 14.

11.3.4.   EXPERIMENTS OF COTTRELL AND WALKER.   Cottrell and Walker[30] measured values of D/μ in a number of polyatomic gases as part of a programme to determine momentum transfer cross sections and to investigate low-energy inelastic resonance scattering.[30,33] The geometry of their apparatus is shown in Figure 11.9. In practice, although their apparatus was designed to enable three values of the inner disk diameter b to be used, only the values 0.526 and 1.026 cm were employed. The agreement between the results obtained with the two modes of division was found to be within the limits of experimental error. In effect, equation 11.7 was used to analyze the results, but although the experimental error is not stated it seems unlikely that, with the geometry used, the differences between the values of D/μ derived from equations 11.7 and 11.8 would lie outside experimental error.

The values of $k_1 = (D/\mu)/(\kappa T/e)$ for methane and ethylene, which are

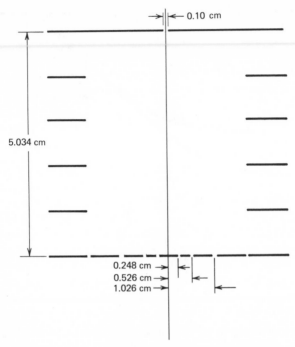

FIG. 11.9.   The dimensions of the apparatus used by Cottrell and Walker.

presented graphically, appear to be superior to those of Cochran and Forester and to the earlier results of Brose and Keyston[34] (for methane) and Bannon and Brose[35] (for ethylene), particularly the values at low E/N, which for each gas approach unity as $E/N \rightarrow 0$. The results are therefore probably the most reliable for these gases, while the values for CD_4, SiH_4 and SiD_4, C_2H_6, C_3H_8, and CH_3OCH_3 are the only ones available.

11.3.5.   EXPERIMENTS OF CROMPTON, ELFORD, AND OTHERS. The experiments of Huxley and Zaazou,[28] Crompton and Sutton,[13] and Crompton, Huxley, and Sutton,[13] and the later analysis by Huxley and Crompton[14] of some of the results of these experiments, left several unresolved problems, for example, the very large discrepancy between the results for air[13,28] and the incompatibility of equation 11.7 with the experimental results. Furthermore, a comparison of the results of many swarm experiments up to this time showed that the discrepancies between the results from various groups, particularly the results for low values of E/N, were often greater than could be accounted for in terms of known sources of experimental error. These facts, together with the increasing demand for transport data of high accuracy for low-energy electrons, prompted a long series of experiments to examine the factors that limit the accuracy of measurements of this kind. This work is described in the papers by Crompton and Jory,[15] Crompton, Elford, and Gascoigne,[36] and Crompton, Elford, and McIntosh.[37] The problems that are common to most if not all such experiments are described in detail by these authors, and since satisfactory solutions were found to almost all of these problems we conclude this section with a summary of their work.

(a) Apparatus

Two apparatuses, designed for somewhat different purposes, were used in these experiments. Figure 11.10 shows the first apparatus,[15] which was designed to enable a wide choice of values of the central collector diameter $2b$ and length h to be made while at the same time maintaining adequate geometrical accuracy. The dimensions were made variable for two reasons. First, experiments to determine D/μ in the presence of attachment, as described in Section 11.4.2, require that measurements be made with at least two values of h. Second, the dimensions best suited for accurate measurement near thermal equilibrium $(D/\mu \rightarrow \kappa T/e)$ are different from those required for measurements in a heavy monatomic gas, where $D/\mu \gg \kappa T/e$ even when E/N is small, as may be shown by using equation 11.8. Rewriting the equation in the form

$$R = 1 - \frac{1}{\left[1 + (b/h)^2\right]^{1/2}} \exp\left(-\lambda h \left\{ \left[1 + \left(\frac{b}{h}\right)^2\right]^{1/2} - 1 \right\}\right) \quad (11.59)$$

we see that, for a given value of b/h, R depends only on the parameter $\lambda h = V/2(D/\mu)$, where V is the cathode-to-anode voltage. When $D/\mu \rightarrow \kappa T/e$, R becomes independent of E/N and the nature of the gas; consequently the ratio then depends only on V. Since V must exceed a certain minimum value if the accuracy of the results is not to be significantly reduced by electrode surface effects, it follows that small values of b and large values of h are required for the current ratios to be approximately equal to 0.5 and therefore measurable with maximum accuracy. Conversely, when D/μ is large for small values of E/N (e.g., in argon[20]), measurements are facilitated by having larger values of b and smaller values of h.

Variable values of b and h were achieved through multiple division of the anode and a screw mechanism to vary its height. By connecting appropriate sections of the anode together, b could be given the values 0.5,

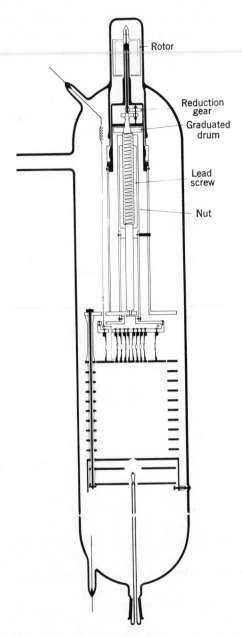

— Rotor

Reduction
gear

Graduated
drum

Lead
screw

Nut

FIG. 11.10. Crompton's and Jory's apparatus[15] for measuring D/μ, which enabled a number of combinations of b and h to be used.

1.0, 1.5, and 2.0 cm, while the anode could be set coplanar with each of the nine guard electrodes to within 0.002 cm, giving integral values of h between 1.0 and 10.0 cm.

Although the apparatus was not designed to operate at low temperature, it was constructed of nonmagnetic metals and glass throughout (apart from the sapphire bearings in the reduction drive and at the lower end of the lead screw) and could therefore withstand baking at moderate temperatures. These precautions ensured that outgassing rates were sufficiently low to allow precise measurements even in the heavy monatomic gases. Details of the mechanical design are given in the original paper.

The second apparatus,[36,37] shown in Figure 11.11, was designed for a more limited application: the determination of values of D/μ at low temperatures, particularly at low values of E/N. It was therefore possible to use a single combination of b and h ($b=0.5$, $h=10.0$ cm), thereby greatly simplifying the construction. A detailed discussion of the choice of values for b, h, the diameter of the guard electrodes, and the diameter of the source hole is given in ref. 36. Apart from the absence of the height adjustment mechanism, there are important differences in the construction of the guard electrodes and the anode, the reasons for which will be described subsequently. Generally these differences are the outcome of attempts to improve the accuracy of the measurements through increasing the uniformity of the electric field within the diffusion chamber. The vacuum envelope of the apparatus was designed for immersion in liquid nitrogen, while the electrode structure was designed to withstand cooling to this temperature without significant distortion.

As will be described subsequently, the results of measurements of this kind are particularly prone to error because of distortion of the receiving electrode. Therefore the construction shown in Figure 11.12 was adopted, in which the segments of the electrode, which are made of copper 0.6 cm thick, are mounted on thin-walled copper cylinders integral with the electrode and vacuum brazed to ceramic cylinders. During cooling the differential contraction between the copper electrode and the ceramic cylinder is absorbed by distortion of the thin-walled cylinder. The success of this technique in preserving the geometry of the electrode was proved by the fact that no short-circuiting between the disk and the annulus occurred at low temperature even though the gap was only 0.005 cm, and that pilot tests using an optical flat showed that the disk and annulus remained undistorted and coplanar at room and low temperature once the whole electrode had come to temperature equilibrium.

In the first series of experiments with these apparatuses a platinum filament made of 0.005-cm-diameter wire was used as the electron source. Temperature stability at room temperature was improved by surrounding

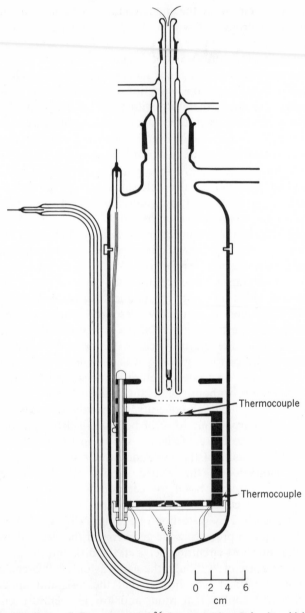

FIG. 11.11. The apparatus of Crompton et al.[36] for measuring D/μ, in which special attention was paid to the uniformity of the electric field.

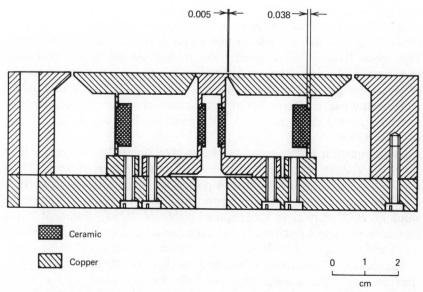

0.005 →|←   0.038 →|←

Ceramic

Copper

0   1   2
cm

FIG. 11.12. The receiving electrode, designed to minimize distortion at low temperatures, which replaced the lower electrode of the apparatus shown in Figure 11.11. (From ref. 37.)

the filament with a water-cooled jacket, or alternatively by immersing the whole apparatus in a water bath. In later experiments[38] the filament was replaced by an americium radioactive source. Americium produces volume ionization mainly by the emission of alpha particles, but the radiation also contains a weak gamma component. Initially it was necessary to compensate for small leakage currents that were produced within the diffusion chamber by the gamma radiation, but subsequently the insertion of a 3-mm lead shield immediately above the source electrode reduced this effect to an insignificant level. The radioactive source is greatly superior to thermionic sources because of its stability and the absence of thermal effects. Its only disadvantage is the low level of ionization at low pressures.

(b) Design Criteria and Determination of Experimental Parameters

Choice of dimensions; dimensional tolerances. One of the main factors that govern the accuracy of the experiments is the choice of the dimensions a, b, and h, and the width of the annular gap separating the central disk and surrounding annulus of the receiving electrode (see Figure 11.1). For a given value of D/μ, equation 11.59 shows that the choice is determined by the range of values over which R can be measured accurately and the range of values of E that can be used. The current-measuring equipment used for these experiments[36] enabled the ratio to be determined to within 0.3% at $R=0.2$ and 0.1% at $R=0.9$, resulting in error limits of $\pm 0.4\%$ for D/μ. For $R \approx 0.5$ the error from this source is reduced to less than 0.1%. The upper limit of E is usually determined by electrical breakdown elsewhere in the apparatus, and therefore for a particular apparatus depends on the gas and on E/N, but the lower limit, which is usually the more restrictive, is determined by the onset of significant errors from contact potential differences and surface effects. It is usually a few volts per centimetre.

Using equation 11.59, it can be shown that when D/μ is greater than about 0.1 V there is no great difficulty in choosing values of b and h that will lead to current ratios within the prescribed range when fields of 10 V cm^{-1} or more are used. The pressure can then be chosen to obtain the value of E/N that gives the required value of D/μ. Additional check measurements can subsequently be made by either increasing or decreasing E and p, keeping E/N constant and the current ratio within the range.

The choice of dimensions becomes more difficult, however, as D/μ decreases, since adequate divergence of the stream occurs only for smaller values of E. On the other hand, E cannnot be reduced indefinitely for the reasons already given, and, consequently, suitable current ratios can be obtained only with small values of b and large values of h. It has been

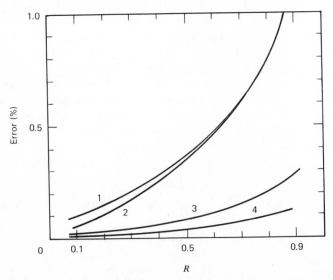

FIG. 11.13.   Curves showing the variation with R of the error introduced by using a source hole of 1-mm diameter (assuming n constant over the hole). Curve 1: $h=2$, $b=0.5$. Curve 2: $h=10$, $b=0.5$; $h=5$, $b=0.5$. Curve 3: $h=10$, $b=1.0$. Curve 4: $h=10$, $b=1.5$. (From ref. 15.)

shown[36] that an adequate, although not ideal, choice of dimensions for measurements near thermal equilibrium at 77 K ($D/\mu \sim 0.01$ V) is $b = 0.5$ cm and $h = 10$ cm. For the sake of simplicity the rest of the discussion will be restricted to an apparatus of these dimensions, although other combinations of b and h are discussed in the references.

The size of the source hole and the accuracy with which it must be located on the axis of symmetry of the apparatus were discussed by Crompton and Jory.[15] Values of D/μ corresponding to given values of R have been calculated taking account of the finite size of the source hole;[15,20] $\Delta(D/\mu)/(D/\mu)$ may then be defined as the difference between any such value and the value calculated from equation 11.59, which assumes a point source. Figure 11.13 shows $\Delta(D/\mu)/(D/\mu)$ plotted as a function of R for various values of b and h, the source hole diameter being taken as 1.0 mm.[15] It can be seen that when $b = 0.5$ cm significant errors occur as R approaches the upper limit of 0.9. A smaller hole would obviously be preferable but cannot ordinarily be used because of the difficulty in obtaining adequate current.

Although the finite size of the source can contribute a significant error, its placement is surprisingly uncritical. An off-axis displacement of the source of 0.005 cm, which is the alignment tolerance for the apparatus shown in Figure 11.11, causes a negligible error ($<0.1\%$) for all values of R within the range.

A compromise also has to be found in deciding on the width of the annular gap in the anode. From equation 11.52 (or 11.53) it can be seen that when $d/h \approx 1$ an error of 0.0005 cm in b (i.e., 0.1%) will lead to an error of 0.2% in the measured value of D/μ. A gap of this width would be necessary, therefore, if the maximum error in D/μ from uncertainty in the effective diameter of the central disk (which is normally taken to the centre of the gap) were to be less than $\pm 0.1\%$. In practice a gap width of 0.005 cm has been used as a compromise between a small uncertainty in b, a reasonably low capacity between the disk and the annulus, and an adequate separation to avoid contact during temperature cycling and consequent damage to the edges of the electrodes. Although the maximum uncertainty with this gap width is $\pm 1\%$, the error is probably considerably less than this because the electrons incident on the gap would be expected to divide approximately equally between the electrodes.

With the form of construction shown in Figure 11.11, it has been shown[36] that, with one exception, dimensional tolerances can be achieved such that the errors in D/μ arising from errors in the geometry of the apparatus are less then 0.1%. The exception is the effective diameter of the central disk. With the optical techniques of measurement that have so far been used, the uncertainty in b arising from the tolerance of ± 0.005 cm on the diameter of the disk and on the diameter of the hole in the annulus leads to a maximum uncertainty of $\pm 0.4\%$. However, experiments with positive ions at low values of E/N (for which D/μ is approximately equal to $\kappa T/e$) suggest that the actual error from this source is considerably less.

Field distortion. Apart from the geometrical factors already discussed, another consideration related to the construction of the apparatus is of equal importance, namely, the uniformity of the electric field within the diffusion chamber. As we have already seen, Warren and Parker[20] concluded that inadequate uniformity of the field in their apparatus was the main cause of the disagreement between their measured and calculated current ratios. This problem is now discussed briefly. A more detailed account will be found in the paper by Crompton et al.[36]

The production of a uniform electric field within a cylindrical chamber of the dimensions commonly used in experiments of this kind is not straightforward because of the relatively small ratio $2c/h$ of the diameter of the chamber to its length. Such a configuration results in the so-called guard electrodes playing an important part in establishing the field rather than serving simply to reduce edge effects, as they would if c/h were very much larger. When small errors occur in the placement of electrodes of the type shown in Figure 11.10, or when the electrodes are slightly distorted, a radial component of the field is introduced. Similar field distortion can arise from the proximity of the vacuum envelope (either metal or glass) when an open guard electrode structure of this type is employed. The presence of such fields within the diffusion chamber has been shown to be a significant source of error.[36]

These considerations led to the development of the electrode structure shown in Figure 11.11. The design is based on the principle that it is better to achieve a high order of uniformity near the axis of the apparatus, where the electron number density is significant, that is, within a cylindrical volume of radius somewhat larger than b, than to attempt to achieve uniformity over the whole volume. This principle rests on the fact that any electrons that are far removed from the axis will almost certainly arrive at the annular electrode regardless of some nonuniformity near the cylindrical boundary. The potential distribution within an apparatus of the dimensions shown has been analyzed by Crompton et al.,[36] using a formula given by C. A. Hurst,[39] and has been shown to lead to a field of adequate uniformity over a cylinder of surprisingly large radius. Figure 11.14, which shows a cross section through equipotential surfaces associated with one module of the guard electrode system, illustrates this point. It may also be seen from the figure that the equipotential surfaces that cut the electrodes midway along their length are plane surfaces; consequently the cathode and anode may be located at any of these planes without disturbing the

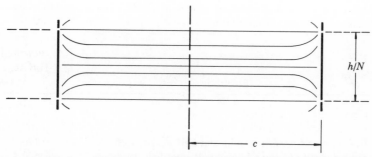

FIG. 11.14. Section through the equipotential surfaces associated with one module of a guard electrode system with N elements of the type shown in Figure 11.11. The equipotential plane surfaces passing through the gaps, and the surface midway between them, can be seen.

field distribution. A structure formed in this way has the advantage that the electrodes are of substantial thickness, which ensures that they can be machined accurately and are not prone to subsequent distortion, and that the volume contained by them is well shielded from the effects of external electric fields.

A less obvious source of field distortion is departure of the surface of the collecting electrode from that corresponding to a true geometrical plane. Radial fields arising from relatively abrupt changes in contour are more likely to occur in the vicinity of the annular gap. Moreover it is in this region that such fields have maximum effect since they cause a spurious distribution of electrons about the gap. The extreme sensitivity of the current ratios to field distortion in this vicinity is demonstrated by the fact that in one apparatus a small lip about 0.0005 cm in height on part of the inner edge of the annular electrode caused systematic errors of more than 1%. For this reason errors from this source are likely to be more serious than other errors from field distortion that have geometrical inaccuracy as their origin.

Finally, field distortion may arise from contact potential differences and surface effects. Whereas distortion that is caused by geometrical inaccuracy leads to errors in D/μ that, to first order, depend only on R and not on E,[20,36] effects that result from surface phenomena give rise to errors in D/μ that vary approximately in proportion to $1/E$. This characteristic behaviour is the basis of one method of deciding which, if either, of the two types of error is present in a set of measurements. Alternatively experiments may be carried out with both positive ions and electrons, in which case errors from geometrical inaccuracy are found to be similar for both electrons and ions, whereas those due to contact potential difference are of different sign.[36] Surface charging effects are more difficult to isolate and are usually identified through a time dependence of the results.[37]

TABLE 11.3. Values of additional potential which, when applied to a given electrode (see Figure 11.11), change the value of D/μ by 1% when $E = 3$ V cm^{-1}

Electrode	Potential Difference (mV)
Central disk (or annulus)	6
Central disk and annulus	100
Collecting electrode and adjacent half guard electrode	170
Source electrode and adjacent half guard electrode	500
Guard electrode 1	350
Guard electrode 2	500
Guard electrode 5	2500
Set of guard electrodes	130

As might be expected, contact potential differences have the greatest influence when they occur in the vicinity of the annular gap in the collecting electrode. This is illustrated by Table 11.3, which shows the effect of applying abnormal potential differences between various electrodes of the apparatus shown in Figure 11.11. It is evident that unless the potential differences across the surface of the receiving electrode can be made very small they will constitute the largest source of error when E is small. It is also evident that the accuracy of any results taken with field strengths of less than about 1 V cm^{-1} is very questionable since to produce surfaces over which the variation in potential is less than a few millivolts has proved difficult.[20,36]

(c) Factors Determining the Energy Distribution Function

The electron energy distribution and hence the transport coefficients are determined by the ratio E/N, the gas temperature T, and the gas itself. The measurement of each of the four quantities necessary to determine E/N (i.e., h, p, T, and the cathode-to-anode voltage V) was discussed in Section 10.6, where the necessary vacuum and gas-handling techniques were also described. An additional requirement that results from the extreme sensitivity of these experiments to the presence of negative ions is discussed in the following subsection.

(d) Measurement of the Current Ratios and the Effect of Space Charge Repulsion

In order to avoid significant space charge repulsion in the diffusing stream it is necessary to ensure that the electron number density is very low. As a consequence the current density at the receiving electrode is low,

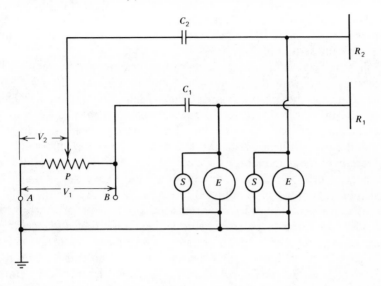

FIG. 11.15. Schematic diagram of induction balances used for precise measurement of current ratios.

particularly when E/N is small, and total currents as small as 10^{-12} A must be used. Since it is frequently necessary to measure the current distribution when 20% or less of the total current is falling on the disk or annulus, the current ratio must be measured with high accuracy. Furthermore, to avoid effects similar to those already described resulting from nonuniformity in potential over the collecting electrode, the potential of the disk and annulus must not depart appreciably from earth potential in the course of the measurement. Hence the method used to measure the current ratios must be capable of measuring with high precision the ratio of two currents, the smaller of which may be as small as 10^{-13} A, while maintaining the segments of the receiving electrode at or near earth potential.

The method that was used in the later experiments, which is described in ref. 36, is a modification of the technique of Crompton and Sutton.[13] Their method employed a pair of integrating induction balances and was developed from the method originated by Townsend[6]; the principle is shown in Figure 11.15. A ramp generator connected to points A and B produces an accurately linear sweep voltage which is applied to the capacitor C_1. A known fraction of the voltage is applied to a second capacitor C_2 by means of the potentiometer P. The capacitors, each approximately 30 pF, were designed to eliminate leakage currents from dielectric soakage[13] and are matched to within 0.01%. With this arrangement two constant displacement currents i_1 and i_2 are generated, the ratio of which is determined simply by the setting of P.

Full advantage of this technique can be taken only by making it an integrating one, with the integration commencing when $V_1 = V_2 = 0$ and terminating just before the ramp voltage reaches its maximum value. To effect the integration a correctly phased signal operates the electromagnetic earthing switches S, unearthing both systems when the voltages are zero and earthing them again near the end of the sweep. Throughout the measurement the potentials of the electrodes R_1 and R_2 are monitored with the electrometers E, and the amplitude of the sweep voltage, and, subsequently, the setting of P adjusted to maintain each system to within 0.2 mV of earth potential. The required accuracy can be obtained without difficulty by using a potentiometer with a resolution and accuracy of better than 0.1%, together with an integration time of the order of 1 min, while the technique ensures that the uniformity of potential of the receiving electrode is adequately maintained throughout the measurement.

Despite the use of low current densities a slight dependence of the current ratios on the magnitude of the total current was observed in some experiments,[37,40] the values of D/μ calculated from the current ratios increasing linearly with the increase of the total current (see Figure 11.16).

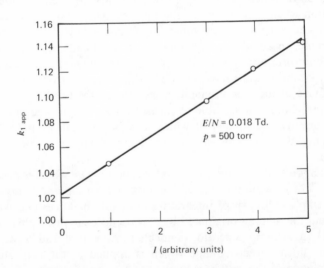

FIG. 11.16.   Typical variation of $k_1 = (De/\mu)/\kappa T$ with total current in nitrogen as observed by Crompton et al. The current dependence was probably caused by oxygen present as an impurity at less than 1 ppm. The maximum current was approximately 3×10^{-12} A.

The effect was greatest at low temperatures, at the lowest values of E/N, and with the highest gas number densities. It was also found that the effect was barely observable at the commencement of a series of experiments, that is, shortly after a new sample of gas had been admitted to the apparatus, but that it steadily increased with time, suggesting that the phenomenon was associated in some way with an increasing level of contamination of the sample. The rate of contamination at room temperature was estimated to be somewhat less than 0.1 ppm hr^{-1} at 500 torr. However, much of the contaminant was condensible and therefore could not have been responsible for the effect at 77 K, where it was most easily observed.[41]

These facts suggested that the phenomenon was caused by the presence of negative oxygen ions in the diffusing stream, three-body attachment to form O_2^- being known to have a significant rate coefficient for very-low-energy electrons.[42] This hypothesis was tested in a series of experiments in which hydrogen was contaminated with traces of oxygen ranging from 0.5 ppm to 3 parts in 10^{10}. Somewhat surprisingly, the current dependence of the results, even at the lowest level, was larger than had previously been observed, indicating that the suggested explanation was not unreasonable.

The effect of space charge on the divergence of a diffusing stream was analyzed by Liley.[43] When the process of attachment accompanies diffusion and drift, the equations that describe electron motion in a steady stream are

$$\text{div}\,\mathbf{j} = -\bar{\nu}_{\text{at}}n \qquad (11.60)$$

and

$$\mathbf{j} = n\mu\mathbf{E} - D\,\text{grad}(n), \qquad (11.61)$$

where $\bar{\nu}_{\text{at}}$ is the attachment collision frequency. The symbols n, $\mathbf{j}$, μ, and D have their usual meanings, and $\mathbf{E}$ is the local electric field, which is the sum of the applied uniform axial field $\mathbf{E}_a$ and the field due to space charge $\mathbf{E}_s$, that is,

$$\mathbf{E} = \mathbf{E}_a + \mathbf{E}_s. \qquad (11.62)$$

In what follows it is assumed that the perturbation to $\mathbf{E}_a$ due to space charge is sufficiently small that f_0 is unaffected by the small variations in $\mathbf{E}$ from point to point in the diffusion chamber.*

---

*Liley's analysis is based on the further assumption that f_0 is unaffected by density gradients, but the solutions that he developed are for the case in which axial diffusion is ignored. Therefore the arguments of Section 11.2 suggest that the results of his analysis are not likely to be seriously affected.

The corresponding equations for the ions are

$$\operatorname{div} \mathbf{j}_I = \bar{\nu}_{at} n \tag{11.63}$$

and

$$\mathbf{j}_I = n_I \mu_I \mathbf{E} - D_I \operatorname{grad}(n_I), \tag{11.64}$$

where the subscript I denotes that the symbol applies to the negative ions.

When the expression for $\mathbf{j}$ from equation 11.61 is used in equation 11.60, and the corresponding substitution for $\mathbf{j}_I$ is made in equation 11.63, the following equations for n and n_I, expressed in cylindrical coordinates, are obtained:

$$\frac{1}{\rho} \frac{\partial}{\partial \rho} \left(\rho \frac{\partial n}{\partial \rho} \right) + \frac{\partial^2 n}{\partial z^2} - 2\lambda \frac{\partial n}{\partial z} - 2\alpha_a \lambda n$$

$$= \frac{2\lambda}{E_a} \left\{ \frac{1}{\rho} \frac{\partial}{\partial \rho} [\rho(nE_{sr})] + \frac{\partial}{\partial z}(nE_{sz}) \right\} \tag{11.65}$$

$$\frac{1}{\rho} \frac{\partial}{\partial \rho} \left(\rho \frac{\partial n_I}{\partial \rho} \right) + \frac{\partial^2 n_I}{\partial z^2} - 2\lambda_I \frac{\partial n_I}{\partial z} + 2\alpha_a \lambda_I \frac{W}{W_I} n$$

$$= \frac{2\lambda_I}{E_a} \left\{ \frac{1}{\rho} \frac{\partial}{\partial \rho} [\rho(n_I E_{sr})] + \frac{\partial}{\partial z}(n_I E_{sz}) \right\}, \tag{11.66}$$

where $\lambda = W/2D$, $\alpha_a = \bar{\nu}_{at}/W$, and E_{sr} and E_{sz} are the radial and axial components, respectively, of the electric field due to space charge. The third and final equation is Poisson's equation, namely,

$$\frac{1}{\rho} \frac{\partial}{\partial \rho} (\rho E_{sr}) + \frac{\partial}{\partial z} (E_{sz}) = \frac{(n_I + n)e}{\epsilon}, \tag{11.67}$$

where ϵ is the permittivity of the medium.

The solution of this set of nonlinear equations is difficult and requires an approximate treatment such as that developed by Liley. If the effects of both diffusion and space charge in the axial direction may be neglected, and E/N is small enough to ensure that $\lambda \simeq \lambda_I \simeq eE/2\kappa T$, it can be shown that the current ratio R (as defined in Section 11.2) must satisfy the following equation:

$$\rho \frac{\partial}{\partial \rho} \left(\frac{1}{\rho} \frac{\partial R}{\partial \rho} \right) - 2\lambda \frac{\partial R}{\partial z} = 2\gamma (1 - \beta \exp - \alpha_z z) \frac{R}{\rho} \frac{\partial R}{\partial \rho}, \tag{11.68}$$

TABLE 11.4. Partial tabulation of coefficients a_{10} and a_{11}

ν	$a_{10}(\nu)$	$a_{11}(\nu)$	ν	$a_{10}(\nu)$	$a_{11}(\nu)$
0.00	0	0	1.00	0.2085	0.0542
0.05	0.0326	0.0091	1.20	0.1977	0.0507
0.10	0.0614	0.0170	1.40	0.1824	0.0462
0.20	0.1087	0.0298	1.60	0.1651	0.0413
0.40	0.1707	0.0462	1.80	0.1472	0.0364
0.60	0.2012	0.0538	2.00	0.1299	0.0317
0.80	0.2115	0.0557			

where $\beta = \Phi(1 - W_I/W)$, $\gamma = \lambda I/2\pi\epsilon W_I E_a$, I is the total current, and Φ is the fraction of I carried by the electrons at $z = 0$.

With the boundary conditions adopted by Hurst and Liley[18] and for small values of γ (i.e., when the effects of space charge repulsion are small), equation 11.68 can be solved for R to give the following expression for the fraction of the total current that falls on the central disk ($\rho = b$) of the receiving electrode (cf. equation 11.58):

$$R = \left[1 - \exp\left(-\frac{\lambda b^2}{2h} \right) \right] - \gamma(1-\beta)a_{10} - \frac{20}{7}\gamma\beta a_{11}(1 - \exp -0.7\alpha_a h),$$

(11.69)

where a_{10} and a_{11} are coefficients that are functions of $\nu = \lambda b^2/2h$. Values of the coefficients for various values of ν are listed in Table 11.4. Equation 11.69 is accurate to within about 1% provided that $0 \leqslant \nu \leqslant 2$.

In the limiting cases corresponding to a stream consisting only of electrons or ions ($\Phi = 1$ or 0), equation 11.69 reduces to

$$R = \left[1 - \exp\left(-\frac{\lambda b^2}{2h} \right) \right] - \frac{a_{10}\lambda I}{2\pi\epsilon W E_a},$$

(11.70)

where W is the appropriate drift velocity (i.e., the drift velocity of the electrons or ions).* This equation may be used to account for (a) the linear increase of the apparent values of D/μ for electrons or ions as I is

*This equation is valid for any value of λ, since a solution can be obtained without the restriction $\lambda \approx \lambda_i$ ($\approx eE/2\kappa T$) when only one species of charged particle is present.

increased and (b) the experimentally observed magnitude of this effect for K^+ ions in hydrogen at room temperature. It also predicts that, for an electron stream free of negative ions, space charge effects will be unobservable under nearly all experimental conditions provided that currents of the order of 10^{-12} A are used.

Equation 11.69 can be used to show that, even when negative ions contribute only slightly to the total current, the divergence of the stream is largely determined by the effect of space charge. Moreover, the effects observed by Crompton et al.[40,41] can be explained quantitatively on the basis of space charge repulsion by O_2^- ions if reasonable estimates are made for the various parameters in the equation. However, further experimental work is required to test the theory more adequately.

(e) Results

The techniques that have been described in this section have been applied to normal hydrogen and parahydrogen, D_2, and N_2 at 293 and 77 K, and to helium and CO_2 at 293 K.[1] The data from these experiments are given in Chapter 14.

TABLE 11.5. Typical sample of the experimental results for D/μ obtained by Crompton and others using the techniques described in this section. The results are for helium at 293 K[63]

E/p_{293} (V cm^{-1} torr^{-1})	E/N (Td)	D/μ (V) Pressure (torr)						Best estimate of D/μ (V)
		40	80	150	200	400	500	
0.007	0.02124						0.0302	0.0302
0.010	0.0303					0.0338	0.0337	0.0337
0.015	0.0455				0.0404	0.0402	0.0400	0.0400
0.02	0.0607				0.0468	0.0467	0.0468	0.0468
0.03	0.0910			0.0607	0.0607	0.0604	0.0602	0.0604
0.05	0.1517		0.0876	0.0876	0.0879	0.0876	0.0874	0.0874
0.07	0.2124		0.1141	0.1141	0.1151			0.1141
0.10	0.303	0.1533	0.1536	0.1536	0.1543			0.1536
0.15	0.455	0.217(5)	0.216(8)	0.217(0)				0.217(0)
0.20	0.607	0.279(1)	0.278(6)	0.278(9)				0.278(9)
0.3	0.910	0.402	0.399					0.399
0.5	1.517	0.640						0.640
0.7	2.124	0.876						0.876
1.0	3.03	1.241						1.241

Table 11.5 contains some of the results for helium at 293 K. These results are typical of those obtained for the other gases. In every case the experiments were performed with an apparatus for which $b = 0.5$ and $h = 10$ cm. Consequently, although the ratio formula given by equation 11.8 was used to analyze the data, negligible error was incurred from the neglect of anisotropic diffusion. The use of this equation was found to lead to results that were independent of the gas number density (see, e.g., the agreement between the results recorded at different pressures shown in Table 11.5); thus there was no inconsistency in the data so analyzed, as was the case in the experiments of Warren and Parker. It may therefore be concluded that the measures taken to establish a uniform field in the diffusion chamber were successful. The agreement between the results obtained for a given value of E/N when the experimental parameters were changed over a wide range, and the accuracy with which each of the experimental parameters was determined, suggest that the error limits of $\pm 1\%$ claimed for most of the data are reasonable.

11.4. MEASUREMENT OF D/μ IN THE PRESENCE OF IONIZATION AND ATTACHMENT

When negative ions and electrons drift through a gas in an electric field, the ratio W/D is very much larger for the ions than for the electrons except when E/N is so small that both the electrons and the ions are almost in thermal equilibrium with the gas molecules, that is, $(W/D)_{\text{ions}} \cong (W/D)_{\text{electrons}} \cong eE/\kappa T$. It follows that when negative ions are formed by attachment from electrons in a diffusing stream the radial displacements of the ions after formation remain essentially unchanged as they drift towards the anode, whereas the electrons continue to disperse by diffusion. At the anode, therefore, the distribution of current for the mixed stream differs from that for a stream consisting only of electrons, a higher proportion of the current being concentrated near the axis. Furthermore, the detailed analysis given in Section 5.4 shows that when measurements are made with a lateral diffusion apparatus, keeping the value of E/N constant but using different values of N, the values of W/ND calculated from the current ratios are not consistent unless allowance is made for attachment.

For these reasons the influence of negative ions on experiments to measure W/D for electrons was recognized from the outset.[6] Subsequently Bailey,[10] Huxley,[2] and others developed techniques that took advantage of the characteristic behaviour of the two components of the mixed stream to measure attachment coefficients and the values of W/D for electrons in attaching gases.

The effect of ionization on the divergence of an electron stream is less obvious, however, and does not appear to have been recognized until much later. The first systematic treatment was given by Huxley[2] in a paper in which he analyzed the influence of both ionization and attachment and showed that the effect of ionization is to modify the structure of the stream in such a way that, when diffusion is isotropic, the coefficient $\lambda = W/2D$ that normally characterizes its divergence becomes $\eta' = (\lambda^2 - 2\lambda\alpha_i)^{1/2}$, where α_i is the ionization coefficient. Since both λ and α_i are proportional to N, η' is also proportional to N; consequently lateral diffusion experiments yield consistent values of η' when E/N is held constant but the number density is varied. Hence it is not possible to detect the presence of ionization in experiments of this kind, and an analysis of the current ratios that takes no account of ionization leads to consistent but incorrect values of W/D or D/μ.

Experiments to measure D/μ in the presence of attachment or ionization, or in cases in which both processes occur simultaneously, are less accurate than those described in the preceding section, but the data from them are nevertheless of considerable importance. A brief description of the methods that have been used will therefore be given.

11.4.1. BAILEY'S METHOD. The first experiments to determine D/μ and the attachment coefficient simultaneously were made by Bailey[10] in 1925. A more detailed account of these experiments and of subsequent experiments carried out in the same laboratory will be found in ref. 11.

A schematic diagram of Bailey's apparatus is shown in Figure 11.17. Electrons released photoelectrically from S drift towards E_0, together with negative ions that are formed subsequently by attachment. The electrode E_0 contains an aperture 4 mm wide which defines the stream initially. Electrodes E_2 and E_4, which contain apertures of the same width, intercept a fraction of the mixed stream, the remainder falling on E_5. Each aperture is sufficiently long that no electrons reach the extremities; consequently, for the purpose of analysis, each can be regarded as having infinite length. The distance between E_0 and E_2 equals the distance between E_2 and E_4.

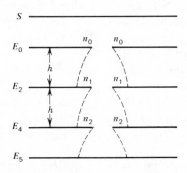

FIG. 11.17.   Schematic diagram of Bailey's diffusion apparatus.

The electrodes E_2, E_4, and E_5 are well insulated from their supports, so that the currents received by each can be measured with the aid of techniques similar to those previously described. The experiment consists in measuring at given values of E/N and N the ratio ξ of the currents received by E_4 and E_2 and the corresponding ratio ζ for E_5 and E_4; the measurements are then repeated at a different value of N, and the results analyzed using the following theory.[10]

If S_2 denotes the ratio of the current passing through the aperture in E_2 to the total current arriving at E_2, and S_4 is the corresponding ratio for E_4, then S_2 and S_4 are related to the measured current ratios ξ and ζ through the equations

$$S_2 = \frac{\xi(1+\zeta)}{1+\xi(1+\zeta)},$$

$$S_4 = \frac{\zeta}{1+\zeta}.$$

(11.71)

Let n_0 and N_0 be the number of electrons and ions, respectively, passing through the aperture in E_0, and let n_1, N_1 and n_2, N_2 be the corresponding numbers for E_2 and E_4. If there is no detachment, $n_0 \exp - \alpha_a h$ electrons arrive at the plane of E_2, while $n_0(1 - \exp - \alpha_a h)$ negative ions are formed between E_0 and E_2, h being the separation of these electrodes.* Assuming that both longitudinal diffusion and the finite width of the apertures (when acting as sources) can be neglected, it can be shown[10] that the fraction $R(\lambda h)$ of the unattached electrons arriving at E_2 that passes through its aperture is given by

$$R(\lambda h) = 1 - \exp\left(-\frac{\lambda b^2}{2h}\right),$$

(11.72)

where $2b$ is the width of the aperture, and $\lambda = W/2D$. It follows immediately that, for the electrons,

$$n_1 = n_0 R(\lambda h) \exp - \alpha_a h$$

$$= a n_0,$$

(11.73)

and

$$n_2 = a^2 n_0,$$

(11.74)

where

$$a = R(\lambda h) \exp - \alpha_a h.$$

(11.75)

---

*We use the symbol α_a in this analysis to be consistent with the neglect of longitudinal diffusion; actually, α_{at} is determined by this method.

Similarly, for the ions,

$$N_1 = N_0 R(\lambda_I h) + n_0 r(1 - \exp - \alpha_a h)$$

$$= N_0 R(\lambda_I h) + c n_0 \tag{11.76}$$

and

$$N_2 = N_1 R(\lambda_I h) + c n_1$$

$$= R^2(\lambda_I h) N_0 + [R(\lambda_I h) c + ca] n_0, \tag{11.77}$$

where r is the (unknown) fraction of the negative ions formed between the electrodes that passes through the aperture in E_2, $c = r(1 - \exp - \alpha_a h)$, and $\lambda_I = W_I / 2D_I$.

Finally, the numbers of electrons and ions arriving at the plane of each electrode are related to S_2 and S_4 (and hence to the measured current ratios ξ and ζ) through the equations

$$S_2 = \frac{n_1 + N_1}{n_0 + N_0}, \tag{11.78}$$

and

$$S_4 = \frac{n_2 + N_2}{n_1 + N_1}. \tag{11.79}$$

The relations expressed by equations 11.73 to 11.79 can be combined to give the following expression for a:

$$a = \frac{S_2[R(\lambda_I h) - S_4]}{R(\lambda_I h) - S_2}. \tag{11.80}$$

Consequently, since $R(\lambda_I h) = R(eEh/2\kappa T)$, it being assumed that the negative ions are in equilibrium with the gas molecules, the value of a corresponding to a given gas number density can be found from the measured values of ξ and ζ. Moreover, from equation 11.75 and from the fact that both λ and α_a are proportional to N, it follows that, if a_1 and a_2 are the values of a determined at N_1 and N_2, then

$$a_1 = R(\lambda_1 h) \exp - \alpha_{a_1} h$$

and

$$a_2 = R(\lambda_2 h) \exp - \alpha_{a_2} h$$

$$= R(m\lambda_1 h) \exp - m\alpha_{a_1} h,$$

where λ_1, α_{a_1} and λ_2, α_{a_2} are the values of λ and α_a at N_1 and N_2, respectively, and $m = N_2/N_1$. We therefore have two equations that can be solved for λ and α_a.

Bailey's method is ingenious in that it avoids the necessity for determining r theoretically, the calculation of this ratio being more difficult than the calculation of $R(\lambda h)$ or $R(\lambda_l h)$ because the ions are generated throughout the volume between the electrodes. However, the method has two shortcomings. In the first place, the apertures must be reasonably wide in order to allow a significant proportion of the electrons and ions to pass through them, and yet no allowance is made in the theory for the finite width when the apertures are treated as sources. Second, diffusion to the edges of the apertures and to the electrodes themselves, particularly by the electrons, is likely to invalidate the assumption that the fraction of electrons or ions passing through an aperture is the same as that falling on the corresponding portion of a geometrical plane. Check measurements made in hydrogen (and therefore in the absence of attachment) showed some anomalies for small values of R, that is, when W/D is small. At the time these anomalies were attributed to the finite length of the apertures, although they may, in fact, have been due to the diffusive effects referred to above.

In a later version of the apparatus[44] provision was made to vary the separation h between E_0 and E_2 and between E_2 and E_4; measurements were then made for a given value of E/N, using a separation h and gas number density N, after which both h and N were increased in the same ratio. The subsequent analysis of the results to determine D/μ and α_a was thereby considerably simplified.

Bailey and his collaborators made a comprehensive study of electron motion in attaching gases, the results of which are summarized by Healey and Reed,[11] but with few exceptions these results have not been checked by more modern experimental measurements based on a more detailed analysis of the mixed stream and better technology. One exception is the set of data for oxygen. Experiments with oxygen were subsequently made by Huxley, Crompton, and Bagot,[45] by Rees,[46] and by Naidu and Prasad.[47] These more recent experiments gave results for D/μ that are in satisfactory agreement with each other although there are discrepancies in the values of α_a/N. The agreement of these results with those of Healey and Kirkpatrick[48] is not good, however, suggesting that the overall experimental error associated with Bailey's technique is large. Consequently, when the results from this group are the only ones available, allowance should be made for the possibility of relatively large errors.

11.4.2. HUXLEY'S METHOD. After the extensive investigations of the Sydney group, no further measurements of this kind were made for a number of years until Huxley's analysis of electron diffusion and drift in the presence of ionization and attachment provided the basis for a more accurate method.

In the original paper[2] it was assumed that diffusion was isotropic and that negligible error was made in neglecting the diffusion of the negative ions. The same assumptions regarding the influence of the cathode and anode boundaries as were described in Section 11.2.2 were adopted in this analysis. With these assumptions, Huxley obtained expressions for the electron and ion number densities within a mixed stream originating from a point source and was thus able to calculate the currents to the various sections of the anode of a slightly modified version of the Townsend-Huxley diffusion apparatus (see Figure 11.18). Since the composition of the stream at the source S is unknown, the diameter of the central disk A_1 is made large enough to ensure that practically all the ions entering at S are collected by the disk. The distribution of electrons and ions falling on the annuli A_2 and A_3 (formed by connecting the three adjacent outer segments of the electrode together) is then unaffected by the initial composition. With this proviso and with the initial assumptions, the ratio $R = i_{A_2}/(i_{A_2} + i_{A_3})$ can be shown to be

$$R = 1 - \frac{\dfrac{h}{d_b}\exp(\lambda h - \eta' d_b) + \dfrac{\lambda h \alpha_a}{\eta'} \displaystyle\int_0^1 \exp(\lambda hs)\left(\exp\left[-\eta' h\left(\dfrac{b^2}{h^2}+s^2\right)^{1/2}\right] - \exp\left\{-\eta' h\left[\dfrac{b^2}{h^2}+(2-s)^2\right]^{1/2}\right\}\right)ds}{\dfrac{h}{d_c}\exp(\lambda h - \eta' d_c) + \dfrac{\lambda h \alpha_a}{\eta'} \displaystyle\int_0^1 \exp(\lambda hs)\left(\exp\left[-\eta' h\left(\dfrac{c^2}{h^2}+s^2\right)^{1/2}\right] - \exp\left\{-\eta' h\left[\dfrac{c^2}{h^2}+(2-s)^2\right]^{1/2}\right\}\right)ds} , \quad (11.81)$$

where b, c, and h are the dimensions shown in Figure 11.18, $d_b = (b^2 + h^2)^{1/2}$, $d_c = (c^2 + h^2)^{1/2}$, $\lambda = W/2D$, $\eta' = (\lambda^2 + 2\lambda\alpha)^{1/2}$, $\alpha = \alpha_a - \alpha_i$, α_a and α_i

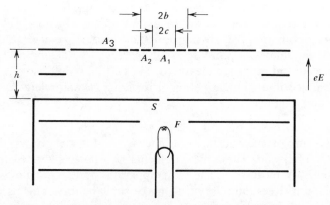

FIG. 11.18. Schematic diagram of a Townsend-Huxley diffusion apparatus for experiments with electronegative gases: $c = 0.5$ cm, $b = 1.0$ cm.

are the attachment and ionization coefficients, respectively, the source is considered to act as an isolated pole source, and it is assumed that a negligible proportion of the electrons and ions arrives at the anode outside A_3.

In a later paper, Hurst and Huxley[49] extended the theory to include lateral diffusion of the ions, so that the method can be applied, in principle, to situations in which it is no longer valid to assume that $W/D \ll W_I/D_I$. The theory given in Chapter 5 is an extension of this work to include anisotropic diffusion.

From an inspection of equation 11.81 the following facts emerge:

1. When $\alpha_a = 0$, the formula reduces to

$$R = 1 - \left(\frac{d_c}{d_b}\right)\exp -\eta'(d_b - d_c). \qquad (11.82)$$

Thus, as already mentioned, measurements of R enable the parameter $\eta' = (\lambda^2 - 2\lambda\alpha_i)^{1/2}$ to be determined, but not λ and α_i separately. Consequently, when ionization but not attachment accompanies drift and diffusion, D/μ can be determined from a lateral diffusion experiment only after α_i has been determined from a growth-of-current experiment.

2. When $\alpha_i = 0$, equation 11.81 becomes a function of the two parameters λ and α_a; that is, because of the form of the equation, R can no longer be expressed as a function of the single parameter η'. Consequently measurements of two current ratios made under different experimental conditions enable both λ (and hence D/μ) and α_a to be determined.

Several experimental procedures are possible, some of which have been used by different workers, as will be described subsequently.

3. When both α_a and α_i are nonzero, the equation becomes a function of the three variables λ, α_a, and α_i, all of which can be determined provided that three current ratios are measured. The fact that both λ and α_i can now be determined in a single experiment is perhaps a surprising result in the light of item 1 above.

Experiments based on this theory were first made by Huxley et al.[45] with the aim of measuring D/μ and α_a/N for oxygen in the range $15 < E/N$ (Td) < 60. These measurements were later extended to smaller values of E/N by Rees.[46] The same apparatus and techniques were used by Crompton, Rees, and Jory[50] to make similar measurements at higher values of E/N in water vapour.

The apparatus used in this series of experiments was described in Section 11.3.5 (see Figure 11.10). The features of the apparatus specifically incorporated for this work were the multiply divided anode and a mechanism for altering the anode-to-cathode separation h through a wide range of values. During the measurements the central disk of the anode was earthed, the distribution of current being found by measuring the ratio of the current received by the smallest annulus to the current received by the remaining insulated portion of the electrode.

Two procedures were used to make the measurements. In the first, measurements were made with two different values of h for given values of N and E. Equation 11.81 then provided two simultaneous equations relating λ and α_a to the current ratios measured at the two heights. The equations could therefore be solved to determine the unknown parameters. When applicable, this procedure has some advantage over an alternative one in which the pressure is varied because, using the apparatus shown in Figure 11.10, the measurements can be made with the same gas sample, thus eliminating possible errors caused by errors in pressure measurements and/or varying levels of impurity in the gas. As shown below, with this method a good separation can be effected between the influence of λ and that of α_a on the spread of the stream.

In the other procedure, different values of N were used for the measurements at the two heights. This procedure had to be adopted because of the limitations on the choice of the experimental parameters that were imposed by the need either to exceed a minimum potential difference between cathode and anode or to maintain the current ratios within certain limits for accurate measurement.

Although the simultaneous equations were solved with a computer in the later work,[50] a graphical method was used initially. This method will now be briefly described because it provides insight into the essential features of the procedure.

Equation 11.81 is first used to construct a set of curves of R as a function of η' for various values of α_a. Figure 11.19 shows the sets of curves for $h = 2$ and 5 cm. From the curves for $h = 2$ cm it can be seen that the current ratios are not greatly affected by attachment unless α_a is large. Consequently a first approximation to η' may be obtained from the current ratio measured at $h = 2$ cm by assuming that $\alpha_a = 0$. This value of η' can then be used to determine a first estimate of α_a from the current ratio measured at $h = 5$ cm, since the distribution of current at this distance from the source is markedly influenced by attachment (see Figure 11.19b). The first estimate of α_a is then used to estimate η' more accurately from the current ratio at $h = 2$ cm, and the iterative procedure continued until satisfactory convergence is obtained, usually after no more than two or three cycles.

After η' and α_a have been determined in this way, λ is calculated from $\lambda = \eta' - [2\lambda\alpha_a/(\eta' + \lambda)] \ (\cong \eta' - \alpha_a$ when α_a is small); D/μ is then calculated from $D/\mu = E/2\lambda$. The values of D/μ and α_a/N may be checked by repeating the measurements, using different gas pressures and possibly different combinations of h.

Naidu and Prasad used the same technique to make measurements in dichlorofluoromethane[51] and oxygen.[47] A schematic diagram of their apparatus, which was designed for concurrent measurements of electron drift velocities using an electrical shutter technique, is shown in Figure 11.20. Electrons were produced photoelectrically from a back-illuminated gold film deposited on a quartz disk. After traversing a region CC' in which they acquired the equilibrium energy distribution, the electrons entered the diffusion chamber through a relatively large source hole of 0.26-cm diameter.

No mechanism was provided for altering the length of the apparatus. Consequently it was necessary to dismantle it between the two sets of measurements, which were made with lengths of either 3.0 and 7.0 or 4.0 and 7.0 cm. It was also found necessary to remove the electrical shutters during the lateral diffusion experiments because they affected the divergence of the stream.

In analyzing the results of their experiments, Naidu and Prasad used two of the assumptions adopted earlier by Huxley, namely, that diffusion is isotropic and that lateral diffusion of the ions is negligibly small compared with that of the electrons. However, equation 11.81 could not be used because of the large diameter of the source hole, and an alternative ratio formula was therefore developed. Unfortunately few details of the derivation of the formula have appeared in the literature, and therefore it is not possible to assess the validity of the ratio formula even when the initial assumptions are justified. Like equation 11.81, the formula would, in any case, require modification to take account of anisotropic diffusion.

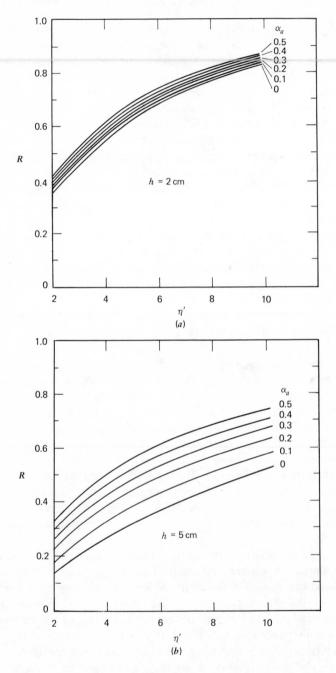

FIG. 11.19. Theoretical current ratio curves; R is plotted as a function of η' for various values of α_a.

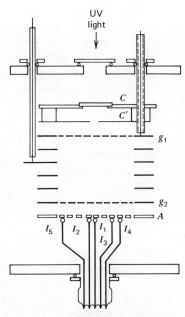

FIG. 11.20.   Schematic diagram of Naidu's and Prasad's apparatus.[51] Electrons were produced by photoelectric emission from the back-illuminated gold-coated surface of C. The grids g_1 and g_2, used for measurements of W with the same apparatus, were removed during the measurements of D/μ. The distribution of current over the anode A was determined by measuring the currents $I_1, I_2, \ldots, I_5$.

Experiments with CCl_2F_2 were made in the range $330 < E/N\,(\mathrm{Td}) < 620$, where both attachment and ionization significantly affect the distribution of current within the diverging stream. However, since the primary aim of the measurements was the determination of D/μ for the electrons, the results were not analyzed to determine α_a and α_i, an alternative procedure being adopted whereby values of D/μ were calculated from the current ratios, assuming values of α_a and α_i taken from other work.[52] Since an analysis of the results taken at two heights and with several gas pressures yielded reasonably consistent values of D/μ, it was concluded that the data for α_a and α_i were of adequate accuracy.

A more detailed analysis was made of the measurement in oxygen with the aim of determining the three parameters D/μ, α_a, and α_i. Several methods are available to obtain sufficient information about the structure of the stream to determine all three coefficients; Naidu and Prasad chose to measure pairs of current ratios at two heights ($h = 4$ and 7 cm). A graphical iterative procedure was then used to find values of the coefficients compatible with the ratios measured at two pressures (3 and 5 torr).

All the work described so far was based on solutions of the steady-state differential equation for n, appropriate to a spatially independent energy distribution function and hence isotropic diffusion. In obtaining these solutions, the boundary conditions described in Section 11.2.2 were adopted. In Sections 11.2.5, 11.2.7, and 11.2.8 it was stressed that, notwithstanding the fact that the energy distribution function is significantly spatially dependent, the semiempirical formula based on these assumptions leads to negligible error in the interpretation of lateral diffusion experiments, provided that the stream is sampled sufficiently far from the source ($eEh \gg \epsilon_{av}$) and close to the axis ($\rho \ll z$). However, when attachment and ionization occur, it is frequently not possible to satisfy these criteria for two reasons:

1. It may be necessary to use low pressures and small values of h in order to inhibit attachment; otherwise the stream will become predominantly ionic near the source, and the accuracy of the measurement of D/μ will be seriously impaired.

2. In order to overcome the problem of the unknown composition of the stream at the source, the current distribution at the anode can be analyzed only beyond a radius c that is large enough to ensure that all the ions entering at the source fall within this radius. The stream cannot, therefore, be sampled as close to the axis as in the experiments described in Section 11.3.

Since the two criteria cannot always be satisfied, there are likely to be errors in the data that have been obtained using equation 11.81 or its equivalents. An estimate of the order of magnitude of the errors arising from the use of the equation may be found by comparing the results so obtained with those derived from the same experimental data using equation 5.62, that is, the ratio formula which results from solving the steady-state differential equation for n assuming anisotropic diffusion.

In the experiments of Huxley et al.[45] the values of D/μ were determined largely from the current ratios measured at $h = 2$ cm, these ratios being only slightly affected by attachment. In order to estimate the errors in the published values of D/μ that results from the use of equation 11.81, therefore, it is sufficient to compare the values of λ calculated from this equation when $\alpha_a = \alpha_i = 0$ (i.e., equation 11.82 with $\alpha_i = 0^*$) with the values of λ calculated from the formula for R which is found from equations 5.62 when $\alpha_a = 0$, that is,

$$R = 1 - \frac{h/d_b' - (1/\lambda_L h)(1 - h^2/d_b'^2)}{h/d_c' - (1/\lambda_L h)(1 - h^2/d_c'^2)} \left(\frac{d_c'}{d_b'} \right) \exp{-\lambda_L(d_b' - d_c')}, \quad (11.83)$$

*Note that with $\alpha_i = 0$ the parameter η' in the equation becomes λ.

where

$$\lambda_L = W/2D_L, \quad d_b'^2 = h^2 + b'^2, \quad d_c'^2 = h^2 + c'^2,$$

$$b' = (D_L/D)^{1/2}b, \quad \text{and } c' = (D_L/D)^{1/2}c.$$

Using the values of D_L/D calculated by Lowke and Parker,[26] it can be shown that, for the experimental conditions that were used (i.e., $R\sim0.5$, $h=2$ cm, $c=0.5$ cm, $b=1.0$ cm), differences in λ (or D/μ) of the order of 5% result from the use of the alternative equations 11.82 and 11.83, the values of D/μ found from equation 11.83 being higher than those found from equation 11.82.

By contrast, differences in the values of α_a obtained from the two analyses arise not from differences between equations 5.62 and 11.81, but as a consequence of the use of the different values of λ to calculate α_a from the current ratios measured at the larger values of h, since the two ratio formulae converge for sufficiently large values of this parameter. The reasons for this will be explained below. As pointed out by Rees,[46] errors of the order of 1% in the values of λ result in errors of about 5% in α_a. It follows that the differences between the values of the attachment coefficient found from the two analyses differ by as much as 20 to 30%, the analysis allowing for anisotropic diffusion yielding the higher values.

As already remarked, in the limiting case $\alpha_a\to0$, equation 5.62 approaches equation 11.83, and significant differences exist between the two ratio formulae (equations 11.82 and 11.83) when h is small. On the other hand, provided that c is not too small, equations 5.62 and 11.81 predict the same values of R when h is sufficiently large. At first sight this result is surprising, although it is not necessary to look far for the explanation. The structure of the stream near the source as predicted by the *pole* solution and *isotropic* diffusion differs significantly from that predicted by the *dipole* solution and *anisotropic* diffusion. In the case of the electrons, the two solutions predict nearly the same distribution at a plane far from the source since lateral diffusion tends to eliminate the influence of the initial distribution. However, such is not the case for the negative ions, since it is assumed that all the ions, including those formed near the source and therefore distributed differently in the two cases, arrive at the plane with the radial displacement they had on formation. The explanation of the similarity of the calculated ratios lies in the fact that, for experimental reasons, the distribution of current is not sampled close to the axis. Consequently, the ions which contribute to the measured currents originate predominantly from electrons that have travelled a considerable distance from the source and whose predicted distributions are therefore essentially the same in the two cases.

A reanalysis of the experimental results of Huxley et al.[45] and of Rees[46] for oxygen has been undertaken by Rees,[53] using equation 5.62 (see Chapter 13). Since the results of Naidu and Prasad were also obtained from an analysis based on the assumption of isotropic diffusion, their data are also subject to errors of uncertain magnitude. It is possible that their published data are less in error from this assumption than is the work of Huxley et al. and Rees since the measurements of Naidu and Prasad were made with somewhat larger values of h. However, no firm conclusion can be drawn without a detailed analysis.

11.4.3. MEASUREMENTS OF D/μ WHEN IONIZATION IS SIGNIFICANT. After the early work of Townsend,[6,11] no further attempt was made to measure D/μ at high values of E/N until investigation in this more difficult regime was reopened by Lawson and Lucas[54,55] and by Crompton et al.[19]

The first of the recent measurements were reported by Lawson and Lucas,[54] who used the apparatus shown diagramatically in Figure 11.21. Electrons were produced by a glow discharge between a spherical copper cathode and electrode 1, the discharge being confined to this region of the apparatus by the glass cylinder C. Electrodes 2 to 5, which contained holes that were carefully placed to prevent line-of-sight paths between the discharge region and the diffusion chamber, were used to establish fields that controlled the current entering the diffusion chamber and to provide a region (between 3 and 6) in which the electrons acquired the appropriate steady-state energy distribution. Electrodes 6 and 7 formed the diffusion chamber, electrode 6 containing a 1-mm hole which served as the electron

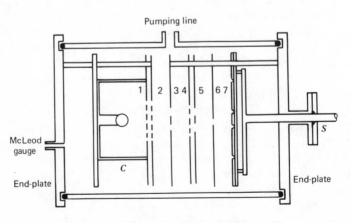

FIG. 11.21. Schematic diagram of Lawson's and Lucas's apparatus.[54]

source. Electrode 7 was divided in the usual manner into a central disk (of diameter 1.5 cm) and surrounding annuli. The electrode could be moved through a total distance of 2.5 cm by means of the shaft S, which entered the apparatus through the Wilson seal in the end plate. During the experiment, gas was continuously pumped and replaced, thus reducing contamination from the elastomers used for the Wilson seal and the O-rings between the end plates and the outer cylindrical envelope. In any case, the hydrogen contained initially about 100 ppm of N_2. Pressures in the range 0.2 to 1.4 torr were measured with a McLeod gauge. The error limit claimed for the pressure measurement was $\pm 2\%$. The experiment consisted in measuring the ratio of the currents to the disk and annulus at a series of values of h between 1.0 and 2.4 cm separated by intervals of 0.2 cm.

The results of these experiments were analyzed using equation 11.81 with $\alpha_a = 0$ and $c = 0$, that is,

$$R = 1 - \left(\frac{h}{d}\right)\exp - \eta'(d-h),\qquad (11.84)$$

where $\eta' = \lambda - \alpha_T$. This formula has the same status as the semiempirical ratio formula given by equation 11.8, that is, it may be derived by assuming that diffusion is isotropic and that the source hole behaves as a point source. Lawson and Lucas claimed to have shown that the formula can be established from a solution of the time-dependent differential equation for n that satisfies the boundary condition $n = 0$ at both cathode and anode, but the derivation contains a fallacy (see footnote on p. 381). Notwithstanding the use of the semiempirical formula, values of η' were obtained that were consistent over the range of experimental parameters used, the reasons for this success being similar to those given in Section 11.2.7.

It should be noted, however, that the use of equation 11.84, rather than a ratio formula based on theory which allows for anisotropic diffusion, may lead to incorrect values of D/μ even though consistent values of this parameter are obtained from an analysis of the experimental results, that is, even when the experimental conditions are such that η' derived with the aid of equation 11.84 is indistinguishable from $\eta(\lambda/\lambda_L)$ derived from equation 11.85 (see below).

From equation 5.55, the ratio formula allowing for anisotropic diffusion is

$$R \cong 1 - \left[1 + \left(\frac{1}{2} - \frac{D_L}{D}\right)\left(\frac{b}{d}\right)^2\right]\left(\frac{h}{d}\right)\exp\left[-\eta\left(\frac{\lambda}{\lambda_L}\right)(d-h)\right],\qquad (11.85)$$

where $\eta = \lambda_L - \alpha_T$. When $D_L/D \sim 0.5$, or when b/d is sufficiently small, this formula reduces to a form indistinguishable from equation 11.84 in so far as the analysis of the experimental results is concerned, although η' is replaced by $\eta(\lambda/\lambda_L)$. When equation 11.84 is used, values of λ (and hence D/μ) are determined from η' by using the relation $\lambda = \eta' + \alpha_T$. However, λ is determined from the modified parameter $\eta(\lambda/\lambda_L)$ that appears in equation 11.85 by using the equation $\lambda = \eta(\lambda/\lambda_L) + \alpha_T(\lambda/\lambda_L)$ (since $\eta = \lambda_L - \alpha_T$). Consequently the values of λ derived from equation 11.84 may require modification to account for the difference between the correction terms α_T and $(D_L/D)\alpha_T$. For the case of hydrogen, the magnitude and sign of these corrections can be seen from the results of Lawson and Lucas, which are reproduced in Figure 11.22. The upper curve in the figure shows the values of D/μ that would be obtained if η' (or $\eta\lambda/\lambda_L$) were to be equated to λ, that is, if the influence of ionization on the spread of the stream were to be ignored. The lower curve shows the values of D/μ corrected for the effect of ionization when diffusion is taken to be isotropic. The correction amounts to about 20% at $E/N = 300$ Td. If the data were to be analyzed using equation 11.85 rather than equation 11.84, the corrected curve would lie in the region bounded by curve 2 and a curve midway between curves 1 and 2, since, for the case of hydrogen in this range of E/N, D_L/D lies between 0.5 and 1.0 (see Chapter 14).

Experiments in hydrogen at high values of E/N were also undertaken

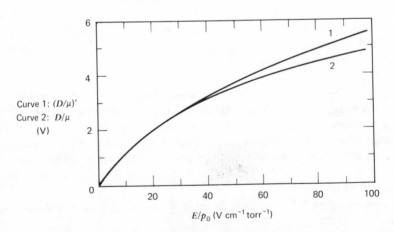

FIG. 11.22.   Lawson's and Lucas's experimental values of $(D/\mu)'$ $(= E/2\eta')$ and D/μ in hydrogen.[54] Here p_0 is the equivalent pressure reduced to 273 K. When analyzed on the basis of equation 11.85, the results lie on a curve between curves 1 and 2 (see text).

by McIntosh.[19] Although there were some similarities between these experiments and those of Lawson and Lucas, significant differences in procedure led to considerable differences in the experimental results.

The apparatus used for these experiments is shown in Figure 11.23. The electrodes, machined from Duralumin and subsequently gold coated, were approximately 21 cm in diameter and contoured to a modified Rogowski profile following the specification given by Bruce.[56] Electrodes of this diameter and shape were chosen with the aim of postponing the onset of electrical breakdown at high values of the product Nh, while at the same

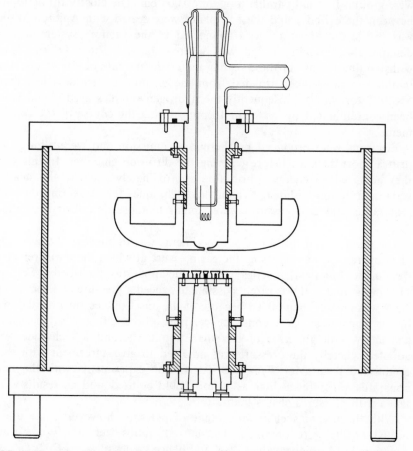

FIG. 11.23. Apparatus used by Crompton et al.[19] for measuring D/μ at higher values of E/N in situations in which Nh approached the values for electrical breakdown.

time preserving a highly uniform electric field near the axis of the apparatus. The anode was divided into three sections, the central disk of diameter 0.7 cm and the surrounding annulus of outer diameter 3.0 cm being mounted on a plate glass support to obtain adequate rigidity and insulation. The central regions of the electrodes were lapped flat, and the entire surface of each electrode was subsequently polished to a high surface finish. A hole of 1-mm diameter in the cathode served as the source of electrons.

The parallelism and spacing of the electrodes (nominally 2.0 cm) were maintained by mounting the electrodes rigidly on two substantial brass end plates, which were separated by a large-diameter glass cylinder whose ends were ground flat and parallel to within 0.003 cm. The closely fitting joints between the cylinder and the end plates were sealed with Apiezon W100 wax, while the apparatus was connected to the vacuum system with a demountable coupling sealed with a pre-outgassed Viton O-ring. Notwithstanding the use of these sealing materials, the rate of rise of pressure in the apparatus when isolated from the vacuum system was less than 5×10^{-6} torr hr^{-1}; consequently the hydrogen samples used would have been contaminated by less than 10 ppm during the course of the experiments.

Electrons were produced by thermionic emission and were allowed to drift a short distance before entering the diffusion chamber. In this way they acquired an energy distribution approximately equal to the steady-state distribution within the chamber. Techniques for measuring the current ratios and gas pressures were similar to those described in Section 11.3.5.

The main aim of this series of experiments was to improve the accuracy of measurement by increasing the gas pressure at which the measurements were made. In an earlier series in which the apparatus described in Section 11.3.5 was used,[57] it was rarely possible to exceed a pressure of 1 torr when E/N was greater than 60 Td because of the onset of electrical breakdown in the apparatus. As a consequence, pressures could not be measured absolutely with an error of less than about 3%, and a technique was adopted whereby the pressure was adjusted to give agreement with well established results for $E/N \leqslant 60$ Td. In the experiments with the new apparatus, pressures as high as 10 torr could be used, and no results were taken with pressures of less than 2 torr.

With the use of such a wide range of pressure, however, some unexpected results were obtained. The current ratios were analyzed, using equation 11.84, to give values of η' and hence values of $k_1' = eE/2\kappa T\eta'$. As already remarked, the values of η'/N (and hence k_1') derived in this way

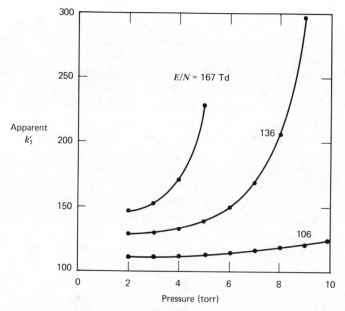

FIG. 11.24.   Observed variation of apparent values of k_1' in hydrogen with gas pressure. The approach to an asymptotic value at low pressure, and the very large increase as Nh approaches the breakdown value, can both be seen. (From ref. 19.)

are expected to be independent of pressure, even though a correction is required to convert them to the Townsend energy ratio $k_1 = (D/\mu)/(\kappa T/e)$ $= eE/2\kappa T\lambda = eE/2\kappa T(\eta' + \alpha_T)$. Figure 11.24 shows that the observed values of k_1' were, in fact, strongly dependent on the pressure, the dependence becoming more marked as E/N increases.* A similar result was noted by Lawson and Lucas as they increased the length of their apparatus while keeping E/N and p constant. They attributed the effect to the influence of secondary processes, and in a later paper[55] modified their analysis to take account of these phenomena (see below).

The results shown in Figure 11.24 indicate an anomalously large divergence of the stream at higher pressures. The explanation lies in the

---

*The origin of this dependence does not lie in the fact that equation 11.84 rather than equation 11.85 was used to analyze the results. If, for example, equation 11.85 had been used to analyze the current ratios corresponding to $E/N = 136$ Td, the value of k_1' at $p = 9$ torr would have been changed by less than 0.5%, and the value at 2 torr by less than 4% (assuming $D_L/D \leqslant 1$).

presence of a distributed source of electrons at the cathode which supplements the primary source, the additional electrons arising from secondary processes at the cathode induced by the incidence of either positive ions or photons on the cathode surface. The very large influence of the supplementary distributed source, as shown in the figure, may be understood when it is remembered that at electrical breakdown each electron leaving the cathode from the primary source produces one electron *at the cathode* as a result of the primary and secondary ionization processes. Thus, at breakdown, the combined strength of the secondary source just equals the strength of the primary source.

By extending the theory given in Section 11.2.3, Hurst and Liley were able to include secondary processes in their analysis of the Townsend-Huxley experiment.[18,19] Thus, to the term $p(\rho,0)$ in equation 11.18 that accounts for the axial component of the primary flux from the source hole, they added two additional terms, $S_I(\rho,0)$ and $S_p(\rho,0)$, to account for secondary electrons produced by the incidence of positive ions and photons, respectively. The equation therefore becomes

$$j_{z+}(\rho,0) = -\zeta_c j_{z-}(\rho,0) + S_c(\rho,0),$$

where

$$S_c(\rho,0) = p(\rho,0) + S_I(\rho,0) + S_p(\rho,0),$$

$$\tag{11.86}$$

$$S_I(\rho,0) = \gamma W\alpha_i \int_0^h n\,dz,$$

$$S_p(\rho,0) = \frac{\delta}{\alpha} W\alpha_i \int_\tau \frac{nz}{4\pi R^3}\,d\tau;$$

γ and δ/α are the positive ion and photon secondary coefficients. In calculating $S_p(\rho,0)$, the volume element $d\tau$ is taken to be situated at a height z above the cathode and a distance R from the point $(\rho,0)$ on the cathode, and the integration is performed over the whole volume between cathode and anode.

When the theory of Hurst and Liley was used to analyze the experimental results, it was found that the results could be adequately accounted for when only the photon effect was included. In this event it can be shown that the current ratio is given by

$$R = b\left[\exp(\eta'-\lambda)h\right]\left[1 - \frac{2\pi\Delta}{\eta'}\left(\frac{[\exp(\lambda-\eta')h]-1}{\lambda-\eta'}\right.\right.$$

$$\left.\left. - \frac{(\exp-2\eta'h)\{[\exp(\lambda+\eta')h]-1\}}{\lambda+\eta'}\right)\right] \times \int_0^\infty I(\xi)\,d\xi,$$

where

$$I(\xi) = \frac{J_1(b\xi)\exp\left\{\left[\lambda-(\eta'^2+\xi^2)^{1/2}\right]h\right\}}{1+\Delta M(\xi)},$$

$$M(\xi) = \frac{-2\pi}{(\eta'^2+\xi^2)^{1/2}}\left(\frac{\exp\left\{\left[\lambda-\xi-(\eta'^2+\xi^2)^{1/2}\right]h\right\}-1}{\lambda-\xi-(\eta'^2+\xi^2)^{1/2}}\right.$$

$$\left. - \frac{\exp\left[-2h(\eta'^2+\xi^2)^{1/2}\right]\left(\exp\left\{\left[\lambda-\xi+(\eta'^2+\xi^2)^{1/2}\right]h\right\}-1\right)}{\lambda-\xi+(\eta'^2+\xi^2)^{1/2}}\right),$$

$$\Delta = \frac{\eta'(\lambda-\eta')(\delta/\alpha_i)}{2\pi},$$

$$\lambda = \frac{W}{2D} \quad \text{and} \quad \eta'^2 = \lambda^2 - 2\lambda\alpha_i. \tag{11.87}$$

Equation 11.87 cannot be evaluated explicity, but it can be computed numerically to any required degree of accuracy by integrating from zero to a sufficiently large value of ξ.[18,19] In practice, R is calculated for given values of E/N and N for an apparatus of known geometry, using published data for α_T and sets of values of λ and δ/α. The values of λ and δ/α that give the best fit to the experimental data for a given value of E/N over the entire range of values of N are then the required values of these parameters.

A test of the agreement between theory and experiment can be made by taking a single value of the secondary coefficient and finding the values of D/μ that give calculated values of R that agree with experiment at each pressure. When the value of the secondary coefficient that gives the best overall fit is found, the constancy of the values of D/μ is a measure of the success of the theory. Table 11.6 shows the results of this test as applied to

TABLE 11.6.   The values of D/μ in hydrogen at $T = 293$ K, $E/N = 136$ Td, determined from the current ratios. At this value of E/N, the single value of δ/α which leads to the best overall fit was found to be 2.82×10^{-3}. The apparent values of $(D/\mu)' = E/2\eta'$ that are obtained using equation 11.84 directly are shown for comparison

p (torr)	2	3	4	5	6	7	8	9
$(D/\mu)'_{app}$ (V):	3.26	3.28	3.36	3.51	3.79	4.24	5.20	7.45
D/μ (V):	3.11	3.11	3.13	3.16	3.18	3.18	3.16	3.11

the results in hydrogen at $E/N = 136$ Td. The maximum disagreement between the values of D/μ determined in this way is less than 3%, despite the fact that the apparent values of $(D/\mu)' = E/2\eta'$ obtained by direct application of equation 11.84 vary by considerably more than a factor of 2.

Figure 11.25 shows the values of the secondary coefficient δ/α plotted as a function of E/N, the values obtained by Morgan and Williams[58] being shown for comparison.

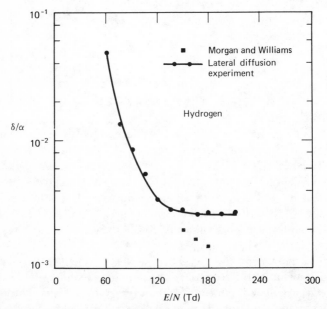

FIG. 11.25.   Variation of the secondary coefficient δ/α derived from the experimental results, assuming a negligible contribution from the positive-ion secondary process. (From ref. 19.)

At higher values of E/N the range of values of N that could be used was restricted. Consequently the analytical technique employed by Crompton et al.[19] could not be used to decide the relative importance of the two possible secondary processes. The values of δ/α for $E/N > 150$ Td that are shown in the figure are therefore the ones that provide the best fit to the experimental data using equation 11.87, but for the reason just stated it was not possible to establish their validity with any certainty. Below $E/N = 150$ Td, however, the data are more reliable since an extended pressure range could be used and because, in any case, the experiments of Morgan and Williams showed that the photon coefficient is an order of magnitude larger than the positive-ion coefficient in this range.

The results of Crompton et al. can be criticized on two counts. In the first place, equation 11.87 was derived from an analysis that did not take detailed account of the spatial dependence of the electron energy distribution or the influence of the boundaries, although some allowance for these phenomena was made semiempirically. Although the self-consistence of the published results suggests that the errors arising from the use of equation 11.87 may be small, it will be necessary to reformulate the theory before any definite conclusions can be reached about the results derived from the current ratios that were measured when secondary processes were significant. Second, even the values of D/μ that were derived from the low-pressure limit of η', that is, from measurements that were essentially uninfluenced by secondary effects, require some modification. Thus, although the current ratio formula (equation 11.84) used in these circumstances may not have been seriously in error, for reasons that have already been discussed, it may be more appropriate to apply a correction of $(D_L/D)\alpha_T$ rather than α_T to the values of η' in order to obtain the values of η and hence D/μ; that is, the data require the same adjustment as those of Lawson and Lucas.

A similar investigation of the effects of secondary processes was carried out by Lawson and Lucas,[55] using the apparatus already described (see Figure 11.21). The variation of η' with electrode separation which they had observed in their first set of experiments[54] for values of $E/N > 300$ Td alerted these authors to the influence of secondary effects, and as a consequence they extended both the theory and the range of their experiments to take account of them.

By following a procedure broadly similar to that of Hurst and Liley, but differing from it considerably in detail, Lawson and Lucas calculated values of R, taking account of secondary electrons released at the cathode by the incidence of positive ions and/or photons.* Their calculations

*The theory was again based on the representation of the electron number density as the sum of the contributions from a set of travelling images and is therefore open to the objection discussed on p.441. For this reason only a broad outline of the procedure will be given here.

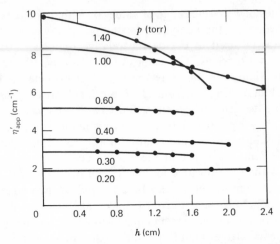

FIG. 11.26. Typical experimental results of Lawson and Lucas for hydrogen;[55] η'_{app} is plotted against electrode separation h at $E/N = 305$ Td for the various pressures used.

showed that, if equation 11.84 were to be used to derive η' from the current ratios in circumstances in which secondary processes play a significant rôle, the apparent values of η' for a given E/N so derived would be dependent on both h and N. The explanation of the dependence on h is similar to the explanation for the dependence on N that has already been given: since the number of secondary electrons, positive ions, and photons produced in the gap is a function of the product Nh, the number of electrons produced *at the cathode* by the secondary process (i.e., the strength of the secondary source unaccounted for in equation 11.84) is a function of h as well as of N.

In the original paper, graphs of the theoretical variation of the apparent values of η'/N with h and N are given. Since the graphs may require modification, however, they are not reproduced here.*

Figure 11.26 shows the observed variation of η'_{app} with h and the gas pressure p at $E/N = 305$ Td. From the figure it can be seen that η'_{app} is not directly proportional to N but that it increases more slowly than N. Thus $(D/\mu)'_{app}$ $(= E/2\eta'_{app})$ increases with increasing N in accordance with the results in Figure 11.24. Furthermore, it can be seen that the effect becomes more pronounced as the electrode separation h increases (cf. values at $p = 0.20$ and 1.40 torr and $h = 0.4$ and 1.6 cm).

For the reasons given above, one would expect the values of η'_{app}/N determined from the current ratios measured at a given value of h to approach the true value of η'/N as N approaches zero, that is, as the

*See footnote on p.449.

number of secondary electrons produced in the gap (and hence the strength of the secondary source at the cathode) approaches zero. Such would not be the case if the electrons were not in quasi equilibrium everywhere in the gap, or if the effects of spatial variation of the energy distribution function due to density gradients were to significantly influence the current ratios, but neither of these effects was included explicitly in the theory (nor were they included in the theory of Hurst and Liley). However, Lawson and Lucas did not make use of this fact but adopted a procedure whereby the current ratio was measured using a range of pressures and several values of h, after which a double extrapolation of the results was made, in the following way, to obtain the true values of $(D/\mu)'$.

By using data such as those shown in Figure 11.26, values of $\eta'_{app}(p,h)$ corresponding to $h=0$ were first obtained by suitable extrapolation. The values of $2\eta'_{app}(p,0)/E$ (i.e., $[(D/\mu)'_{app}]^{-1}$ corresponding to $h=0$) were then plotted as a function of p, as shown in Figure 11.27. Finally these curves were extrapolated to zero pressure to find the values of $(D/\mu)'$ corresponding to $p=0$, $h=0$.

Two other sets of measurements were made. In the first of these the total current to the anode was measured as a function of electrode separation. Equation 5.53 was then used to determine α_T. Second, with the largest electrode separation possible (2.4 cm), the breakdown voltage was measured and used to determine the values of the secondary coefficient. Finally, the values of λ and hence D/μ were determined using the relation

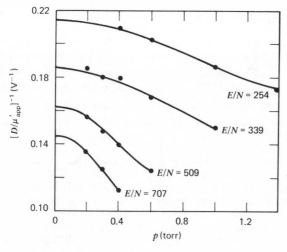

FIG. 11.27. Values of $[(D/\mu)'_{app}]^{-1}$ in hydrogen corresponding to $h=0$, plotted as a function of p for various values of E/N (Td). (From ref. 55.)

$\lambda = \eta' + \alpha_{\mathrm{T}}$. Data were published for D/μ, α_{T}, and the secondary coefficient for values of E/N as high as 1000 Td.

Although the results of Lawson and Lucas are in any case open to the same objection as that made in relation to their earlier results, as well as in relation to those of Crompton et al. (i.e., they were obtained from an analysis that assumed the spatial invariance of the energy distribution function and hence isotropic diffusion), the validity of their work can be further questioned because of the fallacy in the theoretical derivation referred to on p.449.

From the discussion of this section it is evident that further work is required to establish acceptable data for D/μ at high values of E/N, although some reliance can be placed on the data of Crompton et al., obtained under conditions where secondary effects are unimportant. Even in this instance, however, some uncertainty remains. In the first place the influence of secondary process makes it impossible to determine by experiment alone the conditions under which the use of an asymptotic ratio formula (e.g., equation 11.85*) is justified. Thus it is impossible to use the invariance of η/N with N at constant E/N as signifying the validity of the formula, since secondary processes lead to a strong dependence of the apparent values of η/N on N at large values of N, the condition under which the formula might otherwise have most validity. Second, the exact magnitude of the correction to be applied for the effect of primary ionization (see p.441) is unknown, since values of the ratio D_L/D have been neither measured nor calculated for $E/N > 30$ Td.

11.5.   DIRECT MEASUREMENT OF THE DIFFUSION COEFFICIENT D

11.5.1.   CAVALLERI'S METHOD. Before the work of Cavalleri, no successful method had been developed for the accurate measurement of D as a function of E/N, although several attempts to determine thermal diffusion coefficients accurately[59,60] had met with some measure of success. Further reference to these methods will be made later in this section. Cavalleri's method[4] depends for its success on a novel procedure for

---

*As is the case when there is no ionization, the asymptotic limit of this formula is the same as the asymptotic limit of the formula for an isolated pole source, i.e., $R = 1 - (h/d') \exp[-\eta(d'-h)]$ (see equation 5.50). However, as expected, unlike the case of no ionization, the values of D/μ determined from the current ratios using the asymptotic limits of the ratio formulae are dependent on whether or not isotropic diffusion is assumed. If, for simplicity, we consider only the formulae based on the pole source, the two expressions are $R = 1 - (h/d) \exp[-\eta'(d-h)]$ (i.e., equation 11.84) and the one given above. The correction to be applied to the "spreading factor" η' in the first instance is α_{T}, while the correction to the spreading factor $(D_L/D)\eta$ in the second instance is $(D_L/D)\alpha_{\mathrm{T}}$ (see p.442). The quantities η' or $(D_L/D)\eta$ and α_{T} are, of course, those determined directly by experiment.

detecting electrons that does not rely on the detection of electric charge;[59] in some respects it is similar to the method used by Breare and von Engel for the measurement of W (see Section 10.4).

The principle of the method can be understood by referring to Figure 11.28, which is a schematic diagram of the original apparatus used by Cavelleri. The cylindrical glass envelope of diameter $2R$ contains two plane and parallel metal electrodes positioned normal to the axis of the cylinder. The electrodes are separated by a distance H. Both H and R are accurately determined during the construction of the diffusion cell. A side

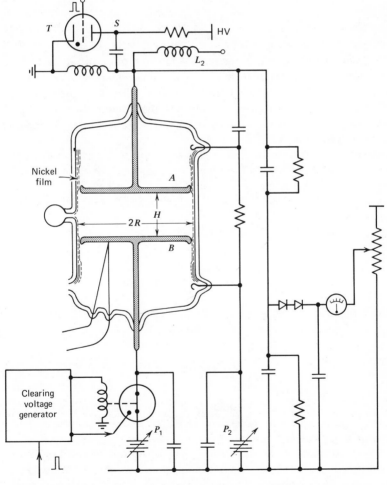

FIG. 11.28.   Schematic diagram of the apparatus used by Cavalleri[4] for the measurement of D by electron density sampling. The pulsed thryatron T is used to ground the point S thus producing the sampling pulse.

arm of small diameter is situated midway between the electrodes and terminates in a thin-walled spherical bulb, which provides access for a roughly collimated soft X-ray beam directed along the mid-plane of the cell.

For simplicity we consider first the application of the method to the measurement of thermal diffusion coefficients. To initiate a measurement, a pulse of X-rays of several microseconds duration is triggered, causing volume ionization of the low-pressure gas sample in the cell. In the absence of all processes other than pure diffusion, the diffusion equation governing the electron density $n(\rho, z, t)$ within the cell at subsequent times is

$$\frac{\partial n}{\partial t} = D\nabla^2 n. \tag{11.88}$$

The general solution of this equation is the sum of an infinite set of modes, but only the fundamental mode contributes significantly after a time interval of the order of the time constant τ_{11} for this mode (see Section 5.5). Consequently, for $t \gg \tau_{11}$ the density distribution within a cylindrical volume of the dimensions shown in Figure 11.28 is closely represented by equation 5.67, that is,

$$n(\rho, z, t) = A_{11} \left[\exp\left(-\frac{t}{\tau_{11}}\right) \right] J_0\left(\frac{c_1 \rho}{R}\right) \sin\frac{\pi z}{H}, \tag{11.89}$$

where A_{11} is a constant, $c_1 = 2.405$, and

$$\tau_{11} = \frac{1}{D\left[(\pi/H)^2 + (c_1/R)^2\right]}. \tag{11.90}$$

It follows that the total number of electrons remaining in the cell after t seconds is

$$N_e = A \exp\left(-\frac{t}{\tau_{11}}\right), \tag{11.91}$$

where A is a constant; consequently, if it is possible to record the decay of N_e with time, equations 11.91 and 11.90 can be used in turn to determine first τ_{11} and then D.

The problem of determining the decay of N_e was solved by Cavalleri in the following way. After a time interval of t_1 ($> \tau_{11}$) has elapsed following the initial X-ray pulse, a damped RF signal (the "sampling pulse") is applied between the electrodes. The initial amplitude of the pulse is large

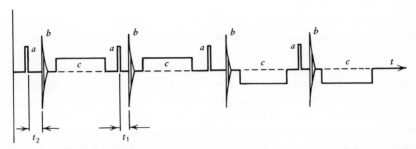

FIG. 11.29. Complete duty cycle: *a* represents X-ray pulse; *b*, sampling pulse; *c*, clearing field; t_1 and t_2 the two different alternate delays between the ends of X-ray bursts and sampling pulses. (From ref. 4.)

enough to cause each electron in the cell to initiate a localized electron avalanche. Consequently, since electronic excitation of the atoms accompanies ionization, each avalanche produces a small pulse of light, and the total light output from the cell is proportional to the number of electrons remaining in the cell at time t_1. The sequence (i.e., X-ray pulse followed by sampling pulse) is then repeated, this time with a delay of t_2 ($> t_1$). The relative magnitude of the light signals is recorded at a series of delay times, using the technique described below, so that the complete decay curve can be determined.

The diffusion time constants for the electrons are typically some tens of microseconds, and the delay times are therefore usually less than a millisecond. Hence repetition rates of a few hundred hertz are possible provided that some provision is made to remove the positive ions that are produced in large numbers by the sampling pulse but that diffuse about 100 times more slowly than the electrons. If these ions are not removed, the concentration of ions in the cell may build up to the point where ambipolar diffusion significantly affects the diffusion time of the electrons. At the end of each cycle, therefore, the clearing voltage generator (Figure 11.28) applies an appropriate potential to the lower electrode, thus creating within the diffusion cell an electric field that rapidly removes the ions. In order to avoid polarization effects that might be produced if the ions were always swept to the same electrode, the output of the generator is reversed after one complete cycle, that is, after each pair of X-ray pulses. The complete duty cycle is illustrated in Figure 11.29.

The number of electrons and ions produced by the X-ray pulse must be kept small, again in order to avoid ambipolar diffusion. Consequently the number of electrons remaining in the cell when the sampling pulse is triggered is usually no more than a few tens and may even average less than one at large time delays. Each measurement is therefore subject to a large statistical fluctuation, and it is necessary to repeat the cycle of measurement many thousands of times to obtain adequate accuracy. The technique used by Cavalleri for measuring and recording the relative populations was as follows.

The diffusion cell was enclosed within a light-tight box on one wall of which was mounted a photomultiplier tube. The axis of the tube lay in the mid-plane of the cell and was perpendicular to the direction of the X-ray beam. The signal from the photomultiplier, suitably amplified, drove an amplitude-to-time converter that produced a pulse whose duration was proportional to the amplitude of the signal. The pulse held open a gate to admit a number of oscillations from a quartz oscillator, the number being proportional to the length of the pulse and hence to the amplitude of the original signal. Two additional gate circuits were used to address the signals alternately to two counters in such a way that one recorded a total count proportional to the sum of the pulse amplitudes corresponding to t_1 while the other recorded the sum for t_2. For some measurements a technique for improving the signal-to-noise ratio was used whereby the signals were superimposed on a rectangular pulse.[4] A block diagram of the detection system and the system used for triggering the thyratron sampling pulse generator is shown in Figure 11.30.

Careful attention was paid to the problems of reducing contact potential differences within the apparatus to a minimum and of compensating for those which could not be eliminated. As in all experiments, this problem becomes particularly difficult as $E/N \to 0$; in some ways the problem is more troublesome in this experiment than in lateral diffusion and drift velocity experiments because of the higher pressures that may be used in the latter. In an attempt to produce surfaces of uniform work function surrounding the diffusion space, a layer of nickel was evaporated onto the stainless steel electrodes and cylindrical glass wall. The resistive film on the wall terminated in the planes of the electrodes; beyond these planes the nickel layer was covered with a conducting layer of tin oxide. The conducting layers were provided with suitable terminations to enable each to be connected to the appropriate electrode.

Notwithstanding these precautions, it was necessary to compensate for a small residual potential difference (~ 10 mV) between the electrodes, and a larger potential difference (~ 100 mV) between the electrodes and the cylindrical wall. Compensation was made by adjusting P_1 and P_2 (Figure 11.28) until maximum diffusion times were recorded.[4]

A typical set of experimental results is shown in Figure 11.31. It is immediately obvious that the initial part of the curve is described by equation 11.91. However, the portion of the curve corresponding to longer time delays indicates that more electrons remain in the chamber at these

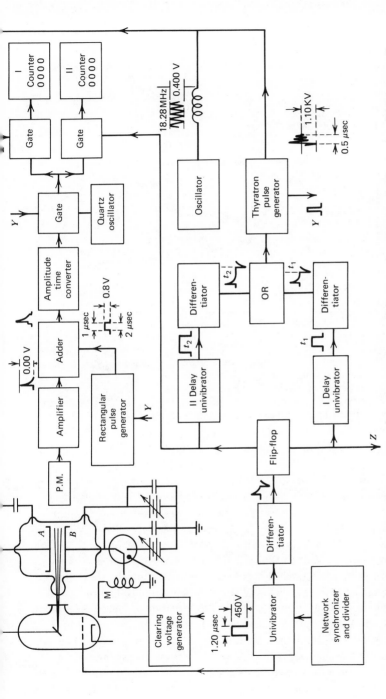

FIG. 11.30. Block diagram of Cavalleri's experiment.[4] At appropriate times the synchronizer actuates the X-ray tube, which is normally cut off, applies the sampling pulse to the electrodes of the diffusion chamber, and finally applies the clearing voltage, the sequence of events being shown in Figure 11.29. For measurement with $E/N \neq 0$, an oscillator supplies a 16-MHz signal to the electrodes with amplitude variable between 0 and 400 V. The "measuring chain" begins with a photomultiplier (P.M.) and ends with the two pulse counters. At the end of the experiment the ratio of the counts is equal to the ratio $\overline{N}_{e1}/\overline{N}_{e2}$ of the mean values of the numbers of electrons remaining in the chamber after time delays t_1 and t_2.

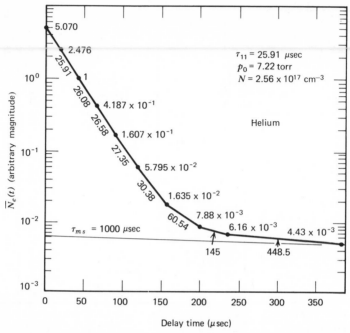

FIG. 11.31. Typical semilogarithmic plot of mean electron population $\bar{N}_e(t)$ (arbitrary magnitude) versus delay time, obtained by Cavalleri for helium at room temperature. The linearity of the initial portion of the curve, corresponding to loss by diffusion, can be seen, as well as the linear portion at large delays due to ionization by metastables. The time constants derived from various sections of the curve are indicated. (From ref. 4.)

times than can be accounted for in terms of a simple diffusion process. The explanation is as follows.

During the initial ionization by the X-ray pulse, metastable helium atoms are produced as well as electrons, positive ions, and photons. The time constant for the decay of the mestastables, due to a number of de-excitation processes, is typically of the order of 1 msec. When a trace of impurity having an ionization potential lower than one or more of the metastable levels is present in the cell, it is possible for the metastables to ionize the impurity molecules, the rate of production of electrons by this process decreasing exponentially as the metastable population decays. Provided that the level of impurity is low, the number of electrons present in the cell at short time delays is scarcely affected by this process. At long time delays, on the other hand, the difference in the time constants for diffusion and metastable decay results in the virtual disappearance of the initial electron population before the metastable population has been greatly decreased. Therefore the decay of the electron population is then largely governed by the decay of the metastable population, with the result that the decay curve plotted on a semilogarithmic scale exhibits further linear portions at large times (see Figure 11.31).

Let N, N_{im}, and n be the number densities of the gas molecules, impurity molecules, and electrons, respectively. Then the equation for the electrons becomes

$$\frac{\partial n}{\partial t} = D\nabla^2 n + KN_{im}, \tag{11.92}$$

where $K = \Sigma \beta_s N_{ms}$, N_{ms} is the number density of the sth type of metastable, and β_s is the rate constant for collisional de-excitation of the metastables by the impurity molecules. When equation 11.89 is used, equation 11.92 becomes

$$\frac{\partial n}{\partial t} = -\frac{n}{\tau_{11}} + N_{im} \sum_s \beta_s N_{ms}.$$

The metastable number densities N_{ms} decay according to the law

$$N_{ms} = N_{ms0} \exp\left(-\frac{t}{\tau_{ms}}\right)$$

where

$$\tau_{ms} = \frac{1}{D_{1ms}\left[(\pi/H^2) + (c_1/R)^2\right]/N + \gamma_s N + \delta_s N^2 + \beta_s N_{im}}.$$

and D_{1ms}, γ_s and δ_s are, respectively, the diffusion coefficient and the two- and three-body de-excitation rate coefficients at unity density for the sth metastable. Consequently, when ionization by the metastables is significant, equation 11.91 is replaced by

$$N_e = A \exp\left(-\frac{t}{\tau_{11}}\right) + \sum_s B_s \exp\left(-\frac{t}{\tau_{ms}}\right), \tag{11.93}$$

where A and B_s are constants. The form of this equation satisfactorily accounts for the shape of the curve shown in Figure 11.31.

So far the description of the experiment has been limited to the case $E/N = 0$, but an essential feature of the method is its applicability to nonzero values of E/N. For these measurements a continuous-wave RF field is generated between the electrodes by applying a suitable signal to electrode A by means of the decoupling coil L_2. If the sinusoidal field is represented by $E = E_0 \sin \omega t$, it can be shown that the electron energy distribution is the same as that resulting from the application of a steady field $E = E_0/2^{1/2}$, provided[4] that $\omega/\nu \ll 1$ (see equation 9.53).

When using the RF heating field, it is necessary to make a small allowance for the loss of electrons immediately adjacent to the electrodes since these electrons are swept into the electrodes by the field. The electrode separation H in equation 11.89 is then replaced by $H - 2s$, where $s = 2W_{max}/\omega$ and W_{max} is the drift velocity corresponding to E_0.[4]

Notwithstanding the importance of Cavalleri's method in other areas, the most important application of the method is likely to be to situations in which electron attachment and ionization are significant processes. As we have already seen, both processes add considerably to the complexity of the measurement of D/μ by lateral diffusion techniques, and they generally lead to a loss of accuracy. Cavalleri's method has two main advantages when applied to these more difficult situations:

1. The method of detection discriminates between electrons and negative ions, so that accurate measurements can be made even when the mean electron energy approaches $3\kappa T/2$, the condition under which lateral diffusion experiments become inaccurate.

2. D can be determined directly when ionization and attachment occur, in contrast to lateral diffusion experiments, in which a correction factor must be applied to the parameter η before D/μ can be calculated (see Section 11.4).

When ionization by metastables can be neglected but attachment and ionization are significant processes, the diffusion equation for n becomes (equation 5.42 with $W = 0$ and $D_L = D$)

$$\frac{\partial n}{\partial t} = D\nabla^2 n + (\bar{\nu}_i - \bar{\nu}_{at})n, \tag{11.94}$$

where $\bar{\nu}_i$ and $\bar{\nu}_{at}$ are the ionization and attachment collision frequencies, respectively. The solution of this equation is again represented by an equation of the form of equation 11.89 but with τ_{11} given by

$$\tau_{11} = \frac{1}{D_1\left[(\pi/H)^2 + (c_1/R)^2\right]/N - (\bar{\nu}_{i_1} - \bar{\nu}_{at_1})N}, \tag{11.95}$$

where D_1, $\bar{\nu}_{i_1}$, and $\bar{\nu}_{at_1}$ refer to the respective parameters reduced to unit number density. It can be seen immediately that the measurement of τ_{11} at two pressures yields two simultaneous equations that can be solved to give $D_1 = ND$ and $\bar{\nu}_{i_1} - \bar{\nu}_{at_1} = (\bar{\nu}_i - \bar{\nu}_{at})/N$. Since $\bar{\nu}_i = \alpha_i W$ and $\bar{\nu}_{at} = \alpha_a W$, $(\alpha_i - \alpha_{at})/N$ can be determined using values of W taken from drift measurements.

It should be noted that, unlike lateral diffusion methods, this method cannot be used to determine α_i/N and α_a/N separately when ionization and attachment occur simultaneously since it is capable of determining only the net rate of production or loss resulting from these processes.

Several factors determine the choice of the operating parameters, for example, the gas pressure and the frequency of the RF heating field. These factors are discussed thoroughly in the original paper. For a given value of E/N, the pressure must be chosen within limits that are determined, on the one hand, by the need for short thermalization times (in order to ensure that the high-energy electrons produced by the initial ionization rapidly attain an equilibrium energy corresponding to the value of E/N) and, on the other, by the need to ensure a finite value of τ_{11}. For the second criterion to be satisfied it can be seen from equation 11.95 that p must be chosen to ensure that $D_1[(\pi/H)^2+(c_1/R)^2]/N > (\bar{\nu}_{i_1}-\bar{\nu}_{at_1})N$. Similarly, the angular frequency ω must b5 small enough to ensure that ω/ν is small, yet sufficiently large to prevent the ripple on the electron energy (due to relaxation of the energy as E passes through zero) from becoming too large.[4]

Cavalleri's original work was followed by the experiments of Gibson, Cavalleri, and Crompton,[61] in which particular attention was paid to the measurement of thermal diffusion coefficients. Initial attempts to repeat Cavalleri's results at small values of E/N ran into unexpected difficulties which were traced to nonuniform surface potentials within the diffusion chamber. In preparing for the experiments, the resistance of the nickel film was purposely reduced in order that the clearing voltage could be used and therefore reasonably high repetition rates. Under these conditions spuriously large values of ND were recorded as $E/N \rightarrow 0$, suggesting that electrons were being removed from the chamber by local fields produced by contact potential differences. The effects of the stray fields could not be compensated for by adjustment of the DC potentials applied to the electrodes and to the cylindrical walls. It was therefore assumed that there were considerable variations in potential over each of the three surfaces. It was also found that when the resistance of the nickel film was made extremely high, and no clearing voltages were applied to the electrodes, the measured values of ND were more nearly in accord with the expected ones.

These preliminary results could be satisfactorily accounted for by assuming that the resistance of the layer covering the inner surfaces of the cell had been made so high that, if any nonuniformity of potential had existed initially, the electrons and ions attracted differentially to these areas could have remained for an appreciable time, thus nullifying the initial nonuniform potential distribution. Accordingly a cell was constructed entirely of glass with electrodes clamped to the outside for the application of the sampling pulse.

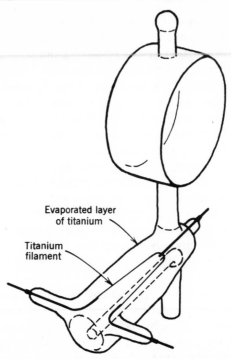

FIG. 11.32. The Cavalleri diffusion cell used by Gibson et al.[61] The cell has no internal metal electrodes. The getter cell mounted below the diffusion cell was used to maintain a high level of gas purity.

The insulated cell proved to be very successful. With repetition rates of a few hertz to allow sufficient time for the ions to diffuse to the walls, reproducible results were obtained for a range of values of sampling pulse frequency, indicating that the rate of replenishment of charge on the walls was sufficient to maintain a uniform distribution of potential. However, a difficulty arose with the new apparatus that had not formerly been evident. Despite rigorous outgassing of the cell and of the valve used to isolate it from the vacuum system, the results recorded at long time delays indicated the presence of a significant level of impurity, resulting in electron production by metastable ionization. Moreover the electron population at long delays generally tended to increase with time, indicating an increase in the level of contamination, and to depend quite markedly on the mean level of ionization generated by the sampling pulse. This last effect was attributed to the removal of impurity ions to the walls during the breakdown by the sampling pulse, an effect observed many years earlier by Townsend and MacCallum.[62]

The fact that the original experiments had not been affected by an increasing level of contamination suggested that the metal film covering the inside surfaces had acted as a getter. A similar effect was achieved with the glass cell by inserting a small getter trap between the isolating valve and the cell (see Figure 11.32), it being assumed that the most likely source of outgassing was the metal isolating valve. With this arrangement it was found that the electron population decayed exponentially with a single time constant for delay times up to and exceeding $5\tau_{11}$. Furthermore there was no evidence of contamination even after the same sample of gas had been used for a series of experiments extending over several months.

The low repetition rates needed when the clearing field cannot be used necessitates long experimental runs to reduce the statistical error to a reasonable level; typically each combination of delay times requires several hours to reduce the rms error to less than 0.5%. Consequently it is desirable to have some degree of automation built into the experiment, and the original electronic system was therefore replaced by one based on a small control and data-logging computer.

A block diagram of the electronic logic is shown in Figure 11.33. On

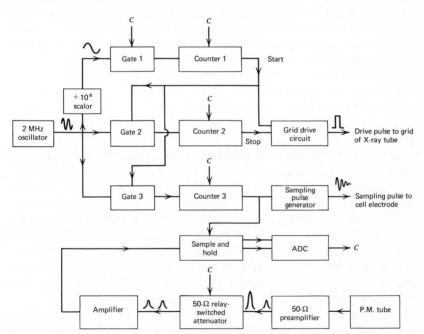

FIG. 11.33. Block diagram of electronic logic used by Gibson (Proceedings of the Decus-Australia 1972 Symposium, p. 55). The points marked C are connections to the computer.

instruction from the computer, gate 1 opens, admitting a 200-Hz signal to counter 1, which has been preset by the computer to a count that controls the duration of the measuring cycle. When the count reaches the preset value, a signal is sent to the control grid of the X-ray tube to initiate the X-ray pulse, and simultaneously gates 2 and 3 are opened to admit signals directly from the 2-MHz oscillator to counters 2 and 3. These counters are also preset by the computer to values that control the duration of the X-ray pulse (counter 2) and the delay times (counter 3). Counter 3 is alternately set to values that yield the two delay times t_1 and t_2.

When the count in 2 reaches the preset value, the X-ray pulse is terminated. Subsequently at time t_1 the preset value in counter 3 is attained, triggering the sampling pulse generator. Finally, on the completion of the count in 1 for the second time, the first measuring half-cycle is completed, and the counters are cleared and reset for the commencement of the next half-cycle.

Simultaneously with the triggering of the sampling pulse, an enabling pulse is sent to the "sample and hold" unit, which remains inactive until this time. The unit then accepts the signal from the light output detection chain and holds a signal which corresponds to the maximum amplitude of the pulse to allow sufficient time for the analogue-digital converter to convert the signal to digital form. The digital signal is then addressed to the appropriate memory, which accumulates a total count proportional to the sum of the pulse amplitudes corresponding to the particular delay time.

The accuracy of the experiment was improved by inserting a set of switched attenuators between the preamplifier and the main amplifier with the switching arranged so that the larger of the two signals (i.e., the one corresponding to t_1) was attenuated but the other was not. By arranging the combination of delay times and the attenuation factor so that the mean heights of the two pulses at the input to the main amplifier were approximately equal, any nonlinearity in the detection system past this point was, to first order, eliminated.

The results from this method have so far not been numerous, but they are nevertheless significant. Figure 11.34 shows a comparison of the values of ND in helium determined by Cavalleri[4] with those derived by combining the results for D/μ and W obtained by Townsend and Bailey and by Crompton, Elford, and Jory.[63] The agreement with the results of Crompton et al. is to within the combined experimental error over the common range of E/N; the agreement with those of Townsend and Bailey, which extend to higher values of E/N, is not as satisfactory.

A measurement of the thermal value of ND in helium at 293 K was made by Gibson et al.[61] in order to compare it with the value calculated from the cross section that was derived by Crompton et al.[64] from drift

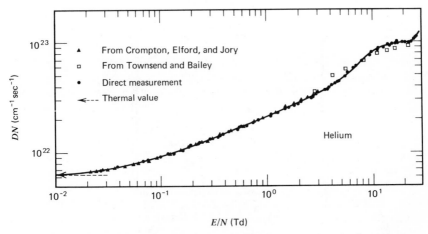

E/N (Td)

FIG. 11.34.   Values of ND measured by Cavalleri, plotted as a function of E/N. Also shown are the values obtained from values of W and D/μ measured by Townsend and Bailey (*Phil. Mag.* **46**, 657, 1923) and by Crompton et al.[63] (From ref. 4.)

and diffusion data obtained at 77 K. The significance of this result is discussed in Chapter 13.

11.5.2.   MEASUREMENTS OF THERMAL DIFFUSION COEFFICIENTS BY OTHER METHODS. Although superseded by Cavalleri's more versatile technique, the earlier experiments of Cavalleri, Gatti, and Principi[59] and Cavalleri, Gatti, and Interlenghi[65] are of historical interest. These experiments were based on the technique for gas amplification which has already been described and which was first proposed by Cavalleri, Gatti, and Redaelli.[66] The first group of authors used a glass cell similar in many respects to the one employed by Gibson et al., the significant difference in the technique being the use of a [210]Po alpha-particle source to produce the initial ionization. The sequence of events following the initial ionization, as well as the techniques for detecting and recording the signals corresponding to two separate delay times, were similar to those already described. However, because the initial ionizing events occurred at random, it was necessary to initiate the sequence with the passage of each alpha particle. A pair of photomultiplier tubes used in coincidence served to detect the alpha particles and thus to start each measurement.

The only difference in principle between the experiment of Cavalleri, Gatti, and Interlenghi[65] and the final form of the experiment as described in the preceding section was the absence of the RF heating field. However, the details of the diffusion chamber were somewhat different, the plane electrodes consisting in this instance of Ni-Cr layers evaporated on the

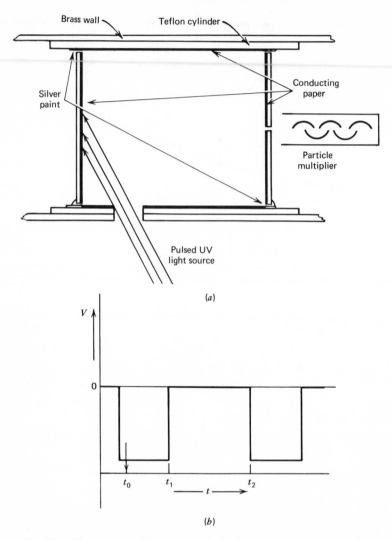

FIG. 11.35. The Nelson–Davis time-of-flight technique for measuring thermal diffusion coefficients: (*a*) schematic diagram of the apparatus; (*b*) sequence of events.

inside faces of the glass end plates. In the light of subsequent experience,[61] it seems doubtful whether a diffusion cell of this type can be used for the measurement of thermal diffusion coefficients when high precision is required.

An entirely different approach was taken by Nelson and Davis,[60] who devised a time-of-flight technique which was a development of the method of Wagner, Davis, and Hurst[3] for measuring longitudinal diffusion coefficients (see the next section). Figure 11.35 shows the principle of the method. The diffusion chamber was a cylindrical volume 25 cm in diameter and 30 cm long, defined by a Teflon cylinder and two metal end plates, the anode containing a small sampling hole. The interior surfaces of the chamber were covered with conducting paper. Electrical connections were made between the end of the paper cylinder and the end plates by means of metallized bands painted on the cylinder to coincide with the positions of the plates. The application of a potential difference to the plates then established a uniform potential drop across the cylindrical wall, and hence a uniform field within the volume, provided that the surface potential was uniform over the surface of the paper. Electrons were produced photoelectrically from the cathode, and those passing through the exit hole were detected by a particle multiplier. A differentially pumped system was capable of maintaining a low pressure in the particle multiplier when the pressure in the diffusion chamber was as high as 100 torr. The throughput of gas helped to maintain adequate gas purity against the unavoidable outgassing of the nonmetallic components. Gas pressure was monitored and controlled by a capacitance manometer, which enabled pressures to be held constant to within 0.005%.

The sequence of events is shown in Figure 11.35b. A UV light pulse occurs at time t_0 shortly after a negative-going square wave voltage pulse has been applied to the cathode. The square wave pulse terminates at t_1, leaving the diffusion chamber field-free for the time interval t_1 to t_2 (the "dwell" time). At t_2 a second square wave pulse is applied. The time t_1 is chosen so that the electron group is brought to the centre of the chamber by the drift field produced by the "take-in" pulse. During the dwell time the centre of mass of the group remains at rest, but the group disperses further by diffusion. Finally, with the application of the "take-out" pulse at t_2, the group drifts to the anode, where the arrival time spectrum is recorded by the particle multiplier and time-of-flight analyzer (see Section 10.5).

The theory of the method can be summarized as follows.[67] If the distribution of electrons in a one-dimensional stationary or drifting group is referred to a set of axes moving with the centroid of the group, the number density is given by

$$n(z',t) = \frac{N_0}{(4\pi D_L t)^{1/2}} \exp\left(-\frac{z'^2}{4D_L t}\right), \qquad (11.96)$$

where it is assumed that the group begins as a highly concentrated distribution at time $t=0$ with N_0 electrons per unit area, and the influence of the cathode and anode on the group is ignored.* Let the full width of the group be defined as the distance $\delta z'$ between the planes at which $n = n(0, t)/e$. Then the half-width $\sigma = \delta z'/2$ is given by $\sigma = (4D_L t)^{1/2}$, and it is clear that, in principle, the determination of σ at any given time t after the initiation of the group enables the longitudinal diffusion coefficient to be determined (see Section 11.6.1).

In the present experiment, the group first spends a time τ in the take-in field, during which its half-width becomes $\sigma_1 = (4D_L \tau)^{1/2}$. During the dwell time τ_D the group disperses further, the half-width becoming Σ_2. Let $\sigma_2 \equiv (4D\tau_D)^{1/2}$ be the half-width that the group would have had if it had been a delta-function distribution at the beginning of the dwell time rather than the distribution of half-width σ_1. Then it can be shown that $\Sigma_2 = (\sigma_1^2 + \sigma_2^2)^{1/2}$.[67] Similarly, if the take-out time is equal to the take-in time τ and $\sigma_3 \equiv (4D_L \tau)^{1/2}$ by analogy with σ_2, the final half-width of the group is

$$\Sigma_3 = \left(\sigma_1^2 + \sigma_2^2 + \sigma_3^2 \right)^{1/2} = 2(2D_L \tau + D\tau_D)^{1/2}. \qquad (11.97)$$

It should be noted that D_L is the value of the longitudinal diffusion coefficient when the drift field is applied, whereas D is the thermal isotropic diffusion coefficient.

In practice, measurements are made of the distribution of arrival times at the plane $z = h$ rather than of the spatial distribution at time $t = 2\tau + \tau_D$. In the absence of the decay of the pulse during the time in which it is being sampled, Σ_3 could be found simply from $\Sigma_3 = W\delta t/2$, where $\delta t/2$ is $|t_1 - t_m|$, t_m is the time at which the number density in the group is a maximum, and t_1 is the time at which the density falls to $1/e$ of the maximum value. Because of the decay, however, a correction must be applied to the measured value of $\delta t/2$ in order to apply the simple formula. It is also necessary to make corrections for the broadening of the arrival time distribution due to the finite width of the group at time t_0 and to electronic fluctuations, and for distortion of the distribution arising from nonrandom sampling. The procedure for applying these corrections is discussed fully in the original papers.[3,60,67]

Let $\delta t_0/2$ be the corrected half-width found from the arrival time distribution such that $\Sigma_3 = W\delta t_0/2$; then it follows from equation 11.97 that

$$\delta t_0^2 = \frac{16}{W^2}(2D_L \tau + D\tau_D). \qquad (11.98)$$

---

*The theory of the method was originally derived on the assumption of isotropic diffusion, but the formula used to analyze the experimental results (equation 11.99) is unaffected by this assumption. During the dwell time, $D_L = D$.

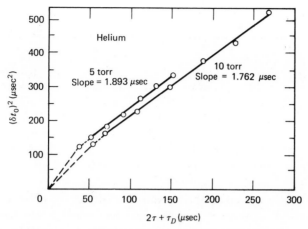

FIG. 11.36. Typical set of diffusion data [squared full width $[(\delta t_0)^2]$ versus drift time (2τ) + dwell time (τ_D)] for helium, obtained by Nelson and Davis.[60]

Consequently, since the total drift time is held constant during an experiment, a plot of δt_0^2 versus τ_D should yield a straight line of slope M, where M is related to D through the equation

$$M = \frac{16D}{W^2}. \qquad (11.99)$$

Figure 11.36 shows the curves obtained by Nelson and Davis for δt_0^2 versus $2\tau + \tau_D$ in helium. These are typical of the results for most of the gases that were studied. The lowest point on each curve corresponds to $\tau_D = 0$. This point can be used to calculate D_L for the value of E/N which applies during the take-in and take-out times. The next point is taken with a dwell time of the order of 10 μsec, during which the electrons attain equilibrium with the gas molecules after the removal of the drift field. Since D_L is changing during this time, one would not expect this point to lie on the straight line that is obtained for increasing values of τ_D. The rest of the curve is well represented by equation 11.98.

Table 11.7 summarizes all the experimental data obtained for helium[67]. The number in parentheses after each value of the slope M signifies the number of values of τ_D that was used to determine the slope; S_M is the fractional standard deviation of the experimental points from the straight line.

In general the differences between the values of Dp derived from different sets of experimental data exceed the standard deviation of the results within each set. For example, the first two results in the table differ by about 8%, whereas the standard deviations are only 1.8 and 2.4%.

TABLE 11.7.   Summary of experimental data for the thermal diffusion coefficient in helium ($T \sim 300$ K)

p (torr)	$\sim E/p$ (V cm^{-1} torr^{-1})	M (μsec)	S_M (%)	$W \times 10^{-5}$ (cm sec^1)	Dp (cm^2 μsec$^{-1\cdot}$torr)
5	0.48	1.893 (7)	1.79	6.255	0.232
10	0.24	1.763 (6)	2.43	4.383	0.212
10	0.24	1.818 (6)	1.71	4.350	0.215

Nelson and Davis examined their results for the effects of residual stray contact potential differences, but they present evidence to suggest that such effects could not explain these discrepancies. Inadequate gas purity is unlikely to have led to significant error, even though the purity of the gas within the diffusion chamber probably did not match the figures quoted for the gas as supplied. Even in the cases of neon and argon the impurity level would probably have had to exceed 1 part in 10^3 or 10^4 to affect the results significantly. When $E/N \neq 0$, the principal effect of molecular impurities on swarm measurements in these gases is to reduce the mean electron energy as a consequence of inelastic collision processes, usually with a corresponding increase in W and decrease in D. However, in the present instance, the thermal distribution cannnot be affected by any level of molecular impurity, and any effect on D arises from a change in the average momentum transfer cross section q_m. In general q_m is larger for the molecular impurity than for neon and argon, and the presence of impurity would therefore be expected to reduce D. Since outgassing from the paper liner is the most likely source of contamination, one would expect lower values of D to be recorded at the lower pressures, where contamination is greatest. In neon the results at 20 torr were, in fact, higher than those at 40 torr. Since the results for neon appear to have been unaffected by impurity, it may be safely concluded that the purity of the gas in all the experiments was adequate. The origin of the random errors in the results, particularly those for neon, is therefore unknown at the present time.

Nelson and Davis present results for nine gases and a comprehensive comparison of their data with values obtained using other techniques. Where the experiments were conducted with a wide range of parameters, the spread of the data usually lies within 5 to 10%. An exception is the case of neon, where the variation is as much as 80%. Since the value of Dp for this gas is considerably higher than for any of the others investigated, the value of δt_0 recorded under similar conditions would have been larger in neon than in the other gases and the intrinsic accuracy of the measurement therefore higher. Hence it seems that the source of the error in neon is associated in some way with the rapid variation of q_m with energy in this energy range (see Chapter 13) and with its very low value.

11.6. MEASUREMENT OF THE LONGITUDINAL DIFFUSION COEFFICIENT D_L

11.6.1. EXPERIMENTS BASED ON THE ANALYSIS OF ARRIVAL TIME SPECTRA. The time-of-flight technique developed by Hurst and his colleagues[3,68-72] for the measurement of diffusion coefficients was originally intended as an alternative to the long-established lateral diffusion technique. At the time of its development the paradox associated with the interpretation of the Townsend-Huxley experiment remained unresolved, although, as discussed in Section 11.2, explanations that assumed special conditions at the cathode and anode of the diffusion chamber had been proposed. Crompton and Jory[15] had argued that, notwithstanding the paradox, the experiments could be carried out under conditions which ensured that the interpretation of the results was unequivocal; furthermore there was considerable experimental evidence to support their argument. The paradox, nevertheless, remained, and the protagonists of the time-of-flight technique held that their new method possessed several advantages over the old, not the least of which stemmed from the belief that it was unnecessary to allow for the influence of the boundaries in developing the theory of the method. This view arises, however, from an oversimplified concept of the rôle of the boundaries in experiments of this kind (see Section 10.5).

The first results from time-of-flight experiments that related to gases for which there was a large body of data from lateral diffusion experiments certainly did not help to clarify the situation. Thus the results for hydrogen obtained by Hurst and Parks,[72] using a G-M detector and relatively high pressures, and those of Wagner and Davis,[68-70] who employed a differentially pumped system with a particle multiplier detector and relatively low pressures, were in reasonable agreement but were only about one half of the values obtained from lateral diffusion experiments. In a later set of experiments Wagner et al.[3] made a number of refinements to the experimental and analytical techniques and obtained data for many more gases. All of these data showed the same trend as the preliminary results in hydrogen. In the case of argon, for example, the values of D/μ were found to be as small as one seventh of the value obtained from lateral diffusion experiments.

In this situation, it was fortunate that the data for helium provided clear-cut evidence to prove either that there was a gross error in the interpretation of the arrival time spectra in terms of diffusion coefficients (which seemed unlikely because of the thoroughness of the investigation and the fact that the results extrapolated to the correct thermal value) or that a new transport coefficient was being measured. Crompton and Jory[74] had shown earlier that the momentum transfer cross section derived from their measured drift velocities in helium could be used to calculate values of D/μ that were in agreement with experimental data from lateral diffusion experiments to within the experimental error. Since the drift velocities measured by Wagner et al. were in close agreement with those of Crompton and Jory, the same argument could be applied to the data of Wagner and his coworkers to show that their measured values of D/μ were quite inconsistent with their drift data if it were assumed that diffusion was isotropic. Perhaps for this reason Wagner et al. came to the following tentative conclusion: "Possibly the diffusion coefficient is a tensor quantity; thus there is a true difference when measurements are made in the two directions with respect to the electric field." This was the first suggestion, backed by experimental evidence, for the anisotropic diffusion of low-energy electron swarms. Wannier[75] had predicted some years earlier that heavy ions drifting and diffusing in an electric field would exhibit the phenomenon, but this was attributed to extremely nonuniform populations of the shells in velocity space. Since it was known, in the case of electrons, that the shells were nearly uniformly populated, no anisotropicity in the diffusion of electron swarms was expected. The evidence from the Oak Ridge National Laboratory for this phenomenon for electrons has been responsible for significant advances in the theory of electron transport in the last few years (see Chapter 4).

Much of the detail of the experiments and their analysis was given in Section 10.5 and need not be repeated here. We therefore restrict our discussion to that part of the analysis which is concerned with the derivation of longitudinal diffusion coefficients (or the ratio D_L/μ) from the arrival time spectra.

We consider, as before, the spectrum that results from N_0 electrons per square centimetre being released simultaneously from the cathode. The number of electrons $E(t)\Delta t$ that arrive at the detector a distance h from the cathode between t and $t+\Delta t$ is given by (equation 10.9)

$$E(t)\,\Delta t = N_0 a W\,\Delta t (4\pi D_L t)^{-1/2} \exp\left[-\frac{(h-Wt)^2}{4D_L t} \right] \quad (11.100)$$

where, as before, it is assumed that there is no loss of electrons by attachment and that the influence of the electrodes may be ignored. If t_m is the time at which $E(t)$ is a maximum, the drift velocity can be calculated to first order from

$$W \simeq \frac{h}{t_m} \qquad (11.101)$$

provided that $\beta = D_L/Wh \ll 1$.

Let t_1 be the time $(>t_m)$ at which the value of $E(t)$ is $1/e$ of its value at t_m. Then by substituting into equation 11.100 it can be shown that

$$t_1 \simeq \frac{h}{W}(1 + 2\beta^{1/2}) \qquad (11.102)$$

provided that $2\beta^{1/2} \ll 1$. It follows that, if we define δt as $\delta t \equiv t_1 - t_m$, then

$$\delta t \simeq 2\left(\frac{hD_L}{W^3}\right)^{1/2}, \qquad (11.103)$$

and that, if δt is determined from the arrival time spectrum, D_L can be found from

$$D_L \simeq \frac{h^2 \delta t^2}{4t_m^3}, \qquad (11.104)$$

(cf. equation 11.98 with $\tau_D = 0$, $2\tau = t_m$, and $\delta t_0 = 2\delta t$).

In Section 10.5 it was shown that, with the assumptions used to derive equation 11.100, a more accurate relation between W and t_m is

$$W \simeq \frac{h}{t_m}(1 - \beta). \qquad (11.105)$$

Similarly, Wagner et al. stated without proof that, to the same order of approximation, equation 11.104 is to be replaced by*

$$D_L \simeq \frac{h^2 \delta t^2}{4t_m^3}(1 - 2\beta). \qquad (11.106)$$

*Equations 11.105 and 11.106 were stated in a somewhat different form by Wagner et al. as

$$W = \frac{h}{t_m}(1 + \beta' - \cdots)^{-1} \quad \text{and} \quad D_L = \frac{h^2}{t_m}\beta'(1 + 2\beta' + \cdots)^{-1}$$

where $\beta' = (\delta t/2t_m)^2$. Since, from equation 11.103, $\delta t \simeq 2(hD_L/W^3)^{1/2}$, it follows that $\beta' \simeq \beta$ and that these formulae are equivalent to the ones given above.

If the foregoing theory is to be applied without modification, three criteria must be satisfied:

1. The half-width of the electron group at time $t=0$ must be negligible compared with the half-width at $t=t_m$.
2. The broadening of the group due to causes other than diffusion, for example, electronic fluctuations, must be negligible.
3. Every electron arriving at the detector must have an equal probability of being recorded, that is, there must be no discrimination against electrons at the rear of the distribution because of the dead-time of the detection system.

In practice none of these criteria can be met adequately, and failure to satisfy any one can have a serious effect on the analysis of the data to determine D_L. This is particularly the case in the high-pressure experiments[71,72] because of the narrow width of the electron groups. For these experiments, therefore, a deconvolution analysis was developed to unfold the true arrival time spectrum from the spectrum resulting from the combination of all three effects and the diffusive process. This technique is described fully by Hurst and Parks,[72] but the description is not reproduced here because in the later low-pressure experiments[3] it was possible to use a more straightforward procedure. Since the low-pressure experiments are inherently more accurate for determining D_L (but not W), only the method of analyzing them will be given.

Let $T_0(t)$ denote the distribution resulting from the effects discussed under criteria 1 and 2 above, that is, the distribution of arrival times that would result if there were no diffusion. The value of $T_0(t)$ was found by determining the arrival time distribution when the system was evacuated, in which case it may be assumed that there is negligible delay between the emission of the electrons from the cathode and their arrival at the detector. The distribution then arises solely from the width of the initial light pulse and from fluctuations in the electronic system and is therefore the required distribution.

The effect of $T_0(t)$ on the final distribution is to produce a smeared function of $E(t)$, which is the convolution integral

$$E'(t) = \int_0^\tau E(t) T_0(\tau - t)\, dt. \qquad (11.107)$$

Since $E'(t)$ is the distribution that is actually measured, we require a method of finding D_L from $E'(t)$ rather than $E(t)$, as $E(t)$ cannot be determined other than by deconvolution.

Let us denote by D'_L the value of D_L that is determined directly from

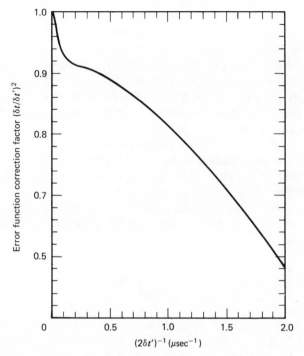

FIG. 11.37. The correction factor $(\delta t/\delta t')^2$ plotted as a function of $(2\delta t')^{-1}$ for the apparatus of Wagner et al.[3]

equation 11.104 for a given value of $\delta t'$, where $\delta t'$ is defined for $E'(t)$ in the same way as δt for $E(t)$. Now, if the independent variable in equations 11.100 and 11.107 is changed from t to $\zeta = t - t_m$, it follows that a given value of δt completely specifies $E(\zeta)$, and hence $E(\zeta')$ and $\delta t'$, since $E(\zeta)$ is a Gaussian distribution under the assumptions used to derive equation 11.102, and $T_0(t)$ is the same for all convolutions. But, from equation 11.104, $D_L = h^2\delta t^2/4t_m^3$ and $D'_L = h^2\delta t'^2/4t_m^3$; consequently

$$D_L = D'_L\left(\frac{\delta t}{\delta t'}\right)^2.$$

Thus, if the ratio $\delta t/\delta t'$ corresponding to a given value of δt (or $\delta t'$) is known, the values of D'_L calculated directly from the measured half-widths using equation 11.104 can be corrected to give the true values of D_L. The procedure is, therefore, to choose a set of values of δt and the corresponding Gaussian functions $E(\zeta)$, to convolute these functions with $T_0(t)$ to form the corresponding functions $E'(\zeta)$, and hence to find the set of values of $\delta t'$. The curve of $(\delta t/\delta t')^2$ versus $(2\delta t')^{-1}$ can then be plotted and used to determine the required correction factor in each case. The curve obtained by Wagner et al. for their apparatus is shown in Figure 11.37.

There remains the problem posed by the failure to meet the third criterion: correcting for the error arising from nonrandom sampling. Because, in general, the dead-time of the detection system exceeds the width of the arrival time spectrum, the experimental conditions must be chosen to ensure that, as nearly as possible, no more than one electron arrives at the detector for every group released from the cathode. Now, the full arrival time spectrum must be built up from the data recorded from a large number of events F, during each of which n_i electrons per square centimetre are released from the cathode but only a small fraction reaches the detector. It follows that when $E(t)$ results from data accumulated over the F events the number N_0 in equation 11.100 must be replaced by $N_0 = \sum_{i=1}^{F} n_i$.

Let

$$\bar{n}_c = \sum_{i=1}^{F} \frac{an_i}{F}$$

be the average number of electrons entering the detector per event. Then, unless $\bar{n}_c \ll 1$, there is some possibility that more than one electron will arrive at the detector per event and that the assumption of random sampling will be violated. The effect of such a violation may be investigated quantitatively in the following way.

If it is assumed that the dead-time of the detector is very much larger than the width of the spectrum, that is, that only one electron can be detected per event, the number $E''(t)\Delta t$ of electrons actually detected between t and $t + \Delta t$ in F events is simply the number $E(t)\Delta t$ which passes through the sampling hole multiplied by the probability $P_0(t)$ that no electron has been recorded up to time t in each event. Thus

$$E''(t) = P_0(t)E(t). \tag{11.108}$$

Using an argument based on Poisson statistics, Hurst and Parks[72,73] showed that

$$P_0(t) = \exp\left[-\int_0^t \bar{n}_c \frac{E(t)}{N_0 a} dt \right],$$

whence

$$E''(t) = E(t)\exp\left[-\int_0^t \bar{n}_c \frac{E(t)}{N_0 a} dt \right], \tag{11.109}$$

and we see immediately that, as $\bar{n}_c \to 0$, $E''(t) \to E(t)$.

Although $\bar{n}_c$ cannot easily be obtained by experiment, it can be shown [73] that this value can be simply calculated from the total number of electrons C recorded in F events, using the relation

$$\frac{C}{F} = 1 - \exp - \bar{n}_c. \tag{11.110}$$

The importance of allowing for nonrandom sampling, which Hurst and Parks called "Poisson distortion," was shown in Figure 10.14, in which $E''(t)$ is plotted for a given $E(t)$ (corresponding to a particular choice of h, W, and D_L) and a set of values of $\bar{n}_c$. When $\bar{n}_c$ is small, $E''(t)$ lies almost on top of $E(t)$; as $\bar{n}_c$ increases, however, the effect is both to displace and narrow the distribution. The effect of the distortion is much more pronounced on the values of D_L derived from the spectra than on the values of W.

By the use of equations 11.101 and 11.104, apparent values of W and D_L were calculated directly from the half-widths of the curves shown in the figure without correcting for the effect of the distortion. These data are given in Table 11.8.

Even when, on the average, only one electron arrives at the detector in each event, the error in D_L that would arise from neglect of the dead-time is 7%. For comparison, the error in W is only 0.5%. When $\bar{n}_c = 100$, the error in W is still less than 4%, but D_L is underestimated by about a factor of 6.

TABLE 11.8. The changes in the apparent values of W and D_L resulting from "Poisson distortion". (From ref.72.)

C/F	$\bar{n}_c$	t_m (μsec)	δt (μsec)	W (cm μsec^{-1})	$D_L \times 10^3$ (cm^2 μsec^{-1})
0	0	28.420	0.561	0.950	2.49
0.0488	0.05	28.410	0.561	0.950	2.49
0.0952	0.10	28.400	0.561	0.951	2.50
0.3935	0.50	28.340	0.554	0.953	2.46
0.6322	1.00	28.270	0.537	0.955	2.32
0.9180	2.50	28.120	0.465	0.960	1.77
0.9650	3.35	28.070	0.432	0.962	1.66
0.9930	5.00	27.990	0.387	0.965	1.24
1.0000	10.00	27.860	0.324	0.969	0.882
1.0000	50.00	27.590	0.239	0.976	0.477
1.0000	100.00	27.500	0.217	0.982	0.411

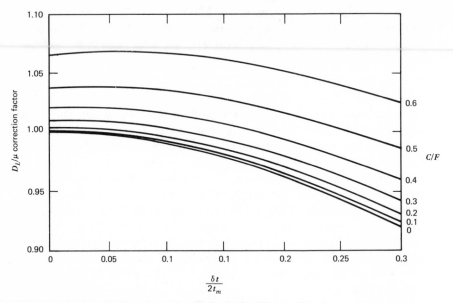

FIG. 11.38.   The correction factor for D_L/μ due to Poisson distortion and the approximation used in equation 11.104, plotted as a function of $\delta t/2t_m$ ($\cong \beta^{1/2}$). C/F is the average number of electrons counted in each event. (From ref. 3.)

In practice[3] a method of correcting for Poisson distortion was devised that was similar to the one already described in relation to the broadening of the spectra due to the finite duration of the light pulse and instrument fluctuations. In preparing curves of the correction factor, the effect of the higher-order terms in equation 11.106 was also included in order that a single correction could be applied to the value of D_L calculated directly from the simple formula 11.104. Both the correction for Poisson distortion and that for the distortion from the simple Gaussian form due to the fact that the electron density is sampled in time rather than space [which is the origin of the higher-order terms in $(\delta t/2t_m)^2$ or β] can be expressed as functions of $\delta t/2t_m$. Figure 11.38 shows the overall correction for both these effects plotted as a function of $\delta t/2t_m$, the curve for $C/F=0$ corresponding to the correction arising solely from the neglect of higher-order terms in the simple formula.

By using the curves shown in Figures 11.37 and 11.38, corrections can be applied to the values of D_L that are calculated directly from the measured values of δt and t_m with the aid of equation 11.104. The true values of D_L can thus be found without recourse to an elaborate analysis of each spectrum.

The significance of the results of these experiments was commented upon at the outset of this section. A summary of the data will be found in Chapter 14.

11.6.2. EXPERIMENTS BASED ON THE BRADBURY-NIELSEN TECHNIQUE. An alternative method of measuring D_L is based on an analysis of the widths of the maxima in the current-frequency curves obtained using a Bradbury-Nielsen electrical shutter apparatus. The method was first referred to by Lowke and Rees[76] in relation to the measurement of diffusion coefficients in water vapour and is based on an earlier analysis by Lowke.[77]

Employing the same assumptions as were used in developing the theory in the preceding section (i.e., that the travelling groups are uninfluenced by the electrodes and that the diffusive flux is small compared with the flux due to drift), Lowke obtained the following expression for the intantaneous current transmitted by the sampling shutter (see also equation 5.1(b) and Appendix 10.2):

$$i = \text{const} \cdot \sum_{m=1}^{\infty} (mf)^{-1/2} \exp\left[-\frac{hW}{4D_L} \frac{f}{mf_1} \left(1 - \frac{mf_1}{f}\right)^2 \right], \quad (11.111)$$

where $f_1 = W/2h$. For the purpose of this discussion two other assumptions are justified:

1. That D_L/hW is sufficiently small that only the jth term in the summation contributes to the jth maximum in the current-frequency curve.

2. That D_L/hW is sufficiently small that, for each maximum, $\Delta f/f_{\text{max}} \ll 1$, where $\Delta f = f'' - f'$, and f' and f'' are the frequencies at which the transmitted current is $i_{\text{max}}/2$. (An equivalent assumption is made in deriving equations 11.102 and 11.103.)

With these assumptions equation 11.111 becomes (for the jth maximum)

$$i_j \cong i_{j_{\text{max}}} \exp\left[-\frac{hW}{4D_L} \frac{f}{jf_1} \left(1 - \frac{jf_1}{f}\right)^2 \right]. \quad (11.112)$$

Let f'_j and f''_j be the frequencies on either side of jf_1 for which $i_j = i_{j_{\text{max}}}/2$, and let the "resolving power" $\mathbf{R}$ be defined as $\mathbf{R} = jf_1/(f''_j - f'_j)$. Then, from equation 11.112, f'_j and f''_j are the roots of the equation

$$\exp\left[-\frac{hW}{4D_L} \frac{f}{jf_1} \left(1 - \frac{jf_1}{f}\right)^2 \right] = \tfrac{1}{2}.$$

After reduction, this equation becomes

$$\left(\frac{f}{jf_1}\right)^2 - \left(2 + \frac{4D_L}{hW}\ln 2\right)\left(\frac{f}{jf_1}\right) + 1 = 0.$$

It follows that

$$\mathrm{R}_j^{-1} = \frac{f''_j - f'_j}{jf_1} = \left[\left(2 + \frac{4D_L}{hW}\ln 2\right)^2 - 4\right]^{1/2}$$

$$\cong 4\left(\frac{D_L \ln 2}{hW}\right)^{1/2}, \tag{11.113}$$

since $D_L/hW \ll 1$.

The method appears to offer some advantage over the procedure described in the previous section, although little use has been made of it. Thus the method is free from errors resulting from Poisson distortion and electronic fluctuations. An additional source of error, however, might be the effect of field distortion in the vicinity of the electrical shutters, although this is expected to be small, particularly if square wave gating signals are used. The effect of the finite width of the electron groups generated at the first shutter may be allowed for by measuring the resolving power with increasing values of j, while maintaining the other

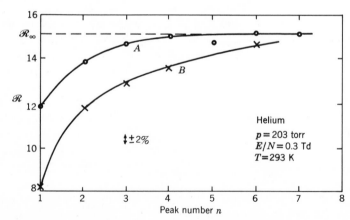

FIG. 11.39.   Variation of the measured resolving power R in helium with the initial width of the electron groups.[78] The resolving power increases as the amplitude and frequency nf_1 of the AC shutter gating signal increases. Curve A, 30 V peak-to-peak; curve B, 15 V peak-to-peak.

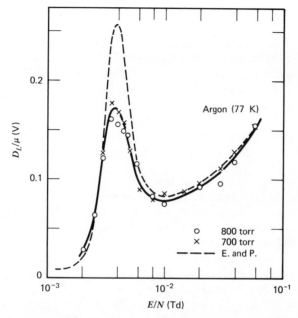

FIG. 11.40.   Values of D_L/μ in argon at 77 K measured with the Bradbury-Nielsen resolving power method. The theoretical curve (E. and P.) was calculated from the theory of Parker and Lowke,[26] using Engelhardt's and Phelps's cross section (*Phys. Rev.*, **133**, A375, 1964). (From ref. 79.)

experimental parameters constant. Since the width of the groups is inversely proportional to j, R_j should approach a limiting value corresponding to the true value of the resolving power as j increases. Figure 11.39 shows a typical set of results taken in helium.[78] The values of D_L/μ calculated from values of R_∞ are in excellent agreement with the values published by Wagner et al.[3]

The only data that have been published from this technique are the results of Robertson and Rees[79] for argon at 77 K. These data, shown in Figure 11.40, were taken in order to check the somewhat striking behaviour of D_L as a function of E/N at low values of E/N, as predicted by Lowke and Parker.[26] The maximum in the curve is a consequence of the rapidly changing slope of the q_m versus energy curve for energies below the Ramsauer minimum. The shape of the curve is in excellent agreement with the theory, although the maximum at $E/N \sim 3 \times 10^{-3}$ Td is considerably lower. As yet, the origin of this discrepancy has not been adequately explained.

REFERENCES

1. See, for example, M. T. Elford, in *Case Studies in Atomic Collision Physics,* Vol. 2, Chap. 2, North Holland Publishing Company, Amsterdam, 1971.

2. L. G. H. Huxley, *Aust. J. Phys.,* **12**, 171, 1959; see also refs. 16, 19, and 45.

3. E. B. Wagner, F. J. Davis, and G. S. Hurst, *J. Chem. Phys.,* **47**, 3138, 1967.

4. G. Cavalleri, *Phys. Rev.,* **179**, 186, 1969.

5. J. S. Townsend, *Proc. Roy. Soc. A,* **80**, 207, 1908.

6. J. S. Townsend, *Electricity in Gases,* Oxford University Press, London, 1915.

7. F. B. Pidduck, *A Treatise on Electricity,* 2nd ed., Cambridge University Press, London, 1925, p. 474.

8. J. H. Mackie, *Proc. Roy. Soc. A,* **90**, 69, 1914.

9. J. S. Townsend and H. T. Tizard, *Proc. Roy. Soc. A,* **88**, 336, 1913.

10. V. A. Bailey, *Phil. Mag.,* **50**, 825, 1925.

11. See, for example, R. H. Healey and J. W. Reed, *The Behaviour of Slow Electrons in Gases,* Amalgamated Wireless Ltd., Sydney, 1941.

12. L. G. H. Huxley, *Phil. Mag.,* **30**, 396, 1940.

13. R. W. Crompton and D. J. Sutton, *Proc. Roy. Soc. A,* **215**, 467, 1952; R. W. Crompton, L. G. H. Huxley and D. J. Sutton, *Proc. Roy. Soc. A,* **218**, 507, 1953.

14. L. G. H. Huxley and R. W. Crompton, *Proc. Phys. Soc. B,* **68**, 381, 1955.

15. R. W. Crompton and R. L. Jory, *Aust. J. Phys.,* **15**, 451, 1962.

16. J. Lucas, *J. Electr. Control,* **17**, 43, 1964.

17. L. G. H. Huxley, *Aust. J. Phys.,* **25**, 43, 1972.

18. C. A. Hurst and B. S. Liley, *Aust. J. Phys.,* **18**, 521, 1965.

19. R. W. Crompton, B. S. Liley, A. I. McIntosh, and C. A. Hurst, *Proceedings of the Seventh International Conference on Phenomena in Ionized Gases, Beograd, 1965,* Vol. 1, Gradevinska, Knjiga Publishing House, Beograd, 1966, p. 86.

20. R. W. Warren and J. H. Parker, *Phys. Rev.,* **128**, 2661, 1962.

21. J. H. Parker, *Phys. Rev.,* **132**, 2096, 1963.

22. J. L. A. Francey, *J. Phys. B,:At. Mol. Phys.,* **2**, 669, 1969; **2**, 680, 1969.

23. E. A. Desloge and R. D. Mitchell, *Aust. J. Phys.,* **23**, 497, 1970.

24. E. A. Desloge, *Statistical Physics,* Holt, Rinehart, and Winston, New York, 1966.

25. J. J. Lowke, *Proceedings of the Tenth International Conference on Phenomena in Ionized Gases,* Oxford, 1971, contributed papers, p. 5.

26. J. H. Parker and J. J. Lowke, *Phys. Rev.,* **181**, 290, 1969; J. J. Lowke and J. H. Parker, *Phys. Rev.,* **181**, 302, 1969.

27. L. B. Loeb, *Basic Processes of Gaseous Electronics,* University of California Press, Berkeley, 1955.

28. L. G. H. Huxley and A. A. Zaazou, *Proc. Roy. Soc. A,* **196**, 402, 1949.

29. L. W. Cochran and D. W. Forester, *Phys. Rev.,* **126**, 1785, 1962.

30. T. L. Cottrell and I. C. Walker, *Trans. Faraday Soc.,* **63**, 549, 1967.

31. J. A. Rees, *Aust. J. Phys.,* **17**, 462, 1964.

32. J. H. Parker and R. W. Warren, *Rev. Sci. Instrum.,* **33**, 948, 1962.

33. T. L. Cottrell, W. J. Pollock, and I. C. Walker, *Trans. Faraday Soc.,* **64**, 2260, 1968.

34. H. L. Brose and J. E. Keyston, *Phil. Mag.*, **20**, 902, 1935.

35. J. Bannon and H. L. Brose, *Phil. Mag.*, **6**, 817, 1928.

36. R. W. Crompton, M. T. Elford, and J. Gascoigne, *Aust. J. Phys.*, **18**, 409, 1965.

37. R. W. Crompton, M. T. Elford, and A. I. McIntosh, *Aust. J. Phys.*, **21**, 43, 1968.

38. R. W. Crompton and A. I. McIntosh, *Aust. J. Phys.*, **21**, 637, 1968.

39. C. A. Hurst, personal communication.

40. R. W. Crompton and M. T. Elford, *Proceedings of the Sixth International Conference on Ionization Phenomena in Gases,* Paris, 1963, Vol. 1, p. 337.

41. Electron and Ion Diffusion Unit, Australian National University, Quart. Repts. 16, 1964, and 24, 1966.

42. L. M. Chanin, A. V. Phelps, and M. A. Biondi, *Phys. Rev.*, **128**, 219, 1962.

43. B. S. Liley, *Aust. J. Phys.*, **20**, 527, 1967.

44. V. A. Bailey and J. D. McGee, *Phil. Mag.*, **6**, 1073, 1928.

45. L. G. H. Huxley, R. W. Crompton, and C. H. Bagot, *Aust. J. Phys.*, **12**, 303, 1959.

46. J. A. Rees, *Aust. J. Phys.*, **18**, 41, 1965.

47. M. S. Naidu and A. N. Prasad, *J. Phys. D:Appl. Phys.*, **3**, 957, 1970.

48. R. H. Healey and C. B. Kirkpatrick, - see ref. 11, p. 94.

49. C. A. Hurst and L. G. H. Huxley, *Aust. J. Phys.*, **13**, 21, 1960.

50. R. W. Crompton, J. A. Rees, and R. L. Jory, *Aust. J. Phys.*, **18**, 541, 1965.

51. M. S. Naidu and A. N. Prasad, *Brit. J. Appl. Phys. (J. Phys. D)*, **2**, 1431, 1969.

52. J. L. Moruzzi, *Brit. J. Appl. Phys.*, **14**, 938, 1963.

53. J. A. Rees, personal communication, 1972.

54. P. A. Lawson and J. Lucas, *Proc. Phys. Soc.*, **85**, 177, 1965.

55. P. A. Lawson and J. Lucas, *Brit. J. Appl. Phys.*, **16**, 1813, 1965.

56. F. M. Bruce, *J. Inst. Electr. Eng.*, **94**, 138, 1947.

57. A. I. McIntosh, Ph.D. thesis, Australian National University, 1967.

58. C. G. Morgan and W. T. Williams, *Proc. Phys. Soc.*, **85**, 443, 1965.

59. G. Cavalleri, E. Gatti, and P. Principi, *Nuovo Cimento,* **31**, 302, 1964.

60. D. R. Nelson and F. J. Davis, *J. Chem. Phys.*, **51**, 2322, 1969.

61. D. K. Gibson, R. W. Crompton, and G. Cavelleri, *J. Phys. B; At. Mol. Phys.*, **6**, 1118, 1973.

62. J. S. Townsend and S. P. MacCallum, *Phil. Mag.*, **5**, 695, 1928.

63. R. W. Crompton, M. T. Elford, and R. L. Jory, *Aust. J. Phys.*, **20**, 369, 1967.

64. R. W. Crompton, M. T. Elford, and A. G. Robertson, *Aust. J. Phys.*, **23**, 667, 1970.

65. G. Cavalleri, E. Gatti, and A. M. Interlenghi, *Nuovo Cimento,* **40B**, 450, 1965.

66. G. Cavalleri, E. Gatti, and G. Redaelli, *Nuovo Cimento,* **25**, 1282, 1962.

67. See also D. R. Nelson and F. J. Davis, *Oak Ridge Natl. Lab. Rept.*, ORNL-TM-2222, 1968.

68. E. B. Wagner and F. J. Davis, *Oak Ridge Natl. Lab. Rept.* ORNL-3697, 1964, p. 134.

69. E. B. Wagner and F. J. Davis, *Oak Ridge Natl. Lab. Rept.* ORNL-3489, 1965, p. 112.

70. E. B. Wagner and F. J. Davis, *Oak Ridge Natl. Lab. Rept.* ORNL-4007, 1966, p. 134.

71. G. S. Hurst, L. B. O'Kelly, E. B. Wagner, and J. A. Stockdale, *J. Chem. Phys.*, **39**, 1341, 1963.

72. G. S. Hurst and J. E. Parks, *J. Chem. Phys.*, **45**, 282, 1966.

73. J. E. Parks and G. S. Hurst, *Oak Ridge Natl. Lab. Rept.* ORNL-TM-1287, 1965.

74. R. W. Crompton and R. L. Jory, *Fourth International Conference on the Physics of Electronic and Atomic Collisions, Quebec*, Science Bookcrafters, Hastings-on-Hudson, New York, 1965, p. 118.

75. G. H. Wannier, *Bell System Tech. J.*, **32**, 170, 1953.

76. J. J. Lowke and J. A. Rees, *Aust. J. Phys.*, **16**, 447, 1963.

77. J. J. Lowke, Ph.D. thesis, University of Adelaide, 1962.

78. M. T. Elford, personal communication, 1972.

79. A. G. Robertson and J. A. Rees, *Aust. J. Phys.*, **25**, 637, 1972.

12

EXPERIMENTAL DETERMINATIONS

OF OTHER ELECTRON

TRANSPORT COEFFICIENTS

12.1. INTRODUCTION

Although experimentally determined values of W and D/μ form the main body of data that are used for the cross-section analyses described in Chapter 13, other transport coefficients have been employed in special circumstances either as the primary data or in tests of the validity of analyses based on the coefficients W and D/μ. Apart from W and D/μ, the coefficients for which most data are available are the collision frequency for momentum transfer ν_m and the ionization and attachment coefficients α_T and α_{at}. Of these, the ionization coefficient is perhaps the most important because it can be measured with adequate accuracy at high values of E/N, where data for D/μ either are nonexistent or are of questionable accuracy because of the difficulties described in Chapter 11. However, since we have generally restricted the subject matter of this book to a discussion of electron drift and diffusion at low values of E/N, where we believe the problems of accurate measurement and interpretation are now, on the whole, well understood, we do not include in this chapter a description of the many experimental determinations of α_T that have been made. The reader is referred to the book[1] *Basic Processes of Gaseous Electronics* by L. B. Loeb for a comprehensive account of work in this field up to 1955, and to the papers[2] by Haydon and his group for descriptions of recent high-precision measurements of ionization coefficients.

The direct measurement of ν_m using microwaves requires techniques that are quite different from the DC swarm techniques with which we are primarily concerned. In Chapter 13 we give a brief resumé of the methods that have been used to analyze the results of microwave experiments because they have direct relevance to similar methods developed to analyze the results of DC swarm experiments, but apart from this we have left detailed descriptions of microwave experiments to others. A comprehensive account will be found in another book in this series.[3] On the other hand, a description of the techniques that have been used to determine attachment coefficients falls within the terms of reference of this book because, in many DC swarm experiments which are concerned with gases of practical importance, diffusion and drift are accompanied by attachment and a full discussion of one involves the other.

Before beginning an account of attachment coefficient measurements, however, we describe briefly the measurement of another transport coefficient which was formerly of great importance but is now less often used, namely, the "magnetic drift velocity" W_M.

12.2.   MEASUREMENT OF THE MAGNETIC DRIFT VELOCITY W_M

As pointed out in Chapter 1, the measurement of the deflexion of an electron stream by a magnetic field normal to the drift field formed the basis of the earliest measurements of electron drift velocity. However, it was later realized[4] that there was no simple relation between the angular deflexion and the drift velocity, and still later[5] that it was possible for W and W_M to differ by a factor of 2 or more, although such a disparity was the exception rather than the rule. The use of the magnetic deflexion method as a source of drift velocity data was therefore abandoned in favour of time-of-flight methods, and the work described in the most recent papers[6-8] had as its aim the determination of accurate values of W_M for use as independent checks of the cross sections and energy distribution functions determined from analyses of the data for other transport coefficients.

The theory of Huxley's method for measuring W_M was given in Chapter 8, where it was shown that the ratio of the transverse velocity $\mathbf{W}_x$ (i.e., the velocity normal to the electric and magnetic fields) to the longitudinal velocity $\mathbf{W}_z \cong \mathbf{W}$ can be determined by measuring the distribution of current about a centrally located transverse cut in the anode when a magnetic field is applied parallel to the cut. Figure 12.1 shows a section through the apparatus used by Jory[6] and others.[7,8] Many features of the design are similar to those used in the drift tubes and lateral diffusion apparatuses described in the preceding chapters, but two points require comment. First, the additional complexity of the receiving electrode, caused by the transverse cut, made it difficult to use the form of construction shown in Figure 11.12. As an alternative, the segments of the

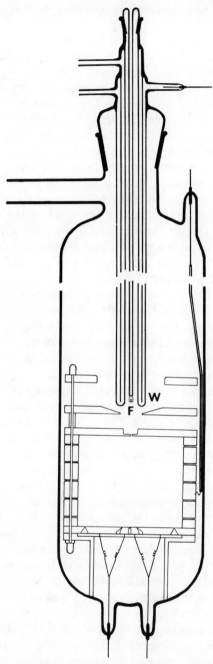

FIG. 12.1. Schematic diagram of the apparatus for measuring the magnetic drift velocity W_M. The transverse cut in the circularly divided lower electrode is normal to the plane of the paper. In this apparatus the electron source is the filament F, which is surrounded with the water cooling jacket W. (From ref. 6.)

electrode, shown in plan view in Figure 1.1b , were mounted on a sheet of plate glass which was itself mounted on the outer ring. This construction, although not suitable for low-temperature measurements, provides good mechanical stability when a number of segments have to be located accurately with respect to each other. The second feature is the water cooling jacket surrounding the filament. It is often necessary to use comparatively low gas pressures in these experiments in order to obtain adequate deflexions of the stream with reasonable values of the magnetic field strength B, as can be shown by using the approximate equation $\tan \theta = BW/E$. Consequently the radioactive sources used in other experiments could not serve to cover the full range of measurements in this instance, and a platimun filament was employed as the electron source. The use of the cooling jacket, through which water at ambient temperature was circulated, helped to remove the heat input from the filament and therefore to improve the temperature stability.

The production of a magnetic field of adequate uniformity over the volume occupied by the electron stream presents some problems. These were solved by using a pair of Helmholtz coils whose design was based on the criteria suggested by Barker.[9] In this way the need to use coils of excessively large diameter and separation was avoided.

With the form of construction shown in Figure 12.1, measurements of W/D and W_M can be made in the same apparatus, provided that alternative anode connexions are used to provide effective anode configurations consisting either of a disk and an annulus or a pair of semicircular electrodes. The values of W/D are required to calculate the values of W_x/W_z, and hence W_M, from the measured current ratios obtained from the magnetic deflexion experiments (see Section 8.6). The experimental procedure is, therefore, to measure first the ratios of the currents to the disk and the annulus at a number of values of E/N in the absence of the magnetic field, and then to measure the ratios of the currents falling on either side of the central cut at the same values of E/N but with suitably chosen values of B. The values of W/D calculated from the first set of measurements are then used in conjunction with the second set to determine the values of W_M. The accuracy of the measurements can be conveniently checked by comparing the results for W_M obtained with different values of B at the same pressure, as well as by comparing results taken at different pressures.

Table 12.1 shows a sample of the results obtained by Jory[6] for nitrogen. In his experiments, any asymmetry in the apparatus was largely accounted for by taking the average value of the results obtained with the magnetic field in opposite directions. Errors from this source are much larger in magnetic deflexion experiments than in experiments to measure W/D.

Table 12.1. Experimental values of W_M (10^5 cm sec^{-1}) in nitrogen at 293 K

E/p (V cm^{-1} torr^{-1})	\<br\>20	\<br\>40	B (gauss)\<br\>60	\<br\>80	\<br\>100	\<br\>120	\<br\>20	\<br\>40	\<br\>60	\<br\>80	\<br\>100	\<br\>20	\<br\>40	\<br\>100	Least Value	Greatest Value	Best Estimate
	p = 200 torr						*p = 100 torr*					*p = 50 torr*					
0.04	3.94	3.94													3.94	3.94	3.94
0.06	4.23	4.27	4.28				4.28	4.24							4.23	4.28	4.27
0.08	4.34	4.41	4.41	4.46			4.40	4.35	4.36						4.34	4.41	4.38
0.10	4.46	4.50	4.49	4.55	4.58		4.49	4.46	4.42	4.39					4.42	4.65	4.4 (8)
0.15		4.76	4.76	4.80	4.85		4.75	4.73	4.70	4.68	4.67	4.82	4.81		4.68	4.85	4.7 (3)
0.20		5.03	5.05	5.07	5.10		5.00	5.00	4.97	4.96	4.96	5.11	5.07		4.96	5.11	5.0 (4)
	p = 20 torr						*p = 10 torr*					*p = 5 torr*					
0.30							5.48	5.45	5.48	5.44	5.43	5.52	5.54	5.56	5.43	5.56	5.5 (0)
0.50		6.20	6.17	6.18									6.26	6.30	6.17	6.30	6.2 (0)
0.70		7.00	7.01	7.01	7.02	7.00							7.06	7.10	7.00	7.10	7.0 (5)
0.90		7.86	7.94	7.90	7.91	7.89	7.84	7.83	7.76	7.76					7.76	7.94	7.8 (5)
1.20		9.36	9.40	9.39	9.37	9.37	9.25	9.25	9.25	9.25					9.25	9.40	9.3 (5)
1.80		12.35	12.45	12.41	12.42	12.38	12.19	12.23	12.26	12.23	12.23				12.19	12.42	12.3
2.50							15.5	15.6	15.6			15.6	15.5		15.5	15.7	15.6
4.00							22.5	22.4	22.6	22.6		22.5	22.5		22.4	22.6	22.5
6.00												31.5	31.6		31.5	31.6	31.6
8.00												40.0	40.2		40.0	40.2	40.1

489

From the table it can be seen that the overall scatter in the data is less than $\pm 2\%$, notwithstanding the large range covered by the experimental parameters p and B for any given value of E/N.

In later experiments with hydrogen and deuterium,[8] some refinements were made to the technique to reduce still further any errors arising from asymmetry. In addition to the obvious error caused by misalignment of the apparatus, errors are caused by contact potential differences across the surface of the anode and by lack of coincidence between the effective centre of the electron source and the geometrical centre of the source hole. Huxley's method is superior to Townsend's original method in that each of these sources of error can be identified and corrected for. By doing so, Creaser[8] was able to reduce the scatter in his data to a level considerably below that of Jory's, but nevertheless the results cannot be regarded as having the same precision as the data for W and D/μ.

Figure 12.2 shows Jory's results for the magnetic deflexion coefficient $\psi = W_M/W$ in nitrogen. These results were selected because they illustrate the point that W_M can be very different from W, and they show that for $E/N < 0.3$ Td the differences between the two quantities are more than

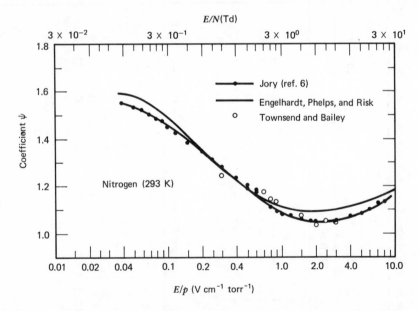

FIG. 12.2.   The variation of the magnetic deflexion coefficient ψ with E/N in nitrogen at 293 K. The experimental values of Townsend and Bailey (*Phil. Mag.*, **42**, 873, 1921) and the calculated values of Engelhardt, Phelps, and Risk (*Phys. Rev.*, **135**, A1566, 1964) are shown for comparison. The same values of W were used in each case to derive ψ from the values of W_M. (From ref. 6.)

50%. This point is further illustrated in Table 12.2, in which calculated values of ψ are shown for combinations of several model cross sections and velocity distribution functions. The numbers in the table are obtained from equations 8.20 and 8.21 (or from equation 8.22) on the assumption that the distribution function can be represented by $f_0(c) = \text{const} \cdot \exp(-c/\alpha)^n$ and the cross section by $q_m = \text{const} \cdot c^r$. The table serves to stress the point that there can be very large differences between W_M and W, while at the same time it explains why reasonable agreement between the two quantities exists in many cases of practical importance, that is, when the cross section is not strongly energy dependent and the distribution function is peaked to a greater extent than the Maxwellian form.

12.3. MEASUREMENTS OF THE ATTACHMENT COEFFICIENTS α_a AND α_{at}

There have been numerous experimental measurements of attachment coefficients based on a number of basically different experimental techniques. Apart from microwave methods, which we omit for the reasons stated at the outset, the various types of experiment can be conveniently divided into two classes: those in which the rate of loss of electrons to form negative ions is determined by observing the spatial variation of the electron number density, and those which rely on a determination of the time variation of the total population of an electron group.

12.3.1. STEADY-STATE METHODS. Within this class of experiment there are three distinct types.

1. Experiments in which the spatial distribution is determined in an apparatus of fixed length. These experiments discriminate between the electrons and ions by taking advantage of their different rates of diffusion.

2. Experiments in which the spatial distribution is found by changing

TABLE 12.2. Values of the magnetic deflexion coefficient ψ corresponding to several model cross sections and distribution functions

			r	
n		-1	0	1
2	(Maxwell)	1.0	1.18	3
4	(Druyvesteyn)	1.0	1.06	1.38
6		1.0	1.03	1.16

the distance between the source and detector, and using a method for detecting the electrons and ions differentially.

3. So-called growth-of-current experiments, which are applicable at higher values of E/N when ionization accompanies attachment.

With the exception of Bailey's method, which we mention first, the methods described below will be seen to belong to one or another of these three types.

(a) Bailey's Method

A description of this method was given in Chapter 11. Like Huxley's method, it has the great advantage that values of D/μ as well as of α_{at} are determined in a single experiment. The essential difference between it and Huxley's method, which is more complex analytically but more straightforward experimentally, lies in the procedure used to allow for the unknown composition of the mixed stream of electrons and ions which enters the apparatus through the aperture in E_0 (see Figure 11.17). In Bailey's method, since the composition of the stream at E_0 was unknown, measurements were made at two different distances from E_0 and advantage was taken of the following facts (see Section 11.4.1):

1. The ratio of the number density of electrons to the number density of ions depends on the distance from the source.

2. The different values of D/W for the electrons and ions result in different fractions of the electron and ion streams falling on the electrodes E_2 and E_4.

Thus Bailey's method is seen to be a combination of types 1 and 2 above.

Bailey and his associates used this method to measure D/μ and α_{at} for a large number of gases. The results of their work are summarized in ref.10. Where there is overlap between this work and more recent investigations, it is generally found that the results obtained using Bailey's method differ considerably from those of the later experiments, which are themselves in reasonably good agreement.[11,27] The question has not been settled as to whether Bailey's method is inherently inaccurate, or whether the discrepancies are due to technical difficulties (such as those associated with gas handling) which would have been difficult to overcome at the time.

(b) Huxley's Method

An account of this method as used in the first instance by Huxley, Crompton, and Bagot,[12] was also given in Chapter 11 and needs little amplification here. The method relies simply on determinations of the spatial distributions of electrons and ions at the anode of an apparatus of fixed length and therefore belongs to type 1. For the purpose of analysis, the stream may be regarded as commencing at the source hole as a pure stream of electrons (see Figure 11.18). This assumption is made valid by the form of division and the dimensions of the anode, which is designed to ensure that the distribution of current is sampled only outside a central disk large enough to intercept all the ions entering the diffusion chamber through the source hole.

A criticism of the method as it applies to the measurement of attachment coefficients, as distinct from measurements of D/μ in the presence of attachment, is its lack of flexibility, for the following reasons. First, it is necessary for $(W/D)_{ions}$ to be very much larger than $(W/D)_{electrons}$ in order to discriminate adequately between the electrons and the ions. Although the theory of the method as developed finally by Hurst and Huxley[13] does not depend on this stipulation for its validity, it is an essential proviso if accurate measurements of the attachment coefficient are to be made. Second, it is necessary for α_{at}/N to be large enough that significant attachment can occur without using high gas pressures. This may be understood by referring to Figure 11.19. In order to measure α_a with reasonable accuracy it is essential that the presence of attachment alter the current ratio significantly, that is, that α_a be sufficiently large that there is good separation between the appropriate ratio curve and the curve for $\alpha_a = 0.$* On the other hand, the use of high gas pressures in order to increase α_a is accompanied by large values of η' (see equation 11.81) and values of R that approach unity. Thus, in order to obtain sufficiently large values of α_a when α_a/N is small, the gas pressure may have to be made so high that the diffusion of the electron stream is negligible and inadequate current arrives at the annuli of the receiving electrode.

In assessing the accuracy of published data obtained with this method the reader should note the modifications to the theory that result from the inclusion of anisotropic diffusion[14] (see Chapters 5 and 11). All the data from the groups at Canberra and Liverpool are subject to some error on this account. For example, in the case of oxygen, the data of Huxley et al. may require corrections as large as 20 to 30%. This is shown by the curves in Figure 12.3 in which the data of Huxley et al., reanalyzed using equation 5.62, are compared with the original data.[14]

Because of its limitations the method has not been largely used. In addition to the original work on oxygen[12,15] and water vapour,[16] the method has been applied by Naidu and Prasad[17] to dichloro-fluoromethane, and again to oxygen but over a considerably larger range

---

*As explained in Section 5.4.3, the attachment coefficient appropriate to the theory of the spreading of a mixed stream of electrons and ions is α_a (see footnote on 496).

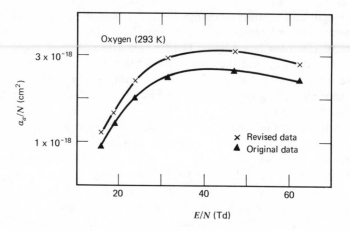

FIG. 12.3. Comparison of the data for α_a/N derived using Huxley's method with and without the inclusion of the correction for anisotropic diffusion (see text).

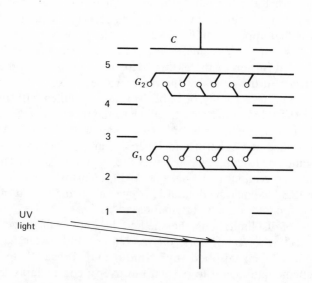

FIG. 12.4. Schematic diagram showing the principle of the method used by Bradbury. The grids G_1 and G_2 are movable and are inserted into the electron and ion stream one at a time.

of values of E/N than hitherto. For $E/N > 60$ Td it was necessary to allow for the influence of ionization on the electron number density within the swarm.* A comparison of the results of the two groups, and of the results obtained by this method with those obtained by other methods, is given in Section 12.3.3.

(c) Methods Employing Electrical Filters

Preliminary experiments based on this technique were made as early as 1929 by Cravath,[1,18] but the first experiments that were free from obvious objection were those of Bradbury.[19] His apparatus is shown schematically in Figure 12.4. The photoelectric source of electrons, the guard electrode structure, and other details of design are similar to those of the drift tubes described in Chapter 10. The essential feature of the apparatus whereby it differs from the types described previously is the provision of the movable shutters G_1 and G_2, which serve as electron filters in this application. Only one of the filters is inserted into the mixed stream of electrons and ions at any one time. When a high-frequency alternating potential difference is applied between the wires of a filter, electrons in its vicinity are swept to the wires, but the more massive ions are only slightly deflected and the majority of them pass through without being captured. Thus, in principle, both the ion current and the current due to unattached electrons can be determined by measuring the current collected by the receiving electrode with and without the high-frequency signal applied to the filter.

In practice, however, as shown by Cravath, the operation of the shutters never approaches the ideal situation described above, since experimental conditions cannot be found such that all the electrons are captured but none of the ions. Accordingly, Bradbury designed his apparatus to have two filters since he argued that, if the two filters behave identically, measurements of the difference between the total current and the ion current at each of the positions G_1 and G_2 should enable errors arising from the imperfect characteristics of the filters to be eliminated, at least to first order. The justification for the procedure is as follows.

Let i and I_0 be the currents measured at C with and without the high-frequency signal. The current I_0 will be somewhat less than the total current of electrons and ions in the stream because electrons are lost by diffusion to the grid wires in each position. Similarly, if it is assumed that

---

*As pointed out in Chapter 11, it is not possible to assess the validity of the theory upon which the analysis of these experiments was based. It appears that the theory may have been developed as an extension of Lucas's analysis of the diffusing stream, which was based on an incorrect argument (see footnote on p.381). It is also possible that, even if this were the case, the analysis as applied to these experiments is not greatly in error since long diffusion chambers (4 and 7 cm) were used.

all the electrons are collected when the RF signal is applied, i will be somewhat less than the ion current at the plane of the filter because some of the ions will be deflected and caught. However, if it is assumed for the moment that these errors are negligible, it follows that $I_1 = I_0 - i_1$ is the current due to free electrons at G_1 and that $I_2 = I_0 - i_2$ is the corresponding current at G_2. Consequently

$$\frac{I_2}{I_1} = \frac{I_0 - i_2}{I_0 - i_1} = \exp - \alpha_a h, \qquad (12.1)$$

where α_a is the attachment coefficient* and h is the separation of the filters. From the form of equation 12.1 it is clear that small errors in I_0, i_1, and i_2 will cancel to first order.

Bradbury presented his results in terms of the "probability of attachment"[19] rather than α_a and thus required values of the electron drift velocity. The values were obtained in the same apparatus, using a somewhat indirect method due to J. J. Thomson.

Bradbury's measurements were restricted to oxygen. Subsequently Kuffel[21] applied the technique with very little modification to both air and water vapour, having first repeated the measurements in oxygen as a check.

Chatterton and Craggs[22] made a systematic investigation of possible sources of error in experiments of this type in an attempt to explain the fairly large differences between the results of these experiments and those from the growth-of-current experiments described in the next subsection. These authors showed that the principal sources of error were losses of electrons between source and collector, due to lateral diffusion, and the unwanted collection of negative ions by the filters, the latter leading to

---

*Here and in the rest of this chapter we make the assumption that has usually been made in developing the theory of these experiments that the effect of diffusion can be ignored when interpreting experiments of this type and therefore assume that the attachment coefficient used in equation 12.1 is $\alpha_a = N \overline{q_{at}(c)c} / W$. The coefficient that would appear in the equation if allowance were made for diffusion is α_{at} (see Section 5.4 and ref. 20). When comparing attachment coefficients obtained in this way with those obtained by Huxley's method, or when comparing attachment rates derived from this type of swarm experiment with those derived from attachment cross sections through the equation

$$N \int_0^\infty q_{at}(\epsilon) \epsilon f(\epsilon) \, d\epsilon = \alpha_a W,$$

it is first necessary to determine the true value of α_a from the value of the coefficient (α_{at}) actually found from equation 12.1. The procedure for making this correction is described in Section 5.4.3.

particularly serious errors. Equation 12.1 is, of course, based on the assumption that there is no loss of ions to either filter, and its use was justified by Bradbury on the grounds that errors due to ion collection would largely cancel. However, Chatterton and Craggs estimated that as much as 10% of the negative-ion current could have been collected in some of Bradbury's experiments. Furthermore, they were able to show that losses of as little as 1 to 2% can lead to very large errors in the measured values of the attachment coefficient when α_a exceeds 0.5 cm^{-1}, and to appreciable errors for all values in excess of 0.3 cm^{-1}. They concluded, therefore, that the accuracy of the data obtainable with this technique is unlikely to exceed that of data from other less complicated methods.

By careful choice of the experimental parameters, Chatterton and Craggs were able to confirm the observation by Chanin, Phelps, and Biondi[23] of the three-body attachment process in oxygen at low values of E/N, and to get reasonable agreement between their results and the ones obtained by other methods at higher values of E/N. However, their work suggests that the technique is of limited value.

(d) Growth-of-Current Experiments

Experiments of type 3 were introduced by Harrison and Geballe[24] and are applicable to the range of values of E/N where ionization and attachment occur simultaneously. The method is an extension of the well known Townsend growth-of-current method of measuring ionization coefficients, in which the current $I(h)$ in a parallel plate discharge gap is measured as a function of electrode separation h while the voltage across the gap is adjusted to maintain a constant value of E/N.[1] When ionization occurs but no attachment, the semilogarithmic plots of $I(h)$ versus h are straight lines whose slopes yield the values of the ionization coefficient α_T (see equation 5.53). When attachment also occurs, however, the plots are no longer straight lines but assume the form shown in Figure 12.5. An analysis of these curves using the following theory enables both α_T and α_{at} to be determined.

Let us assume that the radial distribution of current in the discharge gap is uniform and that we can ignore the effects of the longitudinal number density gradients of the electrons and ions. The second assumption means that we can disregard the influence of the electrodes on the number densities, and also that $\alpha_i \equiv \alpha_T$ and $\alpha_a \equiv \alpha_{at}$. We also assume that the processes of detachment and of negative-ion formation by ion-pair production are negligible, and that the production of electrons by secondary processes can be ignored. The last assumption requires that we restrict our discussion to values of the product ph (gas pressure × electrode separation) that are well below the value at breakdown. From equation 5.43, the

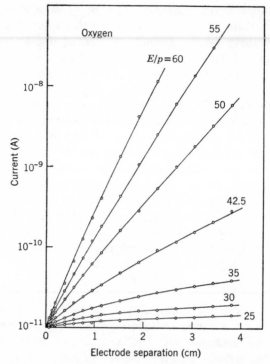

FIG. 12.5. Typical variation of current with electrode separation obtained by Harrison and Geballe[24] for oxygen at a pressure of 11.2 torr. The results are normalized to the same value of j_0. The full curves were fitted to the experimental data using equation 12.5.

appropriate steady-state differential equation for n is seen to be

$$W\frac{\partial n}{\partial z} = (\bar{\nu}_i - \bar{\nu}_{at})n, \qquad (12.2)$$

the solution of which is

$$n(z) = n_0\exp(\alpha_i - \alpha_a)z, \qquad (12.3)$$

where n_0 is the electron number density at the cathode, $\alpha_i = \bar{\nu}_i/W$, and $\alpha_a = \bar{\nu}_{at}/W$.

The corresponding equation for the ions is

$$W_I\frac{\partial n_I}{\partial z} = \bar{\nu}_{at}n,$$

where the subscript I denotes that the quantity refers to ions. This equation is equivalent to

$$\frac{\partial n_I}{\partial z} = n \frac{W}{W_I} \alpha_a$$

$$= n_0 \frac{W}{W_I} \alpha_a \exp(\alpha_i - \alpha_a)z$$

from equation 12.3. On integration, this equation becomes

$$n_I(z) = \frac{n_0 \alpha_a}{\alpha_i - \alpha_a} \frac{W}{W_I} \exp(\alpha_i - \alpha_a)z + \text{const.}$$

When $z = 0$, $n_I = 0$; consequently the constant equals $-\{n_0[\alpha_a/(\alpha_i - \alpha_a)] \times (W/W_I)\}$ and the equation for $n_I(z)$ therefore becomes

$$n_I(z) = \frac{n_0 \alpha_a}{\alpha_i - \alpha_a} \frac{W}{W_I} \{ [\exp(\alpha_i - \alpha_a)z] - 1 \}. \tag{12.4}$$

It follows from equations 12.3 and 12.4 that the current density at the anode is given by*

$$j(h) = eWn(h) + eW_I n_I(h)$$

$$= n_0 eW \left[\exp(\alpha_i - \alpha_a)h + \frac{\alpha_a}{\alpha_i - \alpha_a} \{ [\exp(\alpha_i - \alpha_a)h] - 1 \} \right]$$

$$= \frac{j_0}{\alpha_i - \alpha_a} \{ \alpha_i[\exp(\alpha_i - \alpha_a)h] - \alpha_a \}, \tag{12.5}$$

where j_0 is the current density at the cathode. This equation has been generalized to account for ion-pair production and secondary processes[24] and for detachment.[25]

Equation 12.5 can be rearranged to read

$$j(h) = \frac{j_0}{(\alpha_i/p) - (\alpha_a/p)} \left\{ \frac{\alpha_i}{p} \left[\exp\left(\frac{\alpha_i}{p} - \frac{\alpha_a}{p} \right)ph \right] - \frac{\alpha_a}{p} \right\}. \tag{12.6}$$

---

*It is simpler to evaluate the particle flux at the anode rather than elsewhere in the gap because there is then no necessity to calculate the positive-ion component.

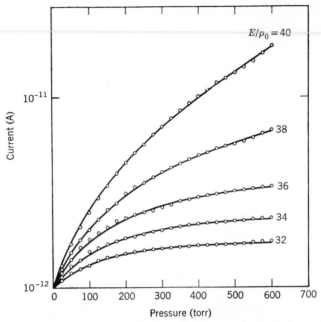

FIG. 12.6. Freely's and Fisher's experimental results for the variation of current with gas pressure in oxygen at various values of E/p_0 (p_0 is the pressure normalized to 273 K) when the electrode separation was held constant ($h = 0.200$ cm). Note the similarity of these curves and those shown in Figure 12.5. (From ref. 26.)

Since α_i/p and α_a/p are functions of E/p, the form of equation 12.6 shows that j/j_0 depends only on the product ph provided that E/p is held constant. Consequently two experimental approaches have been used, that in which the pressure has been held constant and the electrode separation varied, and that in which a constant electrode separation has been employed in conjunction with varying pressure. Figure 12.5 typifies experimental results obtained with the first procedure, and Figure 12.6 shows similar curves obtained with the second method. The equivalence of the two methods is obvious from the figures.

Figure 12.5 illustrates an important limitation of the method which is obvious from the form of equation 12.5. When $\alpha_i \gg \alpha_a$, equation 12.5 approaches the simple exponential form of Townsend's original equation and α_a can no longer be determined by this method. This occurs for $E/p > 60$ V cm^{-1} torr^{-1} ($E/N > 180$ Td) in oxygen. On the other hand, when α_i is small, the current amplification is then so small that it is not possible to determine either coefficient accurately.

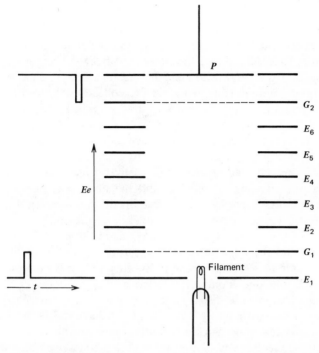

FIG. 12.7. Schematic diagram showing the principle of the method introduced by Doehring. The electric field in the regions E_1G_1 and G_2P is reversed except during the times at which the rectangular pulses are applied.

There have been several investigations of ionization and attachment coefficients in oxygen, the most recent being those of Prasad and Craggs[27] and Freely and Fisher.[26] Sukhum et al.[28] extended the work of Prasad and Craggs in order to investigate the effect of detachment on the earlier measurements and analysis. References to other applications of the technique will be found in the book by Christophorou.[29] From a comparison of the results for the same gas obtained by different workers (see Section 12.3.3) it must be concluded that the method is subject to errors that are as yet not understood.

12.3.2. TRANSIENT METHODS

(a) Doehring's Method

A new method of measuring attachment coefficients was developed by Doehring, whose apparatus is shown schematically in Figure 12.7. Appropriate potentials were applied to the electrodes between the grids G_1 and G_2 to ensure a uniform field in the drift space between the grids. As in the original "four-gauze" experiments (see Chapter 10), the filament (electron source) and the electrode E_1 were held at a suitable potential to ensure a reverse field in the region $E_1 G_1$ and hence no flow of electrons into the drift space. Similarly, the potential of the collecting electrode P was adjusted to give a reverse field in the region $G_2 P$, thus preventing either electrons or ions from reaching the collector. At the time $t = 0$, a rectangular pulse of 20 to 70 μsec duration was applied to E_1 to reverse the direction of the field between E_1 and G_1 and so admit electrons (and some ions) into the drift space. After a known but variable delay, a similar rectangular pulse was applied to P, thereby admitting to the collector any electrons or ions that had arrived at G_2 at that instant. The sequence was repeated at a frequency of 500 Hz, and the quasi-continuous current I arriving at P was recorded. The experiment consisted in finding the variation of the collected current as a function of the time delay t between the opening of the two shutters. The attachment coefficient was then found from the slope of the log I versus t curve, using the following theory.

Let us assume that the open time of the first shutter is made sufficiently long that, on the one hand, the electrons may be considered to have established a steady-state distribution throughout the drift space and, on the other, any ions formed by attachment will have moved an inappreciable distance in that time. Such a compromise is possible because the drift velocity of the electrons is of the order of 10^3 times that of the ions. It follows from the first assumption that, if we neglect the effect of diffusion, the distribution of electron number density within the drift space is (see equation 12.3)

$$n(z) = n_0 \exp - \alpha_a z, \qquad (12.7)$$

where n_0 is the electron number density at the first shutter. If we assume that the electron drift velocity is, in effect, infinite, electrons are present in the drift space only during the time δt during which the shutter opens, and their distribution during that time is given by equation 12.7.

The differential equation for the number density of the ions is (again neglecting diffusion)

$$\frac{\partial n_I}{\partial t} + W_I \frac{\partial n_I}{\partial z} = \bar{\nu}_{at} n. \qquad (12.8)$$

If we make the realistic assumption that $W_I(\partial n_I / \partial z) \ll \bar{\nu}_{at} n$ during the time interval δt, equation 12.8 becomes

$$\frac{\partial n_I}{\partial t} = \bar{\nu}_{at} n$$

$$= \bar{\nu}_{at} n_0 \exp - \alpha_a z;$$

consequently the distribution of ion number density throughout the drift space when the shutter ceases to transmit and the electrons disperse is

$$n_I(z, \delta t) = \bar{\nu}_{at} \delta t n_0 \exp - \alpha_a z. \tag{12.9}$$

The number density of ions at the sampling shutter is now determined as a function of time during the time interval in which the ions cross the drift space. Since the right-hand side of equation 12.8 is now zero, the equation appropriate to this period is

$$\frac{\partial n_I}{\partial t} + W_I \frac{\partial n_I}{\partial z} = 0$$

with $n_I(z, 0)$ given by equation 12.9. The solution of this equation, which satisfies the initial conditions, is

$$n_I(z, t) = \bar{\nu}_{at} \delta t n_0 \exp - \alpha_a (z - W_I t); \tag{12.10}$$

consequently the current density at the grid situated at $z = h$ is

$$j(h, t) = e W_I \bar{\nu}_{at} \delta t n_0 \exp - \alpha_a (h - W_I t)$$

$$= j_0 \exp \alpha_a W_I t, \tag{12.11}$$

where $j_0 = e W_I \bar{\nu}_{at} \delta t n_0 \exp - \alpha_a h$ is the initial ion current density at the sampling shutter. The theoretical curve of current as a function of delay time is therefore as shown in Figure 12.8a. In practice, the finite duration of the initial burst of electrons and the diffusion of the ions in transit cause some rounding of the curve, as shown in the figure.

When plotted on a semilogarithmic scale (see Figure 12.8b), the curve of $I(t)$ versus t is a straight line whose slope is the product of α_a and W_I. The determination of α_a from the slope of the line therefore requires a knowledge of W_I. There are two possible methods of obtaining this information. The first relies on the fact that some negative ions will be formed before the first grid so that the apparatus can be used as a Tyndall-Powell

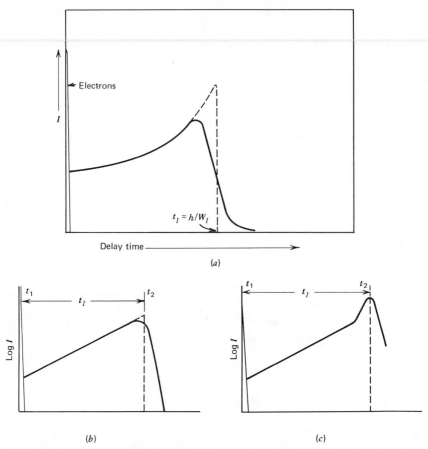

FIG. 12.8. Transmitted-current versus delay-time curves obtained with Doehring's method. (a), linear-linear plot of transmitted current versus delay time; (b), log-linear plot of the same data: the dashed curve would be that obtained if there were no diffusion; (c), curve obtained when some negative ions are admitted by the grid G_1.

"four-gauze" apparatus to determine the velocity of these ions. The second method consists in measuring the time interval t_I between the time t_1 for maximum electron current and the time t_2, that is, the time of arrival of the ions generated at the cathode. But for diffusion the measurement would be straightforward, t_2 then being the time for maximum ion current (see Figure 12.8a); however, the appreciable rounding of the ion current curve makes it difficult to decide exactly which point on the curve corresponds to t_2 (see Figure 12.8b). Fortunately it is possible to adjust the experimental parameters in order to determine t_2 more accurately than might appear possible from the figure. If the open time of the first shutter is small, very few of the ions formed between E_1 and G_1 will enter the drift space. When the shutter is quiescent, there is a high density of ions near E_1 but a low density elsewhere. After the shutter opens the number density of ions at $z = 0$ begins to increase, the density being given by $n_I(0) \equiv n_I(0, t)$ $= n_0 \bar{v}_{at} t$ for $t \ll \delta t$ since there is no contribution initially from those formed near E_1. When, however, $t \sim s / W_I$, where s is the distance between E_1 and G_1, some of these ions can enter the drift space thus contributing to $n_I(0)$. Thus the ion current curve can be changed from the form shown in Figure 12.8b to that in 12.8c. Clearly, the time t_2 is then more closely specified and W_I can be more accurately determined. Depending on the operating conditions, Doehring took t_2 to lie somewhere between the maximum of the curve and the point at which the ion current had fallen to half its maximum value. The error from the uncertainty in t_2 can obviously be reduced by making the gas pressure as high as possible, thus reducing the dispersal due to diffusion.

A modified form of Doehring's technique was used by Chanin et al.[23] for their experiments in oxygen. These experiments revealed the importance of three-body nondissociative attachment in oxygen at low values of E/N, thereby explaining the cause of some of the disagreement between the results of previous workers.

There are three known processes whereby negative ions are formed by low-energy electrons in oxygen.

Nondissociative, radiative attachment.

$$e + O_2 \rightarrow O_2^- + h\nu.$$

The cross section for this process is known to be small compared with the values for the other processes described below, and it may therefore be disregarded.

Dissociative attachment

$$e + O_2 \rightarrow (O_2^-)_{\text{unstable}},$$

$$(O_2^-)_{\text{unstable}} \rightarrow O + O^- + \text{K.E.}$$

This process has a threshold of about 4 eV and is the dominant process when $E/N > 10$ Td provided that the gas pressure is less than about 20 torr (see Figure 12.11).

Nondissociative, collision-stabilized attachment

$$e + (O_2)_{v=0} \rightleftarrows (O_2^-)_{v=m},$$

$$(O_2^-)_{v=m} + (O_2)_{v=0} \rightleftarrows (O_2^-)_{v=m'} + (O_2)_{v=n}.$$

Because the presence of a third body is required to absorb the internal energy of the O_2^- ion and return it to the (stable) ground state, this process depends on N^2 rather than N. At high pressures and low values of E/N it is the predominant mechanism for negative-ion formation.

The apparatus used by Chanin et al.[23] is shown in Figure 12.9. Although similar in principle to Doehring's apparatus, it differs from the latter in the replacement of the pulsed grid electron source by a photocathode illuminated by a pulsed UV light source, and the use of a Loeb shutter in place of the "two-gauze" shutter arrangement employed by Doehring to sample the electron and ion currents. A typical current-delay time curve is shown in Figure 12.10. The absence of the enhanced maximum at $t = t_2$ can be seen. Figure 12.11 shows the results for α_{at}/p as a function of E/p for pressures in the range 7.6 to 54 torr. The pressure dependence below $E/N \sim 15$ Td is clearly in evidence.

Another modification of Doehring's technique was used by Rees[15] to supplement his measurements with the Huxley diffusion method. Using a

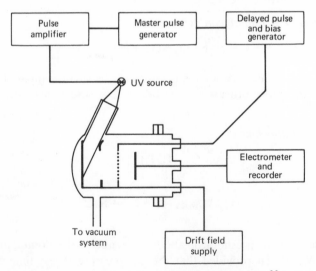

FIG. 12.9. Schematic diagram of the drift tube used by Chanin et al.[23] in their adaptation of Doehring's method.

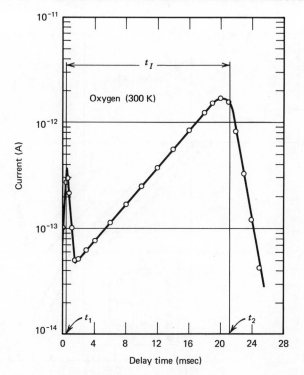

FIG. 12.10. Typical transmitted current versus delay time curve obtained by Chanin et al. with oxygen ($E/N \sim 0.75$ Td, $p = 10.5$ torr, $T = 300$ K). The current peak due to the arrival of the unattached electrons, and the increase of the current with time due to the arrival of the negative ions, can be easily identified. (From ref. 23.)

conventional Bradbury-Nielsen drift tube, Rees repeated the measurements in oxygen, paying particular attention to the ion drift velocity W_I since there were some shortcomings in the measurements by Doehring and by Chanin et al. of this quantity. Doehring's measurements were subject to the normal errors associated with the use of a "four-gauze" technique, that is, errors caused by the poor definition of the source and sampling planes (see Section 10.3.1), and those due to field interpenetration in the vicinity of the shutters. Chanin et al. were able to determine W_I only from curves of the type shown in Figure 12.10. Since they had no means of enhancing the number of ions entering the drift space at time $t = 0$, they had no way of generating the convenient "marker" shown in Figure 12.8. Rees was able to obtain higher precision in his measurements of W_I using both methods. In the first place, as in Doehring's experiment, the negative ions produced before the first shutter provided a marker for the more accurate

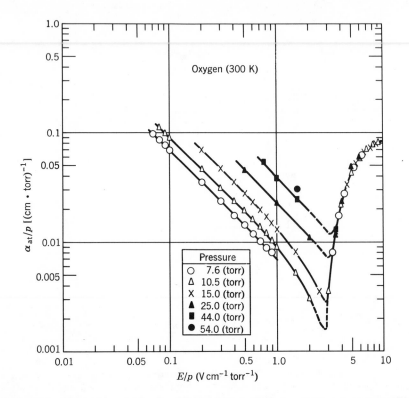

FIG. 12.11. The results of Chanin et al. for the attachment coefficient in oxygen, which clearly showed for the first time the importance of the three-body attachment process at low values of E/p. (From ref. 23.)

determination of t_I when the transient method was used. Second, because of the negligible field distortion within a drift tube terminated with plane shutters, the determination of W_I by the normal time-of-flight method is likely to have been more accurate than Doehring's.

Figures 12.12 and 12.13 show, respectively, typical current versus delay-time curves and a typical current versus frequency curve obtained by Rees. Both the shoulder on the upper curve in Figure 12.12 and the multiple maxima exhibited by the curve in Figure 12.13 are evidence of the presence of more than one ion species.[15] Rees was able to detect four ion species (one of which, at the time, was thought to have been due to an impurity) and to confirm the ion mobilities determined by the transient method with more accurate measurements by the Bradbury-Nielsen method.

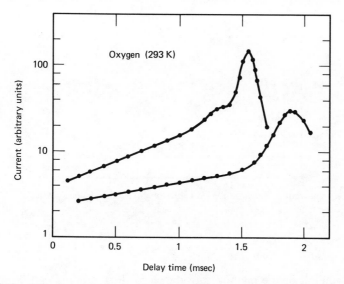

FIG. 12.12. Typical curves of transmitted current versus delay time, obtained by Rees[15] using a Bradbury-Nielsen drift tube with rectangular pulses applied to the grids as in Doehring's method. The first shutter always admitted some negative ions as well as the electrons.

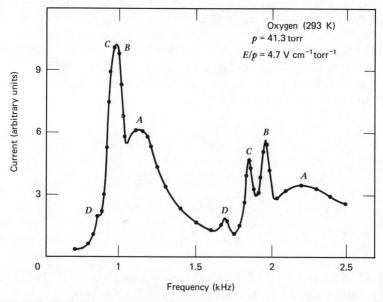

FIG. 12.13. Typical curves of transmitted current versus frequency obtained by Rees[15] when the Bradbury-Nielsen drift tube was used in the normal way with sine wave signals applied to the grids. The "spectrum" shows maxima which can be attributed to four different ion species (A, B, C and D).

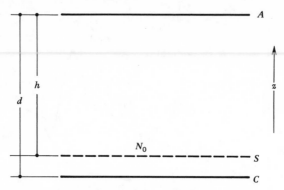

FIG. 12.14. Diagram to illustrate the principles of the Bortner-Hurst method of measuring attachment rates.

(b) Method of Bortner and Hurst

The principle of the method introduced by these authors can be understood with the aid of Figure 12.14. Electrons are released in the plane S by the action of an alpha-particle source which is collimated to confine the particles to that plane. Let h and d be the distances of the plane S and the cathode C respectively, from the anode A, and let N_0 electrons be produced by the passage of a single alpha particle. In the absence of attachment the electrons will move toward the anode with an average velocity $\mathbf{W}$, inducing a current $i = N_0 eW/d$ in the external circuit (see equation 10.2). If the capacity to earth of the anode is C and the time constant of the anode circuit is large compared with the transit time of the electrons, the rate of change of the anode potential is

$$\frac{dV}{dt} = -\frac{N_0 eW}{Cd}, \quad 0 \leqslant t \leqslant t_m,^* \tag{12.12}$$

whence

$$V(t) = -\frac{N_0 eWt}{Cd} + \text{const}$$

$$= V' - \frac{N_0 eWt}{Cd},$$

where V' is the potential of the anode at $t=0$, and $t_m = h/W$. It follows that the maximum amplitude of the (negative) voltage pulse $V_a = V' -$

*Equation 12.12 holds right up to $t = t_m$ since the effects of diffusion are neglected in this analysis.

$V(t_m)$ induced on the anode by the passage of the electrons from S to A is

$$V_a = \frac{N_0 eh}{Cd}.$$

When electron attachment occurs, the form of $V(t)$ is modified. At time t the electrons will have reached the plane $z = Wt$ and their number will have been reduced to $N(t) = N_0 \exp - \alpha_a Wt$. Equation 12.12 is therefore replaced by*

$$\frac{dV}{dt} = -\frac{N_0 eW}{Cd} \exp - \alpha_a Wt, \quad 0 \leqslant t \leqslant t_m,$$

which, on integration, gives

$$V(t) = \frac{N_0 e}{Cd\alpha_a} \exp - \alpha_a Wt + \text{const}$$

$$= V' - \frac{N_0 e}{Cd\alpha_a} (1 - \exp - \alpha_a Wt).$$

Consequently the amplitude of the voltage pulse $V_a(t) = V' - V(t)$ induced on the anode is

$$\left. \begin{aligned} V_a(t) &= \frac{N_0 e}{Cd\alpha_a} (1 - \exp - \alpha_a Wt) \\ &= \frac{V_a}{\alpha_a h} (1 - \exp - \alpha_a Wt), \\ &= \frac{V_a}{\alpha_a h} \left[1 - \exp\left(-\frac{\alpha_a ht}{t_m} \right) \right] \end{aligned} \right\} \quad 0 \leqslant t \leqslant t_m. \qquad (12.13)$$

If the voltage pulse is recorded with the aid of a linear pulse amplifier having a response to a pulse of unit amplitude which is represented by

$$V'(t) = \frac{t}{t_1} \exp\left(-\frac{t}{t_1} \right), \qquad (12.14)$$

*The voltage induced by the negative ions may be neglected with negligible error because $W_1 \ll W$.

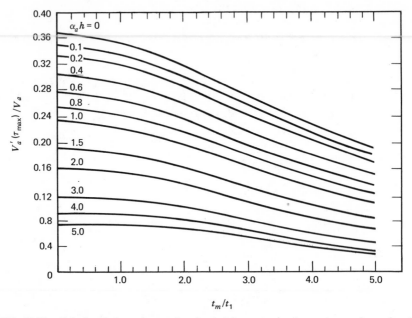

FIG. 12.15. Calculated dependence of pulse height on t_m/t_1 for various values of $\alpha_a h$ (actually $\alpha_{at}h$). (From A. A. Christodoulides and L. G. Christophorou, Oak Ridge National Laboratory Rept. ORNL-TM-3163.)

where t_1 is the differentiating and integrating time constant of the amplifier, the output pulse produced by $V_a(t)$ at a time $\tau > t_m$ is given by [31]

$$V'_a(\tau) = \int_0^{t_m} \frac{dV_a(t)}{dt} V'(\tau - t) \, dt. \tag{12.15}$$

Using the expressions for $V_a(t)$ and $V'(t)$ given by equations 12.13 and 12.14, one may evaluate equation 12.15 for a set of values of $\alpha_a h$ and a series of values of t_m/t_1 thus finding for each pair of values of $\alpha_a h$ and t_m/t_1 the value of $\tau = \tau_{max}$ for which $V'_a(\tau)$ is a maximum. For a given apparatus (i.e., fixed h, d, and t_1) the pulse height $V'_a(\tau_{max})$ is a function of α_a and W. Figure 12.15 shows values of $V'_a(\tau_{max})/V_a$ plotted as a function of t_m/t_1 for various values of $\alpha_a h$. From these curves it may be seen that α_a can be determined from measurements of the ratio of the mean pulse height in the attaching gas to the mean pulse height in a nonattaching gas when t_m is the same in each case, it being assumed that N_0 and t_1 are constants.

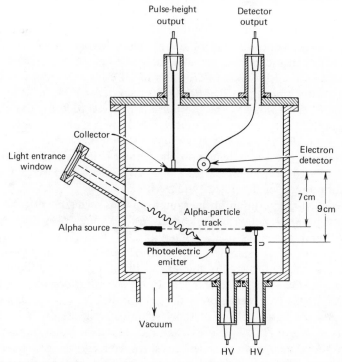

FIG. 12.16.   Apparatus used by Christophorou et al.[32] for measurement of electron attachment and electron drift velocity.

Figure 12.16 shows a schematic diagram of the apparatus used by Christophorou et al.,[32] which was developed from the original apparatus of Bortner and Hurst.[31] The electron source for the pulse height experiments consists of a ring of [239]Pu with a collimating arrangement to ensure that the alpha-particle tracks are confined to the plane of the ring. The plane metal anode is used to collect the electrons and negative ions produced in the ionization chamber and is therefore connected to the input of the pulse amplifier. The values of t_m, and hence the values of W, were found in a separate experiment in which electrons were released from the photocathode by a flash of UV light of short duration incident on the photocathode and were detected by the G-M detector placed behind an aperture in the anode (see Section 10.5). The signal from a photodiode illuminated by the UV light provided the initial time marker.

The experimental procedure for determining α_a was as follows. The pulse heights and values of t_m were first determined for the pure carrier gas* at a given pressure and for a suitable range of values of E/N (and hence t_m). The top curve of Figure 12.15 was thereby obtained. Subsequently a small concentration of the attaching gas (~ 1 part in 10^3 to 10^5) was added to the carrier gas, and the pulse heights and values of t_m were again measured for a series of values of E/N. Let P_{M_1} and t_{m_1} be the pulse height and transit time, respectively, recorded for the mixture at a given value of E/N, and P_{C_1} be the pulse height recorded for the carrier gas when the transit time has the same value. Then $\alpha_a h$ can be determined from the ratio P_{M_1}/P_{C_1} using the curves shown in Figure 12.15, the measurement of the pulse height in the pure carrier gas providing the normalizing factor (i.e., in effect, the value of V_a in equation 12.13 and the amplification factor of the pulse amplifier).

Provided that the addition of the sample gas has negligible effect on the energy distribution function in the carrier gas, the values of α_a/N_A (where N_A is the number density of the attaching gas) determined for different partial pressures of the sample gas should be constant. In this event the values of t_m obtained in the pure carrier gas and in the mixture should also be the same at the same value of E/N. In practice, however, the addition of the sample gas in general alters the energy distribution function, and consequently the values of α_a/N show some dependence on N_A. This is illustrated in Figure 12.17. In this example, in which water vapour and argon are the attaching and carrier gases, respectively, the dependence is particularly strong because of the extreme sensitivity of the energy distribution functions in argon to molecular impurity. Nevertheless it can be seen that the attachment coefficient corresponding to the undisturbed energy distribution function in the carrier gas at the specified value of E/p can be found by linear extrapolation to zero concentration of the sample gas.

Hurst and Christophorou and their collaborators have applied this technique to the examination of a wide range of attaching gases comprising both simple and complex molecules. A major disadvantage of the method arises from the fact that it cannot be applied to the direct measurement of attachment coefficients as a function of E/N in pure gases, since the attachment coefficient measured at a given value of E/N applies to the energy distribution in the carrier gas and not to that for the pure attaching gas. Hence it is not possible to compare directly the results for α_a/N versus E/N obtained by this method with the values derived by the other methods described in this section. Nevertheless the Bortner-Hurst method has found important application in many investigations of attachment cross sections using the procedure described in the next chapter

---

*See ref. 29 for a discussion of the criteria used in the selection of the carrier gas.

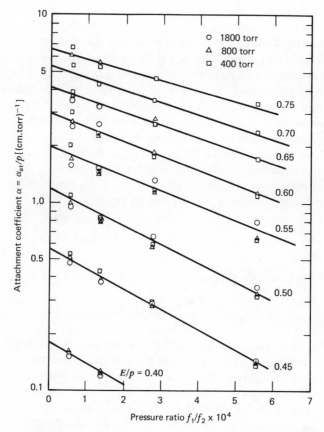

FIG. 12.17.   Attachment coefficient $\alpha = \alpha_{at}/p$ as a function of the ratio of water vapour pressure to argon pressure for various values of E/p. (From ref. 33; see also Section 13.6.)

(Section 13.6). The work is comprehensively reviewed in the book by Christophorou.[29]

(c) Methods Based on the Analysis of Transient Wave Forms

The transient method for measuring drift velocities developed by Herreng, Hornbeck, and Bowe (see Section 10.2) was first applied by Herreng to the measurement of attachment coefficients. An alternative technique was later introduced by Burch and Geballe,[34] but the accuracy of their measurements was poor. Recently Grünberg[35] has applied the same basic principles to measurements at high pressures in oxygen. We describe the essential details of his method because the results from it are the best so far available from experiments of this type and are possibly the best available for oxygen at the present time.

The apparatus used by Grünberg is similar to that described in Sections 10.7.2 and 10.8 and consists of a parallel plate discharge gap with the electrodes machined to a Rogowski profile (see Figure 10.23). Electrons are released from the cathode at time $t=0$ by a pulse of UV light of short duration, and the time variation of the cathode potential $V_R(t)$ (see Figure 10.25) is recorded with a high-speed oscilloscope. For Grünberg's experiments the time constant of the cathode circuit was made large (~ 1 sec) and therefore greatly exceeded the transit time of the negative ions.

The analysis leading to an expression for $V_R(t)$ for $t < t_m$ is essentially the same as that in the preceding section. Thus, if N_0 electrons are released from the cathode initially, $V_R(t)$ is given by (equation 12.13)

$$V_R(t) = \frac{N_0 e}{C d \alpha_a} (1 - \exp - \alpha_a W t), \quad 0 \leqslant t \leqslant t_m,$$

where d is the electrode spacing and $t_m = d/W$. Thus, when $t = t_m$,

$$V_R(t_m) = \frac{N_0 e}{C d \alpha_a} (1 - \exp - \alpha_a d). \tag{12.16}$$

For $t \leqslant t_m$ it is valid to assume that the motion of the negative ions makes no contribution to the current in the gap and therefore to $V_R(t)$. For $t > t_m$, however, the current from the transport of electrons ceases, but a small current continues to flow from the $N_0(1 - \exp - \alpha_a d)$ negative ions produced in the gap. The voltage $V_R(t)$ therefore continues to rise slowly until all the charge carriers (i.e., a number equal to the original number of electrons N_0) have crossed the gap. After a sufficiently long time, therefore, $V_R(t)$ reaches a plateau such that

$$V_R(\infty) = \frac{N_0 e}{C}. \tag{12.17}$$

Figure 12.18 shows typical transient wave forms obtained at two values of E/p. Because the time scale must be chosen so that the plateau can be recorded, the electron transit time occupies a negligible part of the abscissa and the initial rise in $V_R(t)$ appears to be instantaneous. From the oscillograms it is clear that the ratio $V_R(\infty)/V_R(t_m)$ can be determined with considerable accuracy. The attachment coefficient can then be calculated from the equation

$$\frac{V_R(\infty)}{V_R(t_m)} = \frac{\alpha_a d}{1 - \exp - \alpha_a d}. \tag{12.18}$$

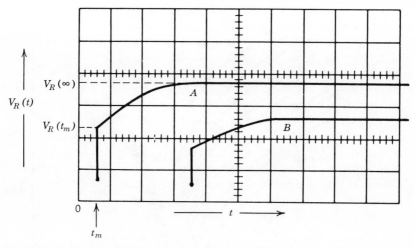

FIG. 12.18.   Typical oscillograms obtained by Grünberg[35] for two different values of E/N. The values of t_m, $V_R(t_m)$, and $V(\infty)$ in relation to curve A are shown.

Grünberg used this method to measure attachment coefficients in oxygen for the pressure range 15 to 880 torr and for values of E/N in the range $0.3 < E/N$ (Td) < 90. The results shown in Figure 12.19 illustrate the remarkable flexibility of the method in its ability to cover a wide range of values of E/N and α_{at} (approximately two orders of magnitude in each case). This feature of the method, coupled with the accuracy claimed for the data (error limits of 2 to 3%), makes it one of the most successful so far developed.

(d) Single-Avalanche Method

These methods are an extension of the ones developed by Raether and his collaborators and described in Section 10.7. They are restricted to high values of E/N and to high gas pressure, where the amplification of the initial number of electrons in the avalanche is large. As a consequence they are beyond the scope of this book, and the reader is referred to the book by Raether[36] for a description of the theory and practice of the methods and the results derived from them. An important extension of this work was made by Frommhold, who accounted for the effects of detachment when interpreting the results of the experiments at higher values of E/N. A comprehensive account of the analysis of the complex situation that exists when ionization, attachment, and detachment occur simultaneously is to be found in ref. 37.

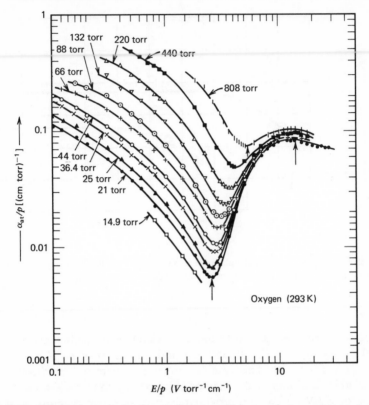

FIG. 12.19. Grünberg's results for the attachment coefficient α_{at}/p in oxygen as a function of E/p and p. (From ref. 35.)

12.3.3. RESULTS AND CONCLUSION. We conclude this section with a comparison of some of the results for oxygen that have appeared in the literature in order to contrast the present state of experimental measurements of attachment coefficients with that of the measurements of drift and diffusion, with which this book is primarily concerned.

Figures 12.20 and 12.21 compare some of the results that have been obtained using the methods described in this section. The overall picture cannot be said to be satisfactory at either high or low values of E/N. Although much of the divergence in the results below $E/N \sim 10$ Td can now be explained in terms of the density dependence of α_a/N due to nondissociative three-body attachment, no simple explanation has been found for the poor agreement between the results obtained in the range $10 < E/N$ (Td) < 100, and it seems likely that deficiencies in the methods or in their application must be responsible for this. Above $E/N \sim 100$ Td,

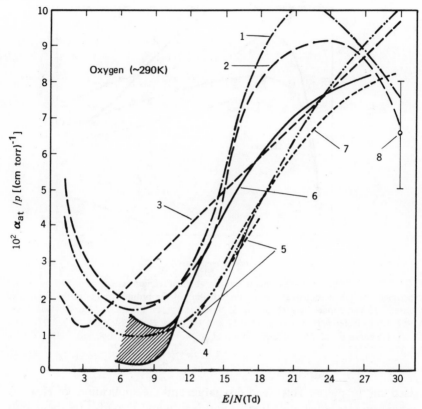

FIG. 12.20. Comparison of attachment data for O_2 at room temperature obtained by
different experimenters. (The ordinate is labelled α_{at}/p, although some of the data are for α_a.
The difference between the two coefficients is clearly unimportant when compared with the
discrepancies between the sets of data.) The shaded area for $E/N < 10$ Td corresponds to the
region in which α_{at}/p varies with pressure. Grünberg's results (Figure 12.19) are omitted for
the sake of clarity, since the range of pressures used by him results in values of α_{at}/p that
span all the data below $E/N \sim 15$ Td. (From H. S. W. Massey, E. H. S. Burhop, and H. B.
Gilbody, *Electronic and Ionic Impact Phenomena*, Vol. 2, 2nd ed., Clarendon Press, Oxford,
1969.) References: curve 1, ref. 19; curve 2, P. Herreng, *Cah. Phys.*, **38**, 7, 1952; curve 3, R. H.
Healey and C. B. Kirkpatrick, from ref. 10; curve 4, ref. 30; curve 5, ref. 15; curve 6, ref. 23;
curve 7, ref. 12; curve 8, ref. 34.

it is claimed that detachment may be responsible for some of the divergence of the results.[28,37]

Almost without exception the methods that have been developed and
used in recent years are open to some objection. At present those which
seem to be most free from criticism are the most recent applications of
Doehring's method by Chanin et al.[23] and Rees,[15] Grünberg's experiments

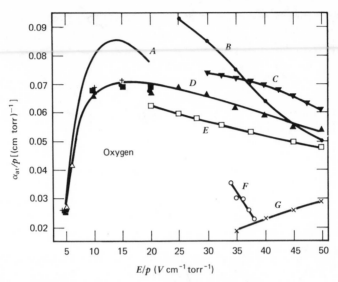

FIG. 12.21. Attachment coefficients in oxygen at room temperature as measured by several experimenters. The comment regarding the labelling of the ordinate made in the caption to Figure 12.20 also applies here. (From M. S. Naidu and A. N. Prasad, *J. Phys. D: Appl. Phys.*, **3**, 957, 1970.) References: Curve *A*, refs. 12 and 15; curve *B*, ref. 24; curve *C*, ref. 27; curve *D*, ref. 17; curve *E*, ref. 26; curve *F*, Dutton et al., *Nature (London)*, **198**, 680, 1963; curve *G*, ref. 37.

based on the pulsed Townsend discharge, and the applications of Huxley's method by Huxley et al.,[12] Rees,[15] and Naidu and Prasad.[17] Of these, only the work of Grünberg is free from obvious objection at the present time. Thus both Chanin et al. and Rees had difficulty in assigning values for the ion drift velocity W_I for use in the calculation of α_{at} from the slopes of their current-delay time curves. The value of the attachment rate $\alpha_{at}W_I$ that may be obtained directly from their measurements is, of course, free from this criticism unless the measurements were significantly affected by ion-molecule reactions. Similarly, the results derived from experiments based on Huxley's method were incorrect as published originally because no account was taken of anisotropic diffusion in the original development of the theory. Rees's work[14] (see p.493) makes it possible to arrive at some assessment of the probable error in these data. His work shows that, because of the problem of correctly interpreting the experimental results when the conditions of the experiments are such that the validity of the asymptotic solutions cannot be verified, the accuracy of data for α_a obtained by this method does not match the accuracy of Grünberg's data. As a technique for measuring attachment coefficients, Huxley's method is, in any case, less satisfactory than the other two because of the limited range of experimental parameters over which accurate measurement can be made. Its greatest utility lies in its application to the measurement of D/μ when accurate values of α_a are known.

We conclude with the comment that the results for oxygen, which has served as a useful standard gas in nearly all investigations, should stand as a warning when assessing the error limits for the data for other gases not subject to such intensive investigation.

REFERENCES

1. L. B. Loeb, *Basic Processes of Gaseous Electronics*, University of California Press, Berkeley, 1955.

2. See, for example, M. A. Folkard and S. C. Haydon, *Aust. J. Phys.*, **24**, 519, 1971; **24**, 527, 1971, and references therein.

3. A. L. Gilardini, *Low-Energy Electron Collisions in Gases*, Wiley-Interscience, New York, 1973.

4. L. G. H. Huxley, *Phil. Mag.*, **23**, 210, 1937; J. S. Townsend, *Phil. Mag.*, **23**, 880, 1937.

5. L. G. H. Huxley, *Aust. J. Phys.*, **13**, 718, 1960.

6. R. L. Jory, *Aust. J. Phys.*, **18**, 237, 1965.

7. R. W. Crompton, M. T. Elford, and R. L. Jory, *Aust. J. Phys.*, **20**, 369, 1967.

8. R. P. Creaser, *Aust. J. Phys.*, **20**, 547, 1967.

9. J. R. Barker, *J. Sci. Instrum.*, **26**, 273, 1949.

10. R. H. Healey and J. W. Reed, *The Behaviour of Slow Electrons in Gases*, Amalgamated Wireless Ltd., Sydney, 1941.

11. A. N. Prasad and J. D. Craggs, *Proc. Phys. Soc.*, **76**, 223, 1960; *Proceedings of the Fourth International Conference on Ionization Phenomena in Gases, Uppsala*, North-Holland Publishing Company, Vol.1, 1960, p.142.

12. L. G. H. Huxley, R. W. Crompton, and C. H. Bagot, *Aust. J. Phys.*, **12**, 303, 1959.

13. C. A. Hurst and L. G. H. Huxley, *Aust. J. Phys.*, **13**, 21, 1960.

14. J. A. Rees, personal communication, 1972.

15. J. A. Rees, *Aust. J. Phys.*, **18**, 41, 1965.

16. R. W. Crompton, J. A. Rees, and R. L. Jory, *Aust. J. Phys.*, **18**, 541, 1965.

17. M. S. Naidu and A. N. Prasad, *Brit. J. Appl. Phys. (J. Phys. D)*, **2**, 1431, 1969; *J. Phys. D: Appl. Phys.*, **3**, 957, 1970.

18. A. M. Cravath, *Phys. Rev.*, **33**, 605, 1929.

19. N. E. Bradbury, *Phys. Rev.*, **44**, 883, 1933.

20. R. W. Crompton, *J. Appl. Phys.*, **38**, 4093, 1967.

21. E. Kuffel, *Proc. Phys. Soc.*, **74**, 297, 1959.

22. P. A. Chatterton and J. D. Craggs, *J. Electron.*, **11**, 425, 1961.

23. L. M. Chanin, A. V. Phelps, and M. A. Biondi, *Phys. Rev.*, **128**, 219, 1962.

24. M. A. Harrison and R. Geballe, *Phys. Rev.*, **91**, 1, 1953, and references therein.

25. H. Schlumbohm, *Z. Naturforsch.*, **16a**, 510, 1961.

26. J. B. Freely and L. H. Fisher, *Phys. Rev.*, **133**, A304, 1964.

27. A. N. Prasad and J. D. Craggs, *Proc. Phys. Soc.*, **77**, 385, 1961.

28. N. Sukhum, A. N. Prasad, and J. D. Craggs, *Brit. J. Appl. Phys.*, **18**, 785, 1967.

29. L. G. Christophorou, *Atomic and Molecular Radiation Physics*, Wiley-Interscience, New York, 1971.

30. A. Doehring, *Z. Naturforsch*, **7a**, 253, 1952.

31. T. E. Bortner, and G. S. Hurst, *Health Phys.*, **1**, 39, 1958.

32. L. G. Christophorou, R. N. Compton, G. S. Hurst, and P. W. Reinhardt, *J. Chem. Phys.*, **43**, 4273, 1965.

33. G. S. Hurst, L. B. O'Kelly, and T. E. Bortner, *Phys. Rev.*, **123**, 1715, 1961.

34. D. S. Burch and R. Geballe, *Phys. Rev.*, **106**, 183, 1957.

35. R. Grünberg, *Z. Naturforsch.*, **24a**, 1039, 1969.

36. H. Raether, *Electron Avalanches and Breakdown in Gases*, Butterworths, London, 1964.

37. L. Frommhold, *Fortschr. Phys.*, **12**, 597, 1964.

13

CROSS-SECTION DETERMINATIONS

FROM ELECTRON TRANSPORT DATA

13.1. INTRODUCTION AND HISTORICAL SYNOPSIS

This chapter is concerned with past and present methods of analyzing the results of swarm experiments although, for reasons that will become apparent, greatest emphasis is given to developments within the last decade.

When the extensive literature of this subject is examined, several major developments are discernible, some of which can be attributed to technological advances that took place shortly before their introduction. Like the experiments themselves, the first analyses were made by Townsend and his collaborators,[1] who were the first to relate the transport coefficients describing the motion of an ensemble of electrons to parameters such as the mean free path and fractional energy loss per collision, which describe the individual collisions between the electrons and gas molecules.

Two remarkable achievements of this early work were the discovery of the energy dependence of the mean free paths of the electrons,[2] that is, their total scattering cross sections, and the fact that the average energy loss per collision in molecular gases greatly exceeds the value predicted on the basis of collisions between hard spheres. As is well known, the discovery of the energy dependence of the scattering cross section—in particular, the pronounced effects observed for the monatomic gases—was made independently by Ramsauer,[3] using a single-collision technique. The work of Townsend and Ramsauer marked the commencement of the complementarity between the so-called swarm and beam methods, which persists to the present day.

The years that followed this pioneering work saw a progressive improvement in the sophistication of the analysis as the basic theory of weakly ionized gases was more fully developed and applied. Comprehensive reviews of the development of the analytical techniques are to be found in the books by Healey and Reed[4] and Loeb[5] and the articles by Allis[6] and Huxley and Crompton.[7].

Notwithstanding their increasing sophistication, all analyses made before the early 1950's were based on somewhat drastic simplifying assumptions,[7] without which little progress would have been possible because of the mathematical complexity of the formulae linking the macroscopic to the microscopic parameters. Attempts were then made to overcome this problem more satisfactorily. One such attempt,[8] based on the then relatively new microwave technology, opened up a field which has been actively pursued ever since.[9] An essential element of the first microwave swarm experiments was the fact that, in many cases, the electron swarm was studied only after it had established an equilibrium energy distribution (Maxwellian) with the gas. In this way the serious problem of estimating precisely the distribution of energies of the electrons in the swarm was overcome. We shall outline these techniques in a little more detail in Section 13.3, although we shall not attempt to discuss them fully. A detailed account is to be found in the book by Gilardini.[9]

The methods used to analyze the results of microwave experiments are notable because they represent the first attempts[8,10] to determine precisely the energy dependence of the momentum transfer cross section rather than the variation of an averaged quantity related to the cross section (see Section 13.2.1). Similar methods were later applied to the results of drift tube and lateral diffusion experiments.[11-14] Although these developments represented an advance, the range of problems to which they could be applied was restricted and the results obtained of limited accuracy (see Sections 13.3 and 13.4).

Finally we come to the developments of the last decade. With the introduction by Frost and Phelps[16] and others[17] of a new analytical approach, the earlier methods, as well as the results derived from them, have become largely obsolete. An essential feature of the new methods is the application of the techniques of numerical analysis to the solution of equation 6.35 for $f_0(c)$ [or the calculation of $f_0(c)$ from equation 3.11] and the subsequent evaluation of the integral expressions for the transport coefficients (e.g., equations 3.14 and 3.27). For this reason the introduction of these methods, which represent the only completely satisfactory solution to a complex problem, had to await the development of high-speed computers.

The application of the new techniques has revolutionized both the scope and the accuracy of the results derived from swarm experiments. The older analyses were rarely capable of giving more than reliable estimates of elastic scattering cross sections. Furthermore they yielded no specific data on the energy dependence of inelastic cross sections, although some gross features associated with inelastic scattering were derived (see Section 13.2.3). On the other hand, the new methods are capable of giving remarkably accurate cross sections for elastic scattering in many instances, together with similar information for inelastic scattering cross sections in certain circumstances.

It is sometimes difficult for those unfamiliar with the field to appreciate how swarm experiments can be ranked with beam experiments as techniques for the detailed study of electron collision processes, the obvious disadvantage being that the energies of the electrons in a swarm are not confined to a narrow band as they are in a beam experiment. Notwithstanding this disadvantage, swarm experiments have features which enable them to make unique contributions, as pointed out in the recent reviews of Crompton[18] and Bederson and Kieffer.[19] In the first place, it seems likely that experiments of this type will always be capable of providing data at very low energies that are more reliable than the data derived from beam experiments. Second, swarm experiments are conducted at relatively high gas pressures, usually in a static gas sample; as a consequence the determination of the number density of the target gas, on which the determination of the absolute magnitude of the cross section depends, presents little difficulty, whereas in other techniques it constitutes a major experimental problem. To offset these advantages the energy resolution of swarm experiments is, apart from exceptional circumstances, greatly inferior to that of beam experiments. The most successful way of attacking many problems concerned with the collisions of low-energy electrons, therefore, is to combine the results from both types of experiment.

In the rest of this chapter we first outline the basic principles of some of the swarm analyses that were used before about 1960 because they provide a useful insight into the physical principles of the techniques currently in use. Subsequently we describe the new methods and some of the results that have been achieved with them.

13.2. BASIC PRINCIPLES AND EARLY METHODS OF ANALYSIS

The problem of determining cross sections from swarm data can be divided into three parts:

1. The derivation of elastic scattering (momentum transfer) cross sections from electron transport coefficients that are measured under conditions in which only elastic scattering occurs, for example, in the monatomic gases at relatively low values of E/N.

2. The derivation of momentum transfer cross sections from transport data measured under conditions in which both elastic and inelastic scattering occur.

3. The derivation of the cross sections for inelastic scattering.

As we shall see, considerable progress was made towards solving the first two problems before the introduction of the new techniques in 1962, but the third remained intractable.

To understand the basic principles of all methods of analysis we return to equations 3.14 and 3.27, which are restated here for convenience:

$$W = \frac{4\pi}{3} \frac{e}{m} \frac{E}{N} \int_0^\infty \frac{d}{dc} \frac{c^2}{q_m(c)} f_0(c) \, dc, \tag{13.1}$$

$$ND = \frac{4\pi}{3} \int_0^\infty \frac{c^3}{q_m(c)} f_0(c) \, dc. \tag{13.2}$$

We also note the following equations for $f_0(c)$.

1. *When elastic scattering only occurs,* $f_0(c)$ is given by (see equation 3.11)

$$f_0(c) = A \exp\left(-\frac{3m}{M} \int_0^c \frac{c \, dc}{\left[Ee/mNq_{m_{el}}(c)c \right]^2 + 3\kappa T/M} \right)$$

$$= A \exp\left(-\frac{3m}{M} \int_0^c \frac{c \, dc}{V^2 + \overline{C^2}} \right), \tag{13.3}$$

where $V = eE/mNq_{m_{el}}(c)c$.

2. *When both elastic and inelastic scattering occur* (and for the purpose of this discussion it is assumed that inelastic scattering is isotropic, that the collision frequency for inelastic scattering is small compared with the collision frequency for elastic scattering, and that superelastic encounters may be neglected), $f_0(c)$ is found by solving equation 6.35, that is,

$$(V^2 + \overline{C^2}) \frac{df_0}{dc} + \frac{3mc}{M} f_0 + \frac{3}{c^3 q_{m_{el}}(c)} \sum_i \sum_k I_{ik} = 0, \tag{13.4}$$

where

$$I_{ik} = \int_c^{c_{1_{ik}}} \left\{ f_0(x) - f_0\left[(x^2 - c_{1_{ik}}^2)^{1/2} \right] \exp - \frac{\epsilon_{ik}}{\kappa T} \right\} x^3 q_{0_{ik}}(x) \, dx,$$

where $q_{0_{ik}}$ is the cross section for the kth transition of the ith inelastic process, ϵ_{ik} is the threshold energy for this transition, $c_{ik} = (2\epsilon_{ik}/m)^{1/2}$, and $c_{1_{ik}} = (c_{ik}^2 + c^2)^{1/2}$.

From the form of equations 13.1 to 13.4 it is clear why, in general, the calculation of $f_0(c)$ for a trial set of cross sections and the subsequent evaluation of the transport integrals cannot be achieved other than by numerical methods.

The analysis introduced by Townsend and used until quite recently avoided any such procedure by making use of the simplification of equations 13.1 and 13.2 that results from the assumption that q_m is a slowly varying function of energy. If it is reasonable to assume that q_m is effectively constant over the range of values of c which contribute significantly to the right-hand side of the equations, it follows immediately that

$$W = \frac{2}{3} \frac{Ee}{mN} q_m^{-1} \overline{c^{-1}} \tag{13.5}$$

and

$$D = \frac{1}{3N} q_m^{-1} \bar{c}. \tag{13.6}$$

Thus, if the relation of $\overline{c^{-1}}$ to $\bar{c}$ is determined by postulating a functional form for $f_0(c)$ (e.g., a Maxwell or Druyvesteyn distribution), equations 13.5 and 13.6 constitute a pair of simultaneous equations for q_m and $\bar{c}$ that can be solved, given the values of W and D for a given value of E/N.

Although the procedure outlined in the preceding paragraph formed the basis of Townsend's analysis, it was developed by him from a somewhat different starting point. Using a free path treatment (see Chapter 7), Townsend found formulae for W and D which, apart from a numerical factor in equation 13.5, were the same as equations 13.5 and 13.6 with the replacement of $(Nq_m)^{-1}$ by l, the mean free path (more properly the mean free path for momentum transfer). A fuller discussion of the formulae derived by Townsend, Huxley, and others and of their application will be found in Healey and Reed[4] and Huxley and Crompton.[7] In general, we limit the discussion in this section to a summary of the most recent applications of similar techniques which avoid the use of computers.[7,20,21]

13.2.1. THE DETERMINATION OF AN EFFECTIVE MEAN VALUE OF $q_{m_{el}}$ AS A FUNCTION OF $\bar{c}$ WHEN ONLY ELASTIC SCATTERING OCCURS: THE MEAN VALUE TECHNIQUE. Let us assume that $V^2 \gg \overline{C}^2$ and that $q_{m_{el}}(c)$ can be replaced by an effective mean value $q^*_{m_{el}}$ in equations 13.1 to 13.3. The equations for W and D are then equations 13.5 and 13.6 with $q_{m_{el}} \equiv q^*_{m_{el}}$. It is to be noted that $q^*_{m_{el}}$ is more properly written as $q^*_{m_{el}}(E/N)$, since it is a mean value of $q_{m_{el}}$ which, for a given value of E/N, satisfies equations 13.5 and 13.6. Thus, although $q^*_{m_{el}}$ is regarded as constant for the purpose of evaluating the integrals at each value of E/N, its value is in general different for each value of E/N.

With the assumptions stated above, it follows immediately that $f_0(c)$ is a Druyvesteyn distribution represented by the equation (see Section 3.7)

$$f_0(c) = A \exp - \left(\frac{c}{\alpha} \right)^4$$

with

$$A = \frac{1}{\pi \Gamma(\frac{3}{4}) \alpha^3}$$

and

$$\alpha = \left(\frac{Ee}{mNq^*_{m_{el}}} \right)^{1/2} \left(\frac{4M}{3m} \right)^{1/4},$$

$\hspace{10cm}$ (13.7)

and that, from the formula on p.78,

$$\overline{c^{-1}} = \left[\frac{\Gamma(\frac{1}{2})}{\Gamma(\frac{3}{4})} \right] \alpha^{-1}.$$

Substitution of this value of $\overline{c^{-1}}$ into equation 13.5 results in the following formula[21] for $q^*_{m_{el}}$:

$$q^*_{m_{el}} = \frac{4}{9} \frac{e}{m} \left(\frac{3m}{4M} \right)^{1/2} \left[\frac{\Gamma(\frac{1}{2})}{\Gamma(\frac{3}{4})} \right]^2 \frac{E/N}{W^2}.$$

$\hspace{9cm}$ (13.8)

Consequently $q^*_{m_{el}}$ can be determined from the value of W measured at each value of E/N. A similar procedure can be used to relate $q^*_{m_{el}}$ to D, but the calculation of $q^*_{m_{el}}$ from measured values of W has been preferred because of the greater accuracy with which W can be measured.

In order to estimate the variation of $q_{m_{el}}$ with electron energy it is necessary to determine at each value of E/N a parameter which characterizes the velocity or energy of the electrons. The parameters that have usually been derived are the root-mean-square velocity $(\overline{c^2})^{1/2}$ or the mean speed $\bar{c}$, although the most probable speed, or the value of c that contributes most to the integral on the right-hand side of equation 13.1, are more logical choices. For this discussion we will follow past practice and use $\bar{c}$.

For the Druyvesteyn distribution, $\bar{c}$ is related to α by $\bar{c} = \alpha/\Gamma(\tfrac{3}{4})$ (see p.78). On substituting the expression for q_m^* given by equation 13.8 into equation 13.7, and then substituting the resulting expression for α into the formula for $\bar{c}$, we obtain

$$\bar{c} = \frac{3}{2\Gamma(\tfrac{1}{2})} \left(\frac{4M}{3m} \right)^{1/2} W. \tag{13.9}$$

Thus, in the special case where only elastic scattering occurs and where the electron energy is greatly in excess of the mean energy of the gas molecules, it is possible to obtain a reasonable estimate of the variation of $q_{m_{el}}$ with energy from the measurement of a single transport coefficient at a number of values of E/N.

13.2.2. EXTENSION OF THE MEAN VALUE TECHNIQUE TO SITUATIONS IN WHICH THE DRUYVESTEYN FORMULA IS INAPPLICABLE: THE METHODS OF TOWNSEND, BAILEY, HUXLEY, AND OTHERS. When inverse collisions between the electrons and molecules significantly affect the electron energy distribution, that is, when $\tfrac{1}{2}m\,\overline{c^2}$ no longer greatly exceeds $\tfrac{3}{2}\kappa T$, or when inelastic collisions are important, equation 13.7 is no longer applicable. However, the *functional form* of $f_0(c)$ may still represent the distribution reasonably well,[22] and the assumption that it does so is a key assumption in much of the older work described in this section. The important difference in the case under discussion is that the constants A and α are no longer simply related to E/N and q_m. For example, in molecular gases, where, in general, inelastic scattering largely controls $f_0(c)$ at all values of E/N, α is usually several times smaller than the value predicted by equation 13.7.

In these circumstances it is no longer possible to eliminate $\overline{c^{-1}}$ from equation 13.5 by using the relation between $\overline{c^{-1}}$, E/N, and q_m that follows from equation 13.7. Therefore an alternative procedure was developed[23] which requires experimental data for W and D/μ as functions of E/N but makes no use of equation 13.7.

From equations 13.5 and 13.6 the following formulae for q_m^* and $\bar{c}$ may be obtained:

$$q_m^* = \frac{1}{3} \left(\frac{2e}{m} \right)^{1/2} (\bar{c}\,\overline{c^{-1}})^{1/2} \frac{E/N}{W(D/\mu)^{1/2}}, \tag{13.10}$$

$$\bar{c} = \left(\frac{2e}{m} \right)^{1/2} (\bar{c}\,\overline{c^{-1}})^{1/2} \left(\frac{D}{\mu} \right)^{1/2}. \tag{13.11}$$

The essential difference between this approach and that described in the preceding section is the estimation of $\bar{c}$ and $\overline{c^{-1}}$ (for substitution in equation 13.5) directly from the measured values of D/μ rather than indirectly from values of W via equation 13.7.

We notice that this procedure introduces an unknown factor into the calculation of q_m^* and $\bar{c}$ that was not present in the method based on equations 13.8 and 13.9, namely, the dimensionless product of the averages $\bar{c}$ and $\overline{c^{-1}}$, whose magnitude depends on the functional form of $f_0(c)$. The appearance of such a factor is to be expected since any specification of the function has been avoided in deriving mean values of q_m and c from equations 13.5 and 13.6. The difficulty is circumvented by postulating the form of $f_0(c)$ and so evaluating the dimensionless factors, it being customary to assume either the Maxwellian or Druyvesteyn form,[7] although other forms have been suggested and used.[4] There is now evidence to suggest that these assumptions are reasonable in the case of some molecular gases,[22] although not for a gas such as argon, where, at low E/N, only elastic collisions occur and where there is a rapid variation of $q_{m_{el}}$ with c.

The representation of the actual distribution by a model is less critical than might appear at first sight. In changing from a Maxwellian to a Druyvesteyn distribution, for example, the factor $(\bar{c}\,\overline{c^{-1}})^{1/2}$ changes by less than 4%; thus, considering the nature of the data which are obtainable by this method, the errors introduced through imprecise knowledge of $f_0(c)$ are relatively unimportant.

For the case of a Druyvesteyn distribution, equations 13.10 and 13.11 reduce to

$$q_m^* = 2.15 \times 10 \quad \frac{E/N}{W(D/.\mu)^{1/2}} \text{ cm}^2 \qquad (13.12)$$

and

$$\bar{c} = 6.44 \times 10^7 \left(\frac{D}{\mu}\right)^{1/2} \text{ cm sec}^{-1}, \qquad (13.13)$$

where E/N is expressed in townsends, W in centimetres per second, and D/μ in volts; for a Maxwellian distribution the numerical factors become 2.23×10^{-10} and 6.69×10^7, respectively.

As has been already remarked, equations similar to these formed the basis of all analyses of swarm measurements until about 1950. Two comments may be helpful to those wishing to evaluate the results of such analyses. First it is important to note that until reliable time-of-flight measurements became available in the mid-1930's, and sometimes even afterwards, the values of W that were used were not true values of W as

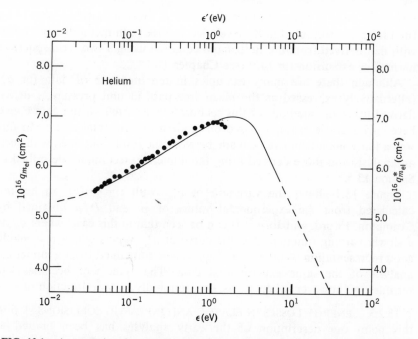

FIG. 13.1. A comparison between the momentum transfer cross section obtained from the accurate analysis and that obtained using approximate formulae. Curve computed from accurate analysis; points are values of q_m^*. (From ref. 21)

defined by equation 13.1 but rather values of W_M as defined by equation 8.18 and determined by magnetic deflexion experiments. The significant difference that can exist between these quantities was not stressed until comparatively recently,[16,24] after detailed analyses by Allis[6] and Huxley[25] of the motions of electrons in crossed electric and magnetic fields. Under certain circumstances failure to allow for the difference between W and W_M can give rise to errors in the values of q_m calculated with the aid of equation 13.10 that are much larger than the error in $(\bar{c}\,\overline{c^{-1}})^{1/2}$ which follows from the approximate representation of the functional form of $f_0(c)$. This is so because the magnetic deflexion coefficient $\psi = W_M/W$ is sensitive to the variation of q_m with c as well as the form of $f_0(c)$ (see Chapter 8). Thus data for q_m^* which are based on values of W_M rather than W are sometimes of questionable accuracy.

The second factor to be borne in mind in assessing the early data is the stage of development of the formula for W, often derived from a free path treatment, at the time at which the analysis was undertaken, for example, whether or not correct account was taken of the distribution of free paths and the distribution of velocities. These matters are discussed fully in the references cited in Section 13.1.† It should be stressed, however, that in its final form the formula for W developed by free path methods is identical with that derived from the Boltzmann equation using a two-term spherical harmonics expansion for $f_0(c)$ (see Chapter 7).

Although there are many examples in the literature of data for q_m^* (alternatively expressed as the mean free path at unit pressure, l or λ), derived using the methods of this section (see, e.g., refs. 4 and 5), we give here only a single example based on modern experimental data.[21] In this way a fair comparison can be made between the results obtained with such an analysis and those derived using the numerical techniques described in Section 13.5.

Figure 13.1 shows the variation of $q_{m_{el}}^*$ with $\epsilon' = \frac{1}{2}m(\bar{c})^2$ in helium, calculated from the experimental values of W and D/μ obtained by Crompton, Elford, and Jory.[21] It can be seen that in this case, where $q_{m_{el}}$ is a slowly varying function of ϵ, the curve of $q_{m_{el}}^*$ versus ϵ' is a reasonably good representation of the curve of $q_{m_{el}}$ versus ϵ obtained from a numerical analysis of the same experimental data. The agreement becomes less satisfactory, however, when $q_{m_{el}}$ is a more rapidly varying function of ϵ.

13.2.3. ENERGY LOSSES IN ELASTIC AND INELASTIC COLLISIONS. Up to this point our description of the early analyses has been limited to techniques that were used to estimate momentum transfer cross sections. As was pointed out in the introduction, no comparable technique was found for estimating the energy dependence of inelastic cross sections, but the importance of inelastic scattering was soon recognized from calculations of another parameter which featured prominantly in the literature, the so-called mean fractional energy loss factor η. The usefulness of this parameter, which, like q_m^*, was calculated from a combination of the transport coefficients, has greatly diminished with the introduction of the techniques described in Section 13.5. Nevertheless it played an important part in the interpretation of a great deal of early work.

Although equations 13.1 and 13.2 are expressed in terms of $f_0(c)$, they can be equally well expressed in terms of averages taken over all possible speeds c, that is,

$$W = \frac{Ee}{3mN} \overline{\left[c^{-2} \frac{d}{dc} \frac{c^2}{q_m(c)} \right]}, \qquad (13.14)$$

$$D = \frac{1}{3N} \overline{\left(\frac{c}{q_m} \right)}; \qquad (13.15)$$

---

†A short but valuable summary is to be found in the introduction to a paper by Cavalleri and Sesta.[26]

furthermore, as discussed in Chapter 7, these formulae can be derived from kinematic arguments without reference to the dynamics of the collisions. There is, therefore, no unique relationship between W and D and the electric field strength E since such relationships require that $f_0(c)$, and hence the averages involving powers of c be known as functions of E. A third equation which provides the required relationship can be found by considering the balance between the rate at which electrons receive energy from the field and the rate at which their excess energy is transferred to the gas molecules. Let the mean fractional energy loss η be defined through the equation

$$\eta = \frac{\overline{\Delta\epsilon}}{\bar{\epsilon}}, \tag{13.16}$$

where $\Delta\epsilon$ is the average energy lost in each encounter by electrons with energy ϵ.† Then the mean rate at which n electrons in the swarm lose energy in collisions is $n \overline{\Delta\epsilon} \overline{\nu_m} = nN \overline{\Delta\epsilon} \overline{q_m(c)c}$, and this must equal the rate at which energy is supplied to the swarm by the field, that is, $neEW$. Thus

$$N \overline{\Delta\epsilon} \overline{q_m(c)c} = eEW$$

$$= \frac{3mNW^2}{\left\{\overline{c^{-2}(d/dc)[c^2/q_m(c)]}\right\}},$$

using equation 13.14. Consequently

$$\eta = \frac{\overline{\Delta\epsilon}}{\bar{\epsilon}} = \frac{6W^2}{(\overline{c^2})\ [\overline{q_m(c)c}]\ \left\{\overline{c^{-2}(d/dc)[c^2/q_m(c)]}\right\}}, \tag{13.17}$$

$$= \left(\frac{6[\ \overline{c/q_m(c)}\ ]}{(\overline{c^2})\ [\overline{q_m(c)c}]\ \left(\left\{\overline{c^{-2}(d/dc)[c^2/q_m(c)]}\right\} \right)^2} \right) \frac{W^2}{(e/m)(D/\mu)}, \tag{13.18}$$

†The parameter η defined in this way should not be confused with $(\overline{\Delta\epsilon/\epsilon})$, which is the average value of the fractional energy loss per collision. For elastic collisions, when $c \gg C$ $(\overline{\Delta\epsilon/\epsilon})$ is independent of ϵ, that is, $f \equiv \overline{\Delta\epsilon/\epsilon} = \Delta\epsilon/\epsilon = 2m/M$. It would perhaps have been preferable if the energy loss factor had been defined in this way rather than as $(\overline{\Delta\epsilon}/\bar{\epsilon})$ (see also the footnote on p.534).

using equations 13.14 and 13.15. The multiplying factor in the large parentheses is dimensionless and depends on both the variation of q_m with c and the velocity distribution function.

Equation 13.18 does not appear explicitly in the literature since, in deriving the formula for η, it has usually been assumed that q_m is constant.[*] With this assumption the equation assumes the simpler form

$$\eta = \frac{3}{2\left[\overline{c^2}\,\overline{(c^{-1})}^2\right]} \frac{W^2}{(e/m)(D/\mu)}$$

$$= \frac{8.53 \times 10^{-16}}{\overline{c^2}\,\overline{(c^{-1})}^2} \frac{W^2}{(D/\mu)}, \qquad (13.19)$$

where W is in centimetres per second and D/μ in volts.

Figure 13.2 shows $\eta/(m/M)$ plotted against D/μ for helium and hydrogen. Although not strictly applicable except in the case of helium at the higher values of E/N, a Druyvesteyn distribution has been assumed in evaluating the dimensionless factor. From the graph it can be seen that in helium η approaches a value of approximately $2.4m/M$, whereas its value in hydrogen is very much larger.

By considering the dynamics of collisions between elastic spheres it can be shown[7,27] that η is given by

$$\eta = S\left(\frac{m}{M}\right)\left(1 - \frac{M\,\overline{C^2}}{m\,\overline{c^2}}\right), \qquad (13.20)$$

where S is a numerical factor depending on the distribution of velocities. The values of S for the Maxwell and Druyvesteyn distributions are 2.66 and 2.40, respectively. The results for helium shown in Figure 13.2 show that the values of η are consistent with this model.[†] On the other hand, the

[*]Equation 13.18 follows immediately from equation 43 of ref. 7 on substituting for k in terms of D/μ.

[†]If we had adopted the alternative energy loss factor f, it could be shown without difficulty that, for constant cross section,

$$f = \frac{3}{2\left[(\overline{c^{-1}})^2 \overline{c^3}/\overline{c}\right]} \frac{W^2}{(e/m)(D/\mu)},$$

that is,

$$f = \left(\frac{\overline{c}\,\overline{c^2}}{\overline{c^3}}\right)\eta.$$

For a Druyvesteyn distribution the dimensionless factor is approximately $1/1.20$. The results for helium shown in Figure 13.2 therefore give $f = 2m/M$ as $\overline{c}/\overline{C} \to \infty$, a better known result.

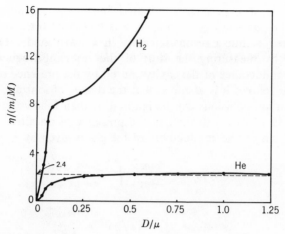

FIG. 13.2.   The energy loss factor in hydrogen and helium. To effect a better comparison, $\eta/(m/M)$ has been plotted.

results for hydrogen show an average energy loss greatly in excess of the value corresponding to elastic collisions. A similar result was obtained by Townsend and Bailey[1] and correctly interpreted by them as evidence of inelastic collision processes. Thus they wrote, "The loss of energy of the electrons, although comparatively small, is very much greater than the loss that would be sustained by a small sphere moving with velocity u in colliding with a large sphere, if both spheres were perfectly elastic. The energy lost by the electron appears as an increase of the internal energy of the molecule such as vibration set up in some of the constituent electrons."

Values of η were determined by Townsend and Bailey and their collaborators for a large number of gases, but, apart from one or two applications,[28] they now have limited use. Tables of values are to be found in refs. 4 and 5, while more modern results for some gases are given in ref. 29. The data published by Hall[30] for H_2 and D_2 are interesting in that they were an early confirmation of the importance of rotational excitation as the dominant energy loss mechanism in diatomic gases.

The warning given at the end of the preceding section applies equally in relation to the early data for η.

13.3.   ANALYSIS OF THE DATA FROM MICROWAVE EXPERIMENTS

Although a detailed account of microwave experiments has not been given in this book, we outline briefly the methods used to analyze the results obtained, because, as has already been mentioned, these methods broke new ground at the time of their introduction in 1951.

In a series of experiments initiated by Phelps, Fundingsland, and Brown,[8,9] the conductivity ratio σ_r/σ_i was measured for a weakly ionized gas contained within a resonant cavity in a wave guide. The ratio was determined by measuring the shift of the resonant frequency and the change in conductance of the cavity caused by the presence of the ionized gas. In what follows it is assumed that the degree of ionization is so slight that electron-ion collisions can be ignored. It then follows from equations 9.18 and 9.19 (with $\omega = 0$) that, in the presence of an alternating electric field $E = E_0 \sin pt$, the conductivity of the gas is given by[31]

$$\sigma = \sigma_r + i\sigma_i = -\frac{4\pi n e^2}{3m} \int_0^\infty \frac{c^3}{\nu + ip} \frac{df_0}{dc} dc$$

$$= -\frac{4\pi n e^2}{3mp} \int_0^\infty \frac{(\nu/p) - i}{1 + (\nu/p)^2} c^3 \frac{df_0}{dc} dc. \qquad (13.21)$$

Consequently,

$$\frac{\sigma_r}{\sigma_i} = -\frac{\int_0^\infty \left\{ (\nu/p) c^3 / \left[1 + (\nu/p)^2 \right] \right\} (df_0/dc) \, dc}{\int_0^\infty \left\{ c^3 / \left[1 + (\nu/p)^2 \right] \right\} (df_0/dc) \, dc}. \qquad (13.22)$$

When all encounters are elastic, $f_0(c)$ is given by equation 9.55:

$$f_0(c) = A \exp - \frac{3m}{M} \int_0^c \frac{c \, dc}{V_{eq}^2 + \overline{C^2}}, \qquad (13.23)$$

where $V_{eq} = eE_0/m[2(\nu_{el}^2 + p^2)]^{1/2}$ and $\nu_{el} = N q_{m_{el}}(c)c$. Thus

$$f_0(c) = A \exp - \frac{3m}{M} \int_0^c \frac{c \, dc}{[e^2 E_0^2 / 2m^2 (\nu_{el}^2 + p^2)] + \overline{C^2}}. \qquad (13.24)$$

Equation 13.22, with the substitution of the appropriate expression for $f_0(c)$, forms the basis of the methods used by Brown and his collaborators.[8–10] In one set of experiments the conductivity is measured under conditions that ensure that the electrons are in thermal equilibrium with the gas molecules, that is, the conductivity is measured when equilibrium has been established, some time after the initial breakdown, and

$V_{eq}^2 \ll \overline{C^2}$. With these conditions $f_0(c) = A \exp - c^2/\alpha^2$, with $\alpha^2 = 2\kappa T/m$, and equation 13.22 then becomes

$$\frac{\sigma_r}{\sigma_i} = -\frac{\int_0^\infty \left\{ (\nu/p)c^4/\left[1 + (\nu/p)^2\right]\right\} \exp(-c^2/\alpha^2)\, dc}{\int_0^\infty \left\{ c^4/\left[1 + (\nu/p)^2\right]\right\} \exp(-c^2/\alpha^2)\, dc}. \qquad (13.25)$$

Two properties of equation 13.25 are immediately obvious:

1. When ν is independent of c, $\sigma_r/\sigma_i N = -\nu/Np$, that is, the ratio is a constant independent of both N and T. In this circumstance a measurement of the conductivity ratio enables ν to be determined without difficulty.

2. When $(\nu/p)^2$ is very much greater or less than unity, that is, the gas pressure is sufficiently high or low, respectively, $\sigma_r/\sigma_i N$ is independent of N regardless of the dependence of ν on c. At intermediate values, however, $\sigma_r/\sigma_i N$ is in general a function of N, becoming most sensitive when $\nu/p \sim 1$. Moreover, the variation of the ratio with N depends on the variation of ν with c. Thus, in principle, the variation of ν with c can be found from an experimental determination of the variation of $\sigma_r/\sigma_i N$ with N over the range of values of N in which $\sigma_r \sim \sigma_i$.

Figure 13.3 shows the variation of $\sigma_r/\sigma_i p_0$ with p_0 (p_0 is the gas pressure normalized to 273 K) which Bekefi and Brown[10] derived from equation 13.25 on the assumption that $\nu = \text{const} \cdot (c/\alpha)^{1.6}$, where α is the most probable velocity at 300 K. The theoretical curve can be seen to fit the

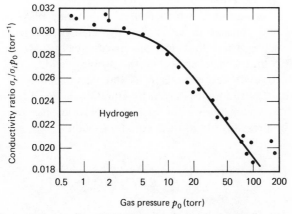

FIG. 13.3.   Conductivity ratio as a function of the gas pressure. ● measurements; the solid line is obtained from theory. (p_0 is the gas pressure normalized to 273 K.) (From ref. 10.)

experimental results (for hydrogen) reasonably well. It should be noted that this representation of v, as well as any results obtained in a similar way, applies only to a relatively small range of velocities in the vicinity of α.

An alternative method which also relies on the electrons being thermalized consists in measuring the conductivity ratio over as wide a range of gas temperatures as possible. If the experimental conditions are chosen to ensure that $(v/p)^2 \ll 1$, equation 13.25 reduces to

$$\frac{\sigma_r}{\sigma_i} = -\frac{1}{\frac{1}{2}\Gamma(\frac{5}{2})\alpha^5} \int_0^\infty \frac{v(c)}{p} c^4 \exp\left(-\frac{c^2}{\alpha^2}\right) dc. \qquad (13.26)$$

Consequently, the representation of the velocity dependence of v by the simple form $v(c)/N = ac^h$ leads to the following expression for the conductivity ratio, which is easily evaluated:

$$\frac{\sigma_r}{\sigma_i} = -\frac{a\alpha^h N}{p} \frac{\Gamma[(h+5)/2]}{\Gamma(\frac{5}{2})},$$

that is,

$$\left.\begin{array}{c} \\ \\ \\ \\ \\ \end{array}\right\} \qquad (13.27)$$

$$\frac{\sigma_r}{\sigma_i N} = -\frac{1}{p}\left(\frac{2\kappa}{m}\right)^{h/2} \frac{\Gamma[(h+5)/2]}{\Gamma(\frac{5}{2})} aT^{h/2}.$$

It follows that measurements of the ratio over a range of values of T enable first h and then a to be determined.

A third set of experiments was designed to yield information about energy losses in situations in which inelastic scattering plays a significant part. The method consists in measuring the change in the conductivity ratio which occurs when the electrons in the cavity are heated by a microwave field that is independent of the probing field.[10]

When all encounters between electrons and gas molecules are elastic, the velocity distribution function is given by equation 13.24. A modified form of this equation which allows for inelastic collisions has been given by Allis.[6,10] In the present notation the equation reads

$$f_0(c) = A \exp - \int_0^c \frac{mc\,dc}{[e^2 E_0^2 / 3mG(v^2 + p^2)] + \kappa T} \qquad (13.28)$$

where $G(c)$ is a factor related to the average energy loss per collision. Equation 13.28 is claimed to be an adequate representation of $f_0(c)$

provided that $G(c)$ is constant or slowly varying.† In the absence of inelastic collisions this equation becomes equation 13.24 with $G = 2m/M$; $2m/M$ is the average energy lost by electrons of a given velocity when scattered isotropically from molecules at rest.[28]

Provided that the experiments are performed with gas pressures low enough to ensure that $\nu^2 \ll p^2$, and provided that $G(c)$ can be regarded as constant, equation 13.28 is a Maxwellian distribution of the form $f_0(c) = A \exp - c^2/\alpha^2$ with

$$\alpha^2 = \frac{2e^2 E_0^2}{3m^2 Gp^2} + \frac{2\kappa T}{m},$$

that is, a distribution characterized by an electron temperature T_e given by

$$T_e = T + \frac{e^2 E_0^2}{3mGp^2\kappa}. \tag{13.29}$$

Thus, if the electron temperature produced by a given field E_0 can be measured, G can be determined.

The experiments described in the preceding section provide, in effect, a calibration curve of $\sigma_r/\sigma_i N$ versus T_e for the apparatus. Therefore, provided that the heating field does not raise T_e beyond the maximum temperature to which the gas was heated in the first set of experiments, T_e can be determined from the measured conductivity ratio and G then found from equation 13.29.

Figure 13.4 shows the significance of inelastic collisions in experiments of this kind. In this figure the conductivity ratios, measured at several gas temperatures T and hence at several values of T_e, are compared with ratios measured at constant T but with varying values of E_0. These latter measurements are plotted against values of T_e calculated from equation 13.29, assuming that $G = 2m/M$. It can be seen that this assumption leads to values of T_e that are much too large, indicating that the assumed value of G is too small. Agreement between the two sets of measurements can be obtained by using a value of $G = 3.5 \times 10^{-3}$, which is the same order as the value determined from drift and diffusion experiments.[32]

In order to extend the range of the measurements to higher electron energies, an alternative analytical technique was devised which combined the procedures already described. When $(\nu/p)^2 \ll 1$, equation 13.22 can be

†See, however, the comment in the bibliography for Chapter 9, where it is shown that this representation is satisfactory only if $T_e \gg T$.

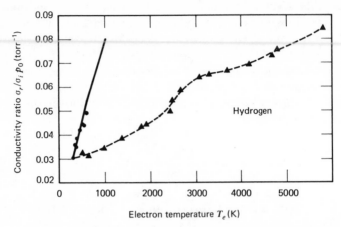

FIG. 13.4. Comparison between the conductivity ratio in H_2 obtained for thermal electrons in heated gas (●) and that obtained when the electrons are heated in a microwave field (——△——△——). The solid line is obtained from theory. (From ref. 10.)

integrated by parts to give

$$\frac{\sigma_r}{\sigma_i} = -\frac{\int_0^\infty (d/dc)[c^3(\nu/p)]f_0(c)\,dc}{3\int_0^\infty c^2 f_0(c)\,dc}. \tag{13.30}$$

If, as before, we replace $\nu(c)$ by $\nu(c) = aNc^h$, this equation becomes

$$\frac{\sigma_r}{\sigma_i N} = -\frac{3+h}{3}\frac{a}{p}\frac{\int_0^\infty c^{h+2} f_0(c)\,dc}{\int_0^\infty c^2 f_0(c)\,dc}. \tag{13.31}$$

Because the measurements now apply to conditions where $T_e > T$, $f_0(c)$ is not the thermal Maxwellian distribution. Bekefi and Brown used equation 13.28 for $f_0(c)$, removing the restriction that G is a constant by giving it a simple power law dependence on c, that is, by putting $G = bc^l$. Substitution of this expression for G in equation 13.28 gives, when $T_e \gg T$,

$$f_0(c) = A\exp\left[-\frac{3b}{l+2}\left(\frac{mp}{eE_0}\right)^2 c^{l+2}\right]. \tag{13.32}$$

Insufficient information is available from measurements of the variation of $\sigma_r/\sigma_i N$ with E_0 to find a, b, h, and l. However, as described in the original paper,[10] it is possible to determine them by combining the microwave results with the results of DC drift and diffusion measurements. The details of the method are not reproduced here because the analysis is complex and the results obtained from it are little more satisfactory than those of the analysis described in the preceding section. Nevertheless, the method of Bekefi and Brown appears to have been the first attempt to obtain more accurate representations of the velocity dependences of ν and G, and as such it was the forerunner of the methods described in the following section for analyzing the results of DC swarm experiments.

13.4. THE ANALYSES OF BOWE, HEYLEN, PHELPS, SHKAROFSKY, AND OTHERS

Using analytical techniques similar to those described in the preceding section, Bowe,[11] Heylen,[12] Phelps and his collaborators,[13,15] and Shkarofsky, Bachynski, and Johnston[14] undertook the analysis of an extensive body of data derived from DC swarm experiments. Shkarofsky et al. have reviewed the basic theory of these analytical techniques as applied to the most general case. We will therefore outline their method of analyzing drift and diffusion data before referring to other procedures which will be seen to be special cases of the more general method.

Following the procedure adopted in the preceding section, we may represent ν and G by $\nu(c) = Nac^h$ and $G(c) = bc^l$. Then, provided that $T_e \gg T$, $f_0(c)$ is given by equation 13.28 with $p = 0$ and $E^2 = E_0^2/2$, that is,

$$f_0(c) = A \exp\left\{-\left[\frac{3}{2(2h+l+2)}\left(\frac{mNa}{eE}\right)^2 b\right]c^{2h+l+2}\right\}. \quad (13.33)$$

The following formulae for W and D/μ are obtained by substituting equation 13.33 into equations 13.1 and 13.2:[14]

$$W = \frac{eE(3-h)}{3mNa}\frac{\Gamma[(3-h)/(l+2h+2)]}{\Gamma[3/(l+2h+2)]}\left[\frac{2(l+2h+2)}{3b}\left(\frac{eE}{mNa}\right)^2\right]^{-h/(l+2h+2)}, \quad (13.34)$$

$$\frac{D}{\mu} = \frac{m}{e(3-h)}\frac{\Gamma[(5-h)/(l+2h+2)]}{\Gamma[(3-h)/(l+2h+2)]}\left[\frac{2(l+2h+2)}{3b}\left(\frac{eE}{mNa}\right)^2\right]^{2/(l+2h+2)}. \quad (13.35)$$

These equations can be written in the forms

$$W = C\left(\frac{E}{N}\right)^{(l+2)/(l+2h+2)},$$ (13.36)

$$\frac{D}{\mu} = D\left(\frac{E}{N}\right)^{4/(l+2h+2)}.$$ (13.37)

Consequently the slopes of the log-log plots of W and D/μ versus E/N can be used to derive for h and l simultaneous equations which can be solved to obtain the values of the indices appropriate to any chosen value of E/N. The constants a and b are then found from the values of W and D/μ at these values.

Shkarofsky et al. applied this analysis to derive velocity-dependent momentum transfer cross sections from measured values of W and D/μ in nitrogen and oxygen as part of a programme to determine collision frequencies in air at high temperature. A comparison of the results for nitrogen with those obtained by using the simpler analysis of Section 13.2.2 is shown in Figure 13.5, the same experimental data having been used in

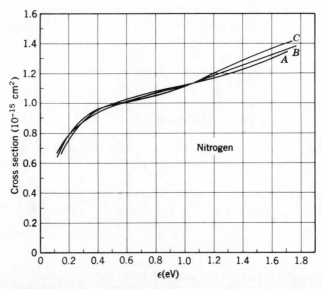

FIG. 13.5. Comparison of momentum transfer cross sections for N_2 derived from swarm analyses, using both the mean value technique and the analysis of Shkarofsky et al. Curves A and B from mean value analysis using Maxwell and Druyvesteyn distributions, respectively; curve C from analysis of Shkarofsky et al. (Based on the corresponding figure in ref. 14.)

each case. A similar order of agreement was obtained in the case of oxygen, from which it was concluded that no great advantage is to be gained from the more exact analysis and that the simpler analysis was adequate to determine the momentum transfer cross section in the minor constitutents of air.

Bowe[11] appears to have been the first author to apply the techniques described in this section, his primary aim being the determination of $q_{m_{el}}(c)$ for electrons in helium, neon, argon, krypton, and xenon from his drift velocity measurements in these gases. For sufficiently low values of E/N, $G = 2m/M$ in these gases, and equation 13.34 is somewhat simplified since $b = 2m/M$ and $l = 0$. Equation 13.36 then reads

$$W = C'\left(\frac{E}{N}\right)^{1/h+1},$$

and the slope of the log-log plot of W versus E/N yields h directly. For example, when $h = 1$ (i.e., $q_{m_{el}} = \text{constant}$), $W = C'(E/N)^{1/2}$ (cf. equation 13.8); when $h = 3$ (i.e., $q_{m_{el}} \propto \epsilon$), $W = C''(E/N)^{1/4}$. Bowe found that his results for helium were well fitted over a reasonably extensive range by assuming $h = 1$, whereas those in argon, krypton, and xenon were fitted over a more limited range by assuming $h = 3$; the results in neon were consistent with $h = 1.314$.

The limitation of analyses of this kind due to the neglect of the thermal term in $f_0(c)$ must be borne in mind. Thus Bowe observed that the ratio $W/(E/N)^{1/2}$ formed from his own drift velocity measurements in helium and from those of Crompton et al.[21] was remarkably constant over an extended range of E/N, and he concluded that $q_{m_{el}}$ was constant over the corresponding energy range. However, a numerical analysis of the same set of experimental results, as described in the following section, shows that $q_{m_{el}}$ varies by about 20% over the energy range in which Bowe's analysis indicates a constant value. It has been shown[21] that this apparent contradiction does in fact originate from the neglect of the thermal term in the approximate treatment, the neglect of the term just offsetting the effect of the variation of $q_{m_{el}}$.

At first sight it may seem surprising that the analysis of Section 13.2.2 leads to a better estimate of the energy dependence of $q_{m_{el}}$ in helium than does Bowe's analysis. The reason is that, in the former analysis, the distribution of velocities is inferred directly from the measured values of D/μ, a technique which enables reasonable estimates of $f_0(c)$ to be made even when the distribution approaches the thermal Maxwellian distribution. It is evident that this advantage offsets the disadvantage that arises from the representation of the cross section by a mean value at each value of E/N rather than by an explicit function of c.

Heylen[12] developed a similar technique but for a somewhat different application, his aim being the determination of elastic scattering cross sections in molecular gases. In two earlier papers Lewis[33] and Heylen and Lewis[34] had attempted to determine energy distributions in hydrogen from an analytical solution of the Boltzmann equation which included terms accounting for inelastic collisions. When rotational and vibrational excitation processes were included, the distribution functions were found to be approximately Maxwellian. On this basis Heylen[12] was able to develop an analytical technique similar to that already described, since he argued that $f_0(c)$ in hydrogen could be adequately represented by the Maxwellian form. With this assumption the distribution function in equation 13.1 could be represented by $f_0(c) = A \exp - c^2/\alpha^2$ with the value of α determined from measured values of D/μ,[32] using the relation $\alpha^2 = (2e/m)(D/\mu)$.

Heylen used a somewhat different representation of the energy dependence of the momentum transfer cross section, namely, $q_m = ac^h e^{-kc}$, the advantage of this form being that more complex variations of the cross section could be represented with appropriate values of h and k. Using this representation and an analytical representation of α versus E/N obtained from the experimental values of D/μ, he derived a formula for W analogous to equation 13.34 but containing three unknowns: a, h, and k. By adjusting these constants to give the best fit between calculated and measured values of W, the energy dependence of q_m could be determined. In a later paper the technique was extended to other gases.[35]

Pack and Phelps[13] developed an analytical technique similar to Bowe's. However, they represented the collision frequency by a series expansion in c and avoided the difficulty of calculating the velocity distribution function by basing their analysis on the temperature dependence of the zero-field mobility. An advantage of their technique is its applicability to molecular gases, in which the presence of inelastic collisions not only makes an accurate representation of $f_0(c)$ difficult but also introduces the energy loss factor as an additional unknown.

Following Pack and Phelps, we replace q_m^{-1} in equation 13.1 by $q_m^{-1}(c) = \Sigma_j b_j c^{1-j}$, and we note that, since $T_e = T$, $f_0(c)$ can be replaced by the Maxwellian distribution $f_0(c) = \pi^{-3/2}\alpha^{-3}\exp - c^2/\alpha^2$ with $\alpha = (2\kappa T/m)^{1/2}$. It follows immediately that $\mu N = W/(E/N)$ is given by

$$\mu N = \frac{e}{m} \sum_j \frac{\Gamma[(5-j)/2]}{\Gamma(\frac{5}{2})} b_j \alpha^{-j}$$

$$= \sum_j B_j \left(\frac{2\kappa T}{m}\right)^{-j/2}. \tag{13.38}$$

The procedure is therefore to make measurements of the drift velocity at a number of temperatures with values of E/N approaching zero. A linear dependence of W on E/N at sufficiently low values of E/N is taken to indicate that the electrons are in thermal equilibrium with the gas molecules.† The zero-field mobilities are then found for each temperature from the slopes of the W versus E/N curves in the linear region. The curve of $(\mu N)_{E/N \to 0}$ versus T is fitted with a series expansion of the form of equation 13.38, using a reasonable set of values of j and appropriate coefficients B_j. The cross section is then given by

$$q_m^{-1}(c) = \frac{m}{e} \sum_j \frac{\Gamma(\tfrac{5}{2})}{\Gamma[(5-j)/2]} B_j c^{1-j}. \tag{13.39}$$

A typical set of experimental data is shown in Figure 13.6 together with analytical curves based on equation 13.38, which fit the data. Although

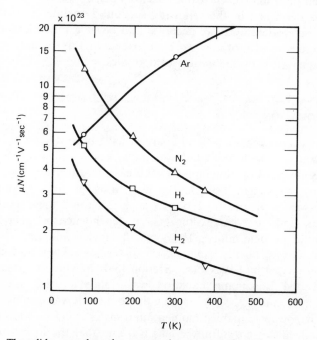

FIG. 13.6. The solid curves show the power series representations chosen to approximate the measured values of μN obtained by Pack and Phelps (shown as ○, △, □, and ▽). (From ref. 13.)

---

†This assumption is not justified when $j = 0$ ($\nu = $ constant). However, in this event, μN is also independent of temperature and both j and b_j are then specified.[15]

measurements were also made in neon, the W versus E/N curve did not show a linear portion even at the lowest values of E/N; from this it was concluded that thermal equilibrium had not been achieved. Therefore the analytical technique could not be applied to these results.

Pack, Voshall, and Phelps[15] applied the same technique to measurements in Kr, Xe, D_2, CO, CO_2, H_2O, N_2O, and NH_3. For CO_2 and N_2O the zero-field mobilities were found to be independent of temperature, a result which leads to a unique cross section. For other gases the polynomial fit to the data is not unique, because of the limited number of temperatures used, and in most cases the authors give a number of alternative cross-section curves which fit their data equally well. Another drawback of the technique is the limited range of energies over which the results are valid. For the temperatures used in these investigations the range was of the order of 0.003 to 0.08 eV.

With the exception of the results for H_2O, N_2O, and NH_3, the same experimental data have been reanalyzed by Phelps and others,[36-39] using the analysis described in the following section. The data from the later analysis are considerably more accurate and cover a much larger energy range than the results of the work described in this section, which is therefore superseded by the more recent studies.

13.5. ANALYSES BASED ON NUMERICAL SOLUTIONS OF THE BOLTZMANN EQUATION

In the preceding section several analyses have been described in which some attempt was made to include the influence of inelastic collisions on the velocity distribution function. In each case an analytical form for the distribution was sought because the evaluation of $f_0(c)$ by numerical techniques was impracticable,† but in no sense could these analyses be regarded as methods whereby the energy dependence of inelastic cross sections could be determined. The work of Heylen and Lewis[42] on the monatomic gases probably came closest to success. This work was based on the earlier paper by Lewis[33] (see Section 13.4), in which an approximate solution to the Boltzmann equation was obtained using the Jeffreys method of approximation. In obtaining the solution it was assumed that no energy exchange accompanied momentum transfer in elastic collisions (i.e., that the molecules were infinitely massive) and that the lowest threshold for the inelastic processes was the one corresponding to electronic excita-

---

†Some authors, for example, Allen[40] and Barbiere,[41] did undertake this formidable task to a limited extent but were able to do so only after making somewhat drastic assumptions with regard to the inelastic terms in the Boltzmann equation. For a review of the work up to about 1955 see ref. 5.

tion. As the original calculations were made for hydrogen, it was not surprising that these assumptions led to somewhat unsatisfactory agreement between theory and experiment. In the later work, Heylen and Lewis were able to allow for energy exchange in elastic collisions; moreover, they used their analysis to calculate energy distributions and transport coefficients in the monatomic gases, where the second assumption is realistic. However, although the results of their calculations were found to be in reasonable agreement with experiment, the method has the limitation common to the techniques already described, namely, the restriction imposed by an analytical representation of q_m.

These problems were satisfactorily overcome for the first time by Frost and Phelps[16] and Carleton and Megill,[43] who used digital computers to obtain numerical solutions to an appropriate form of the Boltzmann equation, thereby avoiding many of the approximations that were a necessary part of all previous work. The approach of the two groups was somewhat different. Carleton and Megill had as their aim the calculation of energy distribution functions for electrons in air in the presence of an AC electric field and a magnetic field normal to it. It was assumed that the electron temperature was sufficiently high that inverse collisions (both elastic and inelastic) between the molecules and electrons could be ignored. In making their calculations, these authors took as input data the best available cross sections for elastic and inelastic collisions for the constituents of air. No attempt was made to modify these cross sections in the light of subsequent comparisons between experimentally observed quantities and corresponding values calculated from the distribution functions.

Frost and Phelps, on the other hand, developed their analysis as an extension of earlier techniques for extracting cross-section data from the results of swarm experiments. Since one of their objectives was to determine the threshold behaviour of the cross sections for rotational excitation for which the excitation energies are small, their analysis was developed to include inverse and superelastic collisions. The procedure used by them was to assume an initial set of cross sections, based on available data from theory and experiment, in order to calculate the energy distribution functions over a wide range of values of E/N. To this point their procedure was similar to that of Carleton and Megill but somewhat more comprehensive. From the distribution functions computed with the trial set of cross sections, data for a number of transport coefficients were calculated and compared with the results of experiment, after which adjustments were made to the assumed cross sections in an attempt to obtain better agreement between calculated and experimental values. When the agreement was judged to be adequate, the final cross sections formed a set that was consistent with experiment, although it was not necessarily a unique set.

Phelps and his collaborators analyzed the results of swarm experiments for a number of molecular gases using this technique.[16,36,37,39] The same technique was applied by Crompton, Gibson, and McIntosh[44] to the special case of parahydrogen at low temperature. An alternative method was developed by Lucas.[45] His method was based on the Gauss-Seidel iterative technique and was applied to calculations at high values of E/N, where collisions of the second kind can be neglected. Gibson[46] combined the original technique of Frost and Phelps with Lucas's technique to develop a method which, like the original one, is capable of covering the full range of E/N but which has some advantages over the original method.

Frost and Phelps[38] also applied their technique to the much simpler case of the monatomic gases at low values of E/N, where inelastic collisions can be ignored. An identical technique was developed by Crompton and Jory,[17,47] while Hoffmann and Skarsgard[48] used a similar method to analyze the results of their microwave experiments.

In the rest of this section we discuss numerical solutions as applied first to the simpler situation of elastic scattering only and second to the more general case. No attempt is made to give a comprehensive survey of the results of these analyses although illustrative examples are presented in the course of the discussion. For comprehensive accounts of the many applications to which the techniques have been put, the reader is referred to the original papers and to the reviews by Phelps,[49] Crompton,[18] Bederson and Kieffer,[19] and Golden et al.[50]

13.5.1. ELASTIC SCATTERING. In the absence of inelastic scattering, the calculation of the transport coefficients for a given energy-dependent momentum transfer cross section reduces to a straightforward process of two stages. The first step is the calculation of $f_0(c)$ for a series of values of E/N by numerical integration of equation 13.3. The second step is the calculation of the transport integrals, for example, thoses for W and ND, also by numerical integration. It is often more convenient to express the relevant relations in terms of electron energy rather than electron speed, especially when, as in the next section, inelastic scattering is included in the analysis. The equations for $f(\epsilon)$, W, ND, and D/μ are then (Section 6.8)†

$$f(\epsilon) = A \exp\left\{ - \int_0^\epsilon \frac{d\epsilon}{\left[ME^2 e^2 / 6mN^2 q_{m_{el}}^2(\epsilon)\epsilon \right] + \kappa T} \right\}, \quad (13.40)$$

---

†From this point on, we follow normal practice and denote $f_0(\epsilon)$ simply by $f(\epsilon)$ since only the isotropic term in the distribution function is referred to.

where $f(\epsilon)$ is normalized through the equation $\int_0^\infty \epsilon^{1/2} f(\epsilon) \, d\epsilon = 1$,

$$W = -\frac{e(2/m)^{1/2}}{3} \frac{E}{N} \int_0^\infty \frac{\epsilon}{q_{m_{el}}(\epsilon)} \frac{df}{d\epsilon} \, d\epsilon, \tag{13.41}$$

$$ND = \frac{(2/m)^{1/2}}{3} \int_0^\infty \frac{\epsilon f(\epsilon)}{q_{m_{el}}(\epsilon)} \, d\epsilon, \tag{13.42}$$

$$\frac{D}{\mu} = -\frac{1}{e} \frac{\int_0^\infty \left[\epsilon f(\epsilon) / q_{m_{el}}(\epsilon) \right] d\epsilon}{\int_0^\infty \left[\epsilon / q_{m_{el}}(\epsilon) \right] (df/d\epsilon) \, d\epsilon} \tag{13.43}$$

Since the only unknown in these equations is $q_{m_{el}}(\epsilon)$, a cross section can, in principle, be determined by comparing calculated and experimental values for a single transport coefficient. However, since it is not possible to establish by a general theoretical argument the uniqueness of cross sections determined in this way, a comparison of calculated and experimental values for another transport coefficient provides a valuable check of the uniqueness and accuracy of the cross section in each case. Under some circumstances the use of data for more than one transport coefficient may be the only way of establishing the cross section in the first instance.[53]

Where possible, cross sections have been derived primarily from drift velocity data because the experimental results for W appear to be intrinsically more accurate than the data for other transport coefficients. However, as pointed out by Frost and Phelps,[38] when $D/\mu \gg \kappa T/e$, the values of D/μ are generally twice as sensitive to variations in $q_{m_{el}}(\epsilon)$ as are the values of W,† so that, given data of equal accuracy, the cross section is more accurately determined from the data for D/μ.

In carrying out the iterative procedure it is important to know where adjustments should be made to the cross section and how large they should be, in order to improve the agreement between calculated and experimental values. The principle underlying the methods of adjustment that have been used can be understood by referring to Figure 13.7 and to Section 13.2.2. The figure shows, plotted to the same scale, the momentum transfer cross section in helium and the energy distribution functions for two values of E/N. It can be seen that, at each value of E/N, there is a

---

†See equations 13.36 and 13.37 with $l = 0$.

relatively narrow range of energies within which the number of electrons per unit energy interval [i.e., $\epsilon^{1/2}f(\epsilon)$] is significant. The swarm can therefore be regarded as a probe of comparatively poor energy resolution which can be made to scan the entire energy range simply by varying E/N (and possibly T).† For this reason, it is the magnitude of the cross section (and its energy dependence) within a restricted range that determines the transport properties of the swarm at any given value of E/N. When there is a discrepancy between calculated and measured transport coefficients over a certain range of values of E/N, the objective is to find the corresponding range of values of ϵ over which an adjustment to the cross section should be made and to determine how large the adjustment should be.

A method which suggests itself is one based on the mean value technique described in Section 13.2.2, where it was shown that the energy-

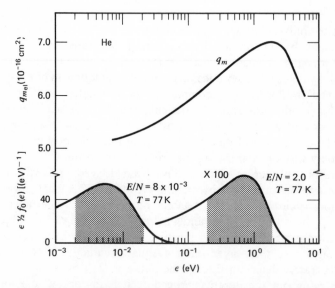

FIG. 13.7.   Electron energy distribution functions in helium at 77 K and at $E/N = 8 \times 10^{-3}$ and 2.0 Td; $\epsilon^{1/2}f_0(\epsilon)\,d\epsilon$ is the fraction of electrons in the swarm with energies between ϵ and $\epsilon + d\epsilon$. Approximately 80% of the electrons in the swarm have energies in the region shown shaded. The momentum transfer cross section derived by Crompton, Elford, and Robertson[51] is shown in the upper section of the figure. (From M. T. Elford, *Case Studies in Atomic Collision Physics*, E. W. McDaniel and M. R. C. McDowell, eds., North Holland, Chap. 2.)

†Although the energy spread of the electrons in the swarm is large at high energies, it compares favourably with the typical spread of a well defined beam at low energies. For example, most of the electrons in a swarm whose mean energy is approximately 15 meV have energies within the range 0 to 30 meV.[18]

dependent cross section $q_m(\epsilon)$ can be replaced by an effective mean value $q_m^*(E/N)$ at each value of E/N. In order to represent q_m^* as a function of a parameter related to the energy, the parameter $\epsilon' = \frac{1}{2}m(\bar{c})^2$ was used previously,[21] but a parameter which represents more accurately the "effective energy" is the energy ϵ^* at which the integrand of equation 13.41, that is, $[\epsilon/q_{m_{el}}(\epsilon)](df/d\epsilon)$, has its maximum value, since it is the electrons with energies close to this value which contribute most to the integral. It is also this group of electrons which gains most energy from the field and hence transfers most energy to the gas molecules in collisions.[44] This can be seen most easily by considering the case in which the thermal motion of the molecules can be neglected, that is when

$$f(\epsilon) = A \exp\left[-\int_0^\epsilon \frac{6mN^2 q_{m_{el}}^2(\epsilon)\epsilon}{ME^2 e^2} \, d\epsilon \right].$$

It follows that

$$\frac{\epsilon}{q_{m_{el}}(\epsilon)} \frac{df}{d\epsilon} = -\frac{6mN^2 q_{m_{el}}(\epsilon)\epsilon^2}{ME^2 e^2} f(\epsilon) \propto \nu_{m_{el}}(\epsilon)\epsilon \, \epsilon^{1/2} f(\epsilon).$$

Since the average energy lost in an elastic collision is proportional to the initial energy, and $\epsilon^{1/2} f(\epsilon) d\epsilon$ is the number of electrons with energy between ϵ and $\epsilon + d\epsilon$, the integrand is proportional to the rate at which electrons in this class lose energy in collisions.

From the preceding arguments it follows that q_m^* versus ϵ^* is a good approximation to $q_{m_{el}}(\epsilon)$, provided that the variation of the cross section is not too rapid. The procedure for using q_m^* as a fitting parameter is therefore to calculate q_m^* [or simply $(E/N)/W(D/\mu)^{1/2}$] from both the measured and calculated transport coefficients, in order to determine the ratios of the two values at appropriate values of ϵ^*, and then to make the corresponding corrections to the input cross section at energies $\epsilon = \epsilon^*$.

Although this procedure serves as a useful tool for adjusting the cross sections as part of the iterative process, it would not be expected to be universally successful because of the assumptions used in its development. It becomes less useful as the energy dependence of the cross section becomes more pronounced; it is of little value, for example, in adjusting the cross section in the vicinity of the Ramsauer-Townsend minimum for the heavier monatomic gases. On the other hand, it finds application at other energies for these gases, and it can be applied equally well to adjust the input values of $q_{m_{el}}(\epsilon)$ for the molecular gases even though, as described in the next section, the overall analysis is then considerably more complex.

It is important to remember, however, that the validity of the final cross section is unrelated to the validity of the assumptions used in deriving the formula for the fitting parameter. The final test of the cross section is the fit of the calculated transport data to the measured data, and its validity rests on the accuracy of the data and the assessment of the uniqueness. These matters are discussed in more detail subsequently.

An alternative fitting parameter was introduced by Frost and Phelps[16] and named by them the "effective elastic collision frequency," v_m. Defined as $v_m = (e/m)(E/N)/W$, this parameter equals the actual collision frequency for momentum transfer when the collision frequency is constant. Like the parameter q_m^*, it becomes ineffective in situations where the assumption on which it is based is seriously violated.

We turn now to the important question of the accuracy and uniqueness of cross sections derived in this way. This problem has been dealt with in detail for only two gases, helium[19,51] and neon.[52] Because, at the time of writing, there are more experimental data to confirm the validity of the helium cross section, we use helium to illustrate the general discussion of this section.

If we accept, for the moment, the validity of the theory on which the analysis is based, the general comment can be made that the accuracy and uniqueness of the cross section appear to depend ultimately only on the accuracy of the experimental data, on the extent of the data (i.e., the range and number of experimental points within a given range), and on the accuracy of the fit between the calculated and experimental curves. The *uniqueness* of the cross section depends essentially on the magnitude of the scatter of the primary experimental data and therefore on the accuracy with which it is justified to attempt to match calculated and experimental values, while the *overall accuracy* of the cross section is governed by the systematic errors in the original data.

We can illustrate the first point by describing the conditions under which it is possible to assert that the cross-section curve for helium is the smooth curve of Figure 13.7, rather than a curve exhibiting a resonance structure within the energy range shown.[21] Figure 13.8 shows a possible resonance at about 0.5 eV, while Figure 13.9 shows the effect which the superposition of such a resonance has on the calculated drift velocities, the round points corresponding to the fractional differences between the calculated and experimental values of W. Also shown on Figure 13.9 are the deviations of the measured values from those calculated using the original curve. It can be seen from the figure that the smooth curve would not have produced acceptable agreement between calculated and experimental values had the experimental values been influenced by the resonance structure. It is therefore possible to discount such a structure with

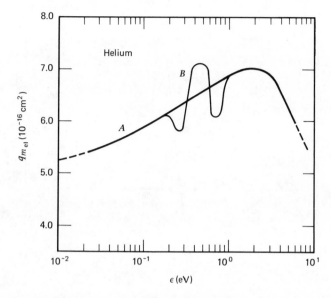

FIG. 13.8. The cross sections used to investigate the effect of fine structure in the momentum transfer cross section in helium. Curve A, derived by Crompton, Elford, and Jory; curve B, curve A but incorporating resonance. (From ref. 21.)

certainty on the basis of these experimental results. On the other hand, had the experimental scatter been $\pm 2\%$ rather than less than $\pm 0.3\%$, no such assertion would have been possible. We note also, however, that even with the low scatter of these data the possibility of a resonance of similar shape but with less than 25% of the amplitude cannot be dismissed, while further calculations would be required to determine how narrow the resonance would have to be to be undetectable.

Figures 13.10 and 13.11 further illustrate the problems of accuracy and uniqueness. Curve 2 of Figure 13.10 starts 2% lower than the standard curve at low energies, crosses the curve in the medium energy range, and rises to 2% above the standard curve in the vicinity of the maximum. The reflection of this variation in the calculated values of W is shown in Figure 13.11, together with the actual agreement between calculated and experimental values. It can be seen that smooth variations of this kind must lie well within a band formed by drawing parallel curves $\pm 2\%$ from the original cross-section curve.

It is to be noted, however, that the degree of uniqueness diminishes towards the high- and low-energy ends of the curve because there are insufficient electrons in the swarm at these energies to influence the measured transport coefficients at the lowest or highest values of E/N at

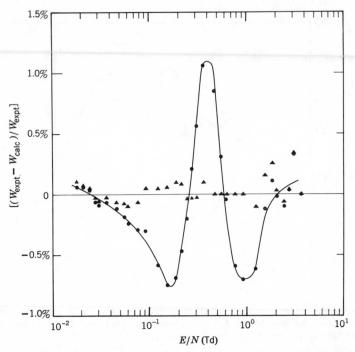

FIG. 13.9. The differences between experimental and calculated drift velocities, using the cross sections of Figure 13.8: ▲ using cross section A, ● using cross section B.

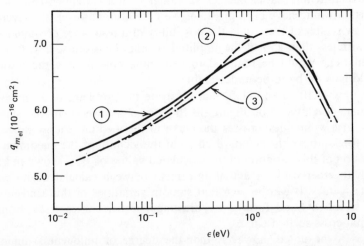

FIG. 13.10. Trial cross sections used to demonstrate the accuracy and uniqueness of the swarm analysis for $q_{m_{el}}$ in helium (taken from ref. 19): curve 1, cross section of Crompton et al.;[21] curve 2, 2% lower than curve 1 at 2×10^{-2} eV, 2% higher than curve 1 at 2 eV; curve 3, 2% lower than curve 1 at all energies.

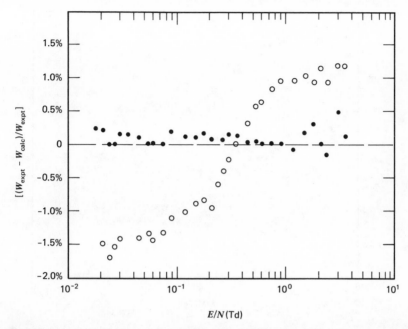

E/N (Td)

FIG. 13.11. A comparison of experimental and calculated drift velocities, using the cross sections shown in Figure 13.10: ● using the cross section of curve 1; ○ using the cross section of curve 2. (From ref. 19.)

which the measurements were made. By applying tests similar to those just described,[51,52] it has been possible to specify the energy range over which the cross section is valid to any required accuracy.

The effect of systematic errors is more straightforward and is also illustrated in Figures 13.10 (curve 3) and 13.12. From the figures it can be seen that an overall systematic error of 1% in the measured drift velocities [due, e.g., to errors in the determination of E, h, or N (see Chapter 10)] gives rise to an overall error of about 2% in the cross section.† When the systematic errors are functions of E/N, the situation is not so straightforward but resembles the case illustrated by curve 2 of Figure 13.10 and Figure 13.11.

Although it is possible to determine an energy-dependent cross section for a monatomic gas by using experimental results for one transport coefficient only, a comparison of calculated and experimental data for another transport coefficient always provides a valuable check of the

†This is consistent with the prediction from equation 13.34 with $l=0$ and $h=1$ (i.e., q_m = constant), in which case $W \propto a^{-1/2}$. It can also be seen from equation 13.41 that, as $D/\mu \to \kappa T/e$, W becomes proportional to q_m^{-1}, that is, the sensitivity of the method improves, as shown in Figure 13.12.

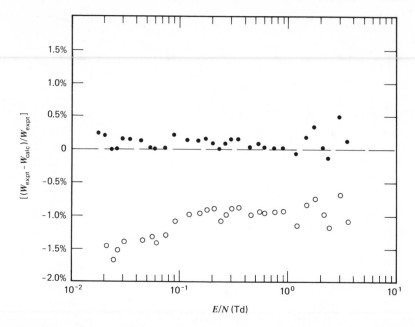

FIG. 13.12.   The results of a test to show the overall accuracy of the helium cross section: ●
as in Figure 13.11; ○ using curve 3 of Figure 13.10. (From ref. 19.)

validity of the cross section and may prove to be essential to determine its
uniqueness. In the case of helium, tests of the kind already described have
established the uniqueness of the cross section derived from drift data
alone, and the comparison therefore serves simply as a cross check. In the
other monatomic gases the presence of the Ramsauer-Townsend minimum
may make it essential to use other transport data to establish a unique
cross section.[53]

Table 13.1 shows a comparison of the experimental data in helium with
the data calculated using the cross section shown in Figure 13.7. Although
the agreement in the case of the drift velocity data does no more than
reflect the low scatter in the experimental results and the accuracy of the
fit, the agreement for the values of D/μ is a reassuring confirmation of
the experimental accuracy, the analytical procedure, and the validity of the
cross section. A comparison of the same type using data for the magnetic
drift velocity W_M provides similar confirmation.[21]

An additional check, which also supports the validity of the theory on
which the analysis is based,† has recently been made by Cavalleri and by
Gibson et al. (see also Section 11.5.1 and refs. 4 and 61 of Chapter 11). The

†See also the paper by K. Kumar and R. E. Robson, *Aust. J. Phys.*, **24**, 835, 1971.

TABLE 13.1.   Comparison of experimental and computed values of W and D/μ for electrons in helium (293 K).

The calculated values were obtained using the cross section shown in Figure 13.7.

E/p_{293} (V cm^{-1} torr^{-1})	E/N (Td)	$10^{-5}\,W$ (cm sec^{-1}) Experimental	Calculated	D/μ (V) Experimental	Calculated
0.006	0.01820	0.418	0.418		
0.008	0.0243	0.533	0.533	0.0313	0.0314
0.010	0.0303	0.637	0.637	0.0337	0.0339
0.015	0.0455	0.863	0.864	0.0400	0.0403
0.020	0.0607	1.052	1.053	0.0468	0.0469
0.05	0.1517	1.840	1.839	0.0874	0.0876
0.10	0.303	2.668	2.667	0.1536	0.1533
0.15	0.455	3.280	3.280	0.217 (0)	0.217 (1)
0.20	0.607	3.789	3.788	0.278 (9)	0.279 (7)
0.4	1.214	5.33	5.33	0.520	0.523
0.6	1.820	6.55	6.53	0.755	0.762
0.8	2.43	7.57	7.58	0.996	1.000
1.0	3.03	8.57	8.54	1.241	1.244

cross section shown in Figure 13.7 was derived from an analysis of drift data in helium at 77 K in which, as already described, the electron temperature was in general greatly increased by the electric field. For example, the energy dependence of the cross section in the vicinity of 0.03 eV was determined from the drift data for electron swarms whose mean energy was approximately four times the thermal value. It is possible to confirm the accuracy of the cross section in this energy range by comparing the value of D at 293 K as calculated using the cross section of Figure 13.7 with the experimental value of D determined by Cavalleri's method. The important difference between the two situations is the fact that the electron velocity distribution function is completely specified in the thermal case so that no assumption need be made regarding its theoretical representation, as was necessary in the theory on which the analysis of the drift data was based. The agreement of the calculated and experimental values to within the experimental error ($\sim \pm 1\%$) provides a satisfying overall cross check of both theory and experiment.

13.5.2.   ELASTIC AND INELASTIC SCATTERING. In most of the work that has so far been published on the analysis of electron swarm data to derive both elastic and inelastic cross sections, the assumption has been made that the total inelastic cross section is small compared with the elastic cross section. With this assumption and the omission of the time and spatial dependence terms equation 6.50 becomes†

$$\frac{d}{d\epsilon}\left\{\left[(eE)^2\frac{\epsilon}{Nq_m(\epsilon)}+\frac{6m}{M}\kappa T\epsilon^2 Nq_m(\epsilon)\right]\frac{df}{d\epsilon}\right.$$

$$\left.+\frac{6m}{M}\epsilon^2 Nq_m(\epsilon)f(\epsilon)+3N\sum_{ik}\frac{m^2}{2}I_{ik}\right\}=0,\qquad(13.44)$$

where (see equation 6.46)

$$I_{ik}=\frac{2}{m^2}\int_\epsilon^{\epsilon+\epsilon_{ik}}\left[f(y)-f(y-\epsilon_{ik})\exp\left(-\frac{\epsilon_{ik}}{\kappa T}\right)\right]q_{0_{ik}}(y)y\,dy,$$

$q_{0_{ik}}$ is the cross section for the kth transition of the ith inelastic process, and ϵ_{ik} is the threshold energy for this transition.

We can express this equation for $f(\epsilon)$ in two ways. From equation 6.49 we note that $dI_{ik}/d\epsilon$ can be expressed as

$$\frac{dI_{ik}}{d\epsilon}=\frac{2}{m^2}\left\{\left[(\epsilon+\epsilon_{ik})q_{0_{ik}}(\epsilon+\epsilon_{ik})f(\epsilon+\epsilon_{ik})-\epsilon q_{0_{ik}}(\epsilon)f(\epsilon)\right]\right.$$

$$\left.+\left[(\epsilon-\epsilon_{ik})q_{0_{-ik}}(\epsilon-\epsilon_{ik})f(\epsilon-\epsilon_{ik})-\epsilon q_{0_{-ik}}(\epsilon)f(\epsilon)\right]\right\},$$

where $q_{0_{-ik}}$ denotes the cross section for a superelastic collision associated with a transition of the ikth type. Consequently equation 13.44 becomes

$$\frac{(eE)^2}{3}\frac{d}{d\epsilon}\left[\frac{\epsilon}{Nq_m(\epsilon)}\frac{df}{d\epsilon}\right]+\frac{2m\kappa T}{M}\frac{d}{d\epsilon}\left[\epsilon^2 Nq_m(\epsilon)\frac{df}{d\epsilon}\right]$$

$$+\frac{2m}{M}\frac{d}{d\epsilon}\left[\epsilon^2 Nq_m(\epsilon)f(\epsilon)\right]$$

$$+\sum_{ik}\left[(\epsilon+\epsilon_{ik})Nq_{0_{ik}}(\epsilon+\epsilon_{ik})f(\epsilon+\epsilon_{ik})-\epsilon Nq_{0_{ik}}(\epsilon)f(\epsilon)\right.$$

$$\left.+(\epsilon-\epsilon_{ik})Nq_{0_{-ik}}(\epsilon-\epsilon_{ik})f(\epsilon-\epsilon_{ik})-\epsilon Nq_{0_{-ik}}(\epsilon)f(\epsilon)\right]=0.\quad(13.45)$$

†The equation in the form of equation 13.44 also implies isotropic scattering in inelastic collisions (see Chapter 6).

This equation is identical with the form of the Boltzmann equation given by Frost and Phelps,[16] who extended Holstein's and Margenau's results to include collisions of the second kind (see Chapter 6).

An alternative form of the equation, which forms the basis of the numerical solution proposed by Frost and Phelps, is obtained by using the fact that equation 13.44 can be integrated once to give

$$\left[(eE)^2 \frac{\epsilon}{Nq_m(\epsilon)} + \frac{6m}{M} \kappa T \epsilon^2 N q_m(\epsilon) \right] \frac{df}{d\epsilon} + \frac{6m}{M} \epsilon^2 N q_m(\epsilon) f(\epsilon)$$

$$+ 3N \sum_{ik} \frac{m^2}{2} I_{ik} = 0. \tag{13.46}$$

As shown in Chapter 6, this is equivalent to the statement that the total point flux across a spherical surface in velocity space is zero. If we now change to the normalized variable $z = \epsilon/\kappa T$ (which requires the transformation $z' = y/\kappa T$ in the original expression for I_{ik}) and divide throughout by $N(\kappa T)^2 q_0 6m/M$, where q_0 is the value of q_m at some reference energy, we obtain

$$\left\{ \frac{M}{6m} \frac{(eE)^2}{N^2 q_0 q_m(z)(\kappa T)^2} + z \left[\frac{q_m(z)}{q_0} \right] \right\} z \frac{df}{dz} + z^2 \frac{q_m(z)}{q_0} f(z)$$

$$+ \sum_{ik} \frac{M}{2m} \int_z^{z+z_{ik}} [f(z') - f(z' - z_{ik}) \exp - z_{ik}] \frac{q_{0_{ik}}(z')}{q_0} z' \, dz' = 0.$$

On introducing the additional normalized variables $\theta = q_m(z)/q_0$ and $\eta_{ik}(z) = M q_{0_{ik}}(z)/2mq_0$, and the constant (for a given set of experimental conditions) $\alpha = (M/6m)(eE/Nq_0\kappa T)^2$, the equation reduces to the relatively simple form

$$\left(\frac{\alpha}{\theta} + \theta z \right) z \frac{df}{dz} + z^2 \theta f + \sum_{ik} \int_z^{z+z_{ik}} [f(z') - f(z' - z_{ik}) \exp - z_{ik}] \eta_{ik}(z') z' \, dz' = 0,$$

$$\tag{13.47}$$

which is equation 3 of ref. 16.

Several methods of solving equation 13.45 or its equivalent, equation 13.47, have been proposed. Some of the methods are suitable only in the region where collisions of the second kind can be neglected with negligible error, so that somewhat simpler solutions can be obtained. Before describing them we note that, once $f(\epsilon)$ has been determined, the problem of calculating the transport coefficients is the same as that described in the preceding section, that is, the evaluation of formulae such as equations 13.41 to 13.43 by numerical integration. We will return later to the problems of how to represent the input cross sections and how best to adjust them in order to obtain a progressive improvement in the agreement between the calculated and experimental values of the coefficients.

(a) Solutions Neglecting Collisions of the Second Kind

When collisions of the second kind can be neglected, equation 13.47 becomes

$$\left(\frac{\alpha}{\theta}+\theta z\right)z\frac{df}{dz}+z^2\theta f+\sum_{ik}\int_z^{z+z_{ik}}f(z')\eta_{ik}(z')z'\,dz'=0. \quad (13.48)$$

A solution of this equation that was proposed by Sherman[54] relies on the relacement of $f(z)$ by $f(z)=v(z)\gamma(z)$, where $\gamma(z)$ is the solution of the equation in the absence of the inelastic terms. Sherman shows that, with certain stipulations, $v(z)\to 1$ as $z\to\infty$. From the definition of $\gamma(z)$ it follows that $\gamma(z)$ is the generalized distribution function given by equation 6.51. Expressed in terms of the normalized quantities, the equation reads

$$\gamma(z)=\exp-\int_0^z\frac{z'\theta^2(z')}{\alpha+z'\theta^2(z')}dz'. \quad (13.49)$$

With the substitution $f(z)=v(z)\gamma(z)$ equation 13.48 becomes

$$\left[\left(\frac{\alpha}{\theta}+\theta z\right)z\frac{d\gamma}{dz}+z^2\theta\gamma(z)\right]v(z)+\left(\frac{\alpha}{\theta}+\theta z\right)z\gamma(z)\frac{dv}{dz}$$

$$+\sum_{ik}\int_z^{z+z_{ik}}v(z')\gamma(z')\eta_{ik}(z')z'\,dz'=0.$$

Since $\gamma(z)$ was defined to make the first group of terms vanish, the equation reduces to

$$\frac{dv}{dz}+\frac{1}{h(z)\gamma(z)}\sum_{ik}\int_z^{z+z_{ik}}v(z')\gamma(z')\eta_{ik}(z')z'\,dz'=0, \quad (13.50)$$

where $h(z) = (z/\theta)(\alpha + \theta^2 z)$. Since $v(z) \to 1$ as $z \to \infty$, integration of this equation gives

$$v(u) = 1 + \int_u^\infty \frac{1}{h(z)\gamma(z)} \sum_{ik} \int_z^{z+z_{ik}} v(z')\gamma(z')\eta_{ik}(z')z' \, dz' \, dz,$$

whence

$$f(u) = \gamma(u)\left[1 + \int_u^\infty \frac{1}{h(z)\gamma(z)} \sum_{ik} \int_z^{z+z_{ik}} f(z')\eta_{ik}(z')z' \, dz' \, dz\right]. \quad (13.51)$$

In the method of calculating $f(u)$ developed by Sherman (the so-called backward prolongation technique), $f(u)$ is set equal to $\gamma(u)$ for all energies above a cut-off energy u_m. We can now write down a system of equations for $f(u)$ for values of u separated by an energy interval Δu small enough to ensure that the integrals can be calculated with sufficient accuracy by the trapezium rule. If the energies are denoted by $u_m, u_{m-1}, u_{m-2}, \ldots, 0$, the first three equations of the set are†

$$f(u_m) = \gamma(u_m),$$

$$f(u_{m-1}) = \gamma(u_{m-1})\left\{1 + \frac{(\Delta u)^2}{4} \frac{1}{h(u_{m-1})\gamma(u_{m-1})}\right.$$

$$\left. \times \sum_{ik} [u_{m-1}f(u_{m-1})\eta_{ik}(u_{m-1}) + u_m f(u_m)\eta_{ik}(u_m)]\right\},$$

$$f(u_{m-2}) = \gamma(u_{m-2})\left\{1 + \frac{(\Delta u)^2}{4}\right.$$

$$\times \sum_{ik} \left[\frac{u_{m-2}f(u_{m-2})\eta_{ik}(u_{m-2}) + 2u_{m-1}f(u_{m-1})\eta_{ik}(u_{m-1}) + u_m f(u_m)\eta_{ik}(u_m)}{h(u_{m-2})\gamma(u_{m-2})}\right.$$

$$\left.\left.+ 2\frac{u_{m-1}f(u_{m-1})\eta_{ik}(u_{m-1}) + u_m f(u_m)\eta_{ik}(u_m)}{h(u_{m-1})\gamma(u_{m-1})}\right]\right\}, \quad (13.52)$$

provided that $u_{m-2} + z_{ik} \geq u_m$.

†Since the integrations are cut off at $u = u_m$, the integrand of the main integral vanishes for $u = u_m$; that is, the second term in the trapezium integration for $f(u_{m-1})$ is zero.

Each equation of the set is linear. Furthermore, the equation for $f(u_{m-1})$ contains $f(u_{m-1})$ as the only unknown. Rearranging the equation, we have

$$f(u_{m-1}) = \frac{\gamma(u_{m-1})\left[1 + \dfrac{(\Delta u)^2}{4}\dfrac{u_m f(u_m)}{h(u_{m-1})\gamma(u_{m-1})}\displaystyle\sum_{ik}\eta_{ik}(u_m)\right]}{1 - \dfrac{(\Delta u)^2}{4}\dfrac{u_{m-1}}{h(u_{m-1})\displaystyle\sum_{ik}\eta_{ik}(u_{m-1})}}.$$

Similarly, once $f(u_{m-1})$ has been determined from the second equation, $f(u_{m-2})$ is the only unknown in the third, and so on. Thus, starting with $f(u_m) = \gamma(u_m)$ at $u = u_m$, the solution can be extended backwards step by step to the origin. In practice, recursion formulae are developed to shorten the calculations.[16]

An alternative method of solution was developed by Lucas.[45] If $f(z)$ in equation 13.48 is represented by $f(z) = A\exp[-\int_0^z D(z')dz']$, the equation becomes

$$-\left(\frac{\alpha}{\theta} + \theta z\right)zDf(z) + z^2\theta f(z) = -\sum_{ik}\int_z^{z+z_{ik}} f(z')\eta_{ik}(z')z'\,dz'$$

whence

$$D(z) = \frac{z^2\theta f(z) + \displaystyle\sum_{ik}\int_z^{z+z_{ik}} f(z')\eta_{ik}(z')z'\,dz'}{[(\alpha/\theta) + \theta z]zf(z)}. \tag{13.53}$$

This equation for D can be solved by a standard iterative technique in which a first approximation to $f(z)$, namely, $f_1(z)$, is inserted into the right-hand side of equation 13.53 and $D_1(z)$ then calculated for a series of values of z. The next approximation to $f(z)$ follows from putting $f_2(z) = A\exp[-\int_0^z D_1(z')dz']$, A being determined in each case by appropriate normalization. The procedure continues until satisfactory convergence is achieved, usually after about thirty cycles. For the initial estimate of $f(z)$, Lucas used a Maxwellian distribution corresponding to an electron temperature estimated from measured values of D/μ, although, as expected, he found the final solution to be independent of the initial choice.

(b) Solutions Including Collisions of the Second Kind

The inclusion of terms accounting for collisions of the second kind considerably increases the complexity of the methods of solution. With the inclusion of these terms equation 13.51 becomes

$$f(u) = \gamma(u) \left\{ 1 + \int_u^\infty \frac{1}{h(z)\gamma(z)} \sum_{ik} \int_z^{z+z_{ik}} \eta_{ik}(z')z' \right.$$

$$\left. \times [f(z') - (\exp - z_{ik})f(z' - z_{ik})] \, dz' \, dz \right\}. \quad (13.54)$$

Using the procedure described in the preceding section, we can derive a set of equations similar to equations 13.52. For convenience the lth member of the set may be represented as

$$f(u_l) = S_0^l f(0) + \cdots + (S_l^l + I_l^l)f(u_l) + \cdots + (S_m^l + I_m^l)f(u_m), \quad (13.55)$$

where the coefficients I and S represent the contributions from the inelastic and superelastic terms, respectively, of equation 13.54.† It can be seen immediately that the term $f(z' - z_{ik})$ in equation 13.54 precludes backward prolongation as a possible technique for solving the equation since, to evaluate $f(u_l)$, $f(u)$ must be known for $u < u_l$. For less obvious reasons, Lucas's method also fails when the S coefficients are other than very small.[46]

Two successful methods of solution have been developed. In the first, due to Sherman and Boyer,[16] the set of linear equations, of which equation 13.55 is one member, is solved by the standard technique of elimination. This method requires a large computer because of the great number of coefficients (m^2, where typically $m \sim 300$) that has to be stored at any one time. An alternative method developed by Gibson[46] is somewhat less demanding because it combines some features of both the backward prolongation and iterative techniques. As in backward prolongation, the equations are solved in order proceeding from the equation for $f(u_{m-1})$. However, when calculating $f(u_l)$ it is necessary to assume a distribution function for all $u < u_l$ since the terms for $u < u_l$ are no longer zero. Initially a Maxwellian distribution corresponding to the value of T_e estimated from the measured values of D/μ is used as the assumed function. As the prolongation proceeds, the value of the normalizing constant of the assumed distribution function is continually adjusted to ensure that the function is continuous at $u = u_l$. With the completion of the first cycle,

---

†In this notation, the first nonzero value of S corresponds to an energy u_l', for which $u_l' - z_{ik} > 0$.

resulting in the first approximate function $f_1(u)$, the procedure is repeated to determine $f_2(u)$, this time using values of $f_1(u)$ for $u < u_l$. For each value of l, $f_1(u)$ is adjusted to ensure that $f_1(u_{l+1}) = f_2(u_{l+1})$. Convergence is usually obtained after three to five iterations.

(c) Fitting Parameters

As was the case for the calculations with elastic scattering, the adjustment of the cross sections is assisted by the use of fitting parameters. For the adjustment of $q_{m_{el}}(\epsilon)$ the parameter q_m^* or ν_m can be used as before. As was shown in Section 13.2.2, the calculation of q_m^* is not greatly affected by the presence of inelastic collisions even though such collisions greatly affect the mean energy of the swarm. This effect is properly accounted for through the term in D/μ which appears in the formula.

A parameter which is sensitive primarily to inelastic collisions can also be found, although its application to the adjustment of the inelastic cross sections is generally less satisfactory because more than one inelastic process can usually occur at a given energy. Figure 13.13 underlines this difficulty, although it also demonstrates how the influence of the various inelastic processes can to some extent be separated. The figure shows graphs (for nitrogen) of the fraction of the total power input to the swarm from the electric field (eEW) which is dissipated in the various collision processes, that is, elastic collisions, rotational and vibrational excitation collisions, and collisions which cause electronic excitation and ionization. From the graphs it is clear that each set of processes is primarily responsible for the energy loss over a certain energy range. It follows, therefore, that, if a parameter can be found which is sensitive to the inelastic cross sections, this parameter can be used as a guide to adjust each group of cross sections in turn.

Employing an extension of the argument whereby q_m^* was derived, Crompton, Gibson, and McIntosh[44] suggested the use of a parameter q_i^*, which was called the *effective inelastic cross section*. If it is assumed that one inelastic cross section q_i is primarily resonsible for inelastic losses at a given value of E/N, and that the energy gain from superelastic encounters may be neglected, the following equation can be obtained from equation 13.45 by multiplying by $(2/m)^{1/2} \epsilon \, d\epsilon$ and integrating over all energies:†

$$e\left(\frac{E}{N}\right)W = \frac{(8m)^{1/2}}{M} \int_0^\infty \epsilon^2 q_m(\epsilon)\left[f(\epsilon) + \kappa T \frac{df}{d\epsilon}\right] d\epsilon + \left(\frac{2}{m}\right)^{1/2} \epsilon_i \int_0^\infty \epsilon q_i(\epsilon) f(\epsilon) \, d\epsilon.$$

$$(13.56)$$

---

† The inelastic terms can be evaluated by changing the variable $\epsilon + \epsilon_i$ to z in the first term and remembering that the lower limit of integration for the second is ϵ_i since $q_i(\epsilon) = 0$ for $\epsilon < \epsilon_i$.

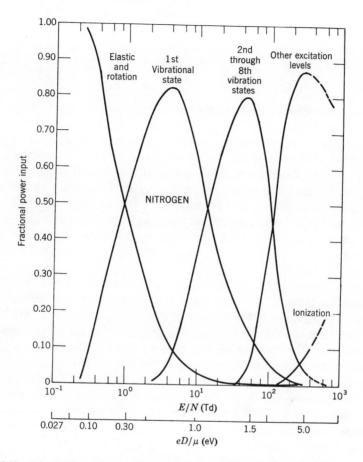

FIG. 13.13.   The distribution between the elastic and inelastic collision processes of the total power input into an electron swarm in nitrogen. (From ref. 37.)

The cross sections $q_m(\epsilon)$ and $q_i(\epsilon)$ are now replaced by the constant effective values q_m^* and q_i^*, respectively, and equation 13.56 is integrated by parts to give

$$e\left(\frac{E}{N}\right)W = \frac{(8m)^{1/2}}{M}q_m^*\left(\int_0^\infty \epsilon^2 f\,d\epsilon - 2\kappa T\int_0^\infty \epsilon f\,d\epsilon\right) + \left(\frac{2}{m}\right)^{1/2}\epsilon_i q_i^* \int_0^\infty \epsilon f\,d\epsilon.$$

As shown in ref. 44, each of the integrals can be expressed in terms of De/μ and a dimensionless factor whose magnitude depends on the form of the energy distribution function. When q_m^* is replaced by equation 13.10, the following expression for q_i^* is obtained:

$$q_i^* = \left(\frac{m\epsilon}{2}\right)^{1/2} \frac{1}{\epsilon_i K_3} \frac{W(E/N)}{(D/\mu)^{1/2}} \left[1 - K_1 \frac{(De/\mu) - K_2 \kappa T}{MW^2}\right], \quad (13.57)$$

where K_1, K_2, and K_3 are combinations of the dimensionless factors which can be evaluated.

The parameter q_i^* is used in the same way as q_m^*, that is, q_i^* is found at each value of E/N both from the measured transport coefficients and from those calculated from the distribution functions determined with a given set of input cross sections. A first-order adjustment to the cross section can then be made by multiplying by the ratio $q_{i_{exp}}^* / q_{i_{calc}}^*$. As before, the "adjustment energy" is taken to be the energy at which the power input to the given inelastic process is a maximum, that is, the value of ϵ for which $\epsilon f(\epsilon) q_i(\epsilon)$ is a maximum [cf. $\epsilon^2 f(\epsilon) q_{m_{el}}(\epsilon)$ for elastic collisions†].

An alternative fitting parameter for inelastic collisions was introduced by Frost and Phelps,[16] namely, the *energy exchange collision frequency* ν_u, defined through the equation $\nu_u/N = W(E/N)/(D/\mu - \kappa T/e)$. When elastic collisions account for a negligible proportion of the energy loss, ν_u is a measure of the product of the inelastic collision frequency and the fractional energy exchange per inelastic collision. Under other circumstances, however, ν_u is approximately equal to the sum of this product and the corresponding product for elastic collisions, since eEW is the total power dissipated in both types of collision. The parameter ν_u therefore becomes less useful when elastic and inelastic losses are comparable, and q_i^* is then to be preferred because it is derived directly from an expression for the power input to inelastic collisions, the power input to elastic collisions having been first subtracted from the total power. However, neither parameter should be regarded as more than an aid in the adjustment of the cross sections. As was the case for q_m^* and ν_m, the accuracy of the final set of inelastic cross sections does not depend on the effectiveness of these devices.

(d) Summary of the Characteristics of Swarm Analyses

As we have already seen, the presence of a number of possible inelastic processes usually makes it impossible to determine a unique set of cross sections when data for only two or possibly three transport coefficients are

---

†The difference arises from the fact that the energy absorbed at each inelastic collision is independent of ϵ, whereas the elastic energy loss is proportional to ϵ.

available. Nevertheless valuable information can be obtained within the limitations described below.

The simplest extension of the case described in the preceding section is that in which elastic scattering is accompanied by only one inelastic process. This situation occurs in the monatomic gases at intermediate values of E/N when electronic excitation becomes significant, and at low values of E/N in parahydrogen when only one kind of rotational transition can be excited. Unique curves for both q_m and the cross section for the inelastic process can then be obtained, factors similar to those already described governing the accuracy and the energy range of the determination. The unique determination of both cross sections is made possible by the fact that $q_{m_{el}}$ enters explicitly into the transport integrals, whereas the inelastic cross section enters only through its effect on the electron energy distribution function.

The situation is not as favourable when more than one inelastic process is present to control the energy distribution. Suppose that there are two such processes with cross sections $q_{0_1}(\epsilon)$ and $q_{0_2}(\epsilon)$ and thresholds ϵ_1 and ϵ_2, and that $\epsilon_1 \ll \epsilon_2$. For the lowest energy swarms, very few electrons have energies greater than ϵ_2; consequently $q_m(\epsilon)$ and $q_{0_1}(\epsilon)$ can be determined uniquely as described above. On the other hand, at higher energies the numbers of inelastic collisions of the two types become comparable. However, an electron loses as much energy in a single collision involving the process with threshold ϵ_2 as it does in many collisions involving the process with the lower threshold. Consequently, collisions in which the energy loss is ϵ_2 play a dominant rôle in determining the energy distribution, and the threshold behaviour of $q_{0_2}(\epsilon)$ can be determined provided that a reasonable extrapolation of $q_{0_1}(\epsilon)$ can be made. We can summarize as follows:

1. Above a transition energy somewhat below ϵ_2 neither inelastic cross section can be uniquely determined from swarm data alone.

2. It is possible to determine $q_{0_1}(\epsilon)$ uniquely from the threshold to the transition energy. Above this energy, uniqueness in its determination is lost unless $q_{0_2}(\epsilon)$ is accurately known from another source.

3. It is impossible to determine $q_{0_2}(\epsilon)$ uniquely for any energy range unless $q_{0_1}(\epsilon)$ is known. On the other hand, if the separation between the threshold energies ϵ_1 and ϵ_2 is sufficiently large, $q_{0_2}(\epsilon)$ can be determined within comparatively narrow limits even though $q_{0_1}(\epsilon)$ is not known with great accuracy.

4. In many gases the situation described in 1 to 3 is a gross oversimplification because there may be many inelastic processes with closely adjacent thresholds, for example, the numerous and closely spaced rota-

tional transitions in the heavier polyatomic gases. In this event little progress can be made unless the energy dependence of the cross sections and their relative magnitudes are known from theory, in which case swarm experiments can be used to normalize the cross sections.

5. The momentum transfer cross section can be obtained with reasonable accuracy at all energies, provided that the total inelastic collision frequency is small compared with that for elastic scattering.

It is not possible to give an adequate summary of all the analyses that have been made, for the most part by Phelps and his collaborators, because different techniques have to be applied to meet the specific situations that arise for each gas. We therefore describe briefly two examples that are representative.

(e) Rotational and Vibrational Excitation in Hydrogen

In order to demonstrate the unique situation which exists for hydrogen, we begin by examining a number of possible ways in which swarm experiments could be carried out in this gas and the consequences to the subsequent analysis to determine the relevant cross sections.

The simplest experiments to conduct are those in normal hydrogen at room temperature. Table 13.2 shows the populations of hydrogen molecules having rotational levels characterized by the rotational quantum number J.

In normal hydrogen at room temperature, four rotational states are significantly populated and four rotational excitation processes are therefore significant. The threshold energies for these processes are 0.044, 0.073, 0.101, and 0.128 eV. In addition, vibrational excitation of the molecule is also possible for energies greater than 0.52 eV. As the mean energy of the electron swarm is increased from the thermal value to about 0.5 eV, seven inelastic processes therefore assume varying importance: the four rotational excitations, two rotational de-excitations, and one vibrational excitation. Even though the cross sections for rotational excitation and de-excitation are effectively coupled by the Klein-Rosseland (see equation 6.48), the situation is sufficiently complex that a unique set of cross sections could not be expected, given data for no more than two or perhaps three transport coefficients. The most effective way of using a swarm analysis in this situation is to test the compatibility of a set of cross sections derived in some other way, or to find the energy dependence for one inelastic cross section, given accurate data for all the others.

The situation can be made considerably less complex simply by lowering the temperature of the gas to 77 K, in which case the populations of the rotational states are those shown in the second row of Table 13.2. Only two

TABLE 13.2. Populations of rotational states in normal hydrogen and para-hydrogen at 293 and 77 K[a]

T (K)	J 0	1	2	3	4	
			population (%)			
293	13.5	67	11.2	7.9	0.3	} normal hydrogen
77	24.87	75	0.13	—	—	
77	99.5	—	0.5	—	—	parahydrogen

[a]In normal H_2 at 77 K the relative populations of the states with even and odd rotational quantum numbers correspond to the populations at room temperature, since it is assumed that no catalyst is present to establish the equilibrium at the low temperature. Similarly, the calculation of the populations in parahydrogen is based on the fact that only states with even quantum numbers are present.

rotational excitations are now significant; furthermore the negligible populations of the rotational states with $J \geqslant 2$ mean that collisions of the second kind become of negligible importance except at the lowest values of E/N. Nevertheless, despite the relative simplicity of this situation, the comparatively small separation of the thresholds of the two rotational transitions means that the energy range over which there is any real separation of the effects of the two types of collision is very limited. The lack of separation is enhanced by the very much larger population of the state with the higher quantum number.

This problem is overcome by using parahydrogen rather than normal hydrogen. The relative populations are then as shown in the third row of the table, and it is apparent that only one inelastic process, the $J = 0 \to 2$ rotational excitation, is of any significance until the swarm has sufficient energy to excite vibrational transitions.

Figure 13.14 helps to illustrate this point quantitatively. Below $E/N \sim 2$ Td, virtually no energy is dissipated in vibrational excitation. Consequently, since only one rotational transition is possible, an analysis of swarm data for $E/N < 2$ Td allows a unique determination of this cross section to be made, as well as of the cross section for momentum transfer. Furthermore, the negligible energy exchange in superelastic collisions for $E/N \gtrsim 10^{-1}$ Td means that these collisions can usually be disregarded, with consequent simplification of the analysis.

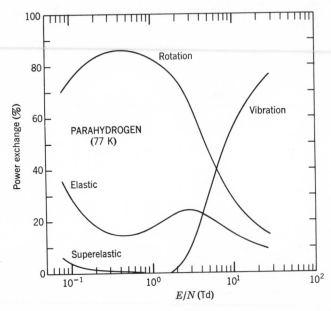

FIG. 13.14. The distribution of the total power input into an electron swarm in parahydrogen between elastic, rotational, and vibrational collision processes. (From ref. 44.)

These arguments suggested low-temperature drift and diffusion experiments in parahydrogen[56] and their subsequent analysis[44] to determine the $J = 0 \to 2$ cross section $q_{r(02)}$. Figure 13.15 shows the cross sections q_m and $q_{r(02)}$, together with the vibrational cross section $q_{v(01)}$, that were found to be compatible with the experimental results for W and D/μ. For energies up to about 0.3 eV the calculated transport coefficients were insensitive to $q_{v(01)}$; consequently $q_{r(02)}$ is uniquely determined up to this energy. The question of the accuracy and uniqueness above 0.3 eV is best answered with the aid of Figure 13.16, in which the rotational cross sections that are compatible with two extreme vibrational cross sections are shown. The upper curve shows the rotational cross section which is required in order to compensate for the complete absence of vibrational excitation. The lower curve is the cross section that is compatible with the vibrational cross section obtained by Ehrhardt et al.,[57] using a beam technique. The improbable shape of the curve for $q_{r(02)}$ above 0.4 eV suggests that the vibrational cross section is too large. With these somewhat extreme forms for $q_{v(01)}$, the error bounds for $q_{r(02)}$ are approximately ±5% up to 0.3 eV, ±10% at 0.4 eV, and ±30% at 0.5 eV.

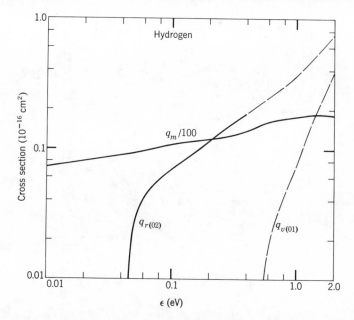

FIG. 13.15. Elastic and inelastic collision cross sections derived from the analysis of transport coefficient measurements in parahydrogen. The unbroken portion of the curve for $q_{r(02)}$ shows the region in which the cross section was determined uniquely. The dashed portion was obtained by extrapolation (see text). The vibrational cross section $q_{v(01)}$ is the vibrational cross section consistent with $q_{r(02)}$. (From ref. 18.)

Figure 13.16 also serves to illustrate some of the points listed in the summary of the characteristics of swarm analyses. Thus it can be seen that the vibrational cross section must be known with good accuracy near threshold if the rotational cross section is to be determined accurately beyond an energy somewhat below $\epsilon_{v(01)}$. This is so because an error in $q_{v(01)}$ requires a compensating adjustment in $q_{r(02)}$ so that $\epsilon_{r(02)}\Delta q_{r(02)} \cong \epsilon_{v(01)}\Delta q_{v(01)}$. Since the cross sections above threshold are of the same order, but $\epsilon_{v(01)} \sim 10\epsilon_{r(02)}$, small errors in $q_{v(01)}$ require large adjustments to $q_{r(02)}$ to compensate for them. Conversely, the threshold behaviour of the vibrational cross section derived from the analysis is comparatively insensitive to the extrapolation of the rotational cross section in the region where $q_{r(02)}$ can no longer be determined uniquely.

A test similar to that described in Section 13.5.1 can be applied to determine the overall accuracy of $q_{r(02)}$. Figure 13.17 shows the fractional changes in W and D/μ that result from a 5% change in $q_{r(02)}$. As is to be expected from the graphs in Figure 13.14, the maximum effects are

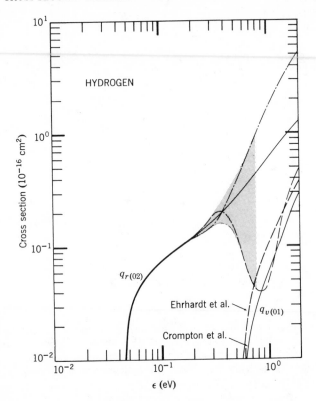

FIG. 13.16. Various combinations of q_r and q_v that are consistent with swarm data for parahydrogen: —, combination based on an extrapolation of q_r, using Henry's and Lane's cross section reduced 3%; — —, q_r curve consistent with the vibrational cross section of Ehrhardt et al.;—·—, q_r curve required if q_v is assumed to be zero everywhere. The shaded area indicates the rapid loss of uniqueness of q_r. (From ref. 44.)

observed at $E/N \sim 1$ Td, where the changes are 1.5 and 2%, respectively. Since the measurements are alleged to be accurate to within about 2%, a reasonable estimate of the error limit of the cross section is about $\pm 5\%$.

For the reasons already given it is not possible to derive a unique vibrational cross section from the swarm analysis alone. However, the wide separation of the threshold energies $\epsilon_{r(02)}$ and $\epsilon_{v(01)}$ and the close agreement between the theoretical cross section of Henry and Lane[58] and the rotational cross section derived from the swarm analysis in the region of uniqueness[44] enables the threshold behaviour of $q_{v(01)}$ to be determined within surprisingly close limits. From threshold to about 0.3 eV the energy dependence of the swarm-derived cross section $q_{r(02)}$ agrees closely with that of the theoretical cross section. It is therefore reasonable to extrapo-

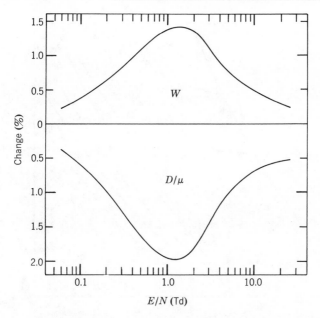

FIG. 13.17.   The changes in W and D/μ produced by a 5% change in $q_{r(02)}$. (From ref. 44.)

late the cross section to higher energies, using the theoretical cross section modified by a factor which gives agreement with the swarm-derived cross section at lower energies. This requires a reduction of the theoretical cross section by about 3%. It is then possible to analyze the swarm data for $E/N > 2$ Td to determine $q_{v(01)}$, the data above $E/N = 10$ Td being the most sensitive to this cross section. Figure 13.18 shows the cross section (labelled CGM) derived in this way.

Although the accuracy of the vibrational cross section cannot be assessed as precisely as that of the rotational cross section, reasonable error limits can be assigned by assuming that the rotational cross section is known within certain limits (say $\pm 50\%$) for energies in excess of 0.4 eV, the energy below which $q_{r(02)}$ may be considered to have been uniquely determined. The results of such an assessment are shown in Figure 13.18, where the assumed limiting forms of $q_{r(02)}$ are shown as dashed curves together with the vibrational cross sections that lead to the best overall fit with the experimental data. A similar argument was used by Crompton, Gibson, and Robertson[59] to determine what they considered to be an absolute upper limit to the vibrational cross section. From the figure it can be seen that, as foreshadowed on p.571, the vibrational cross section can be determined within comparatively narrow limits even when the rota-

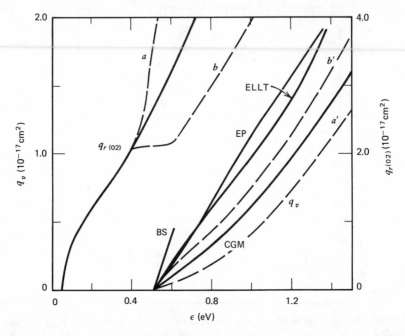

FIG. 13.18. The assignment of error limits to the vibrational cross section in hydrogen derived from swarm experiments. The unbroken curve $q_{r(02)}$ and the q_v curve labelled CGM are the rotational and vibrational cross sections, respectively, determined by Crompton et al.[44] The limiting curves a' and b' for q_v correspond to the assumed limiting rotational cross sections a and b (see text). Also shown are vibrational cross sections near threshold as determined in other experiments: EP, Engelhardt and Phelps;[36] ELLT, Ehrhardt et al.;[57] BS, Burrow and Schulz, *Phys. Rev.*, **187**, 97, 1969.

tional cross section is not accurately known at higher energies.

The cross sections $q_{r(02)}$ and $q_{v(01)}$ derived from the analysis of the parahydrogen data can now be used to derive the cross section $q_{r(13)}$ from an analysis of experimental data for normal hydrogen at low temperature.[46] Since the population of molecules with $J = 1$ is three times the population with $J = 0$ (see Table 13.2), and the energy loss in the $J = 1 \rightarrow 3$ transition is greater than that in the $J = 0 \rightarrow 2$ transition, the $J = 1 \rightarrow 3$ transition is the dominant rotational energy loss mechanism above E/N ∼0.5 Td.[46] Consequently, an analysis of the swarm data above this value enables $q_{r(13)}$ to be determined with reasonable accuracy, even allowing some inaccuracy in $q_{r(02)}$. However, if both $q_{r(02)}$ and $q_{v(01)}$ can be regarded as having been reliably established from the parahydrogen analysis, $q_{r(13)}$ remains as the only unknown, and it can therefore be derived with the same accuracy as $q_{r(02)}$. The final set of cross sections compatible with experimental results for normal hydrogen is shown in

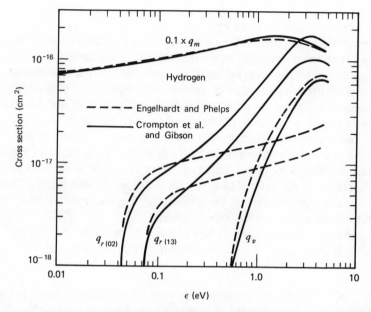

FIG. 13.19. Comparison of the H_2 cross sections found by Crompton et al. and Gibson (unbroken curves) with those of Engelhardt and Phelps (broken curves). (From ref. 46.)

Figure 13.19, together with that derived by Engelhardt and Phelps[36] from a similar analysis of the experimental data of Pack and Phelps,[13] Bradbury and Nielsen,[60] and Warren and Parker.[61] The arguments in favour of the cross sections derived by Crompton et al.[44] and Gibson[46] are given in refs. 46 and 59.

The analysis of swarm data in hydrogen is complicated by two factors which have not been mentioned so far: the presence of collisions in which there is simultaneous rotational and vibrational excitation, and the possibility that the cross sections for vibrational excitation may depend on the rotational state of the molecule. Both questions were examined by Crompton et al.[59] In regard to the first, it was shown that, to within the accuracy of the analysis, the vibrational cross section derived from swarm experiments is the sum of the cross sections for pure vibrational excitation and for simultaneous rotational-vibrational excitation. The influence of the initial rotational state on the vibrational cross section was examined by comparing electron drift velocities in normal hydrogen and parahydrogen in the range of E/N in which W is most strongly influenced by vibrational excitation. Since the drift velocities were found to be virtually identical, it was concluded that, to within the accuracy of the analysis, the *total* vibrational cross section is independent of the initial rotational state, a result which accords with theory.

Although it is possible, in the case of hydrogen, to derive rotational cross sections without recourse to other theoretical or experimental results, the analysis for this gas can be used to demonstrate the effectiveness of another important application of swarm techniques, that is, the testing of cross sections derived by other methods. In the course of the analysis for parahydrogen[44] a number of theoretically derived cross sections were tested for their compatibility with the experimental swarm results. The results of the tests are given in Table 13.3. With the exception of the cross sections calculated by Henry and Lane none of the cross sections resulted in acceptable agreement. Thus in this case, as in others, the method provides a significant test of the validity of the assumptions used in the theoretical derivations of the cross sections.

The analysis was extended to higher energies by Frost and Phelps[16] and Engelhardt and Phelps,[36] who used not only the experimental data for W

TABLE 13.3.   Results of a comparison between calculated and measured values of W and D/μ for various theoretical rotational cross sections. References to the theoretical papers will be found in Ref. 44.

Reference	Type of Calculation	Maximum Deviation (%)
Gerjuoy and Stein (1955)	Born approximation with quadrupole interactions	17
Dalgarno and Moffett (1963)	As above, including non-spherical polarization forces	17
Engelhardt and Phelps (1963)	Dalgarno and Moffett corrected by empirical factor ($\times$ 1.5)	8
Takayanagi and Geltman (1956)	Distorted wave	9
Dalgarno and Henry (1965)	Born approximation and distorted wave	25
Geltman and Takayanagi (1966)	Distorted wave, including short-range nonspherical interaction	11
Lane and Geltman (1967)	Close coupling	8
Henry and Lane (1969)	Close coupling with polarization and exchange	1.4

and E/N at higher values of E/N but also data for the ionization coefficient α_T. Their analysis was therefore concerned with the determination of a consistent set of cross sections for electronic excitation and ionization, as well as the cross sections already discussed. The same problem was examined by Lucas.[45] We will not describe this work here, but we use the investigations of Frost and Phelps and of Engelhardt et al. in nitrogen to illustrate the problems encountered at higher values of E/N. This work also illustrates how other difficulties that occur at low values of E/N in many other gases have been successfully overcome.

(*f*) *Cross Sections for Rotation, Vibration, Electronic Excitation, and Ionization in Nitrogen*

The work of Frost and Phelps[16] and Engelhardt et al.[37] in analyzing swarm data for nitrogen provides examples of most of the problems that are usually encountered in analyzing the data for polyatomic gases over a large range of values of E/N. We therefore give an account of their work without commenting in regard to the impact of later work on it, for example, more recent theories of rotational excitation and new experimental data. Such comment could be made only after further analysis.

Figure 13.13 shows the relative importance of the various collision processes over the range of E/N covered by experimental measurements of the transport coefficients. From the measured values of D/μ the range of values of the characteristic energy $\epsilon_K = eD/\mu$ over which the energy distribution functions are significantly influenced by each process can be inferred. These ranges are shown schematically in Figure 13.20, from which it can be seen that the analysis may be conveniently divided into three regions. These regions are now discussed in turn.

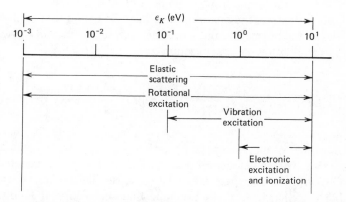

FIG. 13.20. Energy ranges over which the various collision processes significantly affect the electron energy distributions in N_2.

Region A: $\kappa T/e < \epsilon_K < 10^{-1}$ eV. This region can be used to examine the rotational excitation cross sections. The small rotational constant for the nitrogen molecule means that there are many rotational excitations to be considered with thresholds separated by less than 0.001 eV. Since even a thermal distribution at 77 K has a "width" of 20 to 30 meV,[18] many rotational excitation processes contribute to the energy losses of the lowest energy swarms; consequently the unique determination of any of the cross sections is out of the question. In this situation the value of the swarm analysis lies in its ability to verify the energy dependence of the theoretical cross sections and to normalize them.

The large number of cross sections and their low threshold energies introduce two computational difficulties. First, even at 77 K, there are significant fractions of the total population of molecules in many rotational states; collisions of the second kind are therefore significant over the whole E/N range covered by this region, with a consequent increase in the complexity of the analysis. Second, the large number of rotational excitation and de-excitation processes that must be considered as ϵ_K increases means that a computer of excessively large capacity would be required if each cross section were to be represented exactly. Fortunately this can be avoided without serious loss of accuracy because the low threshold energies of the significant cross sections and their close spacing enable them to be represented by a continuous approximation[16,37] when $\epsilon > 0.05$ eV (see Chapter 6).

Using an exact representation of the cross sections at the low-energy end of this range, and the continuous approximation at the high-energy end, Engelhardt et al. found that the rotational cross sections predicted by Gerjuoy and Stein[62] gave an accurate fit to the experimental transport data used by them when a quadrupole moment of $1.04ea_0^2$ was used (a_0 is the Bohr radius). In effect the quadrupole moment is the adjustable parameter which allows for normalization, since the Gerjuoy-Stein theory yields the following formula for the cross section $q_{r(J,J+2)}$:

$$q_{r(J,J+2)} = \frac{(J+2)(J+1)}{(2J+3)(2J+1)} \sigma_0 \left[1 - \frac{(4J+6)B_0}{\epsilon} \right]^{1/2}, \quad (13.58)$$

where B_0 is the rotational constant, $\sigma_0 = 8\pi Q^2 a_0^2/15$, and Q is the quadrupole moment. The value of $Q = 1.04ea_0^2$ found in this way is in satisfactory agreement with the accepted value. The analysis had sufficient sensitivity to enable Engelhardt et al. to conclude that the theory of Gerjuoy and Stein was in better agreement with experiment than were later theories which suggested the inclusion of a polarization correction.

Region B: 10^{-1} eV $< \epsilon_K < 1.4$ eV. As shown in Figure 13.20, vibrational excitation now has to be included in the analysis. In determining the eight vibrational cross sections of importance in this region, two assumptions were made:

1. That the rotational cross sections had been satisfactorily determined from region *A*.
2. That the shapes of the vibrational cross sections were the same as those determined by Schultz.[63]

With these assumptions, agreement was obtained between theory and experiment provided that a "tail" was added to the $v = 0 \rightarrow 1$ cross section extending from threshold to about 1.2 eV (see Figure 13.21), and the total cross section $\sum_v q_{v(v,v+1)}$ was adjusted to give a value of 5.5×10^{-16} cm^2 at 2.2 eV, a value that lay within the spread of Schulz's values for the total cross section as the scattering angle was varied. The threshold behaviour of $q_{v(01)}$ lay outside the range of Schulz's experiments but was investigated theoretically by Chen,[64] who found substantial agreement with the result of the swarm analysis.

Region C: $\epsilon_K > 1.3$ eV. Notwithstanding the fact that experimental data for two additional transport coefficients were available in this region, namely, the photon coefficient δ for the $C^3\Pi_u$ state and the ionization coefficient α_T, the high-energy range was the most difficult to analyze because of the uncertainty regarding a number of excitation processes that had to be included in the analysis. The excitation cross sections that were finally used are shown in Figure 13.22. As was the case for the vibrational cross sections, the initial shapes of these cross sections were taken from Schulz's work. However, it was necessary to adjust both the magnitude and the shape in some instances in order to obtain agreement between calculated and experimental swarm data. At the time of the analysis only three of the processes for which cross sections are shown in the figure had been positively identified; therefore the final cross sections derived from the swarm analysis are effective values in the sense that some of the cross sections may well have been adjusted to their final forms in order to allow for energy losses from processes that had not been observed directly in previous experiments, or to compensate for errors in the ionization cross section. The unusual shape of the cross section with a threshold energy of 14 eV is a case in point.

The ionization cross section of Tate and Smith was used without modification in the analysis. If this cross section were in error, or if another excitation process with a threshold higher than 14 eV were to be identified, a considerable modification of the cross section for the 14-eV process would be necessary, for reasons similar to those given in relation to

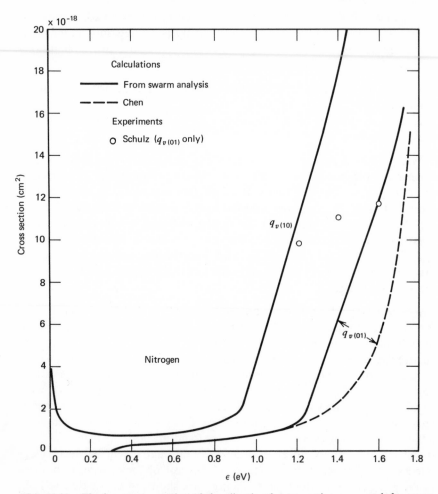

FIG. 13.21.   The low-energy portion of the vibrational cross section $q_{v(01)}$ and the corresponding cross section $q_{v(10)}$ found from the analysis of swarm measurements in N_2. Chen's theoretical curve[64] is identical with the swarm-derived cross section below 1.2 eV. Schulz's results[63] were normalized as described in the text. (From ref. 37.)

the interaction of the rotational and vibrational cross sections in the analysis for hydrogen.

Starting with Schulz's cross section, modifications were made until adequate agreement was obtained with the results for W, D/μ, δ, and α_T. As has been stressed elsewhere, the number of excitation processes and the limited number of transport coefficients for which data are available prevent the final set of cross sections from being unique. The final *total* cross sections that were derived, including the momentum transfer cross

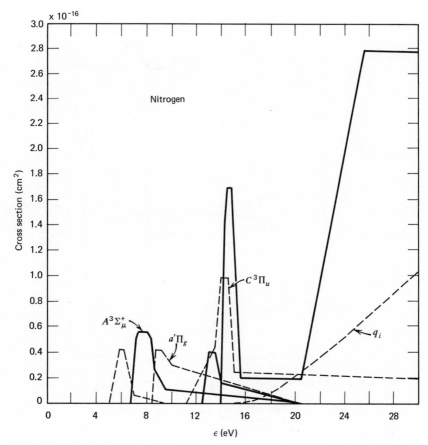

FIG. 13.22.   The final set of excitation cross sections and the ionization cross section q_i used by Engelhardt et al. The full and dashed curves have been used for clarity of presentation. (From ref. 37.)

section, are shown in Figure 13.23. As has been remarked previously, q_m can be obtained without great difficulty throughout the entire range.

Figures 13.24 and 13.25 show the agreement between calculated and experimental transport coefficients that was obtained using the final set of cross sections. Data for the magnetic deflexion coefficient ψ, some typical energy distribution functions, and the results of calculations of the mean electron energy in high-frequency AC fields are also presented in the paper by Engelhardt et al.[37]

Notwithstanding the complexity of the analysis, several important results were obtained from it. We conclude by summarizing them because they

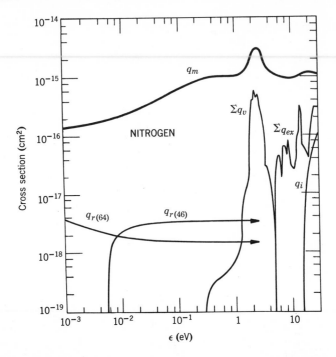

FIG. 13.23. The final momentum transfer cross section q_m, total vibrational cross section Σq_v, total excitation cross section Σq_{ex}, ionization cross section q_i, and typical rotational excitation and de-excitation cross sections $q_{r(46)}$ and $q_{r(64)}$ used by Engelhardt et al.[37] Σq_v represents the total of all vibrational cross sections from $v=1$ to 8; Σq_{ex} represents the total of the cross sections shown in Figure 13.22; q_i is the ionization cross section of J. T. Tate and P. T. Smith (*Phys. Rev.*, **39**, 270, 1932).

are typical of the results of similar analyses carried out by Phelps and his collaborators for a number of gases.

1. The momentum transfer cross section was determined for energies in the range $0.003 < \epsilon < 30$ eV.

2. The Gerjuoy-Stein theory of rotational excitation, when used in conjunction with a quadrupole moment of $1.04 ea_0^2$, was found to predict cross sections that were in good agreement with the swarm results. The only parameter that could be adjusted was the quadrupole moment, the final value of which was in good agreement with experiment.

3. The analysis was used to normalize the set of vibrational cross sections determined by Schulz using a beam technique.

4. The $v=0 \rightarrow 1$ vibrational cross section was shown to have some unexpected structure near threshold.

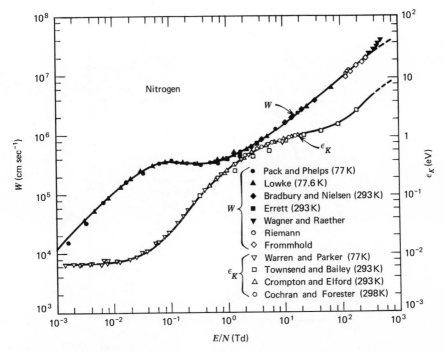

FIG. 13.24. Comparison of calculated values of W and $\epsilon_K = De/\mu$ with experimental values from a number of sources. References to the sources of the data are given in ref. 37, from which the figure is taken.

13.6. BEAM-SWARM TECHNIQUES FOR DETERMINING ATTACHMENT CROSS SECTIONS

The complementary nature of beam and swarm experiments was recognized by Hurst, Christophorou, and others who combined the results of the two techniques to determine absolute attachment cross sections. Most of the recent work of this group of authors has been devoted to the study of electron interactions with organic molecules and therefore lies beyond the scope of this book. For this reason we do not attempt to give a comprehensive coverage of their work or its results, but instead restrict ourselves to a description of the basic principles of the beam-swarm method and its application to two gases. A comprehensive list of references to the work of this group is to be found in the book by Christophorou.[65]

A simplified version of the method was introduced by Hurst, O'Kelly, and Bortner.[66] Using the techniques described in Sections 10.5 and 12.3.2, these authors first measured electron drift velocities in a suitable carrier

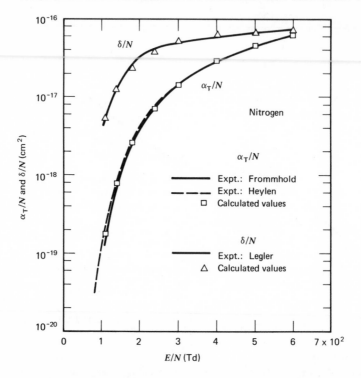

FIG. 13.25. Calculated and experimental results for α_T/N and δ/N in N_2;[37] δ is associated with the $C^3\Pi_u$ state. Calculated values are shown as points, and the experimental results as smooth curves. References to the sources of the experimental data are given in ref. 37.

gas and subsequently determined the attachment coefficient in the same gas containing various concentrations of an attaching gas. In order to be able to refer directly to their published results we use some of the quantities defined by them, even though they differ in some instances from the quantities defined and used elsewhere in this book. We therefore adopt the following definitions:

p	total gas pressure.
$f_1 p$	partial pressure of attaching (sample) gas (in torr).
$f_2 p$	partial pressure of carrier gas (in torr).
N_0	gas number density corresponding to a pressure of 1 torr.
α	attachment coefficient reduced to a sample gas pressure of 1 torr.
α_0	value of α extrapolated to zero concentration of the sample gas, that is, to the situation in which the energy distribution function in the carrier gas is undisturbed by the presence of the sample gas (see below).
$q_a(\epsilon)$	attachment cross section.

According to these definitions of α and f_1, the fraction of the total number of electrons in a swarm captured in drifting a distance of 1 cm parallel to the electric field is

$$\frac{dn}{n} = -\alpha f_1 p, \qquad (13.59)$$

while α is related to the attachment cross section q_a through the equation†

$$\alpha = \frac{N_0(2/m)^{1/2}}{W} \int_0^\infty \epsilon q_a(\epsilon) f(\epsilon)\, d\epsilon. \qquad (13.60)$$

In the original experiments argon was used as the carrier gas in order to investigate the attachment cross section for water vapour. Measurements of α were made for several values of p with various concentrations of water vapour, the ratio f_1/f_2 lying in the range 5×10^{-5} to 6×10^{-4}. The results are shown in Figure 12.17. From the figure it can be seen that α is markedly dependent on the ratio f_1/f_2 but not significantly dependent on p. The fact that α is independent of the number density of the carrier gas implies that the attachment process is dissociative, while the dependence of α on f_1/f_2 at constant p suggests that the energy distribution function is significantly affected by the presence of small concentrations of water vapour. The second conclusion is supported by the observed dependence of the drift velocity on the water vapour concentration.[66] The results shown in the figure also suggest that a linear extrapolation to $f_1/f_2 = 0$ should give values of α that correspond to the situation in which the energy distribution function in the carrier gas is undisturbed by the presence of the sample gas, that is, values of α_0. Only in this situation are the energy distribution functions easily calculable.

If we denote the energy distribution in the pure carrier gas by $f_c(\epsilon)$, α_0 is related to $q_a(\epsilon)$ by equation 13.60 with α replaced by α_0 and $f(\epsilon)$ by $f_c(\epsilon)$. Let us suppose that q_a is a strongly peaked function of energy so that ϵ may be regarded as a constant ($= \epsilon_1$) over the range of ϵ that contributes significantly to the integral Equation 13.60 may then be written as

$$\alpha_0 = \frac{N_0(2/m)^{1/2}}{W} \epsilon_1 f_c(\epsilon_1) \int_0^\infty q_a(\epsilon)\, d\epsilon. \qquad (13.61)$$

It follows that, if the approximations made in deriving equation 13.61 are

---

†We continue to use the distribution function $f(\epsilon)$, which is normalized through the equation $\int_0^\infty \epsilon^{1/2} f(\epsilon)\, d\epsilon = 1$, rather than the distribution function used by Hurst and Christophorou, for which the normalizing relation is $\int_0^\infty f(\epsilon)\, d\epsilon = 1$.[68]

valid, it should be possible to find a value of ϵ_1 such that $\alpha_0 W / \epsilon_1 f_c(\epsilon_1)$ is constant for all values of E/p.

Using values of $f_c(\epsilon)$ for argon calculated by Ritchie and Whitesides,[47] Hurst et al. were able to find a value of $\epsilon_1 = 6.4$ eV which led to remarkably constant values of $\alpha_0 W / \epsilon_1 f_c(\epsilon_1)$. This value and the value 7.7×10^{-18} cm^2 eV for the integrated cross section, which was then obtained from equation 13.61, were in good agreement with the values of the peak energy and integrated cross section published by Buchel'nikova.[67]

In the case of water vapour, argon was an appropriate carrier gas because the dissociative attachment process has a relatively high threshold energy requiring high mean energies in the swarm to obtain appreciable attachment rates. Other carrier gases are more suitable in situations where the attachment process in the sample gas occurs at lower energies. Stockdale and Hurst[68] suggested the use of ethylene to investigate processes with thresholds in the thermal region and nitrogen for the intermediate energy range. For $E/p < 0.2$ V cm^{-1} torr^{-1} the distribution function in ethylene is Maxwellian, corresponding to $T_e = T$. In the case of nitrogen, many of the earlier papers[65] relied on the distribution functions published by Carleton and Megill.[43] However, the use of these distribution functions to calculate the mean energies of the swarms in nitrogen led to discontinuities in the curves of $\alpha_0 W$ plotted against mean energy when a comparison was made of the results obtained with nitrogen and argon as the carrier gases. On the other hand, the use of the distribution functions published by Engelhardt et al.[37] to calculate the mean energies in nitrogen removed the discontinuity,[65] indicating that the technique used to derive them, namely, adjustment of the cross sections to give agreement between calculated and measured transport coefficients, was superior to the method of Carleton and Megill, in which no such adjustment was made.

The usefulness of the method so far described was restricted by the assumption of a sharply peaked attachment cross section. This restriction was removed in the later work of Christophorou et al.,[69] who developed a much more exact method of analysis and used it to combine the unnormalized results for q_a obtained from their own experiments[69] based on the retarding potential difference (RPD) method with the results for $\alpha_0 W$ obtained from experiments similar to those just described. The reader is referred to the original paper for the experimental details.

The raw experimental data from the RPD experiments consist of values of a difference current $I(\epsilon)$ for a series of values of the electron energy; $I(\epsilon)$ is proportional to the cross section at the energy ϵ. In order to transform the curve of $I(\epsilon)$ to a curve representing $q_a(\epsilon)$, it may also be necessary to make a linear transformation of the energy axis to account for the effect of unknown contact potential differences in the electron source.

The cross section $q_a(\epsilon)$ is therefore given by[†]

$$q_a(\epsilon) = KTI(\epsilon), \tag{13.62}$$

where K is the scaling factor and T represents a simple translation of $I(\epsilon)$ along the energy axis. The problem is that of determining K and T.

Let T_j be the jth trial value of T_j, and K_j the corresponding value of K, which remains constant for each trial. On substituting equation 13.62 into equation 13.61 we have, for the ith value of E/p and the jth trial,

$$(\alpha_0 W)_{i_{\text{calc}}} = N_0 \left(\frac{2}{m}\right)^{1/2} K_j \int_0^\infty \epsilon T_j I(\epsilon) f_{ci}(\epsilon) \, d\epsilon.$$

For each trial value of T_j a least-squares fit is made to the experimental data by adjusting K_j so that

$$\sum_i \left[(\alpha_0 W)_{i_{\text{calc}}} - (\alpha_0 W)_{i_{\text{expt}}} \right]^2 = M_j$$

becomes a minimum. In this way the values of K_j and M_j corresponding to each T_j are determined. Finally, the transformation T_j (and the corresponding K_j) is found which makes M_j a minimum.

Figure 13.26 shows experimental values of $\alpha_0 W$ plotted against E/p for *ortho*-chlorotoluene in nitrogen (curve S). Also shown in the figure are the calculated values obtained using $I(\epsilon)$ from the RPD experiment and a set of transformations T_j. From the figure it can be seen that the fit between calculated and experimental values is sensitive to a translation of $I(\epsilon)$ along the energy axis of a few hundredths of an electron volt. The method therefore provides a means of normalizing the cross section not only with respect to magnitude but also with respect to energy. Christophorou et al. gave the name "swarm-normalized beam cross sections" to the cross sections derived in this way.

13.7. CONCLUSION

We conclude this chapter with a few remarks that summarize the very considerable advances that have been made in recent years in the theory and application of swarm experiments.

[†]Christophorou et al. pointed out that the cross section given by equation 13.62 is less sharply peaked than the true cross section because of the energy spread in the electron source. In order to avoid this difficulty $I(\epsilon)$ must be deconvoluted to take account of the energy spread before applying the transformation of equation 13.62.[65]

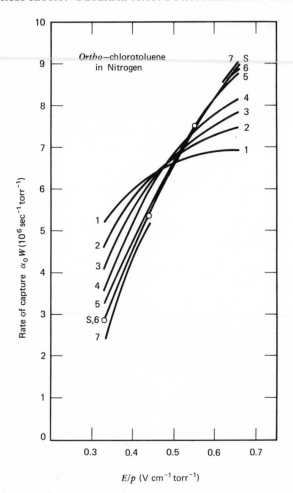

FIG. 13.26. Experimental and calculated rates of capture versus E/p for *ortho*-chlorotoluene in nitrogen for trial functions based on beam experiments. Curve S; swarm experimental rates. Rates calculated from beam data: curve 1, $T_j = 0 \rightarrow 0$ eV; curve 2, $T_j = 0 \rightarrow 0.1$ eV; curve 3, $T_j = 0 \rightarrow 0.2$ eV; curve 4, $T_j = 0 \rightarrow 0.3$ eV; curve 5, $T_j = 0 \rightarrow 0.4$ eV; curve 6 (best fit), $T_j = 0 \rightarrow 0.42$ eV; curve 7, $T_j = 0 \rightarrow 0.46$ eV. (From ref. 69.)

By the end of the 1930's most of the groundwork of the subject had been laid and it might reasonably have been concluded that little else remained to be done. The factors which led to a resurgence of activity were the need for increasingly accurate data for low-energy collision cross sections and an appreciation of the fact that analysis of the results of swarm experiments provides perhaps the only way of obtaining such data. However, these factors in themselves would not have provided the necessary spur, since an overwhelming difficulty which faced the pioneers of the subject was that of interpreting the results of their experiments in terms of the many simultaneously occurring collision processes. As we have seen, this problem has now been solved by obtaining numerical solutions of the transport equations using high-speed computers.

The application of greatly improved analytical methods highlighted the need for experimental data of much greater accuracy, since it became obvious that the uniqueness and accuracy of the derived cross sections were critically dependent on the accuracy of the primary data. Although the technology was available to provide such data, attempts to obtain them soon revealed the inadequacy of some of the assumptions on which the theory of much of the earlier work was based. The full implications of this development have yet to be determined. However, as a result of the progress that has already been made, data for some of the transport coefficients have an accuracy that would have been neither possible nor warranted some years ago.

Notwithstanding the progress that has been achieved in recent years, a limiting factor in some applications of swarm methods is still the lack of experimental data of adequate precision. In others the use of analytical techniques based on insufficiently accurate transport theory inhibits further progress. These problems remain as challenges to both the experimentalist and the theorist.

REFERENCES

1. J. S. Townsend and H. T. Tizard, *Proc. Roy. Soc. A*, **88**, 336, 1913; J. S. Townsend and V. A. Bailey, *Phil. Mag.*, **42**, 873, 1921.

2. A valuable summary of the history of the early work is to be found in a paper by H. L. Brose and E. H. Saayman, *Ann. Phys.*, **5**, 797, 1930.

3. C. Ramsauer, *Ann. Phys.*, **64**, 513, 1921; **66**, 546, 1921; C. Ramsauer and R. Kollath, *Ann. Phys.*, **3**, 536, 1929.

4. R. H. Healey and J. W. Reed, *The Behaviour of Slow Electrons in Gases*, Amalgamated Wireless Ltd., Sydney, 1941.

5. L. B. Loeb, *Basic Processes of Gaseous Electronics*, University of California Press, Berkeley, 1955.

6. W. P. Allis, "Motions of ions and electrons" in *Handbuch der Physik* (Ed., S. Flügge), Vol. 21, Springer-Verlag, Berlin, 1956, p. 413.

7. L. G. H. Huxley and R. W. Crompton, "The motions of slow electrons in gases" in *Atomic and Molecular Processes* (Ed., D. R. Bates), Academic Press, New York, 1962, p. 335.

8. A. V. Phelps, O. T. Fundingsland, and S. C. Brown, *Phys. Rev.*, **84**, 559, 1951.

9. A. L. Gilardini, *Low-Energy Electron Collisions in Gases*, Wiley-Interscience, New York, 1973.

10. G. Bekefi and S. C. Brown, *Phys. Rev.*, **112**, 159, 1958.

11. J. C. Bowe, *Phys. Rev.*, **117**, 1416, 1960.

12. A. E. D. Heylen, *Proc. Phys. Soc.*, **76**, 779, 1960.

13. J. L. Pack and A. V. Phelps, *Phys. Rev.*, **121**, 798, 1961.

14. I. P. Shkarofsky, N. P. Bachynski, and T. W. Johnston, *Planet. Space Sci.*, **6**, 24, 1961.

15. J. L. Pack, R. E. Voshall, and A. V. Phelps, *Phys. Rev.*, **127**, 2084, 1962.

16. L. S. Frost and A. V. Phelps, *Phys. Rev.*, **127**, 1621, 1962.

17. R. W. Crompton and R. L. Jory, *Fourth International Conference on the Physics of Electronic and Atomic Collisions,* Quebec, Science Bookcrafters, Hastings-on-Hudson, New York, 1965, p. 118.

18. R. W. Crompton, *Advan. Electron. Electron Phys.*, **27**, 1, 1969.

19. B. Bederson and L. J. Kieffer, *Rev. Mod. Phys.*, **43**, 601, 1971.

20. L. G. H. Huxley and A. A. Zaazou, *Proc. Roy. Soc. A*, **196**, 402, 1949.

21. R. W. Crompton, M. T. Elford, and R. L. Jory, *Aust. J. Phys.*, **20**, 369, 1967.

22. See, for example, ref. 16; at high values of E/N such a representation may become unsatisfactory, as shown in ref. 37.

23. J. S. Townsend, *Electricity in Gases*, Oxford University Press, London, 1915.

24. R. L. Jory, *Aust. J. Phys.*, **18**, 237, 1965.

25. L. G. H. Huxley, *Aust. J. Phys.*, **13**, 718, 1960.

26. G. Cavalleri and G. Sesta, *Phys. Rev.*, **170**, 286, 1968.

27. A. M. Cravath, *Phys. Rev.*, **36**, 248, 1930.

28. E. W. McDaniel, *Collision Phenomena in Ionized Gases*, John Wiley & Sons, New York, 1964.

29. J. D. Craggs and H. S. W. Massey, "The collisions of electrons with molecules" in *Handbuch der Physik* (Ed., S. Flügge) Vol. 37/1, Springer-Verlag, Berlin, 1959, p. 314.

30. B. I. H. Hall, *Aust. J. Phys.*, **8**, 468, 1955.

31. H. Margenau, *Phys. Rev.*, **69**, 508, 1946.

32. R. W. Crompton and D. J. Sutton, *Proc. Roy. Soc. A*, **215**, 467, 1952.

33. T. J. Lewis, *Proc. Roy. Soc. A*, **244**, 166, 1958.

34. A. E. D. Heylen and T. J. Lewis, *Proceedings of the Fourth International Conference on Ionization in Gases, Uppsala*, Vol. 1, North Holland Publishing Company, Amsterdam, 1960, p. 156.

35. A. E. D. Heylen, *Proc. Phys. Soc.*, **79**, 284, 1962.

36. A. G. Engelhardt and A. V. Phelps, *Phys. Rev.*, **131**, 2115, 1963.

37. A. G. Engelhardt, A. V. Phelps, and C. G. Risk, *Phys. Rev.*, **135**, A1566, 1964.

38. L. S. Frost and A. V. Phelps, *Phys. Rev.*, **136**, A1538, 1964.

39. R. D. Hake and A. V. Phelps, *Phys. Rev.*, **158**, 70, 1967.

40. H. W. Allen, *Phys. Rev.*, **52**, 707, 1937.

41. D. Barbiere, *Phys. Rev.*, **84**, 653, 1951.

42. A. E. D. Heylen and T. J. Lewis, *Proc. Roy. Soc. A*, **271**, 531, 1963.

43. N. P. Carleton and L. R. Megill, *Phys. Rev.*, **126**, 2089, 1962.

44. R. W. Crompton, D. K. Gibson, and A. I. McIntosh, *Aust. J. Phys.*, **22**, 715, 1969.

45. J. Lucas, *Int. J. Electron.*, **27**, 201, 1969.

46. D. K. Gibson, *Aust. J. Phys.*, **23**, 683, 1970.

47. Ritchie and Whitesides developed a similar technique for calculating the energy distribution functions; see ref. 65.

48. C. R. Hoffmann and H. M. Skarsgard, *Phys. Rev.*, **178**, 168, 1969.

49. A. V. Phelps, *Rev. Mod. Phys.*, **40**, 399, 1968.

50. D. E. Golden, N. F. Lane, A. Temkin, and E. Gerjuoy, *Rev. Mod. Phys.*, **43**, 642, 1971.

51. R. W. Crompton, M. T. Elford, and A. G. Robertson, *Aust. J. Phys.*, **23**, 667, 1970.

52. A. G. Robertson, *J. Phys. B: At. Mol. Phys.*, **5**, 648, 1972.

53. J. A. Rees and A. G. Robertson, personal communication, 1972.

54. B. Sherman, *J. Math. Anal. Appl.*, **1**, 342, 1960.

55. A. Farkas, *Orthohydrogen, Parahydrogen and Heavy Hydrogen*, Cambridge University Press, London, 1935.

56. R. W. Crompton and A. I. McIntosh, *Aust. J. Phys.*, **21**, 637, 1968.

57. H. Ehrhardt, L. Langhans, F. Linder, and H. S. Taylor, *Phys. Rev.*, **173**, 222, 1968.

58. R. J. W. Henry and N. F. Lane, *Phys. Rev.*, **183**, 221, 1969.

59. R. W. Crompton, D. K. Gibson, and A. G. Robertson, *Phys. Rev. A*, **2**, 1386, 1970.

60. N. E. Bradbury and R. A. Nielsen, *Phys. Rev.*, **49**, 388, 1936.

61. R. W. Warren and J. H. Parker, *Phys. Rev.*, **128**, 2661, 1962.

62. E. Gerjuoy and S. Stein, *Phys. Rev.*, **97**, 1671, 1955; **98**, 1848, 1955.

63. G. J. Schulz, *Phys. Rev.*, **135**, A988, 1964.

64. J. C. Y. Chen, *J. Chem. Phys.*, **40**, 3507, 1964.

65. L. G. Christophorou, *Atomic and Molecular Radiation Physics*, Wiley-Interscience, New York, 1971.

66. G. S. Hurst, L. B. O'Kelly, and T. E. Bortner, *Phys. Rev.*, **123**, 1715, 1961.

67. I. S. Buchel'nikova, *Soviet Phys.-JETP*, **35**(8), 783, 1959.

68. J. A. Stockdale and G. S. Hurst, *J. Chem. Phys.*, **41**, 255, 1964.

69. L. G. Christophorou, R. N. Compton, G. S. Hurst, and P. W. Reinhardt, *J. Chem. Phys.*, **43**, 4273, 1965.

14

COMPILATION OF

EXPERIMENTAL RESULTS AND

DERIVED CROSS SECTIONS

14.1. INTRODUCTION

In compiling the data for this chapter we have had to restrict the number of gases and the range of E/N for which data are presented. We were also faced with the decision as to whether to present *all* the available data for the gases which were selected or to give an edited summary.

Our choice of gases is perhaps somewhat more arbitrary than our decision as to how to present the data. With one exception, namely, air, we have presented experimental results for the transport coefficients only for gases for which a cross-section analysis of the type described in Chapter 13 is available. An additional reason for our choice is that, since we have been obliged to make a selection, we have presented results for the gases which are primarily of interest to physicists and electrical engineers rather than for those which have been studied for their significance to chemistry or radiation physics. We have attempted to make partial compensation for the omission of a large and important body of material by giving a list of references for other gases at the end of the chapter. The reasons for our choice of the method of presentation are outlined in the following sections.

14.1.1. EXPERIMENTAL DATA FOR TRANSPORT COEFFICIENTS. In presenting the data for the transport coefficients we have avoided the frequently adopted practice of tabulating or graphing all the results available from a number of laboratories, a practice that requires the reader to decide for himself the curve of best fit for these data. Although readers who have a particular reason for doubting the validity of the results presented here will wish to plot the data from the original sources and make their own assessment, we believe that the majority will prefer a method of presentation that gives more ready access to the data. Therefore we have taken the more onerous course of examining the data in detail in order to determine curves which represent what we believe to be the best available data at the time of publication. These curves have then been used to present the data in a form which we hope will be of most value to the reader.

Two alternative methods of presentation have been used. When we have had access to tabulated material, either from papers published in journals or from laboratory reports, we have used these data in a computer interpolation routine to produce tables in which the transport coefficients are listed at values of E/N that are separated by intervals of between 15 and 20%. In the majority of cases the interval is approximately the same as that used in the original investigation. By following this procedure we have been able to produce sets of data in which the transport coefficients are presented at E/N values chosen from a standard set, regardless of the gas or the temperature at which the original measurements were made. This procedure facilitates not only comparison between the results in a particular gas at different temperatures but also comparisons between the results for different gases. Furthermore the results are presented at sufficiently close intervals and with sufficient accuracy that our original curves can be reconstructed by those who need to do so.

In some instances the data were available to us only in graphical form, or are such that they can be represented graphically with sufficient accuracy. In these instances we have presented graphs plotted with sufficient accuracy and reproduced on an adequately large scale that the data can be obtained from them with an accuracy that matches that of the original results. In some cases the graphical representation extends the range of E/N beyond that in which it was thought justified to present the data in tables. For quick reference and comparison the data for W and D/μ in the tables have also been presented in graphical form.

In general, the experimental results have been given only within a range of values of E/N that provides data for cross-section analyses which have made a significant or unique contribution to the knowledge of electron scattering. As has been stressed elsewhere, swarm experiments are particularly valuable in providing data related to low-energy scattering processes, that is, those in which the electron energies are less than a few electron volts, but are less valuable, in general, in situations in which the multiplicity of collision processes makes their interpretation unduly complex. In any case, the cross sections at higher energies which might result from analyses of data at high E/N are generally more reliably measured using single-collision techniques. Furthermore the accuracy of most of the results at high E/N does not compare with that of the results taken at low values of E/N, either because extreme experimental conditions make accurate measurement difficult or because of the problem of relating the measured quantities to the relevant transport coefficients. The interpretation of the results of the Townsend-Huxley experiment under conditions in which primary and secondary ionization are significant is a case in point (see Chapter 11). Nevertheless the data at high E/N are of considerable importance, particularly in relation to electrical breakdown phenomena, and seem certain to continue to attract the attention of experimentalists. Some of the data, particularly the results for the ionization coefficient α_T/N [or the net coefficient $(\alpha_T - \alpha_{at})/N$], have been used in the cross-section analyses from which the results quoted in this chapter were taken. The reader is referred particularly to the papers from the groups at the University of Hamburg,[1] the University of New England,[2] the University of Liverpool,[3] and the University College of Swansea[4] for data in this regime.

Finally, in relation to the transport data, we comment on the paucity of the information provided in this chapter on attachment coefficients. This is due partly to our selection of gases for detailed treatment and partly to our scepticism concerning the value of many of the existing data for this important transport coefficient. In our view the results of Grünberg's work in oxygen (see Table 14.16) are among the few data for electron attachment that will stand for some years, and we have therefore included his results *in toto*. The references given in this chapter contain much of the available data on low-energy electron attachment, and the reader is asked to make the assessment which we ourselves have not attempted.

14.1.2. CROSS SECTIONS DERIVED FROM EXPERIMENTAL SWARM DATA. With one exception, namely, water vapour, the cross sections presented here were obtained using the technique based on a numerical solution of the Boltzmann equation, which was described in Chapter 13 (Section 13.5). In the case of water vapour, for which only the momentum transfer cross section is given, the cross section was derived from the temperature dependence of the zero-field mobility using the technique described in Section 13.4. In this instance the more complex analysis is unnecessary. The energy range over which the cross section has been determined is, of course, restricted because of the small variation in mean electron energy that can be obtained by varying the gas temperature alone.

As was the case with the data for the transport coefficients, we have not included the results of other determinations of the cross sections, either experimental or theoretical, but have given only the cross sections which seem to us at the present time to be the best available *from analyses of the results of swarm experiments.* Comparisons between the data presented here and the results of other investigations are dealt with fully in the papers from which these data were taken.

In some cases, the analysis was based on one or more cross sections taken directly from the results of single-collision experiments, for example, the ionization cross sections. The set of cross sections is then the set which has been found to be compatible with the swarm data. It seems likely that, with few exceptions, more definitive results for inelastic cross sections with thresholds less than a few electron volts can be obtained as more accurate results from swarm experiments and low-energy single-collision experiments become available.

In a few instances it has proved possible to determine the cross sections from an analysis of the experimental swarm data alone. More often, however, the analysis has been based in part on theoretical calculations of the energy dependence of some of the cross sections, using one or more of the physical properties of the molecule (e.g., the quadrupole moment) as adjustable parameters (see Section 13.5.2). As already remarked, other analyses have relied on a combination of this information and experimental data for one or more cross sections derived from single-collision experiments.

It is not our intention to describe the procedure that has been used in each case. The reader is referred to Chapter 13 for a general discussion of the analytical techniques and some illustrative examples, and to the original papers for a detailed description of the procedure adopted for each gas. However, at this point we summarize the formulae that have been derived theoretically for the rotational cross sections since the very large numbers of cross sections that must be included in most analyses prevent them from being tabulated or graphed. Data for the cross sections are therefore sometimes given in terms of one or another of these formulae and the constants that are to be inserted into them.

Formulae of Gerjouy and Stein. These authors[5] derived the following formulae for the cross sections for rotational excitation, based on the assumption that the interaction was dominated by the long-range quadrupole forces:

$$q_{r(J,J+2)}(\epsilon) = \frac{(J+2)(J+1)}{(2J+3)(2J+1)} \sigma_0 \left[1 - \frac{(4J+6)B_0}{\epsilon} \right]^{1/2}, \quad (14.1)$$

$$q_{r(J,J-2)}(\epsilon) = \frac{J(J-1)}{(2J-1)(2J+1)} \sigma_0 \left[1 + \frac{(4J-2)B_0}{\epsilon} \right]^{1/2}, \quad (14.2)$$

where $\sigma_0 = 8\pi Q^2 a_0^2/15$; Q is the electric quadrupole moment in units of ea_0^2, where a_0 is the Bohr radius; and B_0 is the rotational constant. For the rotational transition $J \to J+2$, the energy loss is $\epsilon_J = (4J+6)B_0$, whereas the energy gain in the superelastic transition $J \to J-2$ is $\epsilon_{-J} = (4J-2)B_0$.

In applying these formulae to a particular gas, weighting factors must be used to allow for the different populations of the rotational states. Thus the collision frequency for the rotational transition $J \to J+2$ is

$$\nu_{r(J,J+2)} = N\left[\frac{p_J}{P_r}\exp\left(-\frac{E_J}{\kappa T}\right)\right]q_{r(J,J+2)}(c)c.$$

The factor in the square brackets represents the fraction of the molecules in the Jth rotational state. In this formula E_J is the energy level of the Jth rotational level, given by

$$E_J = J(J+1)B_0;$$

p_J is a factor which allows for the nuclear spin and is given by

$$p_J = (2t+1)(t+a)(2J+1),$$

where t is the nuclear spin, $a=0$ when J is even and $=1$ when J is odd, and $P_r = \sum_J p_J \exp(-E_J/\kappa T)$.

Phelps and his collaborators[6-9] applied these formulae, using the quadrupole moment Q as the only adjustable parameter.

Formulae of Takayanagi. Formulae analogous to equations 14.1 and 14.2 were derived by Takayanagi[10] for the case in which the molecule possesses a permanent dipole moment. These formulae are as follows:

$$q_{r(J,J+1)}(\epsilon) = \frac{(J+1)R_y\sigma_0}{(2J+1)\epsilon}\ln\left[\frac{\epsilon^{1/2}+(\epsilon-\epsilon_J)^{1/2}}{\epsilon^{1/2}-(\epsilon-\epsilon_J)^{1/2}}\right], \qquad (14.3)$$

$$q_{r(J,J-1)}(\epsilon) = \frac{JR_y\sigma_0}{(2J+1)\epsilon}\ln\left[\frac{(\epsilon+\epsilon_{-J})^{1/2}+\epsilon^{1/2}}{(\epsilon+\epsilon_{-J})^{1/2}-\epsilon^{1/2}}\right], \qquad (14.4)$$

where $\sigma_0 = 8\pi\mu^2 a_0/3$, μ is the electric dipole moment (in units of ea_0), R_y is the rydberg (13.6 eV), $\epsilon_J = 2(J+1)B_0$, and $\epsilon_{-J} = 2JB_0$. As before, the cross sections must be multiplied by the weighting factor to allow for the relative populations of the rotational states.

14.2. FORMAT

The data presented in the following sections are grouped in separate sections for each gas, commencing with tabulated data for the transport coefficients, where available, and followed immediately by graphical presentations of the same material or of data for which no tables are given. A similar order is then followed for the cross-section data. When the rotational cross sections used in the analysis are based on theory, the relevant information precedes the graphical presentation of the cross sections.

14.2.1. STATEMENTS OF ACCURACY.

When the data are presented in numerical form, they are, in general, accompanied by a statement of error limits. With few exceptions (which are noted) the limits quoted for the measured transport coefficients are believed to be realistic. In some cases, particularly the measured drift velocities at intermediate values of E/N, we consider them to be rather conservative; in other instances we have widened the limits to allow for the possible effects of systematic errors which, at the present time, cannot be estimated with great accuracy. A case in point is the allowance for the influence of primary ionization in the determination of D/μ by the Townsend-Huxley method. In a very few instances (e.g., the data for oxygen at low E/N) we do not have great confidence in the error bounds ascribed, but included the values on the grounds that an informed guess may be better than no data at all.

The assignment of error limits for the cross sections is considerably more difficult and has been attempted in only a few instances. While we believe that the error limits for the momentum transfer cross sections in the monatomic gases are well founded, the limits for inelastic cross sections are determined more by experience than as the result of an infallible set of tests. The reader is referred to the original papers for background information to support the claims that are made and for statements of the probable accuracy of the cross sections for which no limits have been stated. The review in Chapter 13 gives a general background against which any assessment of uniqueness and accuracy can be made.

14.2.2. REFERENCING.

In this chapter we have departed from the previously adopted practice of listing all the references at the end of the chapter in favour of grouping most of the references with the tables and figures to which they refer. References for the tables are numbered, each number referring to the entry in the table against which it appears in parentheses and to all following entries until superseded. Although this procedure has led to some duplication, our hope is that it will be more convenient for the reader. Section 14.3.12 contains a list of references to gases which are not dealt with in detail, and some general references are included at the end of the chapter. The reference numbers in each table refer only to the references listed below the table in question.

14.2.3.   COMMENTS ON THE VALUES OF ND. Apart from the thermal values of the diffusion coefficient and the data for helium, all the data for ND presented in the tables were calculated by combining the measured values of W and D/μ, as no other data were available. It should be noted that, without exception, all values that were obtained by direct measurement apply to temperatures other than 293 K. Appropriate adjustments are required, therefore, when making comparisons between the data in helium at low values of E/N, or between the directly measured thermal values and the extrapolated values determined from the calculated values of ND_{293}.

Two comments are called for with regard to the error limits stated for these data. We have some reservation about the accuracy claimed for Cavalleri's data for helium (see Table 14.1) and for some of the data from the thermal diffusion experiment of Nelson and Davis. Using the same standard of assessment as has been applied elsewhere, we have concluded that an error limit of $\pm 2\%$ for Cavalleri's results is more justified than the quoted figure of $\pm 0.5\%$. Accordingly, with the author's agreement, we have stated the former figure for these data. We have less feel for the accuracy of the thermal diffusion experiments of Nelson and Davis, since we have had no direct experience with their technique. Our reservation stems from a general assessment of the method and the difficulty in assigning error limits to some of the experimental parameters, as well as from some unexplained randomness in the data (e.g., those for neon). Nevertheless, we do not feel able to amend the error limits given by the authors, which may, in the long run, turn out to be correct.

14.3. TABULATED AND GRAPHICAL DATA

HELIUM	600 – 604
NEON	604 – 607
ARGON	607 – 611
KRYPTON, XENON	612 – 614
HYDROGEN, DEUTERIUM	615 – 626
NITROGEN	627 – 631
OXYGEN	632 – 638
AIR	638 – 640
CARBON DIOXIDE	640 – 645
CARBON MONOXIDE	645 – 647
WATER VAPOR	648 – 651
OTHER GASES	651 – 653

14.3.1 HELIUM.

TABLE 14.1 Experimentally determined transport coefficients for electrons in helium

	77 K	293 K				
E/N (Td)	W (10^5 cm sec^{-1})	W (10^5 cm sec^{-1})	D/μ (10^{-2} V)	ND (10^{21} cm^{-1} sec^{-1})	W_M (10^5 cm sec^{-1})	D_L/μ (10^{-2} V)
0.0				$6.35^{(3)a}$	$6.41^{(4)b}$	
6.0×10^{-3}					6.42	
7.0					6.43	
8.0	$0.347^{(1)}$				6.44	
1.0×10^{-2}	0.412				6.46	
1.2	0.470				6.49	
1.4	0.523				6.55	
1.7	0.596				6.63	
2.0	0.661	$0.453^{(2)}$			6.69	$0.517^{(2)}$
2.5	0.758	0.546	$3.16^{(2)}$	6.90	6.81	0.616	$2.68^{(5)}$
3.0	0.845	0.631	3.36	7.07	7.03	0.709	2.71
3.5	0.923	0.711	3.56	7.24	7.22	0.789	2.84
4.0	0.994	0.786	3.77	7.41	7.40	0.866	2.90
5.0	1.124	0.922	4.20	7.75	7.78	1.009	2.97
6.0	1.239	1.044	4.65	8.09	8.09	1.141	3.22
7.0	1.343	1.156	5.09	8.40	8.38	1.264	3.47
8.0	1.440	1.259	5.53	8.70	8.65	1.374	3.66
1.0×10^{-1}	1.612	1.446	6.45	9.33	9.22	1.567	3.90
1.2	1.767	1.609	7.35	9.85	9.76	1.758	4.43
1.4	1.908	1.758	8.22	10.32	10.17	1.899	4.80
1.7	2.10	1.961	9.58	11.05	10.85	2.11	5.39
2.0	2.27	2.14	10.88	11.66	11.50	2.30	6.08
2.5	2.53	2.41	13.03	12.57	12.50	2.57	6.99
3.0	2.76	2.65	15.21	13.45	13.30	2.83	7.95
3.5	2.97	2.87	17.33	14.23	14.10	3.05	9.45
4.0	3.17	3.08	19.42	14.93	14.90	3.27	10.72
5.0	3.53	3.44	23.6	16.20	16.20	3.66	11.87
6.0	3.85	3.77	27.6	17.35	17.30	3.98	13.59

TABLE 14.1 (*Continued*)

E/N (Td)	77 K W (10^5 cm sec^{-1})	293 K W (10^5 cm sec^{-1})	D/μ (10^{-2} V)	ND (10^{21} cm^{-1} sec^{-1})		W_M (10^5 cm sec^{-1})	D_L/μ (10^{-2} V)
7.0×10^{-1}	4.14	4.07	31.6	18.39	18.34	4.28	
8.0	4.41	4.35	35.6	19.35	19.30	4.57	
1.0×10^0	4.91	4.85	43.4	21.1	20.9	5.09	
1.2	5.36	5.30	51.4	22.7	22.3	5.58	
1.4	5.78	5.73	59.4	24.3	23.7	6.01	
1.7	6.36	6.33	70.9	26.4	25.7	6.58	
2.0		6.86	82.7	28.4	27.7	7.11	
2.5		7.69	102.5	31.5	30.8	8.03	
3.0		8.51	122.7	34.8	33.7	8.82	
3.5		9.28			36.6		
4.0					39.5		
5.0					45.5		
6.0					52.1		
7.0					59.3		
8.0					66.7		

[a]Value measured at 295 K.
[b]Values measured at 300 K.

Sources: 1. R. W. Crompton, M. T. Elford, and A. G. Robertson, *Aust. J. Phys.*, **23**, 667, 1970.

2. R. W. Crompton, M. T. Elford, and R. L. Jory, *Aust. J. Phys.*, **20**, 369, 1967.

3. R W Crompton, D. K. Gibson, and G. Cavalleri, *J. Phys. B: At. Mol. Phys.*, **6**, *1118, 1973*.

4. G. Cavalleri, *Phys. Rev.*, **179**, 186, 1969.

5. M. T. Elford, (personal communication) 1973.

Error limits: W_{77}: ±1%.
W_{293}: ±1%.
D/μ_{293}: ±1%.
ND_{295}: ±2%.
ND_{300}: ±2%.
$W_{M_{293}}$: ±2—±3%.
D_L/μ_{293}: ±3%.

He

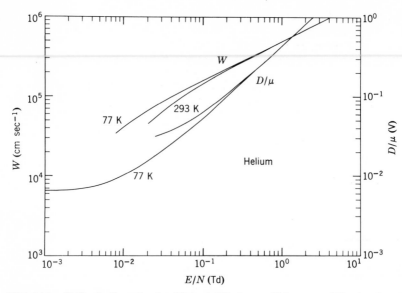

FIG. 14.1. Drift velocity and ratio of lateral diffusion coefficient to mobility for electrons in helium. (From data given in Table 14.1.)

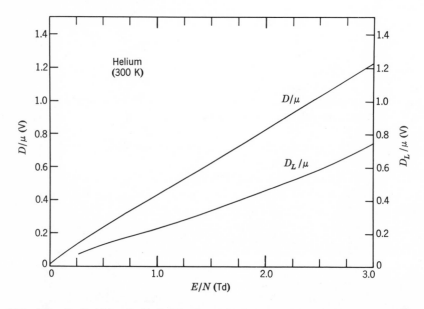

FIG. 14.2. Ratio of longitudinal diffusion coefficient to mobility for electrons in helium with the ratio D/μ plotted for comparison. (From E. B. Wagner, F. J. Davis, and G. S. Hurst, *J. Chem. Phys.*, **47**, 3138, 1967.)

TABLE 14.2. Momentum transfer cross section in helium (from R. W. Crompton, M. T. Elford, and A. G. Robertson, *Aust. J. Phys.*, **23**, 667, 1970)

ϵ (eV)	$q_{m_{\text{el}}}$ (10^{-16}cm^2)	ϵ (eV)	$q_{m_{\text{el}}}$ (10^{-16}cm^2)
0.008	5.18*	0.25	6.27
0.009	5.19*	0.30	6.35
0.010	5.21	0.40	6.49
0.013	5.26	0.50	6.59
0.017	5.31	0.60	6.66
0.020	5.35	0.70	6.73
0.025	5.41	0.80	6.77
0.030	5.46	0.90	6.82
0.040	5.54	1.0	6.85
0.050	5.62	1.2	6.91
0.060	5.68	1.5	6.96
0.070	5.74	1.8	6.98
0.080	5.79	2.0	6.99
0.090	5.83	2.5	6.96
0.10	5.86	3.0	6.89
0.12	5.94	4.0	6.60*
0.15	6.04	5.0	6.26*
0.18	6.12	6.0	6.01*
0.20	6.16		

Estimated error limit $\pm 2\%$ for $0.01 \leqslant \epsilon$ (eV) $\leqslant 3.0$; values with asterisks have estimated error limit of $\pm 5\%$.

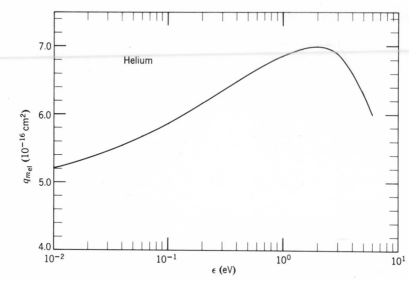

FIG. 14.3. The momentum transfer cross section for electrons in helium for $\epsilon \leqslant 6$ eV. (From data given in Table 14.2.)

14.3.2. NEON.

TABLE 14.3. Experimentally determined transport coefficients for electrons in neon

E/N (Td)	77 K W (10^5 cm sec^{-1})	293 K W (10^5 cm sec^{-1})
1.7×10^{-3}	0.437[1]	
2.0	0.470	
2.5	0.518	
3.0	0.558	
3.5	0.594	
4.0	0.626	
5.0	0.683	
6.0	0.733	
7.0	0.778	
8.0	0.818	
1.0×10^{-2}	0.891	
1.2	0.955	
1.4	1.012	

TABLE 14.3 (*Continued*) **Ne**

E/N (Td)	77 K W (10^5 cm sec^{-1})	293 K W (10^5 cm sec^{-1})
1.7×10^{-2}	1.088	1.022[1]
2.0	1.156	1.094
2.5	1.256	1.199
3.0	1.347	1.293
3.5	1.428	1.383
4.0	1.502	1.463
5.0	1.637	1.590
6.0	1.757	1.715
7.0	1.867	1.826
8.0	1.968	1.933
1.0×10^{-1}	2.15	2.12
1.2	2.32	2.29
1.4	2.48	2.45
1.7	2.69	2.66
2.0	2.89	2.86
2.5	3.19	3.16
3.0	3.46	3.43
3.5	3.70	3.67
4.0	3.93	3.90
5.0	4.34	4.31
6.0	4.69	4.67
7.0		5.00
8.0		5.28
1.0×10^{0}		5.80
1.2		6.25
1.4		6.68
1.7		7.44
2.0		8.48

Thermal diffusion coefficient: ND_{300} 64.8×10^{21} cm^{-1} sec^{-1}[2]

ND_{293} 71.6×10^{21} cm^{-1} sec^{-1}[3]a

[a]Temperature unspecified; taken to be 293 K.

Sources: 1. A. G. Robertson, *J. Phys. B: At. Mol. Phys.*, **5**, 648, 1972.

2. D. R. Nelson and F. J. Davis, *J. Chem. Phys.*, **51**, 2322, 1969.

3. G. Cavalleri, E. Gatti, and A. M. Interlenghi, *Nuovo Cimento*, **40B**, 450, 1965.

Error limits: W_{77}: $\pm 1\%$.

W_{293}: $\pm 1\%$.

ND_{293}: $\pm 4\%$.

ND_{300}: $\pm 4\%$.

Ne

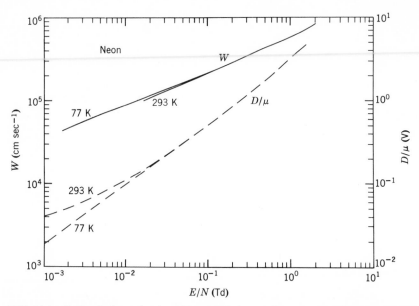

FIG. 14.4.   Drift velocity and ratio of lateral diffusion coefficient to mobility for electrons in neon. (Drift velocity from data given in Table 14.3. No modern experimental data for the ratio D/μ are available; the curve (shown as a dashed line) was calculated from the cross section given in Table 14.4.)

TABLE 14.4.   Momentum transfer cross section for electrons in neon (from A. G. Robertson, *J. Phys. B*: *At. Mol. Phys.*, **5**, 648, 1972).

ϵ (eV)	$q_{m_{el}}$ $(10^{-16}\mathrm{cm}^2)$	ϵ (eV)	$q_{m_{el}}$ $(10^{-16}\mathrm{cm}^2)$
0.03	0.469	0.60	1.402
0.04	0.504	0.70	1.472
0.05	0.536	0.80	1.528
0.06	0.566	0.90	1.580
0.07	0.601	1.00	1.619
0.08	0.636	1.20	1.685
0.09	0.669	1.50	1.753
0.10	0.701	1.80	1.793
0.12	0.754	2.00	1.815
0.15	0.828	2.50	1.860
0.18	0.893	3.00	1.906
0.20	0.930	4.00	1.984
0.25	1.018	5.00	2.070
0.30	1.091	6.00	2.144
0.40	1.225	7.00	2.213
0.50	1.321		

Estimated error limit $\pm 3\%$ for $0.04 \leqslant \epsilon$ (eV) $\leqslant 6.0$.

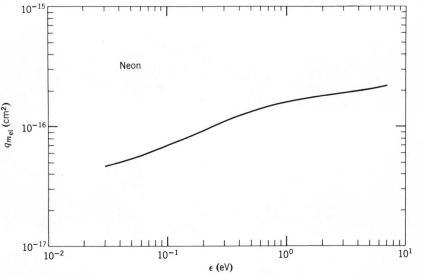

FIG. 14.5. The momentum transfer cross section for electrons in neon for $\epsilon \leqslant 7$ eV. (From data given in Table 14.4.)

14.3.3. ARGON.

TABLE 14.5. Experimentally determined transport coefficients for electrons in argon

E/N (Td)	90 K W (10^5 cm sec^{-1})	293 K W (10^5 cm sec^{-1})
2.0×10^{-3}	0.149[1]	
2.5	0.204	
3.0	0.286	
3.5	0.401	
4.0	0.520	
5.0	0.710	
6.0	0.810	
7.0	0.866	
8.0	0.906	
1.0×10^{-2}	0.961	0.935[1]
1.2	0.994	0.972
1.4	1.031	1.005

TABLE 14.5 (*Continued*)

E/N (Td)	90 K W (10^5 cm sec^{-1})	293 K W (10^5 cm sec^{-1})[t]
1.7×10^{-2}	1.068	1.047
2.0	1.107	1.084
2.5	1.162	1.144
3.0	1.215	1.205
3.5	1.262	1.252
4.0	1.305	1.294
5.0	1.382	1.368
6.0	1.450	1.437
7.0	1.512	1.500
8.0	1.569	1.556
1.0×10^{-1}	1.668	1.654
1.2	1.755	1.741
1.4	1.830	1.820
1.7	1.929	1.918
2.0	2.02	2.00
2.5	2.14	2.13
3.0	2.24	2.23
3.5	2.33	2.31
4.0	2.41	2.39
5.0	2.54	2.52
6.0	2.65	2.63
7.0	2.74	2.73
8.0		2.81
1.0×10^{0}		2.95

Thermal diffusion coefficient ND_{300} 2.85×10^{21} cm^{-1} sec$^{-1(2)}$

Sources: 1. A. G. Robertson, (personal communication) 1972.

2. D. R. Nelson and F. J. Davis, *J. Chem. Phys.*, **51**, 2322, 1969.

Error limits: W_{90}: E/N (Td)$\leqslant 5 \times 10^{-3}$, $\pm 4\%$; $5 \times 10^{-3} \leqslant E/N$ (Td)$\leqslant 10^{-2}$, $\pm 1\% - \pm 4\%$; $10^{-2} \leqslant E/N$ (Td)$\leqslant 10^{-1}$, $\pm 1\% - \pm 2\%$; E/N (Td)$> 10^{-1}$, $\pm 2\%$.

W_{293}: $0.05 \leqslant E/N$ (Td)$\leqslant 1.0$, $\pm 2\%$; E/N (Td)< 0.05, $\pm 3\%$.

ND_{300}: $\pm 1\%$.

Ar

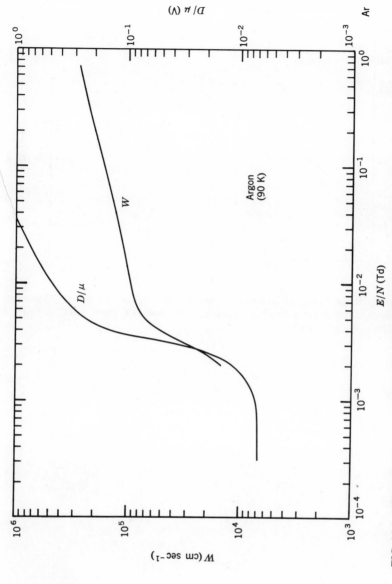

FIG. 14.6. Drift velocity and ratio of lateral diffusion coefficient to mobility for electrons in argon. (From data given in Table 14.5 and R. W. Warren and J. H. Parker, *Phys. Rev.*, **128**, 2661, 1962.)

609

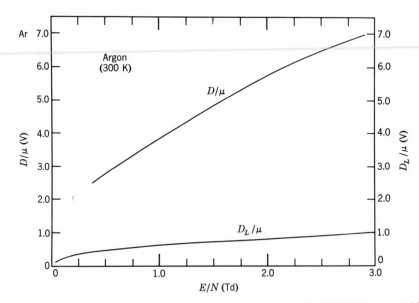

FIG. 14.7.   Ratio of longitudinal diffusion coefficient to mobility for electrons in argon with the ratio D/μ plotted for comparison. (From E. B. Wagner, F. J. Davis, and G. S. Hurst, *J. Chem. Phys.*, **47**, 3138, 1967.)

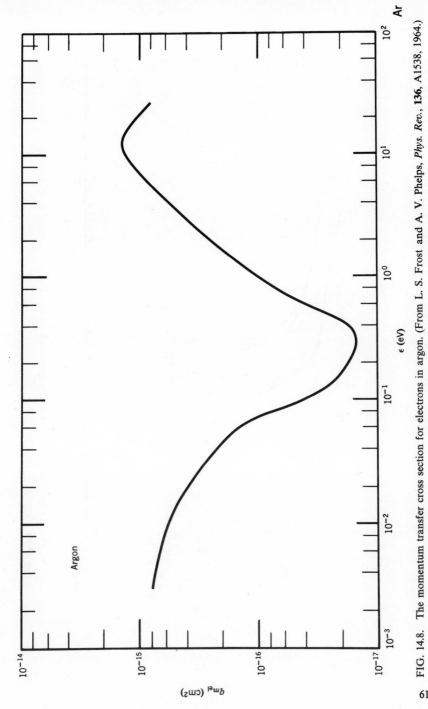

Ar

Argon

ϵ (eV)

$q_{m,el}$ (cm²)

FIG. 14.8. The momentum transfer cross section for electrons in argon. (From L. S. Frost and A. V. Phelps, *Phys. Rev.*, **136**, A1538, 1964.)

14.3.4.   KRYPTON AND XENON.

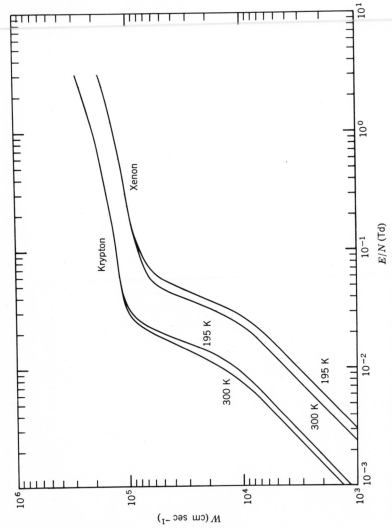

FIG. 14.9.   Electron drift velocities in krypton and xenon. (From J. L. Pack, R. E. Voshall, and A. V. Phelps, *Phys. Rev.*, **127**, 2084, 1962.)

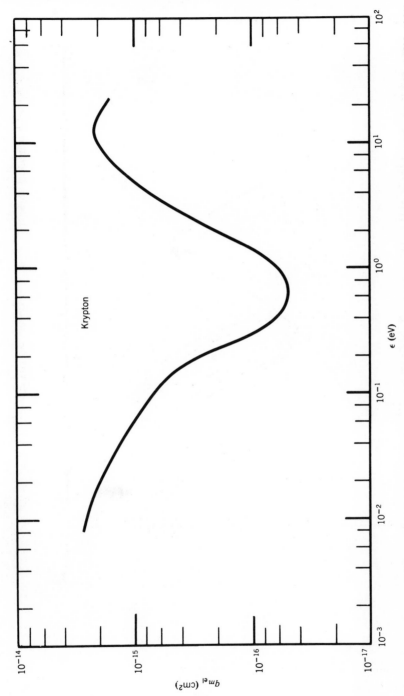

Kr

Krypton

ϵ (eV)

$q_{m\,el}$ (cm²)

FIG. 14.10. The momentum transfer cross section for electrons in krypton. (From L. S. Frost and A. V. Phelps, *Phys. Rev.*, **136**, A1538, 1964.)

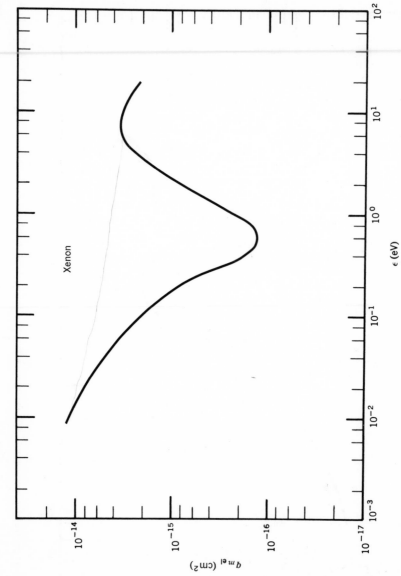

FIG. 14.11. The momentum transfer cross section for electrons in xenon. (From L. S. Frost and A. V. Phelps, *Phys. Rev.*, **136**, A1538, 1964.)

TABLE 14.6. Experimentally determined transport coefficients for electrons in normal hydrogen

E/N (Td)	77 K			293 K			
	W (10^5 cm sec^{-1})	D/μ (10^{-2} V)	ND (10^{21} cm^{-1} sec^{-1})	W (10^5 cm sec^{-1})	D/μ (10^{-2} V)	ND (10^{21} cm^{-1} sec^{-1})	W_M (10^5 cm sec^{-1})
0.0						3.96[7]	
2.0×10^{-3}		0.676[2]					
2.5		0.681					
3.0		0.685					
3.5		0.690					
4.0		0.696					
5.0		0.709					
6.0		0.723					
7.0		0.738					
8.0	0.274[1]	0.755	2.59				
1.0×10^{-2}	0.332	0.790	2.62				
1.2	0.385	0.827	2.65	0.1862[3]			
1.4	0.435	0.867	2.69	0.217			
1.7	0.504	0.926	2.74	0.265			
2.0	0.567	0.986	2.80	0.311	2.59[2]	4.03	
2.5	0.666	1.084	2.89	0.385	2.62	4.04	
3.0	0.758	1.179	2.98	0.459	2.65	4.05	
3.5	0.842	1.268	3.05	0.530	2.68	4.06	
4.0	0.922	1.353	3.12	0.600	2.71	4.07	
5.0	1.071	1.508	3.23	0.737	2.78	4.09	
6.0	1.212	1.649	3.33	0.870	2.85	4.13	
7.0	1.347	1.777	3.42	0.998	2.92	4.17	1.208[8]
8.0	1.478	1.893	3.50	1.124	3.00	4.21	1.346
1.0×10^{-1}	1.725	2.10	3.62	1.369	3.15	4.31	1.604
1.2	1.958	2.29	3.74	1.578	3.29	4.33	1.853
1.4	2.18	2.47	3.84	1.777	3.44	4.37	2.09
1.7	2.49	2.70	3.96	2.07	3.66	4.46	2.41
2.0	2.78	2.94	4.08	2.35	3.87	4.54	2.71
2.5	3.22	3.29	4.23	2.76	4.23	4.67	3.18
3.0	3.61	3.64	4.38	3.13	4.57	4.77	3.61
3.5	3.97	3.98	4.51	3.47	4.90	4.86	3.96
4.0	4.29	4.33	4.64	3.79	5.24	4.96	4.30
5.0	4.86	5.02	4.88	4.33	5.93	5.13	5.03
6.0	5.33	5.72	5.08	4.82	6.63	5.33	5.56
7.0	5.75	6.42	5.27	5.24	7.35	5.50	6.03
8.0	6.11	7.13	5.45	5.61	8.09	5.67	6.47

TABLE 14.6 (*Continued*)

E/N (Td)	77 K W (10^5 cm sec^{-1})	D/μ (10^{-2} V)	ND (10^{21} cm^{-1} sec^{-1})	293 K W (10^5 cm sec^{-1})	D/μ (10^{-2} V)	ND (10^{21} cm^{-1} sec^{-1})	W_M (10^5 cm sec^{-1})
1.0×10^0	6.71	8.60	5.77	6.23[1]	9.53	5.94	7.22
1.2	7.15	10.08	6.01	6.62	11.02	6.08	7.84
1.4	7.50	11.57	6.20	6.92	12.48	6.17	8.35
1.7	7.96	13.80	6.46	7.64	14.65	6.58	9.06
2.0	8.70	15.95	6.94	8.37	16.76	7.01	9.69
2.5	9.36	19.29	7.22	9.13	20.1	7.33	10.63
3.0	9.97	22.4	7.45	9.82	23.1	7.56	11.44
3.5	10.96	25.2	7.89	10.79	25.9	7.98	12.27
4.0	11.60	27.8	8.06	11.47	28.5	8.16	13.04
5.0	12.93	32.5	8.40	12.86	33.0	8.49	14.43
6.0	14.20	36.6	8.66	14.15	37.1	8.76	15.90
7.0	15.42	40.5	8.92	15.35	41.0[5]	8.99	17.40
8.0	16.60	44.0	9.13	16.50	44.7	9.21	18.56
1.0×10^1	18.80	50.6	9.51	18.70	51.1	9.56	20.7
1.2	20.9	56.5	9.84	20.7	57.3	9.88	22.9
1.4	22.8			22.7	63.0	10.21	24.8
1.7	25.5			25.5	71.0	10.65	27.5
2.0	28.1			28.1	78.7	11.06	30.2
2.5	32.2			32.2	91.6	11.79	34.3
3.0				36.6[3]	105.1	12.81	
3.5				40.8	117.9[6]	13.73	
4.0				45.4	133.3	15.12	
5.0				57.1	172.2	19.67	
6.0				69.5[4]	198.2	23.0	
7.0				83.0	219	26.0	
8.0				98.0	237	29.1	
1.0×10^2				128.3	267	34.2	
1.2				165.9	293	40.5	
1.4				194.3	319	44.3	
1.5				213			
1.7					349		
2.0					382		
2.12					396		

Sources: 1. A. G. Robertson, *Aust. J. Phys.*, **24**, 445, 1971.
2. R. W. Crompton, M. T. Elford, and A. I. McIntosh, *Aust. J. Phys.*, **21**, 43, 1968.

TABLE 14.6 (*Continued*)

3. J. J. Lowke, *Aust. J. Phys.*, **16**, 115, 1963.
4. H. A. Blevin and N. Z. Hasan, *Aust. J. Phys.*, **20**, 735, 1967.
5. Electron and Ion Diffusion Unit, Australian National University, Quart. Rept.23, 1966.
6. R. W. Crompton, B. S. Liley, A. I. McIntosh, and C. A. Hurst, *Proceedings of the Seventh International Conference on Phenomena in Ionized Gases, Beograd, 1965*, Vol. 1, Gradevinska, Knjiga Publishing House, Beograd, 1966, p. 86.
7. D. K. Gibson, personal communication, 1972 (result obtained using Cavalleri's method).
8. R. P. Creaser, *Aust. J. Phys.*, **20**, 547, 1967.

Error limits: W_{77}: E/N (Td)$\geqslant 2 \times 10^{-2}$, $\pm 1\%$.

W_{293}: E/N (Td)$\leqslant 25$, $\pm 1\%$; $30 \leqslant E/N$ (Td)$\leqslant 50$, $\pm 2\%$; $60 \leqslant E/N$ (Td)$\leqslant 150$, $\pm 3\%$.

D/μ_{77}: $\pm 2\%$.

D/μ_{293}: $2 \times 10^{-2} \leqslant E/N$ (Td)$\leqslant 6$, $\pm 1\%$; $6 \leqslant E/N$ (Td)$\leqslant 60$, $\pm 2\%$; $60 \leqslant E/N$ (Td)$\leqslant 150$, $\pm 5\%$.

ND_{293}: $\pm 2\%$.

$W_{M_{293}}$: $\pm 3\%$.

TABLE 14.7 Experimentally determined transport coefficients for electrons in parahydrogen

	77 K		
E/N (Td)	W (10^5 cm sec^{-1})	D/μ (10^{-2} V)	ND (10^{21} cm^{-1} sec^{-1})
2.0×10^{-3}		0.683[2]	
2.5		0.689	
3.0		0.695	
3.5		0.699	
4.0		0.705	
5.0		0.716	
6.0		0.730	
7.0		0.745	
8.0	0.274[1]	0.761	2.61
1.0×10^{-2}	0.333	0.795	2.65
1.2	0.387	0.830	2.68
1.4	0.439	0.866	2.71
1.7	0.511	0.924	2.78
2.0	0.578	0.983	2.84
2.5	0.682	1.068	2.91

H₂

TABLE 14.7 (*Continued*)

	77 K		
E/N (Td)	W (10^5 cm sec^{-1})	D/μ (10^{-2} V)	ND (10^{21} cm^{-1} sec^{-1})
3.0×10^{-2}	0.779	1.148	2.98
3.5	0.871	1.223	3.04
4.0	0.959	1.290	3.09
5.0	1.131	1.410	3.19
6.0	1.296	1.507	3.26
7.0	1.457	1.593	3.32
8.0	1.614	1.673	3.38
1.0×10^{-1}	1.913	1.814	3.47
1.2	2.20	1.941	3.55
1.4	2.46	2.06	3.62
1.7	2.83	2.25	3.75
2.0	3.17	2.43	3.85
2.5	3.67	2.73	4.01
3.0	4.11	3.03	4.16
3.5	4.51	3.33	4.29
4.0	4.85	3.64	4.41
5.0	5.42	4.28	4.64
6.0	5.89	4.96	4.87
7.0	6.29	5.65	5.08
8.0	6.62	6.38	5.28
1.0×10^{0}	7.15	7.86	5.62
1.2	7.24	9.35	5.64
1.4	7.67	10.77	5.90
1.7	8.07	13.18	6.26
2.0	8.93	15.39	6.87
2.5	9.60	18.91	7.26
3.0	10.23	22.0	7.50
3.5	11.09	25.0	7.92
4.0	11.73	27.6	8.09
5.0	13.04	32.3	8.42
6.0	14.30	36.5	8.70
7.0	15.52		
8.0	16.70		
1.0×10^{1}	18.90		
1.2	20.9		
1.4	22.8		
1.7	25.6		
2.0	28.1		
2.5	32.2		
2.6	33.1		

TABLE 14.7 (*Continued*)

Sources: 1. A. G. Robertson, *Aust. J. Phys.*, **24**, 445, 1971.
        2. R. W. Crompton and A. I. McIntosh, *Aust. J. Phys.*, **21**, 637, 1968.

Error limits: W_{77}: E/N (Td)$\geqslant 2 \times 10^{-2}$, $\pm 1\%$.
            D/μ_{77}: $2 \times 10^{-3} \leqslant E/N$ (Td)$\leqslant 4 \times 10^{-1}$, $\pm 3\%$; $5 \times 10^{-1} \leqslant E/N$
                  (Td)$\leqslant 6$, $\pm 2\%$.

TABLE 14.8. Experimentally determined transport coefficients for electrons in deuterium

	77 K			293 K			
E/N (Td)	W (10^5 cm sec^{-1})	D/μ (10^{-2} V)	ND (10^{21} cm^{-1} sec^{-1})	W (10^5 cm sec^{-1})	D/μ (10^{-2} V)	ND (10^{21} cm^{-1} sec^{-1})	W_M (10^5 cm sec^{-1})
2.0×10^{-3}		$0.680^{(2)}$					
2.5		0.682					
3.0		0.685					
3.5		0.688					
4.0		0.692					
5.0		0.702					
6.0		0.712					
7.0		0.722					
8.0	$0.280^{(1)}$	0.733	2.57				
1.0×10^{-2}	0.342	0.756	2.59				
1.2	0.401	0.778	2.60				
1.4	0.458	0.801	2.62				
1.7	0.540	0.835	2.65				
2.0	0.618	0.866	2.68	$0.308^{(3)}$	$2.58^{(3)}$	3.98	
2.5	0.743	0.916	2.72	0.383	2.60	3.99	
3.0	0.864	0.962	2.77	0.457	2.63	4.01	
3.5	0.978	1.007	2.81	0.530	2.66	4.02	
4.0	1.087	1.049	2.85	0.601	2.69	4.04	
5.0	1.295	1.131	2.93	0.740	2.74	4.06	
6.0	1.487	1.211	3.00	0.873	2.81	4.08	
7.0	1.666	1.289	3.07	1.002	2.87	4.11	$1.226^{(4)}$
8.0	1.834	1.366	3.13	1.126	2.94	4.14	1.368
1.0×10^{-1}	2.14	1.521	3.26	1.360	3.09	4.20	1.639
1.2	2.41	1.681	3.38	1.577	3.24	4.26	1.883
1.4	2.65	1.840	3.49	1.782	3.40	4.33	2.11
1.7	2.98	2.09	3.66	2.06	3.66	4.44	2.44
2.0	3.25	2.34	3.80	2.32	3.90	4.52	2.72

TABLE 14.8 (*Continued*)

E/N (Td)	77 K			293 K			
	W (10^5 cm sec^{-1})	D/μ (10^{-2} V)	ND (10^{21} cm^{-1} sec^{-1})	W (10^5 cm sec^{-1})	D/μ (10^{-2}V)	ND (10^{21} cm^{-1} sec^{-1})	W_M (10^5 cm sec^{-1}
2.5×10^{-1}	3.64	2.78	4.04	2.70	4.35	4.70	3.15
3.0	3.95	3.23	4.25	3.03	4.80	4.84	3.52
3.5	4.21	3.71	4.46	3.31	5.27	4.98	3.81
4.0	4.43	4.21	4.66	3.56	5.74	5.10	4.10
5.0	4.80	5.21	5.00	4.00	6.74	5.39	4.63
6.0	5.07	6.27	5.30	4.35	7.77	5.63	5.06
7.0	5.30	7.37	5.58	4.65	8.81	5.86	5.41
8.0	5.50	8.48	5.83	4.92	9.86	6.06	5.71
1.0×10^0	5.86	10.66	6.25	5.38	11.90	6.40	6.24
1.2	6.19	12.79	6.60	5.79	13.89	6.70	6.72
1.4	6.52	14.76	6.87	6.19	15.74	6.96	7.14
1.7	7.01	17.51	7.22	6.74	18.30	7.25	7.75
2.0	7.50	19.97	7.49	7.27	20.7	7.50	8.31
2.5	8.28	23.7	7.85	8.10	24.2	7.84	9.21
3.0	9.05	27.0	8.15	8.88	27.5	8.14	10.05
3.5	9.75	30.2	8.41	9.61	30.5	8.37	10.80
4.0	10.4	33.1	8.61	10.24	33.4	8.55	11.52
5.0	11.6	38.5	8.93	11.23	38.8	8.71	12.86
6.0	12.8	43.5	9.28	12.71	43.8	9.27	14.11
7.0	13.9	48.4	9.61	13.77			15.41
8.0	14.9	52.7	9.82	14.80			16.42
1.0×10^1	16.8	61.2	10.28	16.69			18.12
1.2		69.4		18.47			19.98
1.4				20.2			21.6
1.7							23.6
2.0							26.0

Sources: 1. A. G. Robertson, *Aust. J. Phys.*, **24**, 445, 1971.

2. R. W. Crompton, M. T. Elford, and A. I. McIntosh, *Aust. J. Phys.*, **21**, 43, 1968.

3. A. I. McIntosh, *Aust. J. Phys.*, **19**, 805, 1966.

4. R. P. Creaser, *Aust. J. Phys.*, **20**, 547, 1967.

Error limits: W_{77}: E/N (Td) $\geqslant 2 \times 10^{-2}$, $\pm 1\%$.

W_{293}: $2 \times 10^{-2} \leqslant E/N$ (Td) $\leqslant 3.5$, $\pm 1\%$; $4 \leqslant E/N$ (Td) $\leqslant 14$, $\pm 2\%$.

D/μ_{77}: $\pm 2\%$.

D/μ_{293}: $\pm 1\%$.

$W_{M_{293}}$: $\pm 3\%$.

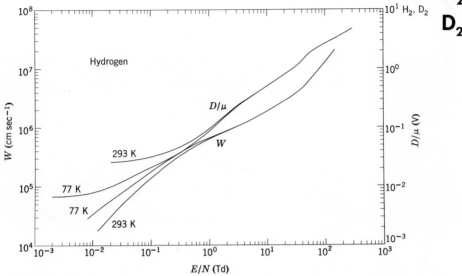

FIG. 14.12. Drift velocity and ratio of lateral diffusion coefficient to mobility for electrons in normal hydrogen. (From data given in Table 14.6.)

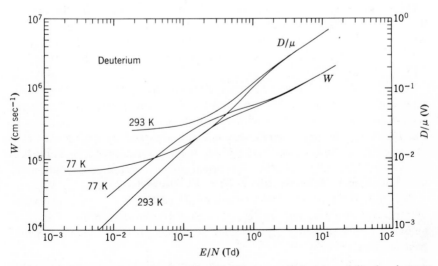

FIG. 14.13. Drift velocity and ratio of lateral diffusion coefficient to mobility for electrons in deuterium. (From data given in Table 14.8.)

621

H₂, D₂

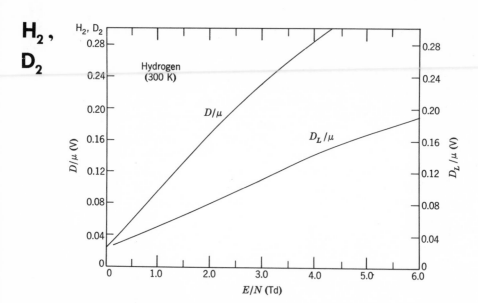

FIG. 14.14. Ratio of longitudinal diffusion coefficient to mobility for electrons in normal hydrogen with the ratio D/μ plotted for comparison. (From E. B. Wagner, F. J. Davis, and G. S. Hurst, *J. Chem. Phys.*, **47**, 3138, 1967.)

Cross sections. For values of E/N of less than 30 Td the energy distribution functions in hydrogen are largely controlled by elastic scattering and collisions which excite rotational and vibrational transitions. The cross sections for momentum transfer and rotational excitation and the sum of the cross sections for vibrational excitation which were obtained from an analysis of the data for W and D/μ in normal hydrogen and parahydrogen in this range of E/N are shown in Figure 14.15 and tabulated in Tables 14.9 to 14.12.

Above $E/N = 30$ Td, vibrational excitation, dissociation, photon excitation, and ionization are the dominant energy loss processes. The cross sections which are consistent with swarm data in this regime are shown in Figure 14.16. The set of cross sections was obtained by taking the cross sections $q_{m_{el}}$, q_d, and q_i (see Figure 14.16) from the results of single-collision experiments and adjusting the vibrational and excitation cross sections to obtain agreement between calculated and measured values of α_i/N.[7]

The vibrational cross section in deuterium (which is not given here) was obtained by assuming that $q_{m_{el}}$, q_d, q_{ex}, and q_i are the same in hydrogen and deuterium and adjusting q_v to obtain agreement between calculated and measured values of α_i/N.

The rotational cross sections that are consistent with low-energy swarm data in deuterium are shown in Figure 14.17. It was not possible to

622

determine these cross sections uniquely, as was done for hydrogen, because of the lack of experimental data for orthodeuterium.

H₂

TABLE 14.9. Momentum transfer cross section for electrons in hydrogen (from R. W. Crompton, D. K. Gibson, and A. I. McIntosh, *Aust. J. Phys.*, **22**, 715, 1969)

ϵ (eV)	$q_{m_{el}}$ $(10^{-16} cm^2)$	ϵ (eV)	$q_{m_{el}}$ $(10^{-16} cm^2)$
0.01	7.3	0.2	12.0
0.02	8.0	0.3	13.0
0.03	8.5	0.4	13.9
0.04	8.96	0.5	14.7
0.05	9.28	0.6	15.6
0.06	9.56	0.7	16.3
0.07	9.85	0.9	17.1
0.08	10.1	1.1	17.7
0.09	10.3	1.4	18.2
0.1	10.5	1.6	18.3
0.11	10.7	1.8	18.2
0.13	11.0	2.0	18.0
0.15	11.4		

Estimated error limit ±5%.

TABLE 14.10. Cross section for the $J = 0 \rightarrow 2$ rotational excitation in hydrogen (from R. W. Crompton, D. K. Gibson, and A. I McIntosh, *Aust. J. Phys.*, **22**, 715, 1969)

ϵ (eV)	$q_{r(02)}$ $(10^{-16} cm^2)$	ϵ (eV)	$q_{r(02)}$ $(10^{-16} cm^2)$
0.439	0.0	0.11	0.079
0.047	0.0185	0.13	0.089
0.050	0.027	0.15	0.099
0.055	0.035	0.20	0.120
0.060	0.042	0.25	0.137
0.065	0.048	0.30	0.160
0.07	0.053	0.35	0.185
0.08	0.060	0.40	0.210
0.09	0.068	0.45	0.236
0.10	0.074	0.50	0.263

Estimated error limit ±5% for $\epsilon \leqslant 0.3$ eV, ±10% at 0.4 eV, and ±30% at 0.5 eV.

TABLE 14.11.   Cross section for the $J = 1 \rightarrow 3$ rotational excitation in hydrogen (from D. K. Gibson, *Aust. J. Phys.*, **23**, 683, 1970)

ϵ (eV)	$q_{r(13)}$ (10^{-16} cm^2)	ϵ (eV)	$q_{r(13)}$ (10^{-16} cm^2)
0.0727	0.0	0.13	0.041
0.0750	0.01	0.15	0.047
0.08	0.017	0.20	0.060
0.085	0.0215	0.25	0.074
0.09	0.025	0.30	0.088
0.095	0.0275	0.35	0.102
0.1	0.0295	0.40	0.118
0.11	0.0335	0.45	0.133
0.12	0.0380	0.50	0.149

See text (page 574) for reference to error limits.

TABLE 14.12.   Cross section for the $v = 0 \rightarrow 1$ vibrational excitation in hydrogen (from R. W. Crompton, D. K. Gibson, and A. G. Robertson, *Phys. Rev.*, **A2**, 1386, 1970)

ϵ (eV)	$q_{v(01)}$ (10^{-16} cm^2)	ϵ (eV)	$q_{v(01)}$ (10^{-16} cm^2)
0.516	0.0	1.5	0.165
0.7	0.019	1.8	0.26
1.0	0.06	2.4	0.405
1.2	0.095	3.0	0.54

See text (page 573) for reference to error limits.

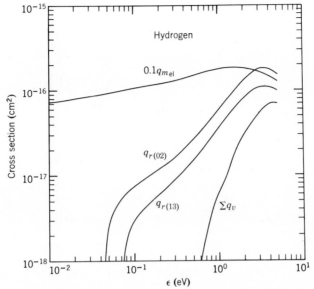

FIG. 14.15. Cross sections for momentum transfer and rotational and vibrational excitation that are consistent with the transport data for low-energy electrons in normal and para-hydrogen. (From D. K. Gibson, *Aust. J. Phys.*, **23**, 683, 1970, and the data given in Tables 14.9 to 14.12.

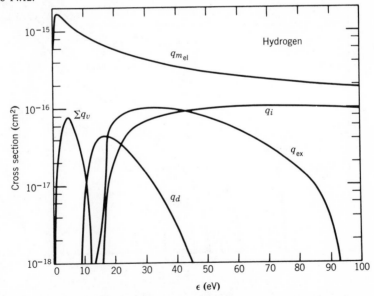

FIG. 14.16. Cross sections for elastic and inelastic scattering that are consistent with the transport data for electrons of higher mean energy in hydrogen. q_d, q_{ex}, and q_i are, respectively, the cross sections for dissociation, photon excitation, and ionization. (From A. G. Engelhardt and A. V. Phelps, *Phys. Rev.*, **131**, 2115, 1963; see also first part of Footnote 10 of A. V. Phelps, *Rev. Mod. Phys.*, **40**, 399, 1968.)

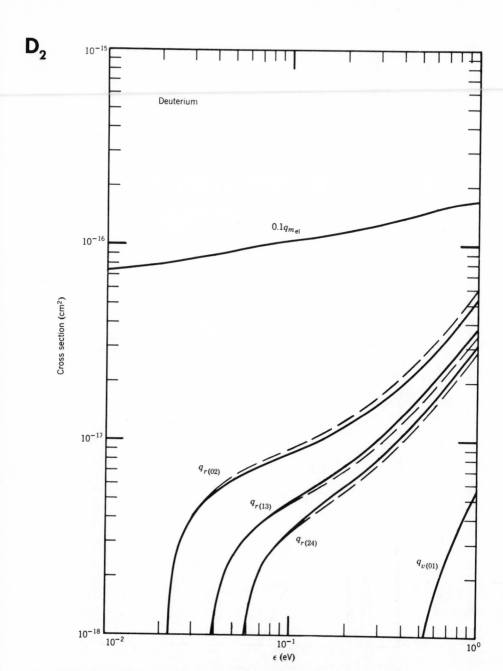

D₂

FIG. 14.17.   Cross sections for momentum transfer and rotational and vibrational excitation for electrons in deuterium. (From D. K. Gibson, *Aust. J. Phys.*, **23**, 683, 1970.)

TABLE 14.13. Experimentally determined transport coefficients for electrons in nitrogen

	77 K			293 K			
E/N (Td)	W (10^5 cm sec^{-1})	D/μ (10^{-2} V)	ND (10^{21} cm^{-1} sec^{-1})	W (10^5 cm sec^{-1})	D/μ (10^{-2} V)	ND (10^{21} cm^{-1} sec^{-1})	W_M (10^5 cm sec^{-1})
1.0×10^{-2}		0.74[2]					
1.2		0.76					
1.4		0.79		0.523[1]			
1.7		0.82		0.630			
2.0		0.86		0.733	2.65[5]	9.72	
2.5		0.94		0.902	2.71	9.77	
3.0		1.02		1.061	2.77	9.79	
3.5		1.11[3]		1.211	2.83	9.80	
4.0		1.20		1.350	2.91	9.82	
5.0		1.38		1.604	3.08	9.88	
6.0	3.47[1]	1.59	9.20	1.810	3.27	9.86	
7.0	3.58	1.83	9.37	1.986	3.48	9.86	
8.0	3.66	2.07	9.46	2.14	3.70	9.90	
1.0×10^{-1}	3.68	2.60	9.57	2.38	4.18	9.92	
1.2	3.67	3.17	9.70	2.54	4.71	9.97	
1.4	3.63	3.77	9.77	2.66	5.28	10.02	
1.7	3.56	4.73	9.90	2.77	6.18	10.08	4.23[6]
2.0	3.49	5.75	10.02	2.87	7.13	10.22	4.32
2.5	3.41	7.46	10.18	2.98	8.69	10.34	4.39
3.0	3.39	9.15	10.34	3.09	10.21	10.50	4.48
3.5	3.42	10.70	10.45	3.19	11.62	10.59	4.55
4.0	3.48	12.25	10.65	3.30	12.97	10.71	4.63
5.0	3.66	14.9	10.90	3.53	15.58	11.00	4.83
6.0	3.84	17.4	11.13	3.75	18.04	11.27	5.03
7.0	4.00	19.8	11.32	3.93	20.4	11.48	5.20
8.0	4.16	22.3	11.59	4.09	22.9	11.71	5.36
1.0×10^{0}	4.47	27.2	12.17	4.43	27.7	12.25	5.61
1.2	4.78	31.8	12.67	4.74	32.3	12.78	5.85
1.4	5.07	36.3	13.15	5.03	36.8	13.22	6.07
1.7	5.54	42.3	13.78	5.49	42.8	13.83	6.41
2.0	6.04	47.5	14.34	5.98	48.0	14.36	6.87
2.5	6.86	54.6	14.98	6.79	55.0	14.94	7.50
3.0	7.72	60.1	15.47	7.67	60.6	15.48	8.33

TABLE 14.13 (*Continued*)

E/N (Td)	77 K W (10^5 cm sec^{-1})	D/μ (10^{-2} V)	ND (10^{21} cm^{-1} sec^{-1})	293 K W (10^5 cm sec^{-1})	D/μ (10^{-2} V)	ND (10^2 cm^{-1} sec^{-1})	W_M (10^5 cm sec^{-1})
3.5×10^0	8.57	64.4	15.78	8.51	65.3	15.88	9.14
4.0	9.42	68.3	16.08	9.33	69.1	16.12	9.92
5.0	11.05	74.3	16.42	10.95	74.4	16.30	11.54
6.0	12.70	79.2	16.76	12.60	79.7	16.73	13.19
7.0		83.2		14.09	83.5	16.82	14.73
8.0		86.7		15.52	87.3	16.94	16.23
1.0×10^1		92.7		18.38	93.2	17.12	19.26
1.2		97.4		20.9	97.9	17.09	22.3
1.4				23.5	101.9	17.13	25.3
1.7				27.3			29.8
2.0				30.9			34.2
2.5				36.5			
3.0				41.7			
3.5				46.8			
4.0				51.8			
5.0				60.9			
6.0				70.3			
7.0				79.0[4]			
8.0				87.4			
1.0×10^2				105.1			
1.2				127.7			
1.4				148.0			
1.7				180.6			
2.0				212			
2.4				252			

Thermal diffusion coefficient ND_{300} 9.47×10^{21} cm^{-1} sec^{-1}[7]

Sources: 1. J. J. Lowke, *Aust. J. Phy.*, **16**, 115, 1963.
2. Taken from curve of best fit to data of R. W. Warren and J. H. Parker, Westinghouse Research Laboratories Sci. Paper 62-908-113-P6, 1962, and source 3.
3. R. W. Crompton and M. T. Elford, unpublished results, 1966.
4. H. A. Blevin and N. Z. Hasan, *Aust. J. Phys.*, **20**, 741, 1967.
5. R. W. Crompton and M. T. Elford, Electron and Ion Diffusion Unit, Australian National University, Quart. Rept. 19, 1965.
6. R. L. Jory, *Aust. J. Phy.*, **18**, 237, 1965.
7. D. R. Nelson and F. J. Davis, *J. Chem. Phys.*, **51**, 2322, 1969.

TABLE 14.13 (*Continued*)

Error limits: W_{77}: ±2%.

W_{293}: $1.4 \leqslant E/N$ (Td)$\leqslant 60$, ±1%; $70 \leqslant E/N$ (Td)$\leqslant 240$, ±2% increasing to ±4%.

D/μ_{77}: E/N (Td)$\leqslant 2 \times 10^{-2}$, ±4%; $2.5 \times 10^{-2} \leqslant E/N$ (Td)$\leqslant 8 \times 10^{-2}$, ±4% decreasing to ±2%; E/N (Td)$\geqslant 10^{-1}$, ±2%.

D/μ_{293}: E/N (Td)$\leqslant 6 \times 10^{-2}$, ±2%; E/N (Td)$\geqslant 7 \times 10^{-2}$, ±1%.

$W_{M_{293}}$: ±3%.

ND_{300}: ±1%.

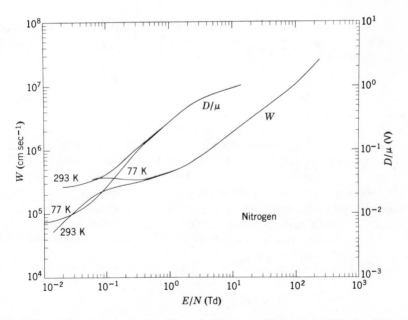

FIG. 14.18. Drift velocity and ratio of lateral diffusion coefficient to mobility for electrons in nitrogen. (From data given in Table 14.13.)

N₂

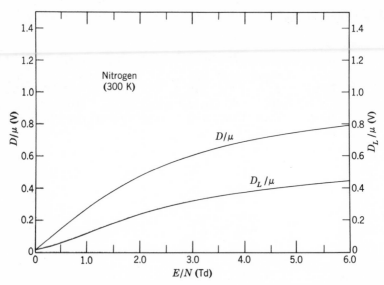

FIG. 14.19. Ratio of longitudinal diffusion coefficient to mobility for electrons in nitrogen with the ratio D/μ plotted for comparison. (From E. B. Wagner, F. J. Davis, and G. S. Hurst, *J. Chem. Phys.*, **47**, 3138, 1967.)

Rotational excitation cross sections. The set of rotational cross sections, of which a typical pair is shown in Figure 14.20, is represented by equations 14.1 and 14.2 with $Q = 1.04ea_0^2$.

Cross sections for momentum transfer and other inelastic processes. The momentum transfer cross section is given in Table 14.14. The determination of the vibrational and electronic excitation cross sections was described in Chapter 13 (Section 13.5.2). The low-energy regions of the cross sections $q_{v(01)}$ and $q_{v(10)}$ are shown in Figure 13.21, and the set of effective excitation cross sections that are consistent with the swarm data is given in Figure 13.22. Figure 14.20 summarizes the cross sections that were obtained from an analysis of the data given in Table 14.13 using experimentally determined values of the excitation coefficient δ/N and the ionization coefficient[8] α_i/N (see Section 13.5.2 and Figure 13.25).

TABLE 14.14. Momentum transfer cross section for electrons in nitrogen (from A. G. Engelhardt, A. V. Phelps, and C. G. Risk, Westinghouse Research Laboratories, Sci. Paper 64-928-113-P4, 1964)

ϵ (eV)	$q_{m_{el}}$ (10^{-16} cm^2)	ϵ (eV)	$q_{m_{el}}$ (10^{-16} cm^2)
1.60×10^{-3}	1.43	1.4×10^{0}	11.45
3.60	1.69	1.6	12.9
6.40	1.94	1.8	16.95
1.03×10^{-2}	2.20	2.0	24.01
2.21	2.94	2.6	29.88
4.00	3.86	3.0	21.63
6.51	4.90	3.6	14.66
1.03×10^{-1}	6.04	4.5	11.52
1.50	7.12	1.0×10^{1}	9.51
3.32	9.34	2.0	12.0
1.00×10^{0}	9.98	3.5	10.5
1.20	10.51	4.0	10.1

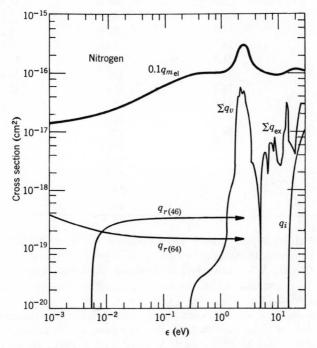

FIG. 14.20. Cross sections for elastic and inelastic scattering that are consistent with electron transport data in nitrogen. (From A. G. Engelhardt, A. V. Phelps, and C. G. Risk, *Phys. Rev.*, **135**, A1566, 1964.)

O₂

14.3.7. OXYGEN.

TABLE 14.15. Experimentally determined transport coefficients for electrons in oxygen

E/N (Td)	W (10^5 cm sec^{-1})	293 K		
		D/μ (10^{-2} V)	ND (10^{21} cm^{-1} sec^{-1})	
6.0×10^{-1}	7.20[1]			
7.0	8.35			
8.0	9.40			
1.0×10^0	11.37			
1.2	12.98	19.5[2]	21.1	
1.4	14.35	20.8	21.3	
1.7	16.17	22.5	21.4	
2.0	17.84	24.4	21.8	
2.5	20.1	27.0	21.7	
3.0	22.0	30.0	22.0	
3.5	23.3	34.2	22.8	
4.0	24.4	38.8	23.7	
5.0	25.7	49.2	25.3	
6.0	26.6	61.0	27.0	
7.0	27.6	72.8	28.7	
8.0	28.7	84.8	30.5	
1.0×10^1	31.4	109	34.3	
1.2	34.6	132	38.1	
1.4	38.2	156	42.5	
1.7	43.6	188	48.2	
2.0	49.4	206	50.9	
2.5	59.9	228	54.7	
3.0	72.3	244	58.8	
3.5		258		
4.0		270		
5.0		290		
6.0		307		

Sources: 1. Values derived from a table prepared by M. T. Elford (personal communication) after graphical assessment of the data of L. M. Chanin, A. V. Phelps, and M. A. Biondi, *Phys. Rev.*, **128**, 219, 1962; R. W. Crompton, and M. T. Elford, *Aust. J. Phys.*, (in press), 1973; and P. Herreng, *Cah. Phys.*, **38**, 7, 1952.

2. J. A. Rees, personal communication, 1972 [based on a reanalysis of the data of L. G. H. Huxley, R. W. Crompton, and C. H. Bagot, *Aust. J. Phys.*, **12**, 303, 1959; and J. A. Rees, *Aust. J. Phys.*, **18**, 41, 1965 (see p.493)].

Error limits: W_{293}: E/N (Td)=1, ±5%; E/N (Td)=1.2 and 1.4, ±2%;
$1.7 \leqslant E/N$ (Td) $\leqslant 9$, ±1%; $9 \leqslant E/N$ (Td) $\leqslant 30$, ±5%.
D/μ: (rough estimate): ±5%.

TABLE 14.16. Experimentally determined attachment coefficients in oxygen at 293 K

O$_2$

E/p (V cm^{-1} torr^{-1})	E/N (Td)	α_{at}/p (cm^{-1} torr^{-1}) at Pressure (torr) of:											
		14.9	21	25	29.75	36.4	44	66	88	132	220	440	880
0.1	3.04×10^{-1}		0.110	0.136		0.178	0.202						
0.12	3.64		0.103	0.122		0.165	0.186	0.232*					
0.16	4.86		0.090	0.107		0.140	0.160	0.215*	0.26*				
0.2	6.07		0.079	0.092		0.124	0.136	0.200	0.235*				
0.3	9.11		0.058	0.072		0.094	0.110	0.155	0.20	0.31	0.40*		
0.4	1.21×10^0		0.046	0.058	0.061	0.074	0.086	0.129	0.165	0.23	0.37*		
0.5	1.52		0.037	0.047	0.053	0.063	0.077	0.107	0.14	0.21	0.30*	0.48*	
0.6	1.82		0.032	0.039	0.045	0.053	0.067	0.093	0.12	0.183	0.27	0.44*	
0.7	2.12							0.082			0.23	0.38*	
0.8	2.43	0.0168	0.023	0.029	0.034	0.043	0.049	0.070	0.95	0.146	0.21	0.36	
0.9	2.73										0.194	0.34	
1.0	3.04	0.0128	0.018	0.023	0.027	0.034	0.041	0.057	0.075	0.120	0.177	0.31	
1.4	4.25	0.0085	0.0121	0.015	0.0177	0.0217	0.0266	0.039	0.050	0.078	0.122	0.16	0.28*
1.8	5.46	0.0060	0.0085	0.0105	0.0122	0.0150	0.0180	0.027	0.034	0.056	0.087	0.115	0.24*
2.2	6.68		0.0061	0.0076		0.0107	0.0129	0.0192	0.025	0.039	0.067		
2.3	6.98		0.0059										
2.4	7.29		0.0056	0.0069									
2.5	7.59		*0.0055*	*0.0067*		*0.0091*	0.0108						
2.6	7.89		0.0059	0.0074		*0.0092*	*0.0105*	0.0154	0.0198	0.03	0.047	0.081	0.18*
2.8	8.50		0.0070	0.0083		0.0097	0.0110	*0.0150*	0.0185	0.0265	0.038	0.072	
3.0	9.11							*0.0150*	*0.0184*	0.0246			0.145*
3.1	9.41												
3.2	9.71							0.0156		*0.0234*	0.035		0.118*

633

TABLE 14.16 (*Continued*)

E/p (V cm⁻¹ torr⁻¹)	E/N (Td)	α_a/p (cm⁻¹ torr⁻¹) at Pressure (torr) of:											
		14.9	21	25	29.75	36.4	44	66	88	132	220	440	880
3.3	1.00×10¹								0.019				
3.4	1.03						0.0140			*0.0236*	0.0333		
3.6	1.09		0.0127	0.0131		0.0149		0.0195	0.0206	*0.024*	*0.033*	0.055	0.100*
3.8	1.15									0.026	*0.0324*	0.052	
4.0	1.21											*0.050*	0.088
4.2	1.27		0.0203	0.0221			0.0225	0.0271	0.0273	0.033	0.035		0.084
4.6	1.40												0.079
4.8	1.46												*0.077*
5.0	1.52		0.035	0.0355			0.037	0.0415	0.0420	0.045	0.047	0.056	*0.076*
6.0	1.82		0.0515	0.0518			0.053	0.056	0.057	0.059	0.060	0.062	0.081
7.0	2.12		0.063	0.0655			0.0658	0.066	0.068	0.070	0.072	0.073	
8.0	2.43		0.072	0.075			0.075	0.072	0.074		0.075	0.077	0.093
9.0	2.73		0.078				0.082	0.080	0.083	0.090	0.086	0.089	0.095
10.0	3.04		0.084	0.084				0.086	0.087	0.094	0.088	0.091	0.099
11.0	3.34										0.090	0.093	0.102
12.0	3.64		*0.084*	*0.085*			*0.086*	0.090	0.093	0.096	0.095	0.094	0.104
13.0	3.95		*0.085*	*0.087*			*0.088*	*0.091*	0.094		0.096	0.096	0.104
14.0	4.25		*0.086*	*0.088*			*0.088*	*0.091*	*0.093*	*0.096*	*0.097*	*0.098*	*0.105*
15.0	4.55							0.090	0.093		0.094	0.098	
16.0	4.86		0.080	*0.088*			*0.086*	0.089	0.089		0.091	0.094	0.102*
18.0	5.46		0.078				0.085*	0.078*	0.088*	0.095*	0.088*	0.084*	0.097*

Source: R. Grünberg, Z. *Naturforsch.*, **24a**, 1039, 1969. The values in italics are the maximum and minimum values of α_{at}/p recorded at a given pressure. Error limits: For pressures less than 100 torr, error limit is usually ±2% to ±3%. Error limit ⩽ ±5% for all values except those marked with asterisk (see table in reference).

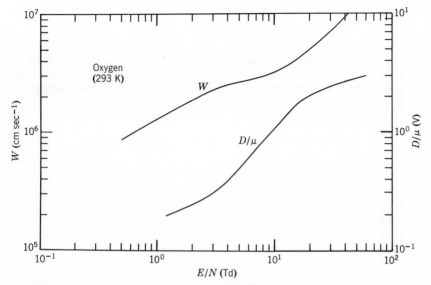

FIG. 14.21. Drift velocity and ratio of lateral diffusion coefficient to mobility for electrons in oxygen. (From the data given in Table 14.15; see also data extending to $E/N = 150$ Td given by M S. Naidu and A. N. Prasad, *J. Phys. D: Appl. Phys.*, **3**, 957, 1970.)

Rotational excitation cross sections. The absence of reliable data for D/μ below about 0.2 eV, which is approximately equal to the threshold of vibrational excitation (0.195 eV), has so far prevented any reliable determination of rotational excitation cross sections for oxygen.[9]

Cross sections for momentum transfer and other inelastic processes. The analysis of swarm data in oxygen is complex and was made more difficult by the limited extent of some of the data and their poor quality. Measurements of the two-body attachment coefficient and of the ionization coefficient [actually $(\alpha_i - \alpha_a)/N$] were used in addition to data for W and D/μ. The resultant set of cross sections is shown in Figures 14.22 to 14.24, the momentum transfer cross section being tabulated in Table 14.17. The reader is referred to ref. 8 for a detailed discussion of the alternative sets of cross sections proposed (see Figure 14.24) and their probable validity.

TABLE 14.17. Momentum transfer cross section for electrons in oxygen (from R. D. Hake and A. V. Phelps, Westinghouse Research Laboratories, Sci. Paper 66-1E2-GASES-P1, 1966)

An alternative cross section, differing by a few per cent from the tabulated values above $\epsilon = 1.0$ eV, was required to fit the transport data when an electronic excitation process with a resonance energy of 4.5 eV was assumed (see *Phys. Rev.*, **158**, 70, 1967).

ϵ (eV)	$q_{m_{el}}$ $(10^{-16}cm^2)$	ϵ (eV)	$q_{m_{el}}$ $(10^{-16}cm^2)$
1.0×10^{-3}	3.0	1.0×10^{0}	7.6
4.0	3.0	1.2	7.9
1.0×10^{-2}	3.0	1.4	7.9
2.0	3.0	1.7	7.6
4.0	3.0	2.0	7.1
6.0	3.0	2.3	6.6
8.0	3.4	2.7	6.0
1.0×10^{-1}	4.4	3.0	5.6
1.2	4.8	3.5	5.3
1.4	5.0	4.0	5.2
1.7	5.0	5.0	5.4
2.0	4.9	6.0	5.8
2.5	4.7	8.0	7.2
3.0	4.6	1.0×10^{1}	8.4
3.5	4.6	1.2	9.0
4.0	4.7	1.5	9.4
5.0	5.1	2.0	9.1
6.0	5.5	3.0	8.6
7.0	6.1	4.0	8.1
8.0	6.8	6.0	7.4
		1.0×10^{2}	6.5

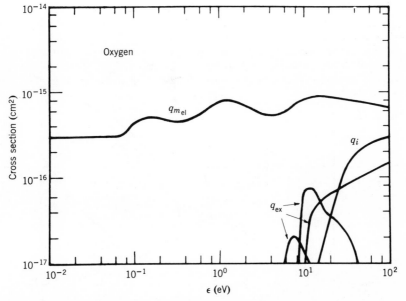

FIG. 14.22. Cross sections for momentum transfer and for electronic excitation (q_{ex}) and ionization (q_i) that are consistent with electron transport data in oxygen. (From R. D. Hake and A. V. Phelps, *Phys. Rev.*, **158**, 70, 1967.)

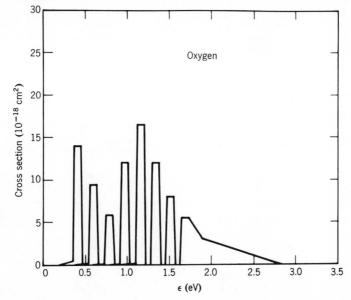

FIG. 14.23. The set of vibrational excitation cross sections which, together with a set of rotational excitation cross sections based on the theory of Gerjuoy and Stein (see text), have been proposed to explain the characteristics of low-energy electron swarms in oxygen. (From R. D. Hake and A. V. Phelps, *Phys. Rev.*, **158**, 70, 1967.)

637

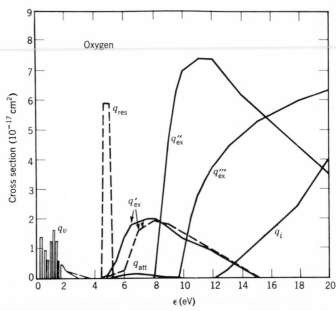

FIG. 14.24. Two sets of inelastic cross sections that are consistent with electron transport data in oxygen: q_{att} is the dissociative attachment cross section, and q'_{ex}, q''_{ex}, and q'''_{ex} are excitation cross sections with thresholds at 4.5, 8.0, and 9.7 eV, respectively. A resonance type of inelastic process, shown by the dashed curve labelled q_{res}, was found to lead to better agreement with measured values of α_a/N while retaining good agreement with the data for W and D/μ. The modified forms of q_v and q'_{ex} are shown as dashed curves. (From R. D. Hake and A. V. Phelps, *Phys. Rev.*, **158**, 70, 1967.)

14.3.8. AIR.

TABLE 14.18. Experimentally determined transport coefficients for electrons in dry, carbon-dioxide-free air

	293 K		
E/N (Td)	W (10^5 cm sec^{-1})	D/μ (10^{-2} V)	ND (10^{21} cm^{-1} sec^{-1})
3.0×10^{-1}	$3.60^{(1)}$		
3.5	3.77		
4.0	4.00		
5.0	4.50		
6.0	4.91		
7.0	5.34	$12.75^{(2)}$	9.73
8.0	5.78	13.96	10.09
1.0×10^0	6.58	16.06	10.57
1.2	7.29	18.17	11.04

TABLE 14.18. (*Continued*) **Air**

E/N (Td)	293 K		
	W (10^5 cm sec^{-1})	D/μ (10^{-2} V)	ND (10^{21} cm^{-1} sec^{-1})
1.4×10^0	7.92	20.7	11.72
1.7	8.76	24.0	12.35
2.0	9.52	27.5	13.09
2.5	10.52	32.9	13.82
3.0	11.36	38.1	14.43
3.5	12.07	42.9	14.78
4.0	12.76	48.1	15.53
5.0	14.11	57.4	16.21
6.0	15.50	65.3	16.87
7.0	16.88	71.7	17.29
8.0	18.26	77.3	17.64
1.0×10^1	21.0	85.9	18.06
1.2	23.8	93.1	18.45
1.4		98.8	
1.7		102.4	
2.0		107.3	
2.5		111.4	
3.0		119.2	
3.5		122.4	
4.0		127.9	
5.0		137.4	
6.0		149.5	
7.0		163.7	
8.0		179.1	
1.0×10^2		213	
1.2		250	

Sources: 1. J. A. Rees, *Aust. J. Phys.*, **26**, 427, 1973.
2. J. A. Rees and R. L. Jory, *Aust. J. Phys.*, **17**, 307, 1964.

Error limits: W_{293}: E/N (Td) $= 3 \times 10^{-1}$, $\pm 5\%$; E/N (Td) $= 4 \times 10^{-1}$ and 5×10^{-1}, $\pm 2\%$; $6 \times 10^{-1} \leqslant E/N$ (Td) $\leqslant 7$, $\pm 1\%$; E/N (Td) $\geqslant 8$, $\pm 2\%$.

D/μ_{293}: $2 \leqslant E/N$ (Td) $\leqslant 120$, $\pm 2\%$.

(No correction for influence of attachment, which could lead to appreciable error below $E/N = 4$ Td. Values below $E/N = 2$ Td are tabulated but should be treated with reservation.)

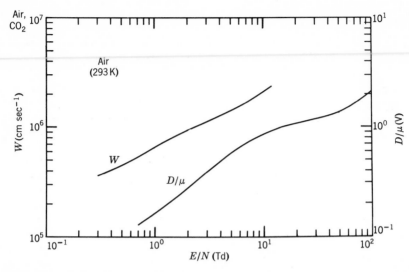

FIG. 14.25. Drift velocity and ratio of lateral diffusion coefficient to mobility for electrons in air. (From data given in Table 14.18.)

14.3.9. CARBON DIOXIDE.

TABLE 14.19. Experimentally determined transport coefficients for electrons in carbon dioxide

	Temperature 293 K		
E/N (Td)	W (10^5 cm sec^{-1})	D/μ (10^{-2} V)	ND (10^{21} cm^{-1} sec^{-1})
3.0×10^{-1}	0.536[1]	2.55[2]	0.456
3.5	0.625	2.55	0.456
4.0	0.714	2.56	0.457
5.0	0.890	2.57	0.458
6.0	1.068	2.58	0.459
7.0	1.246	2.59	0.461
8.0	1.424	2.60	0.463
1.0×10^0	1.781	2.62	0.467
1.2	2.14	2.65	0.471
1.4	2.49	2.68	0.477
1.7	3.03	2.73	0.486
2.0	3.56	2.78	0.496
2.5	4.45	2.88	0.512
3.0	5.37	3.00	0.537
3.5	6.28	3.12	0.561

determine the excitation cross sections. Experimental results for the net coefficient $(\alpha_i - \alpha_a)/N$, as well as for W and D/μ, were used in the analysis. Schulz's data[11-13] for excitation cross sections having thresholds at 3.1, 7.0, and 10.5 eV were used as the starting point for the analysis, the cross sections being normalized to give agreement with the swarm data. The set of cross sections compatible with the measured transport data is shown in Figure 14.28, the momentum transfer cross section being tabulated in Table 14.20. The original paper should be consulted for the provisos that have been made with respect to this set.

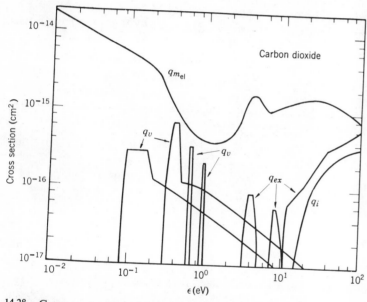

FIG. 14.28. Cross sections for elastic and inelastic scattering that are consistent with electron transport data in carbon dioxide. (From R. D. Hake and A. V. Phelps, *Phys. Rev.*, **158**, 70, 1967; see also revision by J. J. Lowke, A. V. Phelps, and B. W. Irwin, *J. Appl. Phys.*, (in press), 1973.

TABLE 14.19 (*Continued*) **CO₂**

	293 K		
E/N (Td)	W (10^5 cm sec^{-1})	D/μ (10^{-2} V)	ND (10^{21} cm^{-1} sec^{-1})
4.0×10^0	7.20	3.25	0.584
5.0	9.12	3.50	0.639
6.0	11.12	3.84	0.711
7.0	13.24	4.27	0.808
8.0	15.51	4.84	0.939
1.0×10^1	20.6	6.49	1.336
1.2	26.8	9.19	2.05
1.4	34.6	14.53	3.60
1.7	48.7	27.5	7.87
2.0	63.2	43.8	13.82
2.5		73.6	
3.0		99.3	
3.5		121.5	
4.0		139.8	
5.0		171.3	
6.0		197.4	

Thermal diffusion coefficient ND_{300} $\quad 0.486 \times 10^{21}$ cm^{-1} sec$^{-1(3)}$
$\qquad\qquad\qquad\qquad\qquad ND_{293}$ $\quad 0.450 \times 10^{21}$ cm^{-1} sec$^{-1(4)}$

Sources: 1. M. T. Elford, *Aust. J. Phys.*, **19**, 629, 1966.
2. J. A. Rees, *Aust. J. Phys.*, **17**, 462, 1964.
3. D. R. Nelson and F. J. Davis, *J. Chem. Phys.*, **51**, 2322, 1969.
4. Derived from the drift velocity data tabulated above, using the Nernst-Townsend relation.

Error limits: W_{293}: $\quad \pm 1\%$.
$\qquad\qquad D/\mu_{293}$: $\quad 3 \times 10^{-1} \leqslant E/N$ (Td) $\leqslant 14$, $\pm 1\%$; $17 \leqslant E/N$ (Td) $\leqslant 60$, $\pm 2\%$.
$\qquad\qquad ND_{300}$: $\quad \pm 3\%$.
$\qquad\qquad ND_{293}$: $\quad \pm 1\%$.

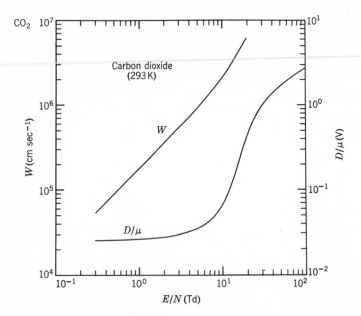

FIG. 14.26. Drift velocity and ratio of lateral diffusion coefficient to mobility for electrons in carbon dioxide. (From data given in Table 14.19 and J. A. Rees, *Aust. J. Phys.*, **17**, 462, 1964.) The electron drift velocity has been shown to be independent of temperature over the range 195 to 410 K (J. L. Pack, R. E. Voshall, and A. V. Phelps, *Phys. Rev.*, **127**, 2084, 1962).

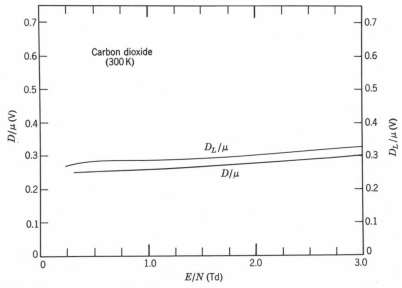

FIG. 14.27. Ratio of longitudinal diffusion coefficient to mobility for electrons in carbon dioxide with the ratio D/μ plotted for comparison. (From E. B. Wagner, F. J. Davis, and G. S. Hurst, *J. Chem. Phys.*, **47**, 3138, 1967.)

Rotational excitation cross sections. Energy losses caused by rotational excitation are so small compared to energy losses in collisions which excite the lowest vibrational state (threshold energy = 0.083 eV) that vibrational excitation dominates even at the lowest values of E/N and T which are experimentally accessible in CO_2. Therefore no definitive data for rotational excitation are available from swarm analyses.

Cross sections for momentum transfer and other inelastic processes. Experimental results for W and D/μ for $E/N < 30$ Td have been used[8] to determine a set of vibrational cross sections based on the energy dependence of the cross sections observed by Schulz.[11] The magnitudes of the cross sections and the momentum transfer cross section $q_{m_{el}}$ were adjusted to give agreement between calculated and experimental swarm data, the final set being shown in Figures 14.28 and 14.29.

A similar procedure, based on data for $E/N > 30$ Td, was used to

TABLE 14.20. Momentum transfer cross section for electrons in carbon dioxide (from R. D. Hake and A. V. Phelps, Westinghouse Research Laboratories, Sci. Paper 66-1E2-GASES-P1, 1966)

ϵ (eV)	$q_{m_{el}}$ (10^{-16} cm²)	ϵ (eV)	$q_{m_{el}}$ (10^{-16} cm²)
1.0×10^{-2}	170	2.5×10^{0}	6.4
2.0	120	2.8	8.5
4.0	85	3.2	12.3
7.0	64	3.6	15.8
1.0×10^{-1}	52	4.0	17.1
1.5	42	4.5	17.0
2.0	34	4.9	14.5
2.5	25	5.2	13.4
3.0	18	5.6	12.7
3.5	13	6.4	10.5
4.2	9.7	8.0	11.7
5.0	7.3	1.0×10^{1}	12.9
6.0	5.7	1.4	14.3
7.0	5.0	1.8	15.6
8.5	4.4	2.8	16.2
1.0×10^{0}	4.1	4.0	14.6
1.25	4.0	5.2	12.7
1.5	4.1	7.5	9.6
1.8	4.4	1.0×10^{2}	8.0
2.2	5.3		

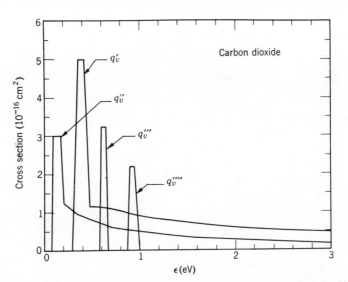

FIG. 14.29. Vibrational excitation cross sections for electrons in carbon dioxide as determined from an analysis of electron transport coefficients. The cross sections q_v', q_v'', and q_v''', and q_v'''' have threshold energies of 0.08, 0.3, 0.6, and 0.9 eV, respectively. (From R. D. Hake and A. V. Phelps, *Phys. Rev.*, **158**, 70, 1967.)

14.3.10. CARBON MONOXIDE. *Transport coefficients*. Graphs showing the variation of W and D/μ with E/N are presented in Figure 14.30, the data for W having been obtained at both 77 and 300 K.

Rotational excitation cross sections. For swarms in which the mean electron energy is less than about 0.1 eV the dominant energy loss mechanism in carbon monoxide is rotational excitation. Hake and Phelps[9] found that the low-temperature, low-E/N results for W and D/μ could be satisfactorily explained by postulating a set of rotational cross sections of the form given by equations 14.3 and 14.4, using a permanent dipole moment of $4.6 \times 10^{-2} ea_0$. A typical pair of the set is shown in Figure 14.31.

Cross sections for momentum transfer and vibrational excitation. The momentum transfer cross section determined from the swarm analysis is given in Table 14.21. Figure 14.31 shows this cross section and the sum of the vibrational cross sections for $\epsilon \leqslant 1$ eV.

645

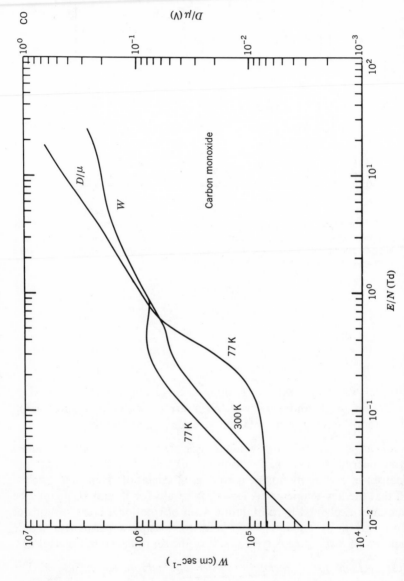

FIG. 14.30. Drift velocity and ratio of lateral diffusion coefficient to mobility for electrons in carbon monoxide. (*W* from J. L. Pack, R. E. Voshall, and A. V. Phelps, *Phys. Rev.*, **127**, 2084, 1962; *D*/μ from R. W. Warren and J. H. Parker, *Phys. Rev.*, **128**, 2661, 1962.)

TABLE 14.21. Momentum transfer cross section for electrons in carbon monoxide (from R. D. Hake and A. V. Phelps, Westinghouse Research Laboratories, Sci. Paper 66-1E2-GASES-P1, 1966)

ϵ (eV)	$q_{m_{el}}$ (10^{-16}cm^2)	ϵ (eV)	$q_{m_{el}}$ (10^{-16}cm^2)
1.0×10^{-3}	40.0	7.0×10^{-2}	6.1
2.0	25.0	1.0×10^{-1}	7.3
4.0	14.0	2.0	10.0
7.0	9.8	4.0	13.5
1.0×10^{-2}	7.8	7.0	15.5
2.0	5.9	1.0	17.0
4.0	5.2		

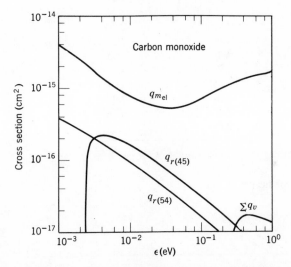

FIG. 14.31. Momentum transfer and inelastic scattering cross sections for energies less than 1 eV that are consistent with electron transport data in carbon monoxide. The cross section Σq_v represents the sum of the vibrational cross sections. Only one pair of the numerous rotational excitation and de-excitation cross sections that determine the energy distribution function at low energies is shown (see text). (From R. D. Hake and A. V. Phelps, *Phys. Rev.*, **158**, 70, 1967.)

H₂O 14.3.11. WATER VAPOUR.

TABLE 14.22. Experimentally determined transport coefficients for electrons in water vapour

	293 K
E/N (Td)	W (10^5 cm sec^{-1})
1.2×10^0	0.286
1.4	0.334
1.7	0.401
2.0	0.468
2.5	0.583
3.0	0.703
3.5	0.822
4.0	0.935
5.0	1.165
6.0	1.396
7.0	1.635
8.0	1.869
1.0×10^1	2.34
1.2	2.80
1.4	3.27
1.7	3.97
2.0	4.67
2.5	6.01
3.0	7.46
3.5	9.26
4.0	11.72

Source: J. J. Lowke and J. A. Rees, *Aust. J. Phys.*, **16**, 447, 1963.

Error limits: ±2%.

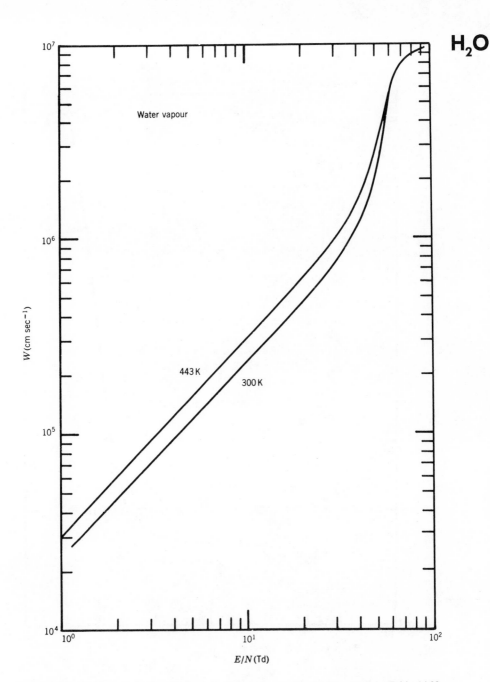

H₂O

FIG. 14.32. Drift velocity for electrons in water vapour. (From data given in Table 14.22 and from J. L. Pack, R. E. Voshall, and A. V. Ph

649

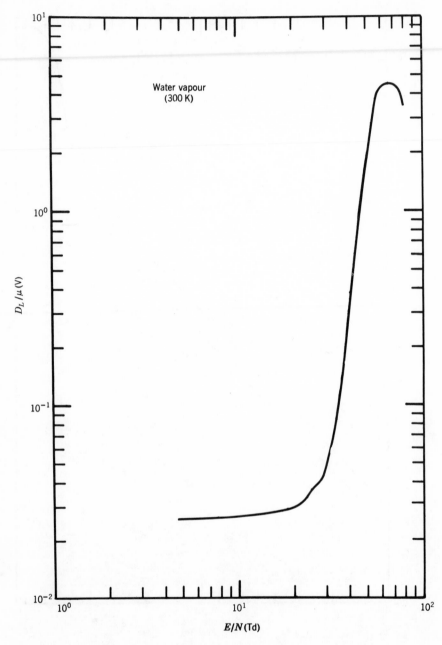

FIG. 14.33. Ratio of longitudinal diffusion coefficient to mobility for electrons in water vapour. (From E. B. Wagner and F. J. Davis, personal communication.)

Cross sections. Because of the lack of reliable data for D/μ for swarms of low mean energy, no swarm analysis has been undertaken to examine inelastic cross sections in water vapour. The momentum transfer cross section shown in Figure 14.34 was determined from the temperature dependence of the zero-field mobility, using the drift velocity data shown in Figure 14.32.

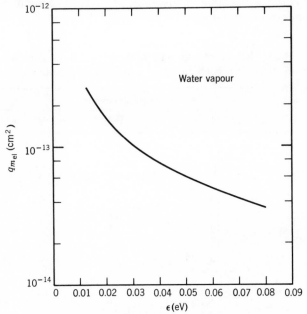

FIG. 14.34. Momentum transfer cross section for electrons in water vapour. (From J. L. Pack, R. E. Voshall, and A. V. Phelps, *Phys. Rev.*, **127**, 2084, 1962.)

14.3.12. EXPERIMENTAL DATA FOR OTHER GASES. Modern techniques have been used to measure electron drift velocities and diffusion coefficients in a large number of gases other than those for which detailed information is given in the preceding sections. We have not attempted to present a comprehensive list of references such as that found in ref. 15 but have restricted the citation to papers concerned with polyatomic gases having comparatively simple structures. For measurements of drift velocities (usually magnetic drift velocities) and the ratio D/μ made before 1941 the reader should consult ref. 14. Early measurements of both W_M and D/μ, as well as of α_a or $\alpha_{c_.}$, in situations in which electron attachment is significant should be treated with some caution.

In addition to the work referred to in this section, there have been many measurements of electron drift velocities and attachment coefficients in more complex polar gases, particularly by the group at the Oak Ridge National Laboratory. For a general coverage of this work and for graphical presentations of much of the work referred to in Table 14.23, the reader should refer to the book by Christophorou.[15]

Table 14.23 contains a list of gases, arranged in alphabetical order, together with references to papers containing experimental measurements of one or another of the transport coefficients. The reference numbers refer to the list at the foot of the table.

TABLE 14.23. References to recent experimental measurements of electron transport coefficients in a number of polyatomic gases

Gas	References for:		
	W	D/μ	D_L/μ
AsH_3	1		
CCl_2F_2	11	11	
CD_4	1	3	
C_2D_2	1		
CF_4	13	13	
C_2F_6	13	13	
C_3F_8	13	13	
C_4F_8	14	14	
C_4F_{10}	13	13	
CH_4	1,6,7	3	6
C_2H_2	2,7		
C_2H_4	2,4,5,6,7,8	3	6
C_2H_6	2,7	3	
C_3H_6	7		
C_3H_8	2,4	3	
C_4H_8	7		
C_4H_{10}	4		
C_5H_{12}	4		
C_6H_6	4		
NH_3	9,10		
N_2O	9,10		
sf_6	12	12	
SiD_4	1	3	
SiH_4	1	3	

Sources: 1. T. L. Cottrell and I. C. Walker, *Trans. Faraday Soc.*, **61**, 1585, 1965.
2. T. L. Cottrell, W. J. Pollock, and I. C. Walker, *Trans. Faraday Soc.*, **64**, 2260, 1967.
3. T. L. Cottrell and I. C. Walker, *Trans. Faraday Soc.*, **63**, 549, 1967.

4. L. Christophorou, G. S. Hurst, and A. J. Hadjiantoniou, *J. Chem. Phys.*, **44**, 3506, 1966. **Misc.**

5. G. S. Hurst and J. E. Parks, *J. Chem. Phys.*, **45**, 282, 1966.

6. E. B. Wagner, F. J. Davis, and G. S. Hurst, *J. Chem. Phys.*, **47**, 3138, 1967.

7. C. R. Bowman and D. E. Gordon, *J. Chem. Phys.*, **46**, 1878, 1967.

8. N. Hamilton and J. A. Stockdale, *Aust. J. Phys.*, **19**, 813, 1966.

9. R. A. Nielsen and N. E. Bradbury, *Phys. Rev.*, **51**, 69, 1937.

10. J. L. Pack, R. E. Voshall, and A. V. Phelps, *Phys. Rev.*, **127**, 2084, 1962.

11. M. S. Naidu and A. N. Prasad, *Brit. J. Appl. Phys. (J. Phys. D)*, **2**, 1431, 1969.

12. M. S. Naidu and A. N. Prasad, *J. Phys. D: Appl. Phys.*, **5**, 1090, 1972.

13. M. S. Naidu and A. N. Prasad, *J. Phys. D: Appl. Phys.*, **5**, 983, 1972.

14. M. S. Naidu, A. N. Prasad, and J. D. Craggs, *J. Phys. D: Appl. Phys.*, **5**, 741, 1972.

REFERENCES

1. See H. Raether, *Electron Avalanches and Breakdown in Gases,* Butterworths, London, 1964, and references therein.

2. See, for example, M. A. Folkard and S. C. Haydon, *Aust. J. Phys.,* **24**, 519, 1971; S. C. Haydon and O. M. Williams, *J. Phys. D: Appl. Phys.,* **5**, L79, 1972; M. A. Folkard and S. C. Haydon, *J. Phys. B: At. Mol. Phys.,* **6**, 214, 1973; S. C. Haydon and O. M. Williams, *J. Phys. B: At. Mol. Phys.,* **6**, 227, 1973.

3. See, for example, M. S. Naidu and A. N. Prasad, *Brit. J. Appl. Phys. (J. Phys. D),* **1**, 763, 1968; N. Sukhum, A. N. Prasad, and J. D. Craggs, *Brit. J. Appl. Phys.,* **18**, 785, 1967; L. E. Virr, J. Lucas, and N. Kontoleon, *J. Phys. D: Appl. Phys.,* **5**, 542, 1972: see also the references listed in these papers.

4. F. Llewellyn-Jones, *Ionization Avalanches and Breakdown,* Methuen and Co., London, 1967.

5. E. Gerjuoy and S. Stein, *Phys. Rev.,* **97**, 1671, 1955; **98**, 1848, 1955.

6. L. S. Frost and A. V. Phelps, *Phys. Rev.,* **127**, 1621, 1962.

7. A. G. Engelhardt and A. V. Phelps, *Phys. Rev.,* **131**, 2115, 1963.

8. A. G. Engelhardt, A. V. Phelps, and C. G. Risk, *Phys. Rev.,* **135**, A1566, 1964.

9. R. D. Hake and A. V. Phelps, *Phys. Rev.,* **158**, 70, 1967.

10. K. Takayanagi, *J. Phys. Soc. Japan,* **21**, 507, 1966.

11. G. J. Schulz, unpublished work.

12. G. J. Schultz, *Phys. Rev.,* **135**, A988, 1964.

13. G. J. Schulz, *Phys. Rev.,* **128**, 178, 1962.

14. R. H. Healey and J. W. Reed, *The Behaviour of Slow Electrons in Gases,* Amalgamated Wireless Ltd., Sydney, 1941.

15. L. G. Christophorou, *Atomic and Molecular Radiation Physics,* Wiley-Interscience, New York, 1971.

AUTHOR INDEX

SUBJECT INDEX